May *both* your problems and your samples in LC/MS be soluble!!

Highest regards,

Ross Willoughby

Ed Sheehan

A Global View of
LC/MS

How to Solve Your Most Challenging Analytical Problems

Ross Willoughby, Ph.D.
Edward Sheehan, Ph.D.
Samuel Mitrovich, Ph.D.

First Edition

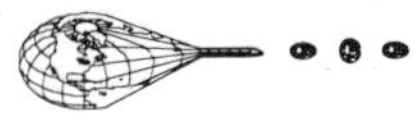

**Global View Publishing,
Pittsburgh, Pennsylvania**

A Global View of LC/MS

How to Solve Your Most Challenging Analytical Problems

By Ross Willoughby, Ph.D.
Edward Sheehan, Ph.D.
Samuel Mitrovich, Ph.D.

Published by:

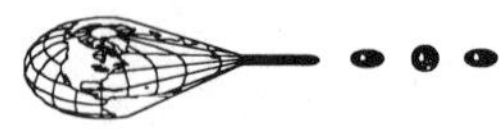

Global View Publishing
Post Office Box 111384
Pittsburgh, PA 15238-9998

First Edition

Printed in the United States of America

Willoughby, Ross.
A global view of LC/MS: How to solve your most challenging analytical problems/
by Ross Willoughby, Ed Sheehan, and Sam Mitrovich - First ed.
Includes bibliographical references and index.
ISBN 0-9660813-0-7 (pbk.)
1. Introduction to LC/MS I. Title
2. Information Sources on LC/MS
3. How to evaluate and buy LC/MS

Contents at a Glance

Dedication

Back Row: (l-r) Brian Chait, Ron Bonner, Tom Covey, Andries Bruins, Matthias Mann, Mark Allen, Craig Whitehouse, Bob Voyksner, Paul Winkler, Ed Sheehan

Front Row: (l-r) Victor Tal'rose, John Fenn, Dick Caprioli, Bruce Thomson, Al Yergey, Ross Willoughby

Group photo of many pioneers in LC/MS taken at the 1996 ASMS Conference in Portland, Oregon.

Dedication

This book is dedicated to the
pioneers in LC/MS.

These stalwart scientists overcame
seemingly insurmountable barriers to
give us one of the important tools we
need to advance our knowledge of the world.

Warning and Disclaimer

It is not the purpose of this manual to reprint all the information that is otherwise available to the authors and/or the publisher, but to complement, amplify and supplement other texts. You are urged to read all the available material, learn as much as possible about LC/MS and to tailor the information to your individual needs. For more information, see the many references in Appendix A.

Any person or organization who decides to use LC/MS as a tool for problem-solving in the laboratory must be willing to invest significant time and effort. LC/MS is not a passive undertaking; rather, it requires significant attention to detail in every aspect of the process; including, training, planning, purchasing, operation, and maintenance.

Every effort has been made to make this guidebook as complete and accurate as possible. However, there **may be mistakes** both typographical and in content. Therefore, this text should be used as a general guide and not as the ultimate source of LC/MS information. Furthermore, this guidebook contains information on LC/MS only up to the date that it was printed.

The purpose of this guidebook is to educate and create context in a complex field. The authors and Global View Publishing shall have neither liability nor responsibility to any person or entity with respect to any loss or damage caused, or alleged to be caused, directly or indirectly by the information contained in this book.

If you do not wish to be bound to the above, you may return this book to the publisher for a full refund.

About The Authors

The authors of this book comprise a biologist, a pharmacologist, and a chemist by education; a developer, an analyst, and a teacher by experience; a liberal, a moderate, and a conservative by politics. It is our hope that this diversity will result in a fair and balanced treatment of this important subject.

Ross C. Willoughby, Ph.D. -

Dr. Willoughby has spent most of the last 20 years in the development of LC/MS interfacing technologies; including, thermospray, APCI, particle beam, continuous flow FAB, and electrospray. He is an analytical chemist by training and was one of the co-inventors of particle beam LC/MS at Georgia Tech. His scientific interests lie in the study of fundamental physical processes associated with LC/MS and instrumentation. Dr. Willoughby has chaired the LC/MS Interest Group of the ASMS for four years and has championed the development of standards and communication within the LC/MS community.

Edward W. Sheehan, Ph.D. -

Dr. Sheehan has spent over 15 years in the pharmaceutical sciences. He is a pharmacologist and pharmacist by training and has a extensive background in applications of LC/MS, particularly in drug analysis and metabolism. His scientific interests lie in the practical application of LC/MS to real world sample analysis and analytical problem solving. He was the recipient of the prestigious Fite-Brackmann Award for his technical writing efforts in support of LC/MS products at Extrel Mass Spectrometry. He is presently an adjunct professor in the School of Pharmacy at the University of Pittsburgh.

Samuel Mitrovich, Ph.D. -

Dr. Mitrovich is a recipient of the prestigious Milken National Educator Award and the Presidential Award for Excellence in Science Education with over 30 years of experience in teaching. He received his graduate degree from the Department of Biological Sciences at the University of Pittsburgh specializing in protein chemistry. He has a strong interest in LC/MS biotechnology applications and development of chemical mapping techniques.

Acknowledgments

The authors would like to acknowledge the support and assistance that we received throughout our three year endeavor of writing this book. We could never have completed this effort without the opinions and advise from many of our colleagues in the field. We would like to thank all the practitioners in the trenches who have related their experiences to us in order that we may provide accurate information about the practical aspects of problem-solving in LC/MS.

We literally interviewed hundreds of lab managers, operators, service technicians, instrument manufacturers, contract labs, and researchers in the field. We appreciate your time and input. We give you a sincere and hardy thanks.

Later in the process, we imposed our extremely "rough" drafts on many of our respected colleagues who are experts in the fields of mass spectrometry and liquid chromatography. We are grateful for their considerate investment of time, effort, and patience on our behalf. The thoughtful criticisms we received have significantly enhanced the quality of this book. These reviewers included Dr. Sebastian Assenza, Gary Astle, Professor Rich Browner, Professor Jack Henion, Dr. Randy Pedder, Dr. Bori Shushan, Dr. David Strand, Dr. Craig Whitehouse, and Dr. Paul Winkler.

Countless others have contributed in many small but very significant ways. We thank you all.

A very sincere and special thanks goes to Mary Ellen Willoughby for reading each and every word on each and every draft. Her patience and dedication to our cause deserves our warmest and most heartfelt appreciation.

Cover by Lori Vinchesi. Photographs by Christine Baxter, Petrakis Photography, and Stewart Harvey.

Preface

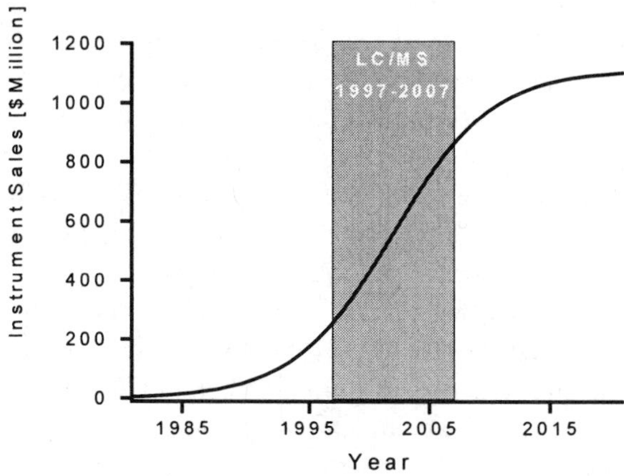

Figure 1- Predicted growth of LC/MS from 1997.[1]

LC/MS Comes Of Age

After 20 years of continuous and rapid development of LC/MS (liquid chromatography/mass spectrometry) technologies and applications, all the pieces are in place to elevate LC/MS into a role as a routine tool for industry, government, and academic analysts to solve many of their most challenging problems. The key to the growth and success of LC/MS as a problem-solving tool is (and will be) in the informing power, reliability, affordability, and availability of commercial systems. The recent seamless integration of separation, detection, and computer technologies, not just the development of LC/MS interfacing technologies, has enabled LC/MS to become a practical problem-solving tool.

Only recently have commercial products and technologies been available in integrated systems engineered for the broader analytical community. The availability of more than a dozen new and affordable systems within the past two years has initiated a rapid growth in this industry (Figure 1). The availability of instrumentation will accelerate the development of new applications; conversely, the development of new applications will accelerate the development of new instrumentation. It is clearly an exciting time in LC/MS.

One aspect of this era of significant growth in LC/MS is the influx of newcomers to the field. The scientists coming into this field will likely be different than the ones who have cut their teeth on LC/MS over the past two decades. Their technical background, experience level, and skill levels will likely be quite different than the veterans of this field. Figure 2 illustrates our view of the distribution of practitioners within LC/MS over the past 20 years versus the next 20. We believe that a trend will occur over the next decade away from the field being dominated by "mass spectrometrists" to a field where LC/MS becomes a routine tool for the entire complement of scientific disciplines.

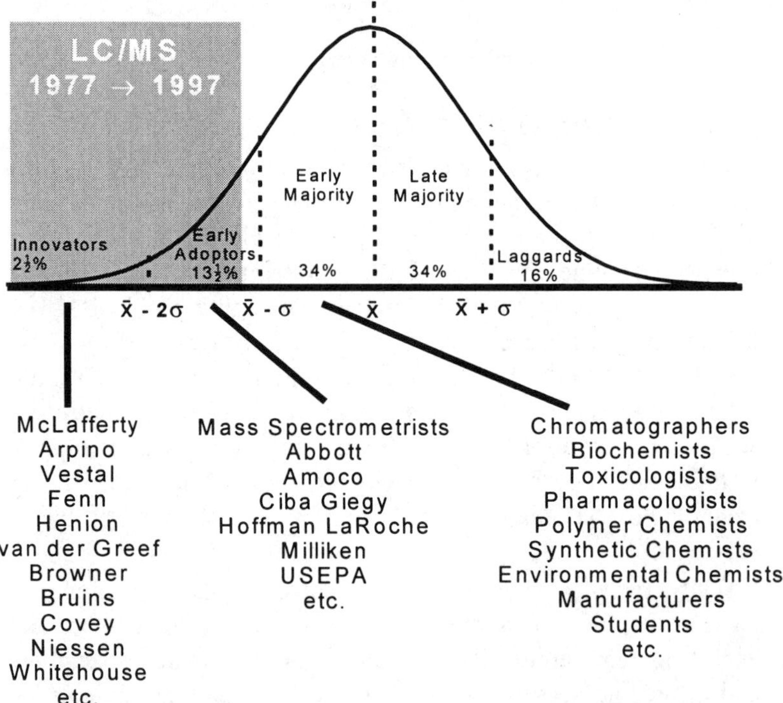

Figure 2- Model for the distribution of practitioners of LC/MS; showing the movement from a handful of innovators in the late 1970s, through early adopters in the present, to the early majority the next decade.[2]

Many benefits will likely come from this (positive) change in the population of practitioners in LC/MS. The analytical power of LC/MS will be broadly applied to problems in nearly all scientific disciplines and touch every aspect of our society. New ideas, new perspectives, and different knowledge-bases will foster further development in this field. Growth is good. Diversity is good.

Unfortunately, rapid growth and diversification do not come without a price. The newcomers to LC/MS are faced with making many choices. They are faced with learning many new techniques, new jargon, and new approaches to solving their problems. They are faced with the uncertainty that is inherent with any change. They risk making the wrong choices. They risk wasting time and money. They risk being consumed with unfamiliar and intellectually demanding technologies. They risk drowning in the mountains of data that are products of LC/MS technologies. They risk missing the opportunity to exploit the data from LC/MS to solve real and substantive problems in their businesses.

Even experienced practitioners of LC/MS have to deal with an ever-expanding applications and technology base. To improve effectiveness, they will need to continually adapt methods, protocols, and possibly, instrumentation, to achieve optimal results.

Why We Wrote This Book

We wrote this book to address the changing needs of this rapidly growing industry. This book aims to provide a guide to the individuals and organizations that are entering or expanding in the field of LC/MS. Our goal is to provide some structure and context to this complex and dynamic field. By providing a guide that focuses primarily on "problem-solving", many of the aforementioned risks associated with rapid growth and change can be averted. This book will help you make the appropriate choices with respect to LC/MS and "your" specific problems. This book identifies the essential sources of information and the resources that are accessible to your organization.

Who Should Read This Book?

We have written this book for any problem-solver in the field of chemical analysis. You may be a newcomer to, or highly experienced in mass spectrometry; this book provides useful information to both.

This book is for managers, group leaders, scientists, and technicians who are considering LC/MS for their labs and need a concise (but comprehensive) treatment of the field in order to better make planning, budgeting, and purchasing decisions.

This book is written for practitioners who require a guide to everyday problem-solving, including standard procedures, protocols, decision trees, flow diagrams, worksheets, checklists, methods, etc.

This book is written for the relatively few current day experts who are likely being besieged with countless questions about LC/MS and wish to direct their inquisitive colleagues to an introductory text with worksheets and evaluation sheets.

We have written this book for students in the collective analytical sciences who are looking for a glimpse of the real world and real world problem-solving. To them we say, "Solving problems is what it's all about.".

This book is not a guide to building your own LC/MS interface or a detailed review about state-of-the-art research topics in LC/MS. If you have interests in these areas, we provide a complete list of texts, conferences, and reviews.

How Is This Book Organized?

A Global View of LC/MS is organized in four **Parts**, with each **Part** intended to assist you in problem-solving. Each new **Chapter** in this book builds upon the previous **Chapter** as part of a global problem-solving model. Each **Part** contains a set of **Chapters** with a specific problem focus. Five **Appendices** are also provided for resource information.

- **Part I** contains three chapters (Chapters 1 through 3) to assist you in determining your technology needs in LC/MS. By the time you finish *Part I,* you should be able to evaluate your personal needs and consider all available alternatives in LC/MS.

- **Part II** contains four chapters (Chapters 4 through 7) which deal with your justification and acquisition of various alternatives, and planning for a successful LC/MS lab. By the time you finish *Part II,* you should be able to acquire and implement the appropriate LC/MS technology to solve your organizational needs.

- **Part III** contains five chapters (Chapters 8 through 12) dealing with practical sample analysis, problem-solving, and methods development. By the time you finish *Part III,* you should be able to identify and solve specific problems with your particular LC/MS technologies.

- **Part IV** provides five Appendices that allow you to access broader resources, general information, and fundamentals of LC/MS.

How To Use This Book

Keep this book in your office, lab, library, or next to your instrument. This book is intended to be a practical guide or manual for getting started in LC/MS, acquiring instrumentation, and productively using LC/MS as a tool to solve your analytical problems. This book allows you to access information relating to LC/MS when you need it. This book is not intended to be read cover-to-cover and put away; rather it should be used like a road map on your summer vacation. Use it when you need it. Keep it handy.

Notes

1 The predicted growth curve for LC/MS instrument sales was estimated by acquiring published and estimated sales values for the LC/MS instrument market through the past decade and applying this information to a classic sigmoidal market growth function as described in Lynch, R., Ross, H. L., Wray, R. D., *Introduction to Marketing*, McGraw Hill: New York 1984.

2 Rogers, E. M., *Diffusion of Innovation*, 4th ed. Free Press: New York 1995. This graph depicts a model gaussian distribution representing the characteristics of a population as new technologies. The far left shows the innovators who are always the first to adopt any new technology. Progressing from left to right, the next group is the early adopters, then early majority. We estimate that the population of LC/MS users in 1997 is in a transition from early adopters, followed by the early majority. This would predict that the majority of LC/MS users have yet to use the technology. The market will likely not saturate for a decade.

3 Gomez, A., Tang, K., "Charge and fission of droplets in electrostatic sprays," Phys. Fluids 6, pages 404-414, (1994)

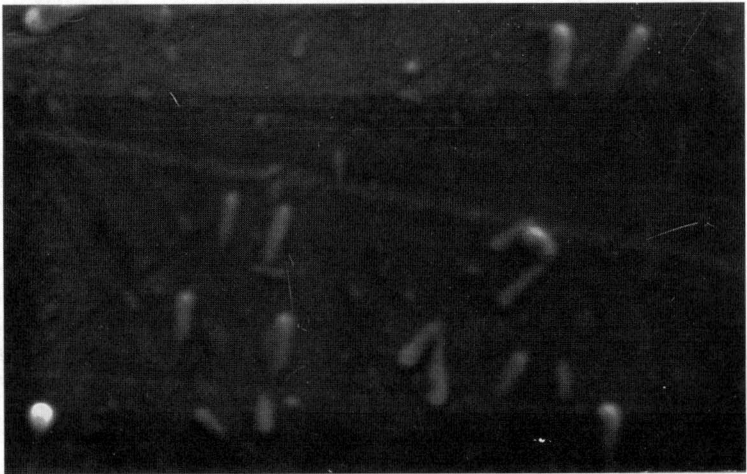

Figure 3- Electron micrograph of dry residue particles from the electrospray aerosol/ion generation process. These micrographs are evidence that the teardrop-shaped particle deformation observed by Gomez and Tang[3] is sustainable and continuous until the liquid droplets crystallize into residue particles in the nm size regime. (1000 ppm solutions of Cytochrome C were sprayed and collected in the first vacuum stage of electrospray). This topic is discussed in Appendix C. (Scale: 0.75 inches = one micrometer).

Detailed Contents

Part I- Is LC/MS Right For You?

Part II- Acquiring LC/MS Capabilities

Chapter 4
Justifying Your LC/MS Acquisition

Chapter 5
Locating Your LC/MS Alternatives

Chapter 6
Choosing Your LC/MS Capabilities

Part III- Solving Problems with LC/MS

Chapter 8

The Prelude to Solving Problems

Chapter 9

Solving Identity Problems

Chapter 10

Solving Quantification Problems

Chapter 11

Solving Target Problems

Part IV- Appendices

Appendix A

Sources for LC/MS Information

Appendix B

Practical Information for LC/MS

Appendix C

Interfaces- How do they work?

Appendix D

Mass Analyzers- How do they work?

Appendix E

Introduction

Problem-solving activity pervades all aspects and levels of human life. There are economic, political, social, and technological problems, all of which can be found in individual, group, organizational, or social levels. And because of the interdependent nature of most problems, they can affect or be affected by other problems at any level of human activity.

Arthur Van Gundy,
in ***Techniques of Structured Problem Solving***[1]

Solving Problems With LC/MS

To fully benefit from the information contained within *A Global View of LC/MS*, you must first have an appreciation for the formal and structured process of problem-solving. This (brief) introduction to "structured" problem-solving describes the general model we use repeatedly throughout this text. This book assumes that most activities relating to LC/MS are problem-solving activities. Therefore, describing LC/MS under this context will allow people to take appropriate actions, make appropriate decisions, and get appropriate results.

There is a large body of literature devoted to structuring the problem-solving process.[2,3,4] One of the best references in this field (and our personal favorite) is *Techniques of Structured Problem Solving* by Arthur VanGundy.[4] We have incorporated many of the problem-solving techniques and methods described in this reference throughout *A Global View of LC/MS*. We encourage our readers (all of which are problem-solvers we hope) to expand your knowledge and understanding of structured problem solving techniques. The rewards will be immeasurable. Also note that structured problem-solving is simply another name for the scientific method.[*5]

[*] The "Scientific Method" is described as "principals and procedures for the systematic pursuit of knowledge involving the recognition and formulation of a problem, the collection of data through observation and experiment, and the formulation and testing of hypotheses."

Structured Problem-Solving With LC/MS

We have chosen a three-stage problem-solving model for this text because it is simple, graphical, and relevant to our topic, LC/MS. This model includes three processes that are essential to successful problem solving: **intelligence**, **design**, and **choice**.[6] The processes associated with these stages are illustrated in Figure 4.

Intelligence is the first and possibly the most important stage in the problem-solving process. We begin the process by searching for information in order to fully characterize the problem. This process is represented by a converging funnel. Why a funnel? Generally, in this stage you are funneling all available information toward a clear problem statement. You also evaluate and analyze your available information in order to formulate a working definition of the problem. The product of this process should be a statement or restatement of the problem. This stage is relevant to all activities in LC/MS. Whether you are a newcomer to LC/MS and must define your technology needs or a seasoned practitioner and must define your experimental needs, intelligence is key to your success.

Design is a problem-solving stage characterized by a diverging funnel. Once your problem is clearly defined, you should seek out as many alternative solutions as practical. But which solutions? The problem-solver is often faced with either no alternatives or too many alternatives, particularly in LC/MS. For example, with so many different technologies in LC/MS, which ones are suited for your specific problem? What criteria should you use to evaluate and select one? Always keep an open mind in the design stage of problem-solving in order to consider all alternatives, even those not readily apparent.

The **Choice** stage involves the converging process of taking all of the solution alternatives generated in the design stage and reducing them down to a single selection. The problem-solver is faced with using their evaluation criteria to reduce the generally larger number of alternatives to a single solution implementation. In LC/MS, this stage is encountered when making appropriate technology selections, when choosing instrumentation, and when making day-to-day experimental decisions. In the best of worlds, the most appropriate solution will be selected at the end of this process.

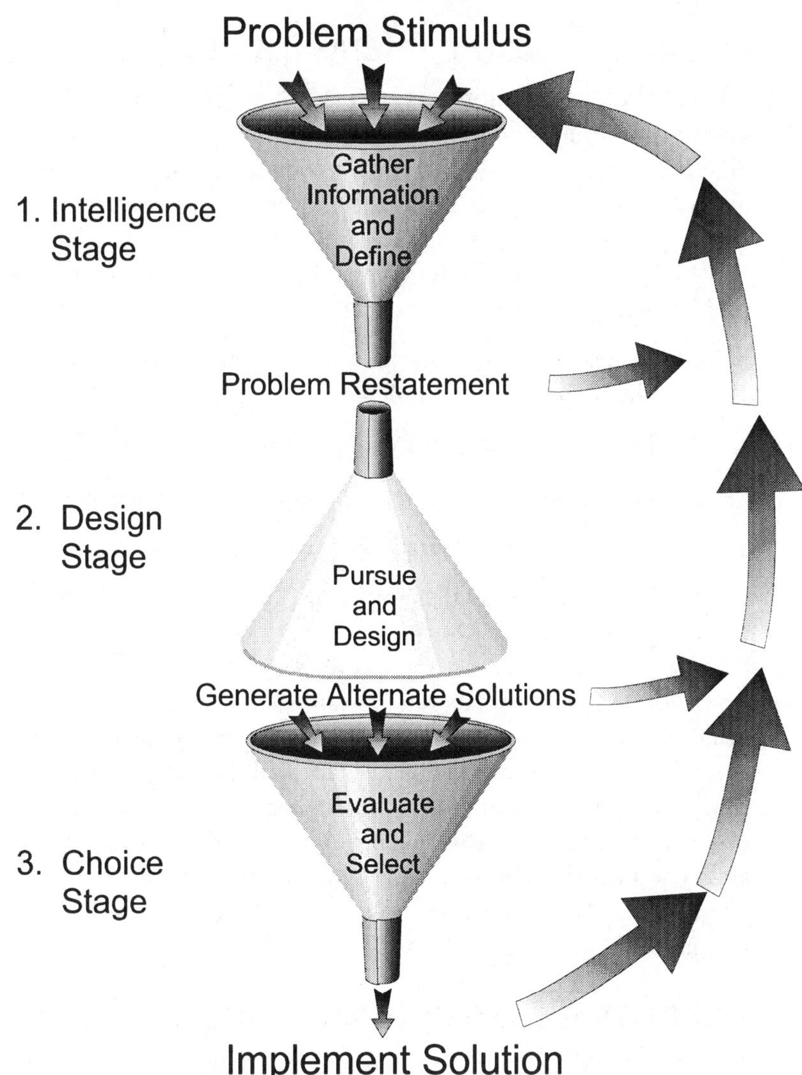

Figure 4- The three stages of problem-solving utilized throughout this book. All problems begin with some stimulus and hopefully end with the implementation of the appropriate solution. In LC/MS, this cyclical process will help the problem-solver focus on the important issues that lead to successful problem solutions.

Defining Your Problems In LC/MS

When do you have a problem? Ask a technician, a scientist, a lab manager and an administrator in your organization to define a given problem. The likelihood of a consistent response is negligible. This is why problem-solving within many groups or organizations can be quite difficult.

The following preconditions must be met in order to begin the problem solving process.[7]

1. A gap must exist between what is and what should be. The term "problem gap" is used for this discussion.
2. There must be an ability to measure this problem gap.
3. There must be an awareness that the problem gap exists.
4. There must be a motivation to close the problem gap.
5. You must have the ability and resources to close the problem gap.

If you are running laboratory samples, the problem gap will usually translate directly into a chemical measurement. The ability to measure the problem gap will be limited by your analytical and instrument technologies. The awareness of problems and how to measure the gap may be also limited by your ability to identify components in your sample. The motivation to close any problem gap may be influenced by economic, safety, regulatory and liability factors.

Benefits Of Structured Problem-Solving

The importance of the problem-solving process is sometimes overlooked in our formal education, but becomes painfully obvious as we enter into the business environment. In most successful laboratories, the analyst is paid to solve problems not run samples. Effective problem-solving skills are a critical component to success at every level of society. A recent symposium at PITTCON'95 concluded that "lack of problem solving skills" was one of the major issues facing analytical chemistry today[8]. Significant loss in time, efficiency, and frequent rework in the laboratory are a result. This is particularly important in industries and applications where accurate and efficient chemical analysis can mean the difference between in-spec or out-of-spec, competitive or non-competitive, on-time or delayed, profit or loss, healthy or ill, even life or death. Why we take the problem-solving process so lightly is an enigma.

The modern-day pharmaceutical industry is a good example of the benefit of combining problem-solving techniques with new measurement technologies, such as LC/MS. A tremendous demand is placed on the entire industry to produce new and effective products at an extremely rapid pace.[9] In order to meet these demands, they are forced to use better tools and better protocols (structured problem-solving). LC/MS and problem-solving go hand-in-hand with applications where speed, information and reliability are prerequisites for success, and in some cases, even survival.

Notes

[1] This is the first statement in the first paragraph of *Techniques of Structured Problem Solving* by Arthur VanGundy. This statement articulates the scope, persuasiveness, and interdependence of problems far better than we ever could.

[2] Simon, H.A. *The New Science of Management Decision*, rev. ed. Prentice Hall, Englewood Cliffs, N.J., 1977.

[3] Richards, T. *Problem-Solving Through Creative Analysis.* Gower Press, Essex, U.K., 1974.

[4] VanGundy, A.B. *Techniques in Structured Problem Solving*, 2nd ed. Van Nostrand Reinhold, New York, N.Y., 1988.

[5] *Webster's Ninth New Collegiate Dictionary*, Merriam-Webster, Springfield, Mass.,1983.

[6] Simon, H.A. op cit.

[7] VanGundy, A.B, op cit.

[8] "The Education and Training of Analytical Chemists," Symposium at PITTCON 1995, Atlanta, Georgia. (Overview of this symposium appeared in *American Laboratory*, July 1995.)

[9] Holmer, Alan F., Pharmaceutical Research and Manufacturers of America, *1997 Industrial Profile*, available on the internet at www.phrma.org. This is an excellent overview of the demands placed on the pharmaceutical industry to get products to the market faster and more efficiently. The increased demand for information about molecular structure in the process of discovering and formulating new drugs point directly at LC/MS as an indispensable tool for meeting these industry demands. The benefit of judicious use of LC/MS will likely be more effective and profitable drugs.

Part I

Is LC/MS right for you?

1. Intelligence

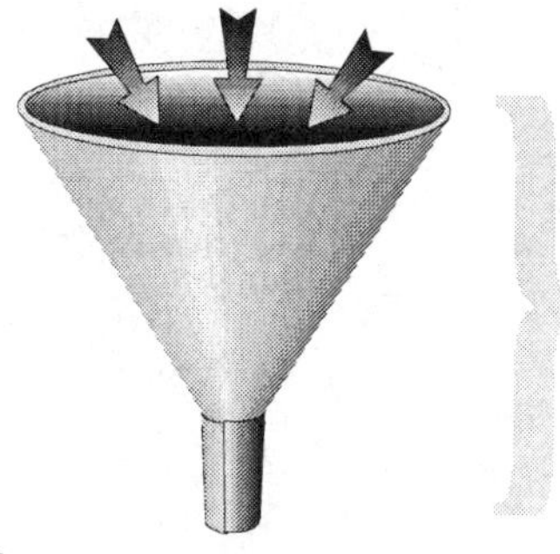

Chapter One- Does LC/MS solve your problems?

What are your needs?

2. Design

Chapter Two- What are your LC/MS alternatives?

Evaluate potential LC/MS technologies

3. Choice

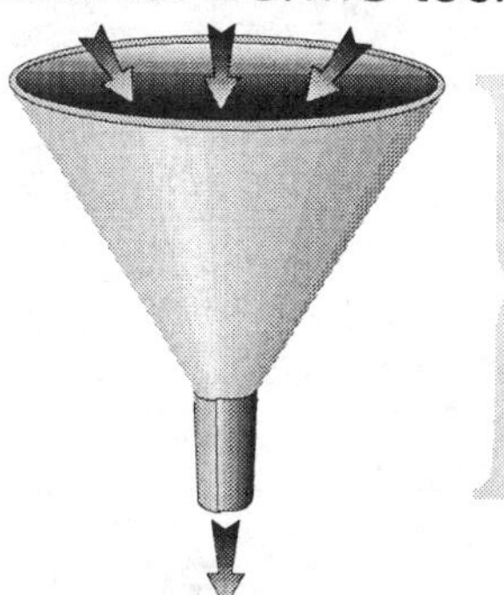

Chapter Three- Evaluating Your Alternatives

Select an appropriate LC/MS technology

Chapter One- Flow Diagram

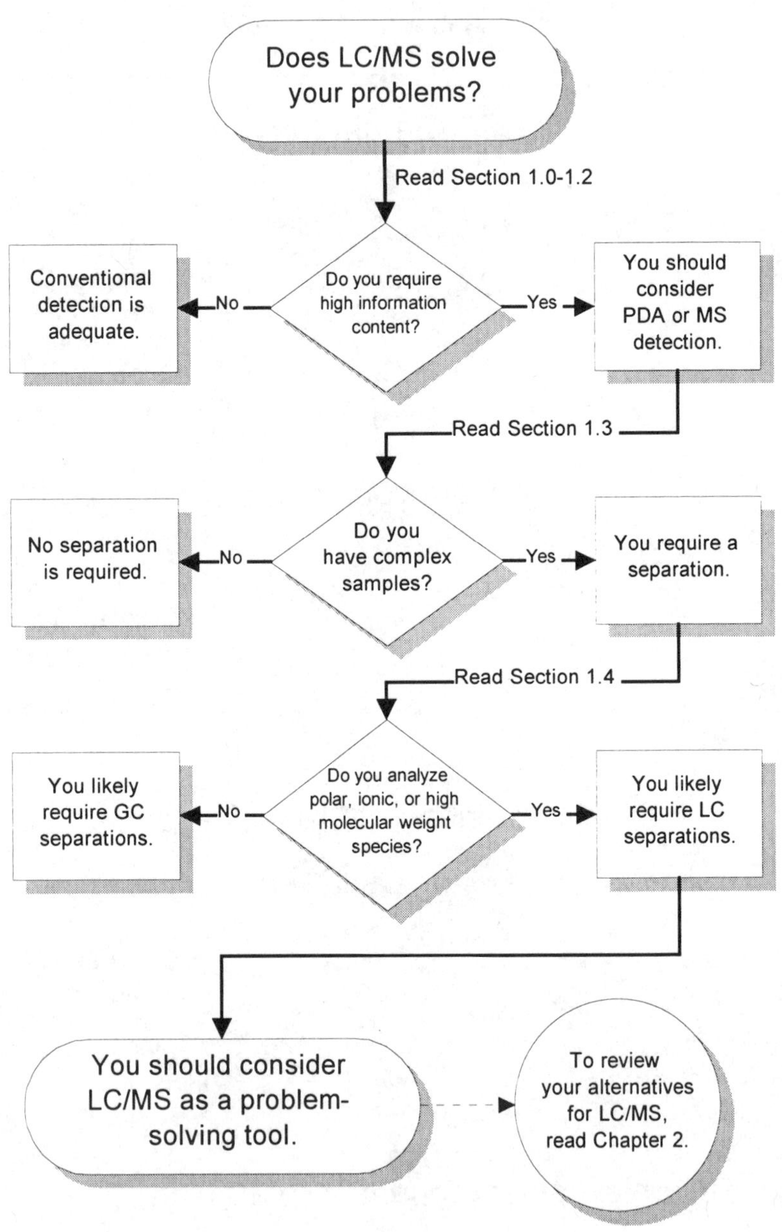

Does LC/MS solve your problems?
Read Section 1.0-1.2
Do you require high information content?
No
Conventional detection is adequate.
Yes
You should consider PDA or MS detection.
Read Section 1.3
Do you have complex samples?
No
No separation is required.
Yes
You require a separation.
Read Section 1.4
Do you analyze polar, ionic, or high molecular weight species?
No
You likely require GC separations.
Yes
You likely require LC separations.
You should consider LC/MS as a problem-solving tool.
To review your alternatives for LC/MS, read Chapter 2.

Chapter One

Does LC/MS Solve Your Problems?

(or, What problems does LC/MS solve?)

Man is a tool-using animal...Without tools he is nothing, with tools he is all.

Thomas Carlyle, in ***Sartor Resartus***[1]

1.0- Getting Started In LC/MS

If the periodic table is the "alphabet" of chemistry, then molecules are "words", and chemical reactions and interactions are "sentences." In this way nature communicates her past, our present, and continually writes scripts for her future. Every chemical system mankind encounters comes with a set of hidden instructions that we need only take the time to read. Unfortunately, we are blind and illiterate without the tools to decipher this language.

One important tool to observe the language of chemistry is liquid chromatography/mass spectrometry (LC/MS). Each letter (element), word (molecule), and sentence (reaction) in nature's library is accessible through LC/MS. LC/MS opens the doors to this library to levels that would make Andrew Carnegie proud.[2] Your start in LC/MS should include a tour of this library.

You must be living in a vacuum (pun intended) for the past few years if you haven't heard about LC/MS. Virtually every technical journal, trade journal, conference, and junk mail advertisement in the chemical industry seems to feature some review of, or announcement of, or product for, or compatibility with, LC/MS. If the continuum of background information about LC/MS hasn't annoyed you too much, you may be interested in finding more information about LC/MS.

To get started in LC/MS we prescribe the following:

1. Learn what your industry is doing with LC/MS. There is nothing wrong with being a follower (especially into the minefields of new technology). Search your trade literature for reviews of LC/MS that are specifically focused on your industry. (*Selected Reviews* are listed in *Appendix A.*)
2. Get to know someone who is already in this field, preferably, someone from your industry who is not a direct competitor. Is someone already doing LC/MS in your company? Develop a relationship with someone who has practical experience but who isn't trying to sell you anything. This person can serve as a sounding board as you get deeper into this technology.
3. Read *Chapters One* through *Four* of this book, *A Global View of LC/MS*. This will provide you with an overview of the field and a glimpse at the practical alternatives that you face with LC/MS as a problem-solving tool.
4. Attend at least one meeting of *the American Society of Mass Spectrometry* (ASMS). This meeting has everything. More than 1,200 papers and posters, covering almost any topic in the field of LC/MS, will be at this meeting. If there is an application for LC/MS in your field, it will likely be presented. If you cannot attend this meeting, review the abstracts on the Internet at **www.asms.org**. In addition, this meeting is a good source of LC/MS contacts within your industry.
5. Subscribe to *LC-GC* or *LC-GC International*. This publication routinely contains a wealth of practical information about LC/MS, including reviews, problem-solving, tips, tricks, market analysis, and products. This magazine is free to users of chromatographic equipment. You can subscribe at (218) 723-9477 or on-line at **www.lcgcmag.com**.
6. Depending on your background and job function, you or someone in your department should attend a course on LC/MS. Most courses are given in one- or two-day sessions. The instructors are generally experts in the field and present a broad and balanced overview of the field. On-site courses are also available from a variety of sources. (See *Appendix A* for

LC/MS Training.) We also provide current course information at our Internet site at **www.lcms.com**.

7. Review Appendix A of this book to familiarize yourself with key resources; including, references, conferences, books, journals, and training.

Dispelling The Myths

Getting started in LC/MS should also involve educating yourself in the "up-to-date" capabilities of LC/MS. Many legitimate concerns about various LC/MS technologies that existed 10 years ago may not apply today. Most of these "myths" no longer exist. Some of the myths include: "*Mass spec is too expensive,*" "*Mass spec is not rugged and reliable,*" or (our favorite) "*Mass spec is only semi-quantitative.*" Today, complete LC/MS systems sell for less than $100,000.[3] Some labs are running more than 1,000 samples per week (per instrument).[4] For many applications, the most accurate quantitative results come from mass spectrometry, particularly when isotopically labeled standards are used.[5]

Other myths we would like to dispel are: "*You need a Ph.D. to run one*." The state of the art in commercial LC/MS provides well engineered and highly integrated instrumentation. Most new instruments are designed for ease-of-use and repetitive/routine analysis. If you are a competent liquid chromatographer, you will be successful with LC/MS if you choose the appropriate system.

We recommend that you keep an open and objective mind about this field. It is changing so rapidly and growing so quickly that reviewers and experts are having a difficult time keeping up with the state-of-the-art. Fortunately, the focus of instrument manufacturers over the past several years has been away from esoteric technologies and more toward reliable, well-engineered LC/MS systems. LC/MS is truly becoming a practical, problem-solving tool.

1.1- Defining Your LC/MS Needs

In the *Introduction* we defined problems in terms of "problem gaps," the difference between what is and what should be. Getting started in LC/MS should also involve taking a closer look at the problems facing your lab and your organization (Your problem gaps!). Are you content with your current analytical techniques or methods? If you aren't satisfied, how would you change them? What improvements would you make to your methods? Do you want faster turnaround, higher throughput, fewer repeats, better accuracy, lower uncertainty, lower liability, purer products, more information.......? You and your organization are best equipped to define your problem gaps.

To help you evaluate and define the particular needs and problems in your laboratory we provide a simple worksheet (page 28). *For a down-loadable 8x11-inch copy of this worksheet (and all other worksheets) visit our Internet site at* ***www.lcms.com***. Take a hard look at where you are now and where you should be. This process of defining the problem gaps in your lab (or organization) may lead you to take a serious look at various alternatives in LC/MS to meet your future needs; conversely, this process may also lead you to conclude that you really don't have an immediate or significant need for LC/MS. Without clearly defining your problem needs, you will have a difficult time weighing your requirements for LC/MS technologies, or any other technology for that matter.

Here are some examples of problem gaps that may be closed using LC/MS technologies:

- Improve the quality of a product by identifying impurities,
- Increase the number of drug candidates by allowing higher throughput of samples and higher information content,
- Increase the efficiency of half-life screening in early animal studies by "N-in-1 dosing,"[6]
- Decrease the decision-time for control of a production process by providing specific information that is traceable to a spe-

cific chemical origin,

- Decrease cost-per-sample in a lab,
- Decrease sample work-up requirements in a lab,[7]
- Decrease labor content for each analysis,
- Increase productivity of a lab,
- Improve troubleshooting capabilities for LC methods,
- Decrease the time to develop methods,[8]
- Increase the defensibility of chemical evidence in forensic analysis,
- Improve the ability to measure disposition of compounds in the environment (e.g. metabolites/degradants of agrochemicals),
- Improve the diagnostic capabilities for given disease states,[9]
- Increase the shelf-life by identifying decomposition processes.

Defining your problems requires that you know your organization, your industry, your industry specific technologies, and your past. Defining your problems requires that you also have some vision of your future. Armed with this knowledge, you are best equipped to characterize your own problems.

Problem definition is a process of assessing your specific problem needs, defining the obstacles you may encounter along your problem path, and defining the constraints under which you are required to operate. *Worksheet 1.0* will help you to identify your problem needs. *Worksheets 1.1* and *1.2* will help you assess your technology needs. *Worksheet 4.0* (*Chapter Four*) will help you to assess your non-technology needs. The more clearly you define your specific needs, the better you will be able to evaluate and select appropriate solutions.

Worksheet 1.0- Defining Your Needs

Part A. Defining your problem gaps.

What are the conditions that currently exist in your lab?..in your organization? (e.g. 10 samples/day)	What should be the conditions that exist in your lab?..in your organization?(e.g. 100 samples/day)
1.________________	________________
2.________________	________________
3.________________	________________
4.________________	________________
5.________________	________________

Part B. How would you measure the difference between your current conditions and what should be for items above?[10] (e.g. higher throughput)

1.________________________________

2.________________________________

3.________________________________

4.________________________________

5.________________________________

Part C. Do you (or could you) have the ability and resources to close these problem gaps?[11] ❑ yes ❑ no

Comment: ________________________________

Part D. Do you or your organization have the motivation to close these gaps?[12] ❑ yes ❑ no

Comment: ________________________________

Summary: The problem-solving process is the process of closing a problem gap; it first requires that you identify your problem, then have some parameter to measure the problem, have the ability and resources to solve the problem, and have the motivation to solve the problem.

Problem Restatement: Can you restate your problems in terms of a chemical measurement? Describe the requirements you anticipate for that measurement.

____

____

Note: Copies of this worksheet can be downloaded from **www.LCMS.com**.

1.2- LC/MS Yields Enhanced Information

It is interesting that experts in the science of problem-solving define the existence of a problem by the ability to measure the problem.[13] In other words, in order to solve a problem, we must be able to measure it. This "measurement" requirement for problem-solving has been demonstrated in many areas of science and technology, but medicine and biotechnology have led the way. The ability to make specific chemical measurements through developments in a wide variety of analytical techniques, including LC/MS, has fueled the rapid growth and advancements in these fields.

In Section 1.1, we ask you to consider the problems in your organization or lab in terms of your specific measurement requirements. Now we ask you to consider the information that you need from your measurements. The information we acquire from our measurement is used to formulate solutions to our problems. The progression from data, to information, to knowledge, to wisdom[14] is appropriately applied to problems in LC/MS. If "measurement" is the prerequisite for a problem to exist, then "information" is the prerequisite for a solution.

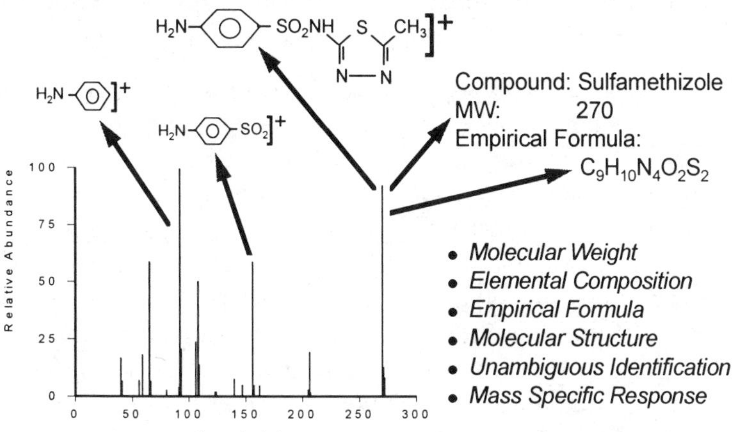

Figure 1.0- The mass spectrum provides more compound specific information than virtually any other detector available to the analyst. (EI of Sulfa-methizole)

The information-rich content of mass spectral detection is the primary reason for considering LC/MS for your specific chemical measurements (Figure 1.0). Mass spectrometry (MS) can provide information not available by any other mode of detection. That means MS can solve problems that no other technology is capable of solving. Table 1.0 itemizes the types of chemical information available from MS detection, along with the benefit that the information provides.

Table 1.0-Chemical Information Derived from MS

Information Type	Benefit
Molecular Weight	Provides Identity
Molecular Structure	Provides Identity
Elemental Composition	Provides Identity
Sequence	Provides Identity
Mass-Specific Response	Enhances Quantification

Figure 1.0 shows a typical mass spectrum of the drug sulfamethizole to illustrate the information content of a single mass spectrum. Labeled are the sources of information from a mass spectrum that lead to compound identification and selective quantification. A high degree of compound specific information is obtainable from virtually any mass spectrometer and any ionization technique. An important criterion for evaluating various detection technologies is specificity. Table 1.1 compares mass spectrometry (MS) to ultraviolet (UV) detection in terms of specificity. Clearly, in problems that require compound specific information, mass spectrometry is the preferred analytical solution.[15] (Evaluating the distinctions between one mass spectrometry technique and another is discussed more fully in *Chapter Three*.)

Hidden Benefits To MS Detection

As stated, the obvious benefit of using a mass spectrometer as a detector is the opportunity it affords to quickly identify sample components. The value of compound specific information is fairly obvious, particularly for the analysis of "unknown" sample components. However, several hidden benefits may be derived from the information rich detection associated with mass spec. These so-called hidden benefits are potentially shorter method development times[16], potentially shorter sample run times[17], and potentially lower cost per sample.

Because of the high compound specificity of mass spectral detection, the method developer can in some cases take short-cuts in developing their chromatographic methods. A method developer can use a unique analyte-specific mass measurement to isolate the response of the analyte from other components in the sample (enhanced selectivity). This allows the developer to lower the requirements for separation that are typical with conventional detectors (e.g. baseline-baseline separations). The mass selectivity allows the developer to remove spectral interferences without compromising their analytical results.

Lowering the requirements for baseline separations with mass spectrometers can also allow the developer to shorten their sample run times. Two-minute analyses are quite common with LC/MS for target assays. Production labs are running more than a thousand sample per week. At these sample rates the cost per sample could become competitive with LC/UV detection, with the added benefit of much lower uncertainty in the analytical result.

Table 1.1- Specificity of Various LC Detectors

Specificity	UV	MS	MS/MS/(MS)
Class Specific	✓	✓	✓
Compound Specific		✓	✓
Moiety Specific			✓

An often overlooked benefit of MS for problem-solving is found in its standardization alternatives (See *Chapter Ten*). The use of analog or labeled standards to accurately monitor sample treatment and effects from chemical or physical interferences gives mass spectral results unparalleled accuracy for complex samples.

Table 1.2 lists examples of analytical problems that can be solved with mass spectrometric detection and the benefit of using MS instead of alternative detectors.

Mass Spectrometer vs. PDA Detection

One important question that you should answer in your problem evaluation is "Do you need compound-specific detection?" If so, you should strongly consider MS for your detection mode. Mass spectrometry has no rival for problems that require a high degree of compound specific information. Figures 1.1 and 1.2 show a comparison of results from photodiode array (PDA) versus mass spectral detection using LC/PDA/MS of an anti-inflammatory drug and its oxidative metabolite. The information contained within the UV traces is limited while the mass spectra contain molecular ions, elemental information, and structural information to allow unambiguous identification of each component. Your ability to identify a component in a mixture is significantly enhanced with MS.

Class-specific detectors such as UV make it difficult to identify complete unknowns. HPLC, using conventional detection methods, are generally developed to confirm the presence of a known or a "target analyte."[18] Compound-specific information is not absolutely required for simple mixtures where the identity of components is not in question and the methods have significant validation data to support the reliability of the results.[19]

Compound-specific detectors such as MS gives you unambiguous identification of sample components. The rich chemical information available from mass spectra allows you to identify unknown components in samples, even when a standard is not available for comparison. MS is ideal for drug metabolism, product stability, natural products elucidation, and forensic analysis; all areas of analysis where structural elucidation and identification is critical.

With MS, you don't have to know what you're looking for before you perform the analysis; with UV, you do. With MS, you can identify compounds as they are produced. This has a significant advantage in areas of research and industry where speed and accuracy of the sample information is critical to solving a problem or making a decision. Drug discovery and diagnostics are two important areas where rapid identification of sample components is key to successful problem-solving.

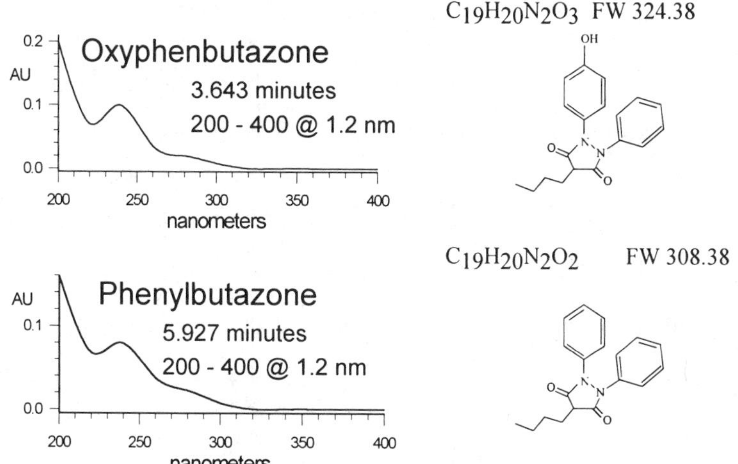

Figure 1.1- Comparison of results from diode array detection of non-steroidal anti-inflamatory (NSAI) drug phenylbutazone and an oxidative metabolite. No significant difference exists.[20]

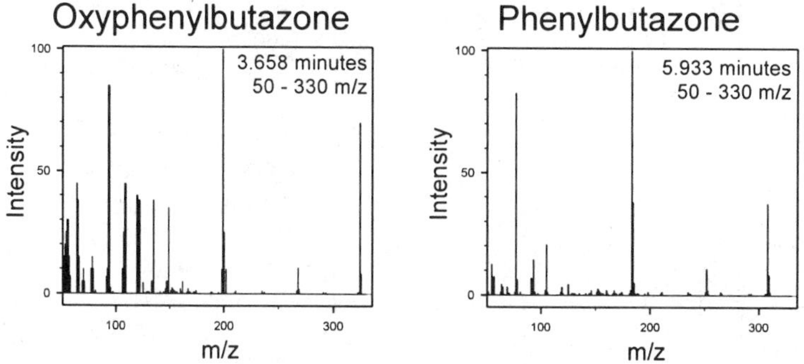

Figure 1.2- Comparison of results from mass spectrometry of the non-steroidal anti-inflamatory (NSAI) drug phenylbutazone and an oxidative metabolite. These electron ionization spectra provide information for unambiguous identification of analytes.[21]

Table 1.2-Problems Requiring Compound-Specific Detection

Problem	Examples	Benefit of MS
Identification of Unknowns	• Purity • Product Integrity • Toxicants • Pollutants • Drug Discovery	• Fast Interpretation • Unambiguous • Yields Mass • Yields Structure • Increased Confidence
Isolation of Unknowns	• Screening Complex Matrices • Screening Natural Products	• Fast Isolation • Yields Unique Responses • Yields Clues for Further Separation
Identification of Target Analytes	• Pharmacokinetics • Broad-Based Environmental Methods • Rapid Screening-Prenatal	• Confirms Presence • Confirms Absence • Unambiguous • Legally Defensible Evidence
Isolation of Target Analytes	• Analyzing Complex Matrices • Analyzing Variable Matrices	• Yields Unique Responses • Yields Unique Responses for Quantification • Simplifies Integration • Allows Internal Standardization
Characterization of Knowns	• Filing an IND • Filing a Patent • Preparing a Certified Reference Standard	• Unambiguous Identification • Traceable to Compound • Strengthens Your Filing Position

Enhanced Identity And Specificity

Table 1.3 lists representative examples of "identity" problems solved in various industries. We present these applications of LC/MS to give you published examples of problems that have already been solved by LC/MS. Table 1.4 provides examples of problems solved by using the "enhanced specificity" of MS. If you are new to the field of MS, you may want to read the references that pertain to your industry to get a better grasp of the state-of-the-art in your field.

For more comprehensive reviews of LC/MS applications, we recommend "LC/MS Update" published by *HD Science* (See *Appendix A* for availability.). Several other texts and review articles also cover various fields of LC/MS and MS quite completely. These are listed in *Appendix A* under *Reviews*.

Table 1.3- Examples of LC/MS for Identity Problems

Industry/ Application	Problems Requiring MS Identification	Solution to Problem
Pharmaceutical Sciences	Identification and characterization of compounds that have stimulative effects on biological processes.	LC/MS/MS analyses of compounds that stimulate DNA synthesis.[22]
Clinical Sciences	Diagnosis of metabolic disorders.	Electrospray LC/MS/MS of amino acids, organic acids and fatty acids.[23]
Biochemistry & Biotechnology	Identification of the molecular weight, sequence and post-translational modifications of biotechnology products.	Characterization of minor impurities from the production of eicosapeptide with electrospray LC/MS.[24]

Table 1.3- Examples of LC/MS for Identity Problems (cont.)

Industry/ Application	Problems Requiring MS Identification	Solution to Problem
Food Chemistry	Identification and confirmation of adulterants/residues in food stuff.	The use of Flow-FAB LC/MS for the detection of tetracycline antibiotics in honey.[25]
Forensic Sciences	Identification of explosives from a blast site.	Identification of military and commercial explosives from post blast residues using thermospray LC/MS.[26]
Environmental Chemistry	Confirm contamination in soils, water, sludge, and air.	The use of particle beam (EI) LC/MS to identify atrazine and its metabolites in soil samples.[27]
Synthetic Chemistry	Determination of molecular weight and elemental composition of synthetic and semi-synthetic compounds.	Confirmation of the coalescence of buckyballs, C_{60} and C_{70}, to higher fullerenes.[28]
Polymer Chemistry	Determination of the molecular weight distribution and polydispersity of polymers.	Use of LC/MS in determining the molecular weight and polydispersity of polyamidoamines.[29]
Petroleum Chemistry	Identification of geochemical markers in the assessment of paleo-environments for the exploration of oil resources.	Use of APCI LC/MS to identify metalloporphyrin pigments from Triassic Serpiano oil shale.[30]

Table 1.4- Examples of LC/MS for Specificity Problems

Industry/ Application	Problems Req. MS Specificity	Solution to Problem
Pharmaceutical Sciences	Determining the number and identity of drug metabolites.	LC/MS (PB & APCI) of a pyrrolo-pyrimidine based antioxidant.[31]
Clinical Sciences	Confirm the presence of bio-markers associated with inherited metabolic disorders in fetal blood or urine.	Rapid screening and quantitation of amino acids, organic acids, fatty acid oxidation disorders with electrospray LC/MS/MS.[32]
Biochemistry and Biotechnology	Confirm the presence and absence of peptides from enzymatic digest.	Use of electrospray to determine the molecular weight of peptides and oligonucleotides.[33]
Food Chemistry	Confirm the presence of adulterants in food stuff.	Detection of orange juice adulteration by means of ICP-MS.[34]
Forensic Sciences	Confirm the presence of explosives from-blast residues.	Thermospray LC/MS in the analysis of explosives from a blast site.[35]
Environmental Chemistry	Confirm the presence of pollutants,	The use of ICP/MS for detection of metal ions.[36]
Synthetic Chemistry	Confirm the extent of synthesis in combinatorial libraries.	The use of MALDI-MS to deduce the sequence of peptide from a combinatorial library.[37]
Polymer Chemistry	Confirm molecular weight, repeating units, end groups, and polydispersity of polymers.	The use of electrospray LC/MS to determine the molecular weight and polydispersity of polyamido-amine.[38]
Petroleum Chemistry	Confirm the presence of production grade crude oil.	The use of APCI LC/MS in the detection of geo-porphyrin pigments.[39]

1.3- LC/MS Addresses Sample Complexity

Another essential consideration for determining your problem needs for LC/MS is your sample complexity. The high information content of mass spectra does not come without some constraints. You cannot interpret your mass spectral information unless it represents a single chemical species. Unfortunately, most of the world is not pure. Typical "real world" samples will vary widely in matrix and analyte complexity. Solving problems with even simple mixtures will usually require a separation to isolate your analytes for detectability and interpretation. Separations are also necessary to remove interferences that affect analyte response and confuse interpretation. Most applications using MS detection will require a separation.

Figure 1.3 illustrates the relationship between information requirements and complexity of samples for solving problems with LC/MS. The shaded area represents the problem-solving domain of LC/MS. This figure is presented in order to rank the various separation and detection technologies for comparison in meeting specific problem needs. This figure may give you a qualitative picture of where LC/MS fits in the scheme of your information needs and your sample complexity.

Analysis with MS requires separations for the following reasons:

- To accurately interpret a mass spectrum, the spectrum must be related back to a single chemical species.
- To accurately quantify an analyte, the mass specific response must be related back to a single chemical species.
- To accurately quantify an analyte, the analyte response must be consistent.

We separate analytes from the matrix to remove spectral interferences that overlap our analyte spectra; and we separate to remove chemical and physical interferences that affect the response of our analytes and standards. Table 1.5 describes the effect of interferences on your analytical results.

Do You Have Complex/Variable Matrices?

The nature of the matrix of your sample may also mandate a separation (e.g. soils, tissues, blood, sludge). Although matrix and complexity are essentially the same problem, one should note that the sample matrix can affect your analytical results beyond the addition of chemical and physical interferences in the mass spectrometer. Your matrix may also contribute to less than complete recovery of your analyte (e.g. analytes bind to proteins, analytes trap inside insoluble sample materials, analytes are not distributed homogeneously within the sample).

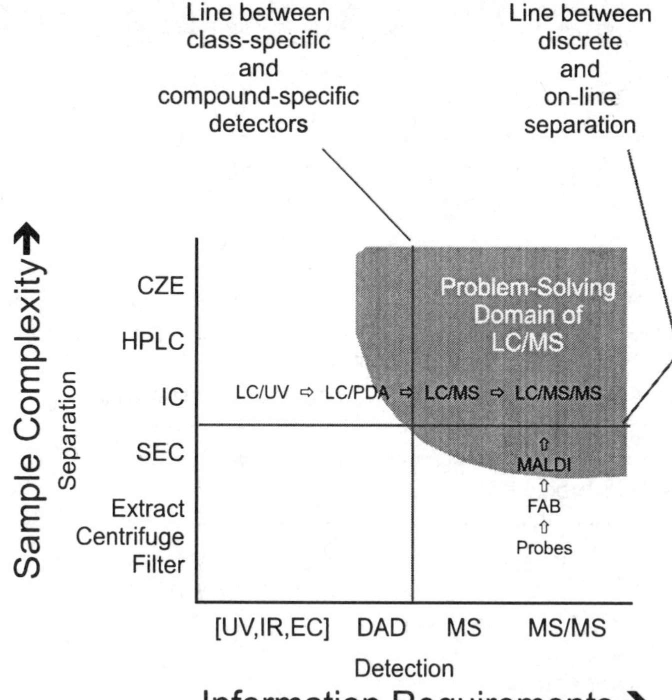

Figure 1.3- Problem-solving for LC/MS covers the domain where samples are too complex for discrete separation techniques and information requirements exceed those of conventional detectors.

The aforementioned effects on analytical response because of spectral and chemical interferences are clearly compounded when you have samples of variable matrix or background composition. Variation in sample matrix creates an uncertainty with method validation. If you have a variable matrix, you will have difficulty finding a representative matrix blank with which to validate your method and ensure reproducibility and consistency of your analytical response. As a general rule, the better your separation, the less likely that you will encounter interference effects.

Table 1.5- Effects of Interferences on Mass Spectrometry

Cause	Effect
Spectral Interferences- Background compounds yield ions that overlap with analyte ions in the mass spectrum.	Prohibits accurate quantification and confuses or negates interpretation and identification of analyte, particularly if these compounds originate in the sample.[40]
Chemical Interferences- Presence of one or more components in the sample may suppress or enhance the response of the analyte through a chemical process.	Degrades the accuracy of quantitative analysis. Interferences may also change the spectrum qualitatively (adduction) and complicate the identification process.[41]
Physical Interferences- Presence of one or more components in the sample may suppress or enhance the response of the analyte through a physical process.	Degrades the accuracy of quantitative analysis.[42]

LC/MS Is Suited For Trace Analysis

Trace analysis is by far the most difficult challenge in chemical analysis. The effects of spectral and chemical interferences are exaggerated when you reduce the level of analyte relative to everything else. Separations are especially important when having to deal with interferents at a concentration many orders of magnitude higher than your analyte. As a rule, we use separations in trace analysis to isolate trace analytes from respective interferences and to concentrate the trace analyte into as narrow a chromatographic band as possible.

Table 1.6 provides examples of separation problems in various industries to illustrate some of the issues relating to isolation of sample components or issues relating to removal or separation of analytes from interferences. Very few industries or applications don't have to deal with separation of complex mixtures on a daily basis.

Higher Selectivity From Separations

The majority of methods employing chromatography and mass spectrometry (GC/MS or LC/MS) use retention time as the initial selection criterion for target analyte identity, before library matching and quantification. Matching target analyte retention time and associated mass spectrum from a sample with those from a target reference standard produce the most reliable results attainable by any analytical technology today (described in detail in *Chapter Eleven*). The separation step adds one additional dimension (time) to identity when you are required to provide unambiguous evidence of the presence or absence of a target component. Note: Target methods prove the presence (and absence) of a target component by their detection (or lack of detection) at a given retention time.

By combining MS with a chromatographic technique, we can effectively separate sample components in both time (chromatography) and mass (mass spectrometry) to isolate, identify, and quantitate virtually any component in any mixture. With this combination, we can literally find the proverbial "needle in a haystack." One of the most compelling "needle in a haystack" examples is told by Professor Hunt and associates from the University of Virginia. He relates the isolation and identification of a single membrane epitope (polypeptide) for mammalian cancer from a cell culture containing literally thousands of other membrane proteins.[43]

What Are The Benefits Of A Separation?

- Separations are required to allow interpretation of single analyte spectra in the presence of other species.
- Separations are required to allow quantitation of target analyte masses in the presence of spectral interferences.
- Separations increase the reliability of the identification of target analytes by using retention time as a selection criterion.

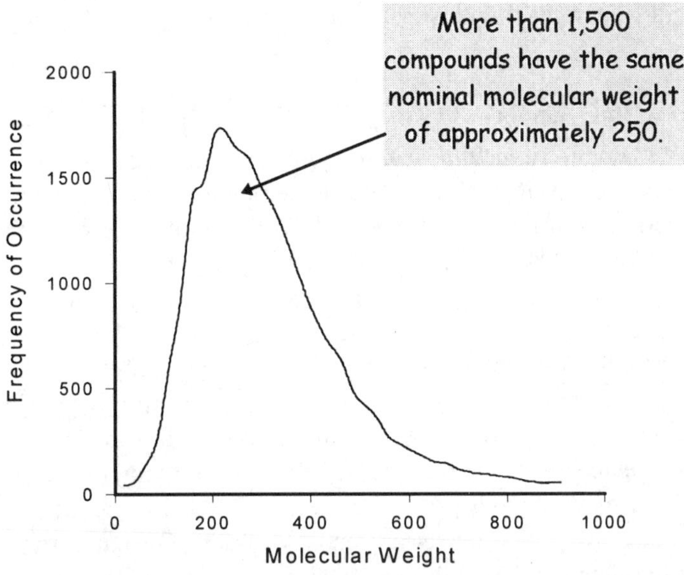

Figure 1.4.- The frequency of occurrence versus molecular weight for the distribution of compounds found in the Wiley Mass Spectral Library (6th Edition) showing the likelihood of isobaric interferences, particularly for masses 100 to 500.[44] LC separation is an effective way to remove these interferences.

Table 1.6- Examples of LC/MS for Separation Problems

Industry/ Application	Problems Req. Separation	Solution to Problem
Pharmaceutical Sciences	Screening for microbial metabolites in culture filtrate brutish.	Determine peak purity and identification by comparison to spectral libraries.[45]
Clinical Sciences	Screening for therapeutic drugs and their metabolites.	Reversed-phase separation, ID and quantitation of antiepileptic drugs and metabolites.[46]
Biochemistry and Biotechnology	Peptide mapping.	Comparison of the peptide retention map of a manufactured protein with the retention map of a reference protein.[47]
Food Chemistry	Screening for and separation of flavoring components.	Separation and identification of fully methoxylated flavones in citrus juices.[48]
Forensic Sciences	Screening for impurities in manufacturing by-products to determine origin.	Analysis of an illicit drug sample; identifying cocaine itself, impurities, and manufacturing by-products.[49]
Environmental Chemistry	Screening for environmental pollutants.	Reversed-phase separation and quantitation of pesticides.[50]

Table 1.6- Examples of LC/MS for Separation Problems (cont)

Industry/ Application	Problems Req. Separation	Solution to Problem
Synthetic Chemistry	Separation of reaction components and homologues from a synthetic reaction.	Scale up to the preparative separation of eosin-Y and its halogenated homologues.[51]
Polymer Chemistry	Separation and analysis of polymer additives.	Extraction and separation of antioxidants and UV absorbers from polypropylene.[52]
Petroleum Chemistry	Separation of geo-markers in heavy crude oil fractions.	LC-GPC were used for the determination of porphyrins in crude oil.[53]

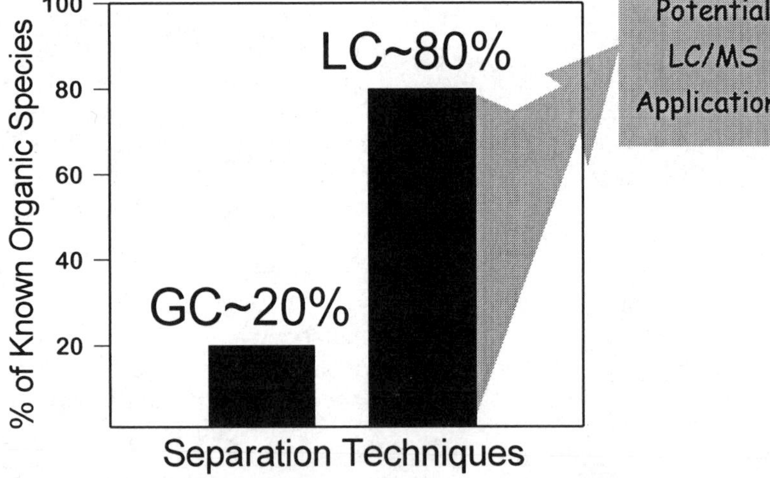

Figure 1.5- "It has been estimated that only 20% of known organic compounds can be satisfactorily separated by GC, without prior chemical modification of the sample" (Snyder and Kirkland, *Introduction to Modern Liquid Chromatography*).[54] LC, on the other hand, applies to a significantly broader population of compounds.[55]

1.4- LC/MS Accommodates Polar Compounds

We cannot adequately answer the question, "Does LC/MS solve your problems?" unless you first identify and define your problems, and the measurements associated with those problems (*Section 1.1*); defines the information requirements of your measurements (*Section 1.2*); defines the complexity of your samples and how it will effect your measurements (*Section 1.3*); and lastly, defines the properties of your samples in order to select the appropriate measurement techniques.

This section will address the last issue, sample properties. The chemical properties of your sample generally affect every aspect of your analytical measurement- from sampling, to storage, to separation, and finally, to detection. LC/MS has received considerable attention in the past several decades because it applies to a very broad range of compounds that vary widely in molecular weight, ionizability, and polarity.[56]

What Compounds Are Amenable To LC/MS?

In general, we can say that all compounds that can be separated with liquid chromatography (or the electrophoretic techniques) are amenable to LC/MS. Figure 1.5 illustrates that the potential application of LC separations is several times that of GC. The general properties of LC/MS sample components are as follows:

- **High Molecular Weight-** LC is the preferred separation technology for most compounds over 250u.
- **Ionic-** LC technologies (including IC and CE in the broadest sense) are the only way to separate species that are permanently ionized in solution.
- **Ionizable-** A large majority of ionizable compounds are separated with LC because of the potential reactivity and instability of these compounds (e.g. most drugs and metabolites).
- **Polar/non-ionic-** Many polar molecules decompose at the temperatures required to introduce them into a gas chromatographic column (e.g. sugars).

Figures 1.6(a) and 1.6(b) give examples of compounds that have properties of high molecular weight, charge, and solubility.

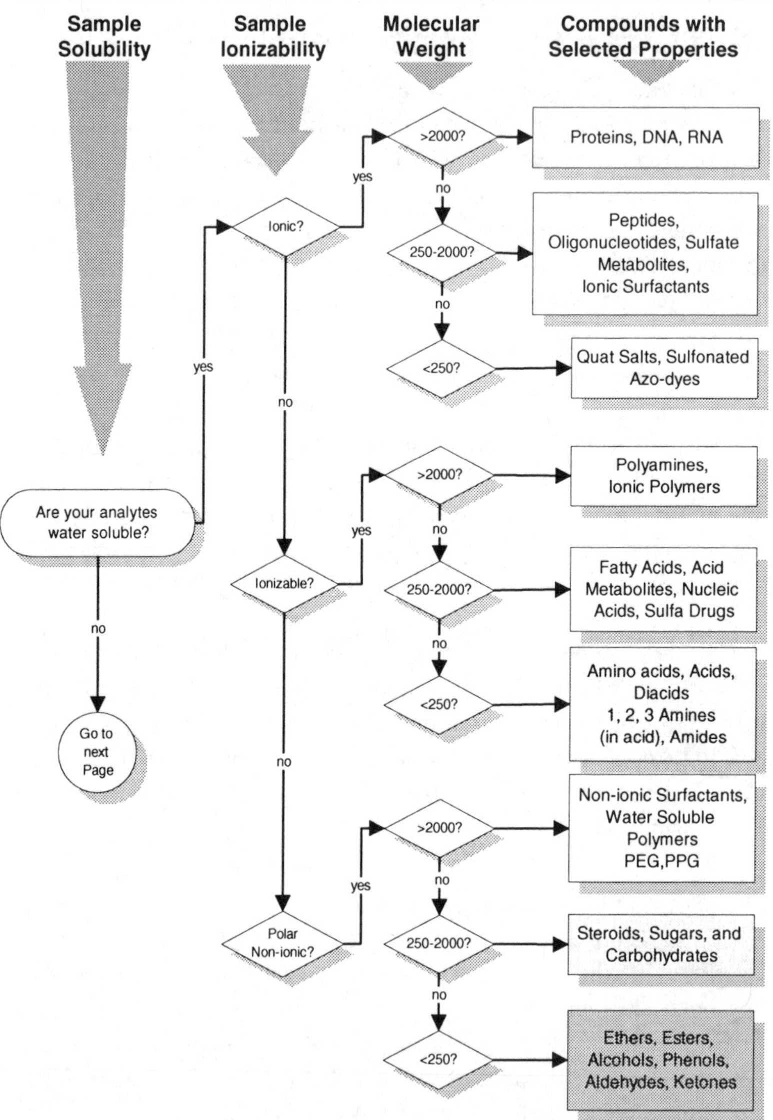

Figure 1.6(a)- Examples of water soluble compounds with different charge and molecular weight properties. These properties can be used to evaluate the appropriate selection of LC versus GC separation techniques.

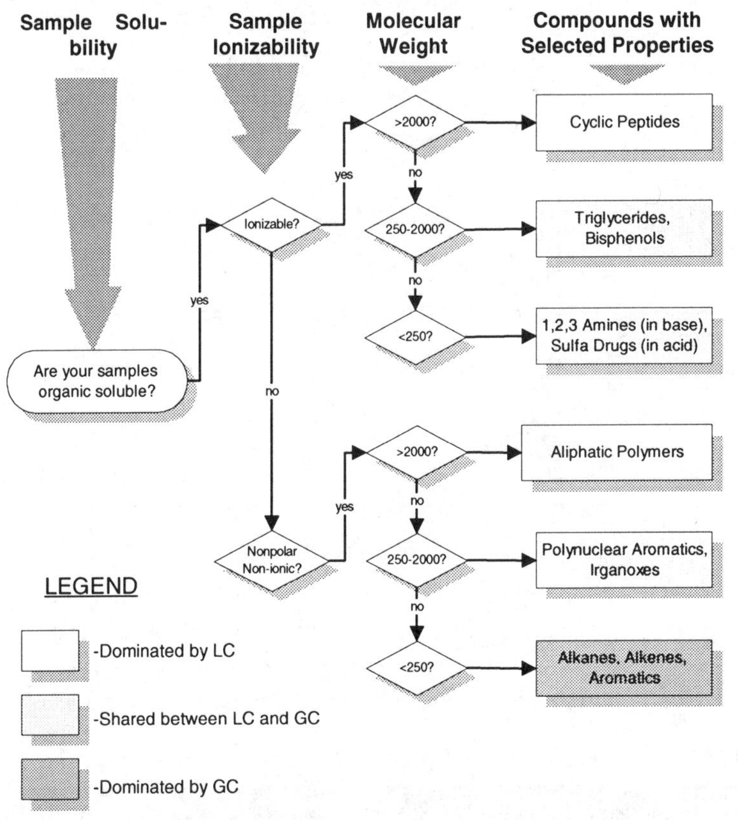

Figure 1.6(b)- Examples of organic soluble compounds with different charge and molecular weight properties.

LC Versus GC Separations

If your lab is already using LC, you most likely have addressed the question of LC versus GC. The applications domain of LC/MS (or LC) consists of compounds that tend to be "nonvolatile and thermally unstable." The boundaries of applicability of LC or LC/MS are not clearly defined in the literature. To add to the confusion, many compounds are capable of separation with both. In Figure 1.7 we have mapped the domain of LC and GC using our own quantitative model based on polarity and molecular weight.[57]

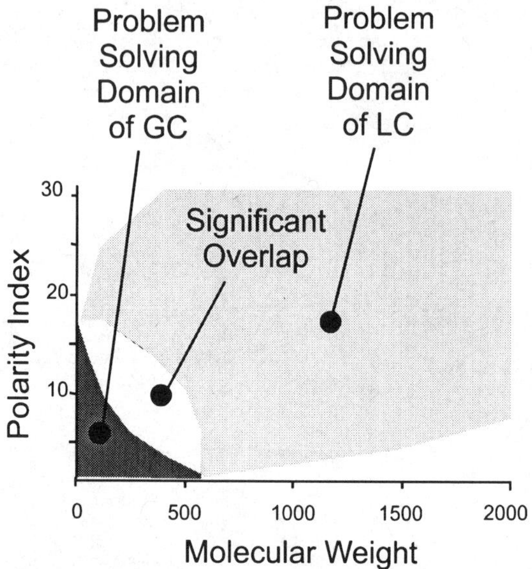

Figure 1.7-Graph showing the analytical domain of GC and LC using polarity versus molecular weight. There is also a significant overlap between GC and LC in the most populated region for molecular mass (100 to 500u).[58]

Table 1.7- Examples of Compd. Classes Analyzed by LC/MS

Industry/ Application	Application	Typical Compounds
Pharmaceutical Sciences	• Metabolic profiling • I.D. and purity • Formulation • Pharmacokinetic profiling	• lipid lowering drug: lifibrol[59] • carotenoids[60] • degradation of cephalosporins[61] • phenylbutazone & metabolites[62]
Clinical Sciences	• Metabolic disorders • Therapeutic drug monitoring • Assay confirmation	• amino acids, organic acids[63] • antiepileptic drugs and metabolites[64] • backup to EMIT assay[65]

Table 1.7- Examples of Compd. Classes Analyzed by LC/MS(cont.)

Industry/ Application	Application	Typical Compounds
Biochemistry and Biotechnology	• Post translational modification • Non-covalent complexes • Nucleic acids • Peptide mapping	• histactophilin modified by myristic acid[66] • margatoxin/peptides[67] • oglionucleotides[68] • hemoglobins[69]
Food Chemistry	• Drug residues • Artificial sweeteners • Flavor components • Fungal metabolites	• doxycycline in bovine tissues[70] • diet sweeteners[71] • flavones in citrus[72] • mycotoxins[73]
Forensic Sciences	• Explosives • Performance-enhancin agents and metabolites • Adulterants in consumer products	• commercial & military[74] • cortiosteroids[75] • trytophan analogs, eosinphilia-myalgia syndrome[76]
Environmental Chemistry	• EPA: Appendix-VIII • Quaternary ammonium salts • Pollutants in hazardous waste sites • Textile dyes	• brucine, PTU, DES[77] • paraquat & diquat[78] • chlorophenoxy herbicides[79] • non-ionic textile dyes[80]
Synthetic Chemistry	• Combinatorial libraries • Reaction products	• nonpeptide angiotensi antagonist[81] • reporting structures in journals[82]
Polymer Chemistry	• Additives • Molecular weight & polydispersity	• antioxidants, Irganox 1010[83] • polyisoprene, polyethylene oxide, and poly-dimethyl-siloxane[84]
Petroleum Chemistry	• Oil exploration • Fractionation of oil	• geo-porphyrins[85] • PAH[86]

The Mass Spectrometry Perspective

Most of the published results with LC/MS originate from the MS community. Much of the emphasis in this community has not been on LC detection; rather, it has been on enhanced ionization capabilities and increased applicability of MS for chemical analysis. LC/MS has, to a large degree, revolutionized MS by opening entirely new worlds for mass analysis that have yet to be exploited. Electrospray has led the way by providing spectra of species that no one dreamed possible only one decade ago. At one time, the traditional mass spectrometry lab would send impure samples back to the synthesis lab for repurification. Now on-line separations allow much higher complexity and less pre-treatment of samples.

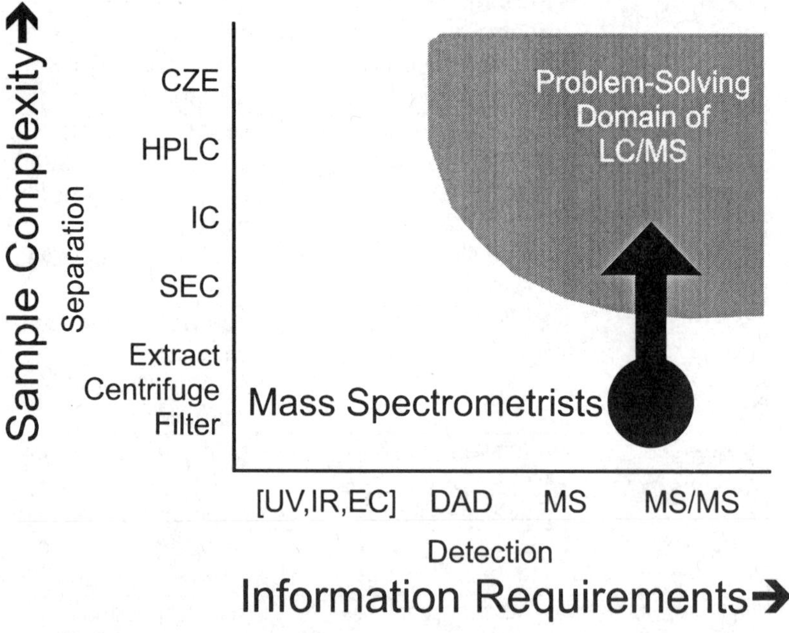

Figure 1.8- A significant number of an estimated 10,000 mass spectrometrists around the world have embraced LC/MS as a tool for problem-solving to deal with more complex samples.[87]

The Chromatography Perspective

The chromatography community is as diverse as science itself. Virtually every scientific discipline has separation problems that require LC. They have been in quest of the "universal" detector for decades with LC/MS receiving the lion's share of the attention. LC/MS technologies have been around for many years, however, the engineering of reliable, low cost, easy-to-use systems (not technology) has precluded broad acceptance of LC/MS by the chromatography community. Finally, the introduction of well-engineered commercial products designed specifically for the needs of the various chromatography communities makes universal detection for LC a reality.

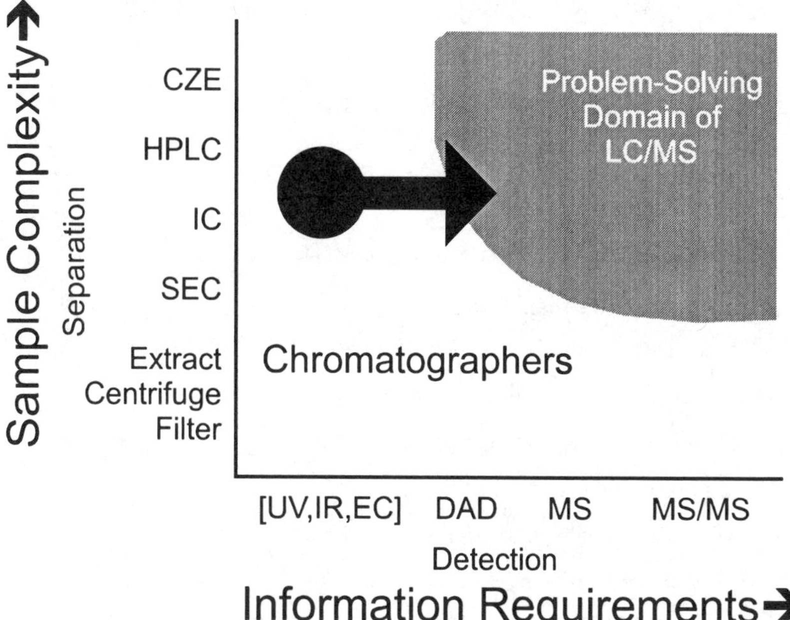

Figure 1.9- An estimated 150,000 chromatographers around the world have yet to use information-rich MS detection as a tool for problem solving, primarily because systems were not available to meet their needs (low cost, reliable, easy to use). This situation has changed with the advent of LC "detector" products.[88]

Worksheet 1.1- Evaluating Your Information Needs

Check the box that is appropriate for your problems.

1. Do you need to identify sample components:	Yes	No
– by determining molecular weight of compound?	❐	❐
– by determining empirical formula of compound?	❐	❐
– by determining structure of compound?	❐	❐
– by determining isomeric structure of compound?	❐	❐
– by determining structural confirmation?	❐	❐
– by determining sequence of compound?	❐	❐
– by producing a noncovalent interactions?	❐	❐
– by producing a legally defensible identification?	❐	❐

2. Do you need to elucidate component structures:	Yes	No
– to determine modifications to known structures attributable to your products and processes?	❐	❐
– to identify degradation pathways for your products or formulations?	❐	❐
– to identify synthetic pathways for your reactions?	❐	❐

3. Do you need to enhance selectivity:	Yes	No
– to isolate a target from a complex mixture?	❐	❐
– to isolate a target analyte from complex matrices?	❐	❐
– to eliminate background and interferences?	❐	❐
– to improve detection limits?	❐	❐
– to improve precision of your measurement?	❐	❐

Note: Copies of this worksheet can be downloaded from **www.LCMS.com**.

If your information requirements have prompted you to check any **Yes** boxes, you should consider mass spectrometry as an alternative detection technology for your analytical problem-solving. *Chapter Two* reviews your MS alternatives. *Chapter Three* discusses evaluation of your MS alternatives.

Worksheet 1.2-Evaluating Your Sample Needs

Check the box that is appropriate for your problems.

1. Do you potentially have complex samples:	Yes	No
– that have more than one component?	❐	❐
– that have multiple sample components with widely varying concentrations?	❐	❐
– that have a variable matrix?	❐	❐
– that are of biological origin? (urine, blood)	❐	❐
– that are limited in amount? (forensic, residue)	❐	❐

2. Do you have chemical interferences that may affect the response or the interpretation of your mass spec results:	Yes	No
– that interact with your analytes in solution?	❐	❐
– that interact with your analytes in the gas-phase?	❐	❐
– that degrade and contaminate your detector?	❐	❐

3. Do you potentially have samples that contain:	Yes	No
– analytes with molecular weight above 500?	❐	❐
– analytes that are ionic?	❐	❐
– analytes that are thermally labile?	❐	❐
– analytes that are highly polar?	❐	❐
– that are limited in amount? (forensic, residue)	❐	❐

Note: Copies of this worksheet can be downloaded from **www.LCMS.com**.

If your separation requirements have prompted you to check any box **Yes** for questions 1 and 2, you will likely require a separation to isolate and identify your analytes. If your checked **Yes** for any box in question three, you will likely require liquid chromatography or electrophoresis to separate your sample components.

Worksheets 1.0, 1.1, and 1.2 are intended to provide a structured approach to evaluating your technology needs for MS, LC, and ultimately LC/MS as they relate to your specific problems. They are not intended to provide evaluation of specific alternatives; that will be done later in *Chapter Three* for specific LC/MS technologies, and in *Chapter Six* for specific LC/MS capabilities.

Discussion Questions And Exercises

1. Identify five problems that require high information content measurement within your organization. Fill out Worksheet 1.0.
2. Compare and contrast the advantages and disadvantages of PDA versus mass spectrometry detection.
3. Name five detection techniques that result in compound specific response. Give a general statement of the analytical utility of each.
4. What is the value of chemical information to your organization? List five examples of how information benefits your organization.
5. What are the pros and cons of separating components in your sample with mass spectrometry (using mass selectivity) versus chromatography (using column selectivity)?
6. Compare and contrast derivatization GC/MS versus LC/MS. Give an example of a preferred application of each.
7. What percentage of the top chemicals produced (on a tonnage basis) are amenable to separation with GC?
8. Why is information about thermally labile, low volatility, and/or high molecular weight molecules important to society?

Notes

[1] This quote was extracted from Edward Yourdon's *Modern Structured Analysis.* Prentice-Hall, Inc.:New Jersey, 1989. Thomas Carlyle, *Sartor Resartus*, Book I, Chapter 4. We use this quote in Chapter One to emphasize our view of LC/MS as a problem-solving "tool", man is ultimately the problem "solver".

[2] The great Scottish-born philanthropist Andrew Carnegie donated funds to establish 2,500 public libraries throughout the United States and Great Britain. He believed the key to social evolution was found in educating the masses. We believe the key to social evolution may be found in

"measuring" the masses.

[3] In early 1997, Waters began offering its Integrity "system" for $99K, including, HPLC pumps, autosampler, diode array, switching module, and mass analyzer. A large number of other so called "LC detector" products are currently available in the $120K to $150K range (see Tables 5.1 and 5.2., *Chapter Five*).

[4] Phoenix International Labs in Montreal publishes its LC/MS sample track record for various drug studies on its Internet site. It averages more than 1,000 samples per week per instrument after running tens of thousands of samples. Its Internet site is: www.pils.com.

[5] Quantitative analysis with mass spectrometry has become the defacto standard for applications that require reliable, legally defensible, interference free data. The majority of published quantification methods involve GC/MS; however, recently many more LC/MS methods have come on-line. For a good overview of quantification in mass spec, read, Boyd, R.D., "Quantitative trace analysis by combined chromatography and mass spectrometry using external and internal standards," *Rapid Commun. Mass Spectrom.* 7, pages 257-271 (1993).

[6] (a) The practice of N-in-1 dosing involves dosing multiple drug candidates into one animal early in the development process. This process has a significant benefit of requiring only one method and one animal to monitor the pharmacokinetics of as many as 16 compounds. The information obtained in these screening experiments gives valuable information about candidate selection and elimination. These experiments could not be done by any technique other than LC/MS. We found only one paper on N-to-1 dosing at ASMS in 1996 (Hamilton, B.H., and Abolin, C.R., "Multiple Simultaneous Quantitative LC/MS - A New Tool to Support Drug Candidate Selection," *Proceedings of the 44th ASMS Conference on Mass Spectrometry and Allied Topics*, page 612, Portland, Oregon, May 12-16, 1996.). (b) In 1997, Timothy Olah and colleagues at Merck Research Laboratories describe the use of LC with tandem MS for determining mixtures of drug candidates as part of a high-throughput bioanalysis involving pharmacokinetics and metabolic stability studies. This approach relies on (1) simple isolation methods, (2) isocratic chromatography, and (3) LC/MS/MS conditions of APCI (atmospheric pressure chemical ionization) and Reaction Monitoring that can be adapted and applied to numerous agents. In a single analysis, they were able to determine the plasma concentration of 12 drug candidates. Using this method the authors were able to screen more than 400 compounds in a six-month period. (Olah, T.V., McLoughlin, D.A., Gilbert, J.D., "The

simultaneous determination of mixtures of drug candidates by liquid chromatography atmospheric pressure chemical ionization mass spectrometry as an in-vivo drug screening procedure," *Rapid Commun. Mass Spectrom.* 11, pages 17-23 (1997).).

[7] Robert Stevenson sites in a review of PBA'96 (Osaka, Japan, August 20-23, 1996) stated that the use of LC/MS/MS was for high throughput analysis in combinatorial chemistry increased sample preparation time by 80%. (Stevenson, R., "The World of Separation Science," *American Laboratory*, News Edition (December 1996).).

[8] Robert Stevenson sites in a review of HPLC '96 (San Francisco, June 16-20, 1996) that it was reported that "Conventional UV methods take two to four months to develop. .....LC/MS/MS reduces the methods development time to two weeks or less" (Stevenson, R., "The World of Separation Science," *American Laboratory*, News Edition, (March 1997).).

[9] Donald Chase, David Millington, and their associates at Duke University Medical Center describe semi-automated electrospray (ES) LC/MS/MS methods, based on isotope dilution, for the analysis of amino acids and acylcarnitines in human whole blood, plasma and urine. Complete metabolic profiles of target compounds were generated in less than one minute per sample. These semi-automated methods will allow the analysis of up to 500 samples per day. They describe one method for the diagnosis homocystinuria and other hypermethioninemias from dried blood spots on newborn screening cards. They predict that utilizing LC/MS/MS will successfully detect hypermethioninemias with very low rates for <u>false positives</u> and <u>false negatives</u>. (Chace, D.H., Hillman, S.L., Millington, D.S., Kahler, S.G., Adam, B.W., Levy, H.L., "Rapid diagnosis of homocystinuria and other hypermethioninemias from newborns' blood spots by tandem mass spectrometry," *Clin. Chem.* 42, pages 349-355 (1996).).

[10] Measuring a problem gap is a precondition of the problem solving process. If your problems can be easily restated in terms of a chemical measurement, then the analytical chemistry component of your problem solution can usually be evaluated and implemented in a straightforward manner. For this text, we want to emphasize the redefinition of problems in terms of chemical measurement. Can we define our problems in terms of our measurements?

[11] Depending on the nature of your problem(s), you may have to develop and/or acquire resources to address your problems. For this exercise,

assume you have approval to hire, train, and purchase the necessary resources to close your problem gap.

[12] Motivation is a key component to successful problem-solving. Without it, the barriers to finding a particular problem solution may be insurmountable. You may be successful in motivating your organization by providing your knowledge of the benefits of closing a problem gap. If you organization isn't motivated to address a problem condition, you will likely face many internal barriers to change.

[13] VanGundy, A.B. *Techniques in Structured Problem Solving*, 2nd ed. Van Nostrand Reinhold: New York 1988.

[14] Monster Under the Bed.

[15] Compound specific information is essential for problems where the identity of a sample constituent is unknown or requires unambiguous confirmation.

[16] Robert Stevenson ... PBA'96, *op cit.*

[17] Robert Stevenson ... HPLC'96, *op cit.*

[18] See Glossary for the definition of target analytes.

[19] For instance, UV methods are quite adequate for bulk drug assays.

[20] Waters, these spectra were provided courtesy of Waters Corporation.

[21] Waters, *op cit.*

[22] Griffiths, W.J., Hjertman, M., Lundsjo, A., Wejde, J., Sjovall, J., Larsson, O., "Analysis of dolichols and polyphenols and their derivatives by electron impact, fast atom bombardment and electrospray ionization mass spectrometry," *Rapid Commun. Mass Spectrom.* 10, pages 663-675 (1996).

[23] (a) Chace, D., *et al., op cit.* (b) Sweetman, L., "New born screening by tandem mass spectrometry (MS-MS)," *Clin. Chem.* 42, pages 345-346 (1996).

[24] Papayannopoulos, I.A., "Use of low- and high-energy collision-induced dissociation tandem mass spectrometry in the identification of unusual amino acids in a semisynthetic polypeptide," *J. Am. Soc. Mass Spectrom.* 7, pages 1034-1039 (1996).

[25] Oka, H., Yoshitomo, I., Hayakawa, J., Harada, K., Asukabe, H., Susuki, M., Himei, R., Horie, M., Hakazawa, H., MacNeil, J.D., "Improvement of chemical analysis of antibiotics. 22. Identification of residual tetracyclines in honey by frit FAB/LC/MS using a volatile mobile phase," *J. Agri. Food Chem.* 42, pages 2215-2219 (1994).

[26] Berberich, D.W., Yost, R.A., Fetterolf, D.D., "Analysis of explosives

by liquid chromatography/thermospray/ mass spectrometry, " *J. Forensic Sci.* 33, pages 946-959 (1988).

[27] (a) Behymer, T.D., Bellar, T.A., Ho, J.S., Budde, W.L., "Method 553: Determination of Benzidines and Nitrogen-Containing Pesticides in Water by Liquid-Liquid Extraction or Liquid-Solid Extraction and Reverse Phase High Performance Liquid Chromatography/Particle Beam/Mass Spectrometry," pages 173-212, IN: Methods for the Determination of Organic Compounds in Drinking Water, Supplement II; US-EPA Report, EPA/600/R-92/129; Order No. PB92-207703, 270 pages, 92 (23), Abstract No. 266,774 (1992). (b) Behymer, T.D., Bellar, T.A., Budde, W.L., "Liquid chromatography/particle beam/mass spectrometry of polar compounds of environmental interest," *Anal. Chem.* 62, pages 1686-1690 (1990).

[28] She, Y., Tu, Y., Liu, S., "C-118 from fullerenols: Formation, structure and intermolecular nC_2 transfer reactions in mass spectrometry," *Rapid Commun. Mass Spectrom.* 10, pages 676-678 (1996).

[29] Kallos, G.J., Tomalia, D.A., Hedstrand, D.M., Lewis, S., Zhou, J., "Molecular weight determination of a polyamidoamine Starburst polymer by electrospray ionization mass spectrometry," *Rapid Commun. Mass Spectrom.* 5, pages 383-386 (1991).

[30] Rosell-Mele, A., Carter, J.F., Maxwell, J.R., "High-performance liquid chromatography-mass spectrometry of porphyrins by using an atmospheric pressure interface," *J. Am. Soc. Mass Spectrom.* 7, pages 965-971 (1996).

[31] Zhao, Z., Koeplinger, K.A., Bundy, G.L., Banitt, L.S., Padbury, G.E., Hauer, M.J., Sanders, P.E., "*In vitro* and *in vivo* biotransformation of 6,7-dimethyl-2,4-di-1-pyrrolidinyl-7H-pyrrolo(2,3-D) pyrimidine (U-89843) in the rat," Drug Metab. Dispos. 24, pages 187-198 (1996).

[32] Sweetman, L., *op cit.* and Chace, D., *et al., op cit.*

[33] Burlingame, A.L., Boyd, R.K., Gaskell, S.J., "Mass spectrometry," Anal. Chem. 68, pages 599R-651R (1996).

[34] Nagy, S., Wade, R. (Eds.), *Methods to detect adulteration in fruit juice beverages, Volume I*, AG Science: Auburndale, 1995.

[35] Berberich, D.W., *et al., op cit.*

[36] Tomlinson, M.J., Lin, L., Caruso, J.A., "Plasma mass spectrometry as a detector for chemical speciation studies," *Analyst* 120, pages 583-589 (1995).

[37] Youngquist, R.S., Fuentes, G.R., Lacey, M.P., Keough, T., "Generation and screening of combinatorial peptide libraries designed

for rapid screening by mass spectrometry," *J. Am. Chem. Soc.* 30, pages 3900-3906 (1995).

[38] Kallos, G.J., *et al., op cit.*

[39] Rosell-Mele, A., *et al., op cit.*

[40] Spectral interferences that originate from a constant background source can usually be eliminated with spectral background subtraction. Spectral interference originating from a sample can only be removed through separation processes. MS/MS is commonly used to remove spectral interferences since fragmentation pathways of isobaric species are usually quite different.

[41] A common misconception is that chemical interferences can be removed with MS/MS. Many labs believe that they can neglect chromatography and isolate their analyte using MS/MS. Unfortunately chemical effects on analyte response carry through multiple stages of mass analysis. This issue will be addressed more fully in *Chapter 10.*

[42] Physical interferences are usually observed in aerosol processes where the transport of an analyte to a detector is suppressed or enhanced. A good example is the carrier effect on particle beam LC/MS, whereby the low analyte concentration signal is enhanced by the presence of a non-volatile buffer (Bellar, T.A., Behymer, T.D., Budde, W.L., "Investigation of enhanced ion abundances from a carrier process in high-performance liquid chromatography particle beam mass spectrometry," *J. Am. Soc. Mass Spectrom.* 1, pages 92-98 (1990).). Another example of a physical interference is the precipitation of analyte at high concentrations in electrospray. The formation of solid particles in electrospray may prevent the analyte from desorbing into the gas phase (Sheehan, E.W., Willoughby, R.C., "Photographic studies of electrospray." *Proceedings of the 41st ASMS Conference on Mass Spectrometry and Allied Topics*, page 770, San Francisco, CA, May 30-June 4, 1993).

[43] (a) Hunt. D.F., Henderson, R.A., Shabanowitz, J., Sakaguchi, H., Michel, H., Sevelir, N. Cox, A.D. Appella, E., Engelhard, V.H., "Characterization of peptides bound to the class I MHC molecule HLA-A2.1 by mass spectrometry," *Science* 255, pages 1261-1263 (1992). (b) Henderson, R.A. Michel, H., Sakaguchi, H., Shabanowitz, J., Appelia, E., Hunt. D.F., Engelhard, V.H., "HLA-A2.1-associated peptides from a mutant cell line: a second pathway of antigen presentation," *Science* 255, pages 1264-1266 (1992).

[44] Palisades Corporation, Ithaca, New York, graciously provided a listing of its Sixth Edition of the *Wiley Data Base.*

[45] Fiedler, H.P., "Identification of new elloramycins, anthracycline-like antibiotics in biological cultures by high performance liquid chromatography and diode array detection," *J. Chromatogr.* 361, pages 432-436 (1986).

[46] Reidman, M., Huber, L., "Automated serum sample analysis using cleanup columns in LDLC," *Am. Clin. Prod.,* pages 8-15 (April 1986).

[47] Garnick, R.L., Solli, N.J., Pappa, P.A., "The role of quality control in biochemistry: An analytical perspective," *Anal. Chem.* 60, pages 2546-2557 (1988).

[48] Sendra, J.M., Navarro, J.L., Izquierdo, L., "C_{18} solid phase isolation and high performance liquid chromatography/ultraviolet diode array determination of fully methoxylated flavones in citrus juices," *J. Chromatogr. Sci.* 26, pages 443-448 (1988).

[49] Lurie, I.S., Moore, J.M., Cooper, D.A., Kram, T.C., "Analysis of manufacturing by-products and impurities in illicit cocaine via high performance liquid chromatography and photodiode array detection," *J. Chromatogr.* 405, pages 273-281 (1987).

[50] Schlett, C., "Multi-residue analysis of crop protection agents and metabolites by HPLC and diode-array detection," *Am. Environ. Lab*, pages 12-17 (December 1990).

[51] Van Leidekerde, B.M., de Leenheer, A.P., "Analysis of xanthene dyes by reverse-phase high-performance liquid chromatography on a polymeric column followed by characterization with a diode array detector," *J. Chromatographia* 528, pages 155-162, (1990).

[52] Newton, I.D. *The separation and analysis of additives in polymers*, Chap 2., pages 8-36, IN: Polymer characterization, (eds.) Hunt, B.J. and James, M.I., Blackie Academic and Professional: Glasgow (1993).

[53] Gratzfeld-Huesgen, A., "Simultaneous determination of porphyrins, aromatics and olefins in crude oil by gel permeation chromatography and multiple signal detection," *LC-GC* 7, pages 836-845 (1989).

[54] Snyder, L.R, and, Kirkland, J.J., *Introduction to Modern Liquid Chromatography*. Wiley: New York (1991). This often quoted ratio has been a major justification for development of LC/MS over the past 20 years.

[55] In addition to covering the applications domain not covered by GC, LC is capable of separating most of the 20% of compounds that are covered by GC.

[56] In fact, it may be an easier task to describe the compounds that are not amenable to LC/MS rather than the ones that are!

[57] A description of this mapping procedure was presented at ASMS in

1995 (Mitrovich, S., Willoughby, R.C., "Mapping Chemical-space: Application and Prediction within LC/MS," *Proceedings of the 43rd ASMS Conference on Mass Spectrometry and Allied Topics,* page 494, Atlanta, Georgia, May 21-26, 1995.) This work has subsequently been submitted to *J. Chem. Education* (1998).

[58] *ibid.*

[59] Sun, E.L., Fennstra, K.L., Bell, F.P., Sanders, P.E., Slatter, J.G., Ulrich, R.G., "Biotransformation of lifibrol (U-83860) to mixed glyceride metabolites by rat and human hepatocytes in primary culture," *Drug Metab. Dispos.* 24, pages 221-231 (1996).

[60] Nelis, H.J., De Leenheer, A. P., "Profiling and quantitation of bacterial carotenoids by liquid chromatography and photodiode array detection," *Appl. Environ. Microbiol.* 55, pages 3065-3071 (1989).

[61] Hayward, D.S., Zimmerman, S.R., Jenke, D.R., "Drug stability testing by monitoring drug and degradate levels by liquid chromatograph," *J. Chromatogr. Sci.* 27, pages 235-239 (1989).

[62] Covey, T.R., Lee, E.D., Henion, J.D., "High speed liquid chromatography/tandem mass spectrometry for the determination of drugs in biological samples," *Anal. Chem.* 58, pages 2453-2460 (1986).

[63] Sweetman, L. *op cit.*

[64] Covey, T.R., *et al., op cit.*

[65] Reidman, M. and Huber, L., *op cit.*

[66] Hanakam, F., Eckerskorn, C., Lottspeich, F., Muller-Taubenberger, A., Schafer, W., Gerisch, G., "The pH-sensitive actin-binding protein hisactophilin of Dictyostelium exists in two isoforms which both are myristoylated and distributed between plasma membrane and cytoplasm," *J. Biol. Chem.* 270, pages 596-602 (1995).

[67] Bakhtiar, R., Bednarek, M.A., "Observation of noncovalent complexes between margatoxin and the K_v 1.3 peptide ligands: A model investigation using ion-spray mass spectrometry," *J. Am. Soc. Mass Spectrom.* 7, pages 1075-1080 (1996).

[68] Limbach, P.A., Crain, P.F., McCloskey, J.A., "Characterization of oligonucleotides and nucleic acids by mass spectrometry," *Curr. Opin. Biotechnol.* 6, pages 96-102 (1995).

[69] Shackleton, C.H.L, Witkowska, H.E., "Characterizing abnormal hemoglobin by MS," *Anal. Chem.* 68, pages 29A-22A (1996).

[70] Riond, J.L., Hedeen, K.M., Tyczkowska, K., Riviere, J.E., "Determination of doxycycline in bovine tissues and body fluids by high performance liquid chromatography using photodiode array ultraviolet-

visible detection," *J. Pharm. Sci.* 78, pages 44-47 (1989).
[71] Di Pietra, A.M., Cavrini, V., Bonazzi, D., Benfenati, L., "HPLC analysis of aspartame and saccharin in pharmaceutical and dietary formulations," *Chromatographia* 30, pages 215-219 (1990).
[72] Sendra, J.M., Navarro, J.L., Izquierdo, L., "C_{18} solid phase isolation and high performance liquid chromatography/ultraviolet diode array determination of fully methoxylated flavones in citrus juices," *J. Chromatogr. Sci.* 26, pages 443-448 (1988).
[73] Frisvad, J.C., Thrane, U., "Standardized high performance liquid chromatography of 182 mycotoxins and other fungal metabolites based on alkylphenone retention indices and UV-VIS spectra (diode array detection)," *J. Chromatogr.* 404, pages 195-214 (1987).
[74] Cappiello, A., Famiglini, G., Lombardozzi, A., Massari, A., Vadalia, G.G., "Electron capture ionization of explosives with a microflow rate particle beam interface," *J. Am. Soc. Mass Spectrom.* 7, pages 753-758 (1996).
[75] Park, S.J., Kim, Y.J., Pyo, H.S., Park, J., "Analysis of cortiosteroids in urine by HPLC and thermospray LC/MS," *J. Anal. Toxicol.* 14, pages102-108 (1990).
[76] (a) Anonymous, "Analysis of L-tryptophan for the etiology of eosinophilia-myalgia syndrome," Centers for Disease Control, *Morb. Mortal. Wkly. Rep.* 39, pages 589-591 (1990). (b) Anonymous, "Pharmaceutical Impurity Identification," .....API LC/MS technology was the first to obtain molecular weight and structural information on the suspected causative agent of Eosinophilia-Myalgia Syndrome, di-L-trypotophan aminal of acetaldehyde, Application Note by PE SCIEX (1990).
[77] Behymer, T.D., Bellar, T.A., Budde, W.L., "Liquid chromatography/particle beam/mass spectrometry of polar compounds of environmental interest," *Anal. Chem.* 62, pages 1686-1690 (1990).
[78] Song, X, Budde, W.L., "Capillary electrophoresis electrospray mass spectra of the herbicides paraquat and diqua," *J. Am. Soc. Mass Spectrom.* 7, pages 981-986 (1996).
[79] Kieber, R.J., Mopper, K., "Determination of picomolar concentration of carbonyl compounds in natural waters, including seawater, by liquid chromatography," *Environ. Sci. Technol.* 24, pages 1477-1481 (1990).
[80] Brown, M.A., Kim, I., Sasinos, F.I., Stephens, R., "Analysis of Target and Nontarget pollutants in Aqueous and Hazardous Waste Samples by Liquid Chromatography/Particle Beam Mass Spectrometry," Chap. 13, pp. 198-214, IN: *Liquid Chromatography/Mass Spectrometry: Applica-*

tions in Agricultural, Pharmaceutical, and Environmental Chemistry, (ed.) M.A. Brown, American Chemical Society: Washington, D.C. (1990).

[81] Brummel, C.L., Vickerman, J.C., Carr, S.A., Hemling, M.E., Roberts, G.D., Johnson, W., Weinstock, J., Gaitanopoulos, D., Benkovic, S.J., Winograd, N., "Evaluation of mass spectrometric methods applicable to the direct analysis of non-peptide bead-bound combinatorial libraries," *Anal. Chem.* 68, pages 237-242 (1996).

[82] Synthesis, (*Journal of Synthetic Organic Chemistry*) Instructions to authors.

[83] Allen, D.W., Clench, M.R., Crowson, A., Leathard, D.A., Saklatvala, R., "Characterization of electron beam generated transformation products of Irganol 1010 by particle beam liquid chromatography mass spectrometry with on-line diode array detection," *J. Chromatogr.* A 679, pages 285-297 (1994).

[84] Belu, A.M., DeSimone, J.M., Linton, R.W., Lange, G.W., Freidman, R.M., "Evaluation of matrix-assisted laser desorption ionization mass spectrometry for polymer characterization," *J. Am. Soc. Mass Spectrom.* 7, pages 11-24 (1996).

[85] Rosell-Mele, A., *et al., op cit.*

[86] Gratzfeld-Huesgen, A., "Simultaneous determination of porphyrins, aromatics and olefins in crude oil by gel permeation chromatography and multiple signal detection," *LC-GC* 7:836-845 (1989).

[87] The value of 10,000 mass spectrometrists world-wide is a crude estimate based on the aggregate of annual instrument sales, installed base, obsolescence rates, and number of labs. This number is derived by the authors from a large number of market sources.

[88] The value of 150,000 chromatographers world-wide is a crude estimate based on the aggregate of annual instrument sales, installed base, obsolescence rates, and number of labs. This number is also derived by the authors from a large number of market sources.

Chapter Two- Flow Diagram

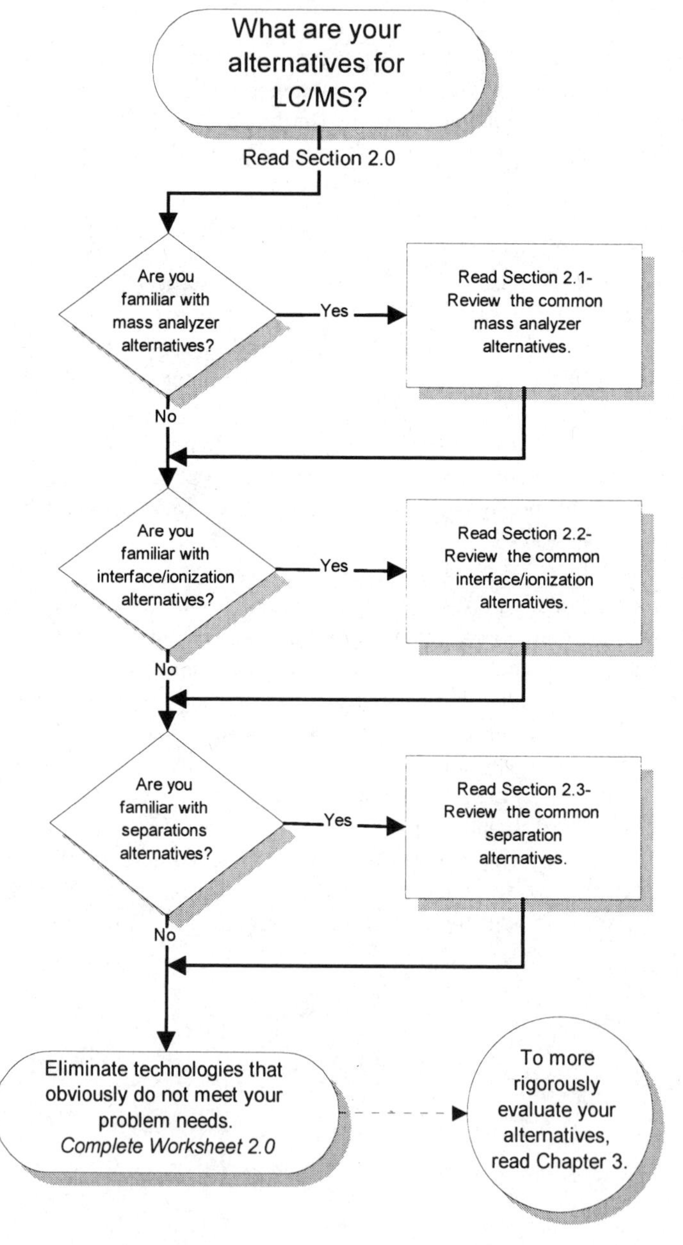

What are your alternatives for LC/MS?
Read Section 2.0
Are you familiar with mass analyzer alternatives?
Yes
Read Section 2.1- Review the common mass analyzer alternatives.
No
Are you familiar with interface/ionization alternatives?
Yes
Read Section 2.2- Review the common interface/ionization alternatives.
No
Are you familiar with separations alternatives?
Yes
Read Section 2.3- Review the common separation alternatives.
No
Eliminate technologies that obviously do not meet your problem needs.
Complete Worksheet 2.0
To more rigorously evaluate your alternatives, read Chapter 3.

Chapter Two

What Are Your LC/MS Alternatives?

(or, Drowning In Alphabet Soup)

The field cannot be seen from within the field.
Ralph Waldo Emerson

2.0- Sorting Through The Maze

The most significant innovations in analytical instrumentation on the market today involve a chain of components that must be optimized to meet the user's analytical needs. In reality, none of the individual components themselves is generally new; what is new is the means by which they are linked. We can think of this as an "analytical chain." In LC/MS, we move from a sample, through a chain of components, to an analytical result (see Figure 2.0). In less conceptual terms we start with a separation stage, move through a linking stage (or conversion phase) and then on to mass analysis. Your analytical results are highly dependent upon each link in this chain.

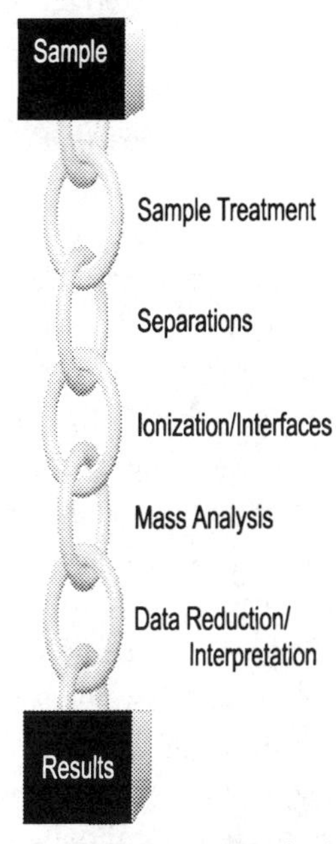

Figure 2.0- The Analytical Chain of LC/MS.

Since LC/MS consists of this chain of individual components as shown in Figure 2.0, it also follows that the quality of any analysis is dependent upon all the components working in concert. If just one component is not suitable for your analysis, you may fail to get the information you need to solve your problem. Fortunately, manufacturers today have removed much of the *art* that surrounded LC/MS and consequently have made the technique relatively seamless and user-friendly.

Nonetheless, differences exist among the available systems. To determine what is best for answering your analytical problems it is necessary to understand each of the components. For example, what is the difference between ES/MS and APCI/MS as compared to FT/ICR/MS, PB/LC/MS and CE/MS, not to mention CCFAB/MS? Sometimes just trying to figure out what each acronym means is a chore let alone understanding their sometimes subtle differences and the significance of variations. Add to this the wide variety of methods available for liquid chromatography and the task truly becomes daunting.

In an attempt to simplify this task, we have divided this chapter into three sections that represent the three major stages of LC/MS seen in Figure 2.1; namely,

1.) separation (*Section 2.3*),
2.) ionization/interface (*Section 2.2*), and
3.) mass analysis stages (*Section 2.1*).

The analytical utility and performance of your LC/MS system will largely be affected by your choice of alternative selected for each of these stages. These sections contain brief summaries (**Overviews**) of each technique. We have limited these two-page **Overviews** to technologies that are currently commercially available. Some of the outdated or esoteric LC/MS technologies are discussed in *Appendix E, Pioneers in LC/MS.*

The **Overviews** presented in this chapter are not intended to present an in-depth theoretical treatment of each technology; rather, they are intended to provide you with the essential information you need to distinguish one technology from the other.[1] You can acquire more in-depth knowledge about each topic by reading *Appendices C* and *D* or reading the cited references in each **Overview** and *Appendix A*.

Also note that we have included ionization and interfacing into a single link of our chain model for LC/MS. A more traditional (and perfectly correct) approach would be to include ionization into the mass analysis section. We felt that coupling ionization with interfacing technologies makes for a more orderly organization of the multitude of alternatives for LC/MS.

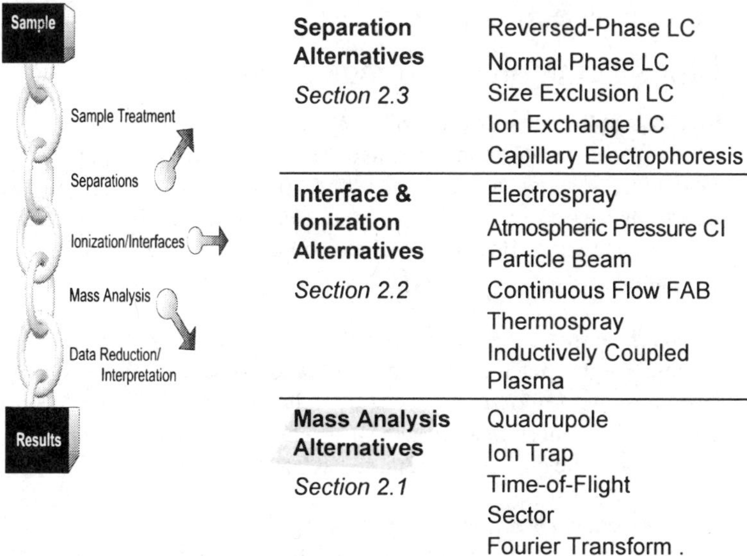

Figure 2.1- The analytical chain of LC/MS showing the large number of potential alternatives for each technology.

When reading this chapter keep in mind the topics we have already discussed in *Chapter One* and how they relate to your particular problems. Does a particular technology alternative meet your information requirements? Does a particular technology alternative address the complexity of your samples? Is a particular technology alternative compatible with your sample types and sample properties? Once you have familiarized yourself with the wide variety of LC/MS technologies available to you, we will discuss how to evaluate the differences and select appropriate technologies in *Chapter Three*. We have provided a worksheet at the end of this chapter to allow you to eliminate obvious mismatches to your problem needs before proceeding to Chapter *Three*. This culling process may save you some time.

2.1-Mass Analyzer Alternatives

To a large extent, the information that you will ultimately derive from LC/MS is determined by the operating conditions and performance specifications of your mass analyzer. For this reason we will review them first. The following are attributes that are common to all mass analyzers:

Things common to all mass analyzers

- **All mass analyzers determine the mass of an ion.**
  No mass spectrometer can determine the mass of an uncharged atom or molecule. The uncharged species must first be ionized (see *Section 3.2*) so that it can be repelled and attracted by the fields generated in the mass analyzer. Basically, a charge on an atom gives us the ability to manipulate it.

- **All mass analyzers determine the mass-to-charge ratio.**
  A mass spectrometer determines the mass-to-charge ratio of an ion. This is usually referred to as *m/z* (mass divided by the charge). Since the charge on most ions created in a mass spectrometer is 1, *m/z* is equivalent to the mass and is reported as such.[2]

- **All mass analyzers measure gas-phase ions.**
  For the mass of an ion to be determined, the ions must pass through the analyzer region of the mass spectrometer on its way to the detector. To do this the ions must be in the gas phase; otherwise they simply would not have the necessary mobility.

- **All mass analyzers must operate at very low pressure.**
  A mass spectrometer must operate at reduced pressures ($< 10^{-4}$ torr) because gas-phase ions must have a free path through the analyzer. The greater the vacuum, the longer the mean free path.

All of the mass analyzer alternatives presented in *Section 2.1* have the means of separating and detecting gas-phase ions. It is the means by which we separate and detect these ions that will determine the information content of a given experiment. Instrumental parameters such as mass resolution, mass accuracy, mass range, sensitivity, scan speed, and MS/MS all reflect the relative capabilities and utility of the various mass analyzer alternatives. These parameters are defined more fully in *Chapter Three, Section 3.1.* For the purposes of this chapter; a brief description of each parameter is presented below:

Resolution describes the ability of a mass analyzer to separate adjacent ions. There are high and low resolution analyzers. High resolution instruments are generally used to isolate ions from isobaric (same nominal mass) interferences in order to make accurate mass determinations. Increased resolution also increases experimental selectivity.

Mass accuracy is the ability of a mass analyzer to assign the mass of an ion close to its true mass. High mass accuracy (exact mass) is usually associated with analyzers that also have high resolution. Exact mass measurements can be used to predict the elemental composition of an analyte.

Mass range is usually defined by the lower and upper m/z value observable by a mass analyzer. If your analyte MW does not fall within the mass range of an instrument, it will not be detected.[3]

Sensitivity is the ability of a particular instrument to respond to a given amount of analyte. This type of mass analyzer has a significant effect on sensitivity as does the mode with which the analyzer is operated.

Scan speed is the rate at which we can acquire a mass spectrum, generally given in mass units per unit time. The scan speed will affect the amount of information we can reasonably acquire with a given mass analyzer and the chromatographic peak width.

Tandem mass spectrometry (MS/MS; or MS^n, n=1,2,3,...) provides the ability to mass-analyze sample components sequentially in time or space in order to improve selectivity of the analyzer or promote fragmentation and facilitate structural elucidation. This technology is a key consideration for mass analyzers, particularly in LC/MS where most of the ionization processes produce exclusively molecular ions, not fragmentation spectra..

The following **Overviews** provide you with a description of how ions are separated with each analyzer; how they are detected and the implications of these processes for solving analytical problems.

Overview: Quadrupole Mass Analyzers (Q)

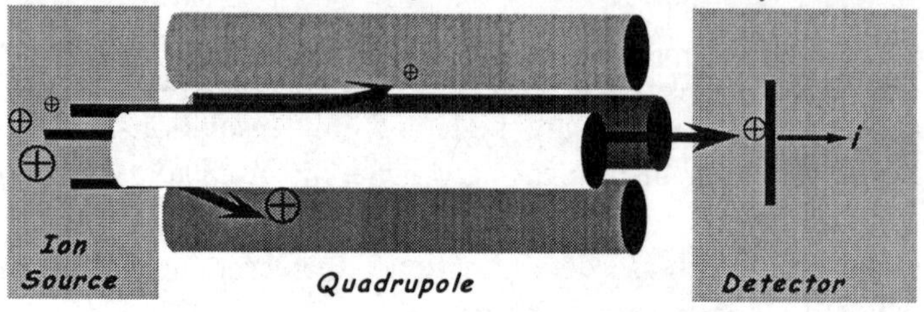

Figure 2.2- Quadrupole Mass Analyzer

Mass Separation

As the name implies, quadrupoles (quads) consists of four precisely parallel rods (called poles) equally spaced around a central axis. Ions are introduced along the axis of the poles as shown in Figure 2.2. The energies of the ions as they transit the quad are only a 5-10 eV and thus quads are considered low energy mass analyzers. By applying precisely controlled voltages (rf & DC) to opposing sets of poles, we can create what is known as a "mass filter." By ramping the voltages on each set of poles a complete range of masses can be passed to the detector. Only ions with a particular mass-to-charge ratio will pass through the filter to be detected at a particular applied voltage. Note that only one mass-to-charge ratio is filtered at any given time with quads. See *Appendix D* for a more thorough explanation.

Structural Elucidation

Quads are used to obtain structural information about sample constituents in two ways. The first, is the "single quad" mode which relies on the ionization process to produce structurally significant fragments. The externally generated fragments are mass-analyzed and a structure is elucidated from the observed masses and intensities. This is the most common operating mode for mass spec and has yielded its greatest utility with electron ionization spectra and the associated EI libraries. Particle Beam LC/MS allows this type of ionization (Section 2.2).

"Triple quad" (MS/MS) allows analysts to fragment their ions within the quadrupole mass analyzer.[4] Significant improvements can be made in specificity and selectivity with the wide variety of triple quadrupole operation modes.

Analytical Utility

Quads are the most common and versatile of all the mass analyzers. Quads have a very long applications record and an extensive history of use with mass spectral libraries. They are particularly well-suited for high throughput and production applications where dirty samples and background contamination may affect the quality and productivity of the results.[5]

Most of the applications with quads occur at unit resolution with mass accuracy at circa 0.1 u. Although quads can operate at upper masses above 4,000 u, they are best suited for applications under 1,000 u. The "triple quad" has become the standard tool in LC/MS for target methods such as those found in clinical assays.

Detection Modes

The most sensitive operating mode of single quads is known as selected ion monitoring (SIM)[6]; and for triple-quads, selected reaction monitoring (SRM). In these non-scanning modes of analysis one monitors a single m/z at maximum collection efficiency and maximum sensitivity.[7] There are typically two orders of magnitude sensitivity difference between scanning and SIM modes.

Summary -Quads

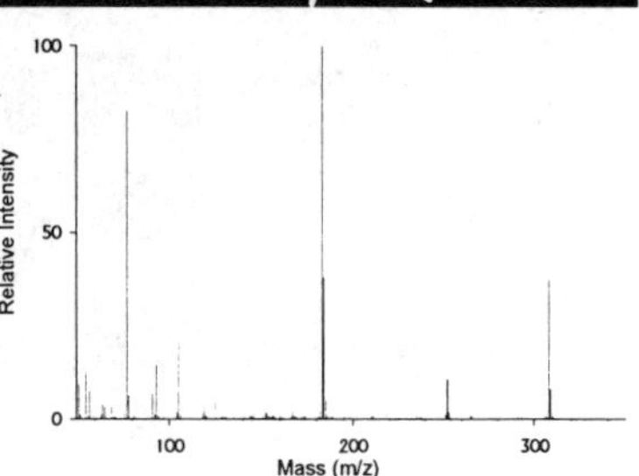

Figure 2.3- Typical histogram (bar graph) ion profile.

Resolution

- Low : unit (1 mass unit width)

Mass Range

- 50-2,000 u

Scan Speed

- 4,000 u/sec max

Vacuum Requirements

- Minimal: 10^{-4}-10^{-5} torr

Variations:

- Single quad- SIM/Scanning
- Triple quad- SRM/Scanning
- Polarity- Positive/Negative Ion
- Hybrids, Q-TOF *(Section 3.1)*

Common LC/MS interfaces:

- API: Electrospray & APCI
- Particle Beam
- Thermospray

Pioneers

- Dawson, Peter H. (1976,1995)[8]

A Good Reference

- Miller & Denton, (1986)[9]

Symbol[10] ═══

Overview: Ion Trap Mass Analyzers (IT)

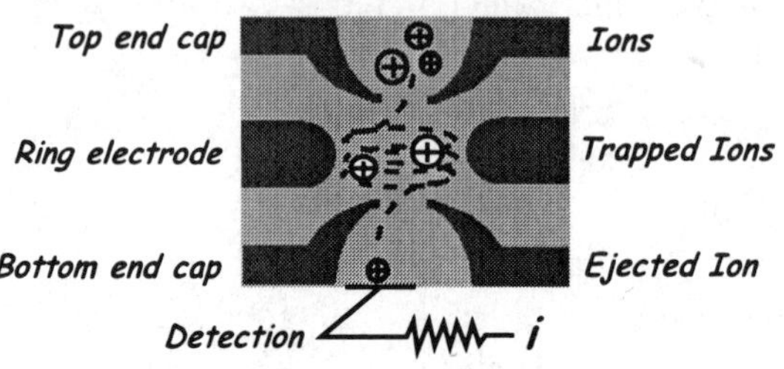

Figure 2.4- Ion Trap Mass Analyzer

Mass Separation

As the name implies "ion trap" mass analyzers work by first trapping ions then detecting them based on their m/z ratios (see Figure 2.4). The ion trap is a variation of the quadrupole mass filter, and consequently is sometimes referred to as a "quadrupole ion trap" or Paul trap. Using the same principles of electrodynamic focusing as quads, the trap contains (or traps) ions in a 3-dimensional volume rather than along the center axis. After trapping, the ions are detected by placing them in unstable orbits, causing them to leave the trap.

An ion trap consists of two end caps and a ring. Helium gas is added to the trap causing the ions to migrate toward the center.[11] The more tightly focused ions are ejected and detected with higher efficiency. Theoretical aspects of the ion trap are in *Appendix D*.

Structural Elucidation

Ion traps allow the trapping of specific ions, followed by collisional dissociation, so that tandem mass spectrometry (MS/MS or MS^n) can be performed. MS/MS with traps are tandem in time. (See *Section 3.1*) This capability provides the analyst with rich structural information. MS/MS capability with ion traps is performed with electrostatic scan functions, requiring sophisticated software control. Current traps have MS^n capability as high as n=10 (n=2 or 3 are most practical).

Ion traps can have either internal or external ion sources; however, a practical implementation of LC/MS on the trap required external sources to be used. The external source allows coupling to atmospheric pressure ionization techniques, APCI and electrospray.

Analytical Utility

Ion traps are simpler in design and cheaper in price than triple quads. It should be emphasized that MS/MS is becoming a prerequisite for many applications in LC/MS where identity (structural information) and high selectivity are required. The trap has supplied these problem-solving capabilities to the broader analytical community primarily because of their reduced price. Popular applications of traps have been found in combinatorial chemistry and drug metabolism.

The commercial implementations of traps have been limited to low resolution (unit), 0.1u mass accuracy, and mass range ca. 2,000.[12] By compromising some of the operating conditions, higher resolution and mass accuracy can be obtained with traps (e.g. Zoom Scan[13]).

Detection Modes

Ion traps are pulsed mass analyzers in the sense that stored ions are released from the trap by applying a bias voltage pulse to one end cap. The ejected ions are collected by an electron multiplier. Because of the storage capacity of the trap, the effective duty cycle of the trap can be quite high. Even so, quads in SIM mode are currently winning the detection limit battle by a slight edge over traps.

Summary-Ion Traps

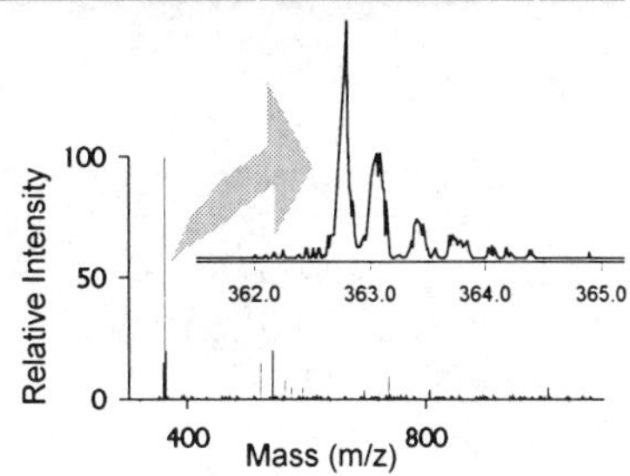

Figure 2.5- ZoomScan™ of mass 362.8 (tryptic fragment).

Resolution
- Low (unit), can operate higher.

Mass Range
- 2,000 u

Scan speed
- 4,000 u/sec

Vacuum requirements
- Low: 10^{-3} torr

Variations:
- MS/MS(MS)
- Polarity-Positive/Negative Ions
- Sources- Internal & External
- Resonance Excitation

Common LC/MS interfaces:
- API: Electrospray & APCI

Pioneers
- Paul and Steinwedel, (1953); Post and Henrich, (1953).

A Good Reference
- March and Hughes, R (1989)[14]

Symbol [15]

Overview: Time-of-Flight Analyzers (TOF)

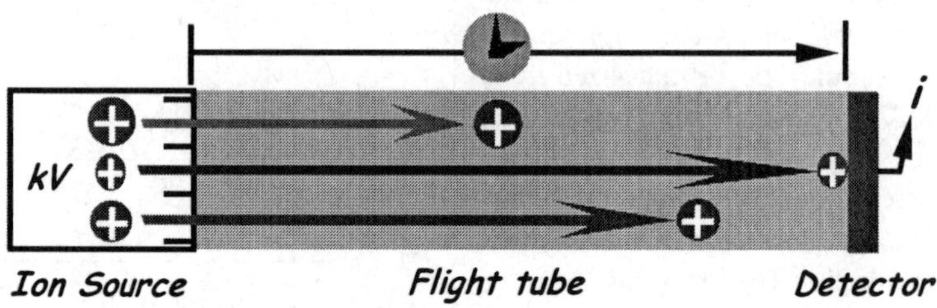

Figure. 2.6- Time Of Flight Mass Analyzer

Mass Separation

Time-of-Flight (TOF) mass analyzers are, in principle, one of the simplest mass analyzers. They are based on the fact that ion velocity is mass dependent. A "linear" TOF (Figure 2.6) consists of simply an ion source, a (field free) flight tube and a detector. A bundle of ions from the ion source region are pulsed (accelerated with a high voltage potential) down the flight tube. Each mass entering the flight tube has a different velocity; small mass ions have higher velocity relative to large mass ions. The initial bundle of ions will separate as they proceed down the flight tube and arrive at the detector at different times. The arrival time is directly related to mass or, in this case, m/z.

A common variation of TOF incorporates an electrostatic mirror (a reflectron)[16] into the field-free region to compensate for energy differences of the initial ions.

Structural Elucidation

In the past, the requirement for MS/MS for structural elucidation in LC/MS has made TOF somewhat "suspect" as a tool for identity and elucidation problems. This was unfortunate since TOF has characteristics that are ideal for structural elucidation problems; namely, high sensitivity in the scanning mode;[17] and high scan speed.[18]

Fortunately, recent developments and commercial product implementations have reversed this concern. The combination of TOF in tandem with other mass analyzers (i.e., quads, sectors) has made TOF a viable contender for a prominent position in LC/MS/MS. For example, quads in tandem with TOF allow high efficiency SIM in the first analyzer (quad) and high efficiency scanning in the second analyzer (TOF) (see *Section 3.1* for details).

Analytical Utility

Unparalleled scan speed, ion collection efficiency, and mass range make TOF useful almost everywhere. The power of TOF is found in its ability to collect a high percentage of all ions generated in the ion source; that is, it is sensitive over the entire mass range. For applications where "complete spectra" are required (identity), the high duty cycles of TOF make it the technology of choice. TOF has no practical upper mass limit and therefore is unrivaled for applications where high molecular weight is important (industrial polymers, biopolymers). TOF is the ideal analyzer for high speed/efficiency separations (CE/MS, CEC/MS).

Recent technology developments have also added high resolution and high mass accuracy to the list of TOF of capabilities.[19] With these developments, TOF becomes extremely versatile, with lower cost linear instruments for high sensitivity and higher cost reflectron instruments for high resolution and mass accuracy.

Detection Modes

The improvement in high speed electronics over the past several decades has allowed TOF to come of age (i.e., sampling rates of 500 MHz). These sampling rates allow thousands of scans per second.

Summary-TOF

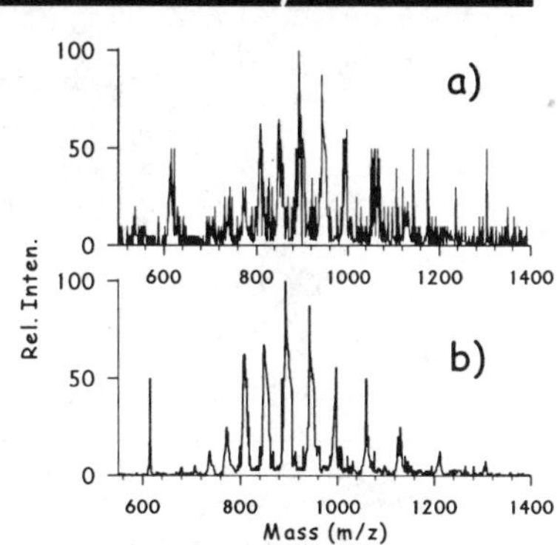

Figure 2.7-ES spectra (a) linear and (b) reflectron TOF.

Resolution
- Can be low (unit mass) or high

Mass Range
- Unlimited

Scan speed
- Very Fast: >10^6 u/sec

Vacuum Requirements
- High: 10^{-7} torr or greater

Variations:
- Linear & Reflectron[20]
- Tandem with quads & sectors
- Post-source decay

Common LC/MS interfaces:
- API: Electrospray & APCI
- MALDI[21]

Pioneers
- Stephens, W.E. (1946); Wiley, W.C. and McLaren, I.H. (1955).[22]

A Good Reference
- Cotter, R.J. (1994)[23]

Symbols[24] Linear Reflectron

Overview: Sector Mass Analyzers (E & B)

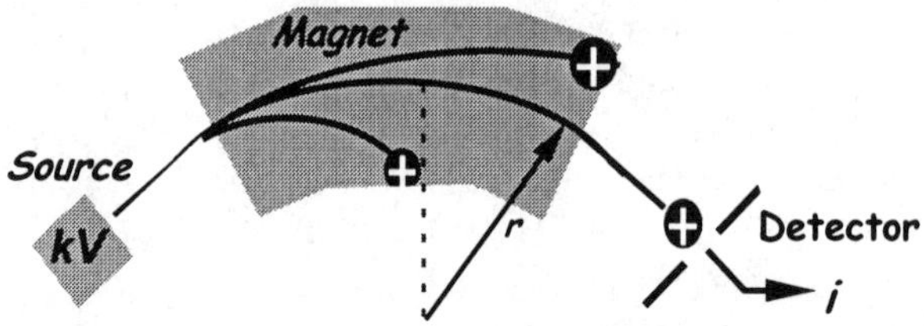

Figure. 2.8- Magnetic Sector Mass Analyzer

Mass Separation

Magnetic sector mass spectrometers (sectors) have dominated high resolution applications in mass spec for the past 30 years, particularly organic analysis.[25] In a sector instrument, ions created in the ion source are accelerated with voltages of 4-8kV into the analyzer magnetic field (Figure 2.8). The radius of curvature in a given magnetic field of the sector is a function of m/z. Ions of different masses can be separated and detected by a fixed detector by varying either the magnetic field or the source voltage to scan the mass range.[26]

To achieve high resolution, magnetic sectors are generally coupled with electric sectors (energy filters) to accommodate variations in initial kinetic energy. Electric sectors (denoted as E) and magnetic sectors (denoted as B) are placed in a variety of tandem configurations as illustrated in *Section 3.1*.

Structural Elucidation

Two approaches to structural elucidation in LC/MS are the use of electron ionization (see particle beam-*Section 2.2*) and MS/MS, with MS/MS being the more common implementation. MS/MS on sectors is actually performed on a wide variety of instrumental configurations, including, EB, BE, EBE, BEB, EBQQ, BEQQ, and BE-TOF (See *Section 3.1* for more details).[27] Each of these hardware configurations has a particular performance characteristic, with the major advantage being high resolution isolation of precursor and/or product ions. This leads to high selectivity and accurate assignment of ion masses.

One major advantage of using sectors for structural elucidation is the ability to produce higher energy collisions leading to greater decomposition of analyte precursor ions and ultimately more structural information.

Analytical Utility

Although other analyzer technologies have encroached on the applications base for sector instrumentation, sectors still have general utility for high resolution and high energy MS/MS applications. Their long track record makes them still a viable alternative for high resolution (up to 100,000) with double focusing instruments (use E and B).

The dynamic range and reliability of sector instruments still makes them a competitive alternative for quantitative analysis. However, sector technology will likely not have a significant role in low cost "LC detector" applications. A primary role of sector technologies in the future may well be in hybrid combinations with quads and TOF.

Detection Modes

Sectors, being a true spectrometer in contrast to quads (which isolate one mass at a time), are capable of separating all ions all the time; unfortunately, detecting them at a single point limits detection to one mass at a time. To improve this situation, and improve sensitivity, multichannel array detectors are used to collect ions over a larger spatial area. The improvement in detection sensitivity with multichannel array detection is greater than 100-fold. (See *Appendix D* for more details.)

Summary-Sectors

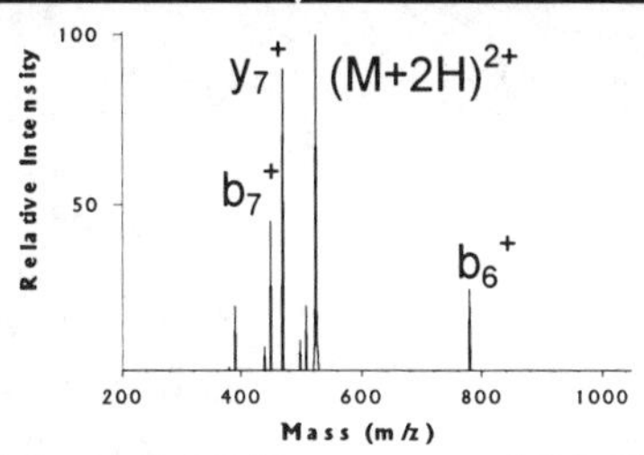

Figure 2.9-B/E spectrum of $(M+2H)^{2+}$ ion.

Resolution
- Single Focusing: Moderate
- Double Focusing: High

Mass Range
- 20,000 u

Scan speed
- Slow

Vacuum requirements
- High: 10^{-7} torr

Variations:
- 1-, 2-, 3-, 4-Sectors
- Tandem with Quads and TOF

Common LC/MS Interfaces:
- API: Electrospray & APCI
- Particle Beam & Thermospray
- Continuous Flow FAB

Pioneers:
- Aston, F.W. (1919); Hipple, J.A. and Fox, R.E. (1946).[28]

A Good Reference
- Desiderio, D.M. (1994); McLafferty (1983) .[29]

Symbol [30]

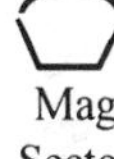
Mag Sector

Electric Sector

Overview: Fourier Transform-MS (FTMS)

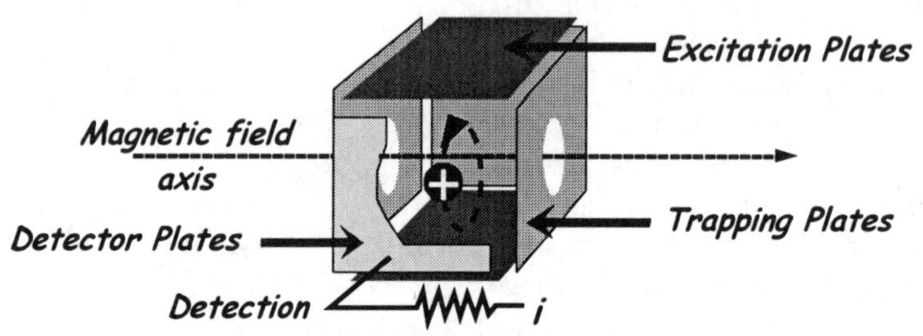

Figure. 2.10- Fourier Transform Mass Analyzer

Mass Separation

The Fourier Transform Mass Spectrometer (FTMS) (also known as ion cyclotron resonance (ICR) MS or the "magnetic ion trap") consists of a cell contained within a high vacuum chamber ($<10^{-6}$ torr) centered in a very high magnetic field (See Figure 2.10).[31] Ions are trapped in the cell by a combination of a magnetic field and electric potentials applied to the trapping plates. In the presence of the magnetic field the ions will take on circular trajectories about the axis of the magnetic field. The frequency of rotation of ions around the axis of the magnetic field is inversely proportional to mass. The frequency of ion rotation is detected indirectly through induced current on the detector plates as the ions pass near the plates. The frequency of ion rotation can be converted to mass through a fourier transform. (*Appendix D*).

Structural Elucidation

FTMS (like quadrupole ion traps) allows the trapping of specific ions, followed by collisional dissociation, so that tandem mass spectrometry (MS/MS) can be performed (See *Section 3.1*). This capability provides the analyst with rich structural information. MS/MS with FTMS is performed by typically adding a collision gas to the cell and allowing the analyte ions to repeatedly collide with gas molecules until analyte molecules dissociate into product ions.[32]

Highly selective excitation and detection waveforms can be applied to the excitation plates. This allows precise control over isolation, dissociation, and detection of sample ions and fragment ions, aiding in interpretation.

FTMS requires an external source for coupling to the LC/MS techniques, such as APCI and electrospray.

Analytical Utility

The combination of high resolution (>500,000), high mass accuracy, simplicity, and versatility have made FTMS the premier research technology in mass spectrometry. Initially, the utility of FTMS was hampered by lack of commercially available LC/MS sources; that condition no longer exists. With an external source for APCI and electrospray, FTMS has demonstrated a broad range of utility for elucidation of structures and identification of unknowns. A significant limitation to the widespread implementation of FTMS is the high cost.

To date, the majority of applications have tended to be in qualitative, not quantitative, analysis. This certainly does not mean that FTMS is not a quantitative tool, rather that its greatest utility thus far has been for identity and structural elucidation problems.

Detection Modes

The unique and indirect ion detection of FTMS offers many advantages. Since analyte ions are not consumed by the detection, a single ion can be measured repeatedly giving FTMS the highest potential sensitivity and the ability to perform more than one experiment with a single ion. Ion losses are not as significant from one experiment to the next.

Summary-FTMS

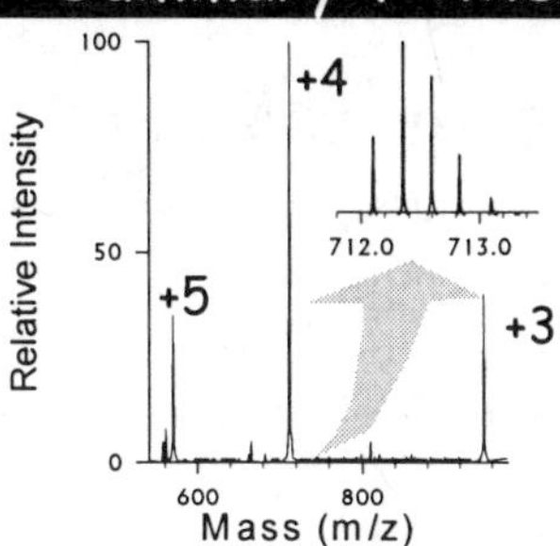

Figure 2.11-ES spectrum of a tryptic fragment, $(M+4H)^{+4}$.

Resolution
- High (<500,000)

Mass Range
- >15,000 u

Scan speed
- Fast (>millisecond)

Vacuum requirements
- High: 10^{-7} torr for unit mass, $<10^{-9}$ torr for high res

Variations
- EI and CI
- Polarity-Positive/Negative Ion
- SWIFT: Stored Waveform Inverse Fourier Transform

Common LC/MS interfaces
- API: Electrospray & APCI
- MALDI[21]

Pioneers
- Marshall, A.[33]

A Good Reference
- Buchanan, M.V. (1987).[34]

Symbol[35]

2.2-Interfacing Alternatives

Interfaces & Ion Sources

A wide variety of commercial alternatives exist for coupling liquid chromatography to mass spectrometry. In this section we have included the processes of "interfacing" and "ionization" because we feel it is difficult to consider your alternatives in LC/MS without addressing both at the same time.[36]

The aggregate of interfacing and ionization in LC/MS must accomplish the same thing for all combinations of LC/MS; that is, all interface/ionization combinations must convert dissolved analyte eluting from a separation system into gas-phase ions at reduced pressure. Table 2.0 itemizes the processes associated with this conversion. An *evaporation process* is required to convert analyte in the liquid phase into the gas phase; a *pressure reduction process* is required to convert sample separated at atmospheric pressure into low enough pressures for mass analysis; and an *ionization process* is required to convert sample into ions (a prerequisite for mass analysis). All interfaces have to deal with these three processes in one manner or another.

Table 2.0- Conversion processes required for interfacing liquid chromatography with mass spectrometry (indicated by arrows).

	LC	Conversion Process	MS
State-of-matter:	Liquid-phase	Evaporation	Gas-phase
Pressure:	Atmospheric	Pressure Reduction	High Vacuum
Charge State:	Neutral (Ionic)	Ionization	Ionic

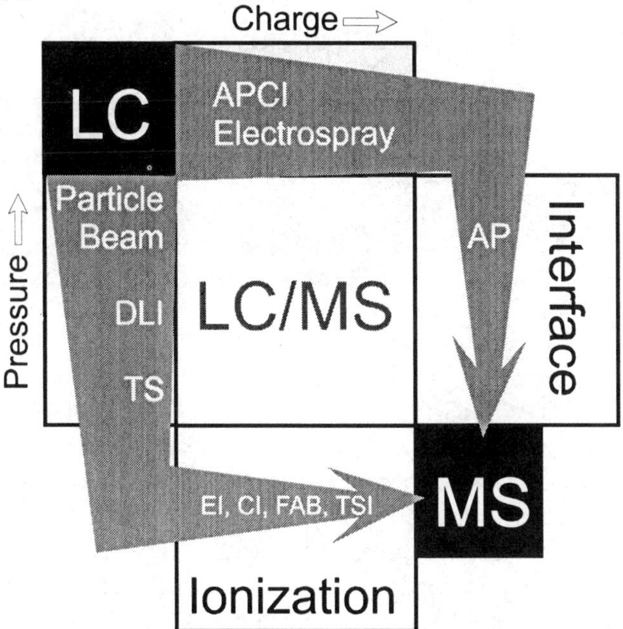

Figure 2.12- Pathways for combining interfaces and ionization in LC/MS.

The utility of any given interface and ionization system will depend upon its respective ability to efficiently transport your analyte from solution into the gas phase and into the vacuum system, either as a charged species or by creating a charged species along the way. In addition, the utility will also be determined by its ability to maintain the structural integrity of the analyte through these conversion processes.

Figure 2.12 illustrates many of the various LC/MS technology alternatives presented in this section. One group of technologies creates ions at atmospheric pressure, then samples ions through so-called atmospheric pressure (AP) interfaces; the other group of interfaces sample analyte into the vacuum system and ionize at reduced pressure. Here we present **Overviews** of the major interfacing technologies that are available to you commercially. Each interfacing alternative generally has more than one ionization alternative. Some of the advantages and limitations of these approaches will be discussed here; conversely, a more in-depth treatment of evaluation and selection of interfacing technologies is presented in *Chapter Three, Section 3.2*.

Overview: Electrospray (ES)

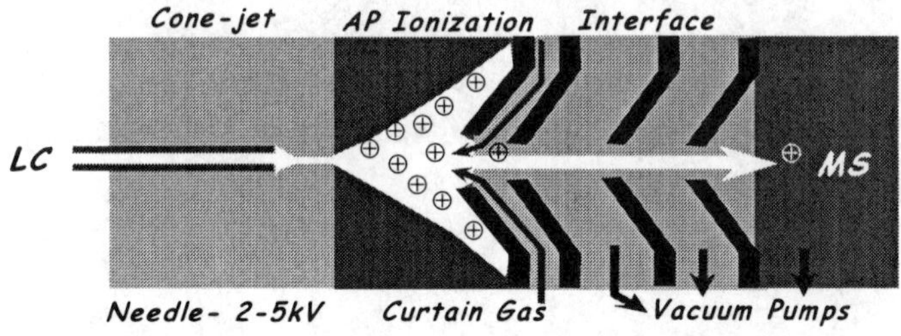

Figure 2.13-The Electrospray Interface

Interfacing

The electrospray (ES) process produces gas-phase ions at atmospheric pressure which are sampled into the vacuum system through a series of sampling apertures separating successive vacuum stages (Figure 2.13). The simplicity of the electrospray design makes it a universal tool for LC/MS interfacing. ES is composed of simply a hollow needle with a high electrical potential through which the effluent flows (1-10 uL/min.). The high field at the tip of the needle produces a cone shaped liquid meniscus from which a spray of highly charged droplets emerges. Subsequent evaporation of the droplets results in ion formation (see *Appendix C*). Conventional ES operates at ca. 1-10 uL/min flow rates. Pneumatic assistance (Ionspray- IS), ultrasonic assistance (Ultraspray- US), and thermal assistance (Turbospray) all allow increases in allowable liquid flow rate.

Ionization

Unlike most ionization processes in mass spectrometry which occur in the gas phase; electrospray ionization (ESI) is the transfer of ions present in a liquid phase into the gas phase. A prerequisite for gas-phase ion production with ESI is that your analyte exists in solution as an ion. The exact mechanism(s) of ion formation continues to be debated; however, the fact remains that virtually any ion in solution can be observed with the mass spectrometer. If a species exists in solution with more than one ionizable site, then ESI will also produce more than one charge in the gas phase. In this way we can observe multiple charged species in the mass spectrum (Figure 2.17). This effectively folds the mass spectrum of a high molecular weight species into the mass range of a typical mass analyzer.

Analytical Utility

Electrospray is revolutionary. Virtually any "ion" in solution is amenable to analysis with electrospray; including, large macromolecules, quat salts, etc. The gentleness of the electrospray process extends the polarity limit of analysis with mass spectrometry far beyond any previous boundary with any previous technique (See Figure 2.16). The production of multiply charged species, especially of proteins, extends the effective range of a mass analyzer to megadaltons.

Considerations

In order to effectively solve problems with ES you must have an understanding of the solution chemistry of your sample. Unfortunately, acid-base equilibrium, redox chemistry, and competitive interaction with other solution species may affect your results. Buffers and high quality separations may be necessary for quantitative response. In addition, to evaluate unknowns you should analyze in both positive and negative ion modes .

Summary -ES

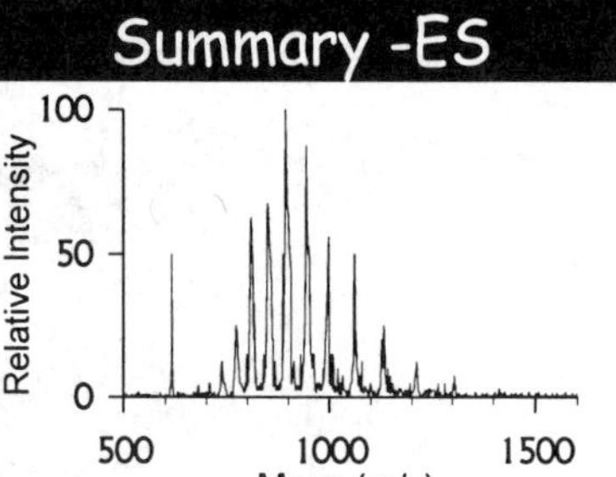

Figure 2.14-ES spectrum of myoglobin.

Information Content
- Molecular Weight
- Structure with MS/MS or CID

General Applicability
- Large ionic molecules
- Small singly charged ions

Ionization Modes
- Polarity- positive/negative.

Variations:
- Ionspray: (IS) <1.0 mL/min [37]
- Ultraspray: (US)<0.5mL/min[38]
- Turbospray: (TS)>0.1mL/min[39]
- Nanospray: (NS) nL/min[40]

Available on MS Types
- All

Pioneers
- Dole et al.(1968)[41]
- Yamashita and Fenn (1984)[42]

A Good Review
- Mann, M. and Fenn, J.B. (1992)[43]

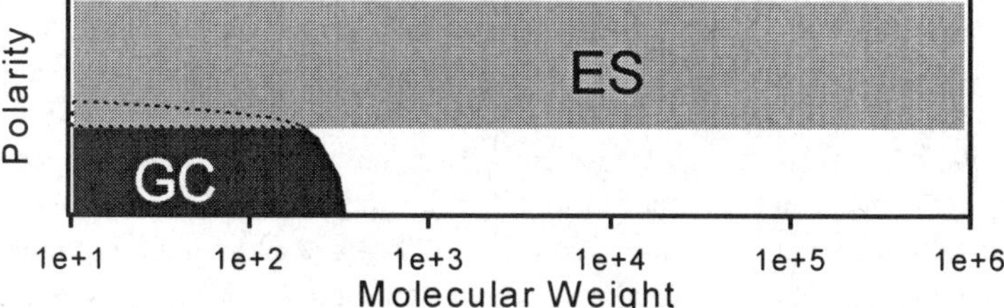

Figure 2.15- The problem solving domain of electrospray.

Overview: Atmospheric Pressure-CI (APCI)

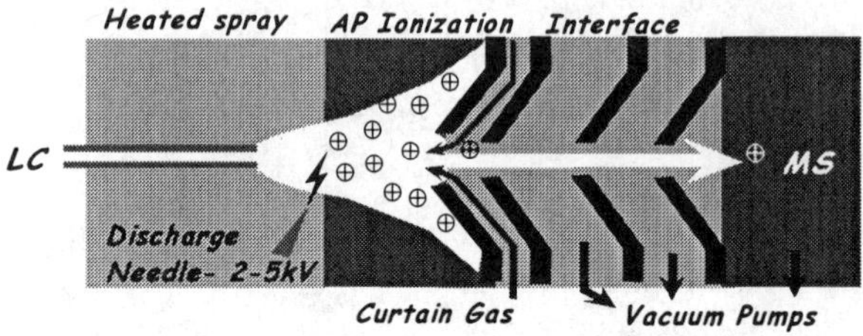

Figure 2.16- The APCI Interface.

Interfacing

Like electrospray, atmospheric pressure chemical ionization (APCI) creates gas-phase ions at atmospheric pressure. As a consequence, the interface for APCI is identical to that of electrospray; namely, the **atmospheric pressure interface** (referred to as API). This allows both APCI and electrospray to be configured on the same instrument, enhancing the problem-solving capabilities of both and should be considered when evaluating your alternatives for LC/MS. The ions produced are sampled through a series of apertures into successive vacuum chambers as shown in Figure 2.16.

With APCI, the effluent from a liquid chromatograph is sprayed into a heated spray chamber (~400°C). The heat rapidly evaporates both solvent and solutes.[44] This high temperature evaporation process does not degrade most drugs and metabolites (some conjugated metabolites require ES).

Ionization

In contrast to electrospray, ionization with APCI occurs in the gas-phase. It is accomplished with a source of electrons introduced on-axis with the heated spray. The electrons are usually supplied by a discharge source although other sources have been used as well such as a ^{63}Ni beta emittor. The APCI ion sources produce a rich plasma of reagent ions resulting from interaction with the electrons. The reagent ions are produced by electron ionization of the source gases; usually air or nitrogen is used. The simplified general reaction involves the ionization of N_2 and O_2 by electron ionization (EI) leading to the formation of hydronium ion-water clusters $H_30^+(H_2O)_n$.[45] At atmospheric pressure there is significant interaction between reagent ions and analyte ions produced by the heated nebulizer. The gas-phase analytes will become and remain protonated if their proton affinity is greater than that of water (see *Appendix C*).

Analytical Utility

APCI is the method of choice for drugs and metabolites. It is probably the most widely used technology for high throughput target applications, particularly in the pharmaceutical industry. The sensitivity, ruggedness, and reliability of APCI puts it ahead of ES for many pharmaceutical applications. APCI appears to be much less susceptible to chemical interferences than ES. The ionization process of APCI is one of the most efficient. Ionization efficiencies can approach 100% under ideal conditions. The only major limitation of all AP ionization techniques is our inability to collect all of the ions generated in the mass analyzers (typically less than 1%).

Considerations

It is important to understand that APCI is extremely sensitive for compounds such as amines that have a high proton affinity. For compounds of lower proton affinity, the response may be diminished.

Summary-APCI

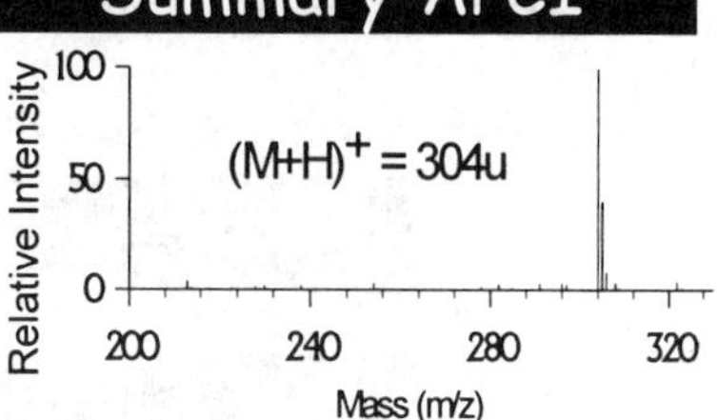

Figure 2.17-APCI spectrum of scopalamine (MW=303).

Information Content:
- Molecular weight <2,000u
- Structure requires MS/MS

General Applicability
- Ionizable, polar and nonpolar molecules

Ionization Modes
- APCI positive/negative

Flow Rates
- 0.5 - 2.0mL/min

Available on MS Types
- All

Pioneers
- Horning, et al (1974)[46]

A Good Reference
- Thompson, B.A., Danylewch, M.I. (1983).[47]

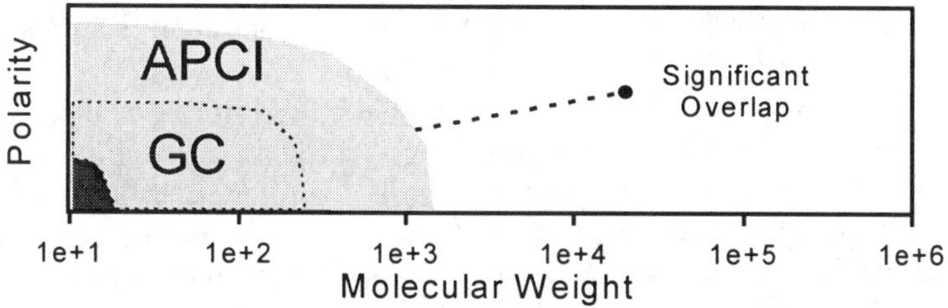

Figure 2.18- The problem solving domain of APCI.

Overview: Particle Beam (PB)

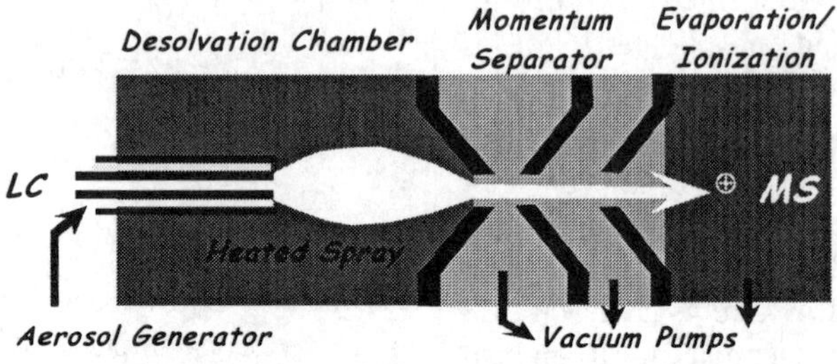

Figure 2.19- The Particle Beam Interface

Interfacing

The particle beam (PB) interface was designed to make LC/MS the liquid-phase counterpart of GC/MS. PB is a classical enrichment interface that is designed to remove mobile phase solvent while transferring the majority of analyte into the mass analyzer.[48] Although there are a number of implementations of PB, they all have the same three key components:

1.) aerosol generator,

2.) desolvation chamber, and

3.) a momentum separator.

The aerosol generator serves to disperse the liquid effluent into a high surface area spray. The desolvation chamber serves to provide space and heat to effect rapid and complete desolvation of the aerosol, producing solvent-depleted solute particles.

The momentum separator serves to direct the aerosol through a series of apertures (at atmospheric pressure) into low pressure ion source of the MS (see *Appendix C*).

Ionization

A beam of solute particles (particle beam) enters the source of the mass analyzer and impinges on the surfaces of the ion source. The ion source walls are heated to several hundred degrees facilitating the "flash vaporization" of the solute particles. Once in the vapor phase, the solute molecules are ionized by conventional means; the most common is electron ionization (EI). However, chemical ionization (CI) and fast atom bombardment (FAB) have also been commercially available. PB introduction has been implemented on a number of versatile instrument platforms that utilize the same ionization source for both GC/MS and LC/MS. This provides PB with several advantages, including the ability to precisely control the chemical ionization processes, and the ability to use the same tuning procedures as are found with GC/MS.

Analytical Utility

In an era when API techniques are dominant there are still a number of problems that are best solved by use of the conventional ionization modes, EI and CI. PB is the only commercial LC/MS interface today that produces EI spectra. The value of EI spectra is the reproducibility of molecular fragmentation and ion abundances. This is ideal for library searching the available EI spectral data bases of most industrially, environmentally and socially significant molecules.

Particle beam technology is ideal for low-cost identity detection for LC. For compounds less than 1,000u PB with EI offers universal detection. These capabilities are not available with other interfacing techniques.

Considerations

PB should not be considered for trace analysis where sample is limited. The sensitivity of PB is generally limited to ppm levels.

Summary-PB

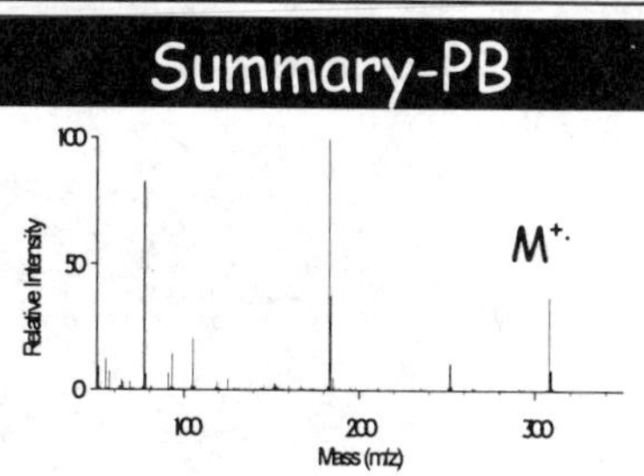

Figure 2.20-EI spectrum of phenylbutazone (MW=308).

Information Content
- Structural Elucidation (EI)
- Molecular Weight (CI)

Variations:
- Thermal, Pneumatic & Ultrasonic aerosol generation

General Applicability
- Small ionic, polar and nonpolar compounds (<1,000u)

Flow Rates
- 0.1-1.5mL/min.

Ionization Modes
- EI, CI, (FAB)

Available on MS types
- Quads & Sectors

Pioneers:
- Willoughby & Browner, (1984); Winkler & co-workers (1986).[49]

A Good Reference
- Behymer, et al. (1990).[50]

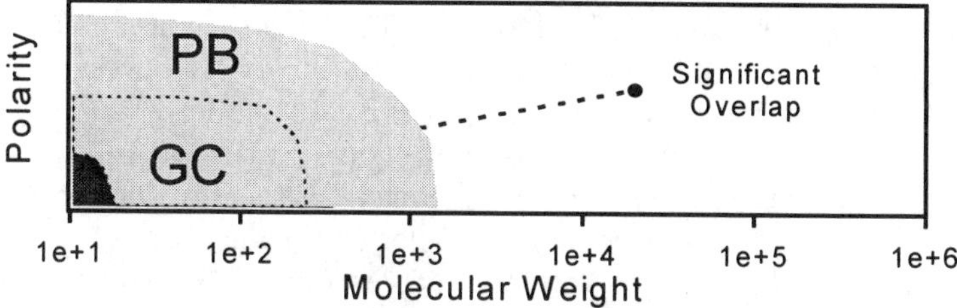

Figure 2.21- The problem solving domain of particle beam.

Overview: Continuous Flow FAB (CFFAB)

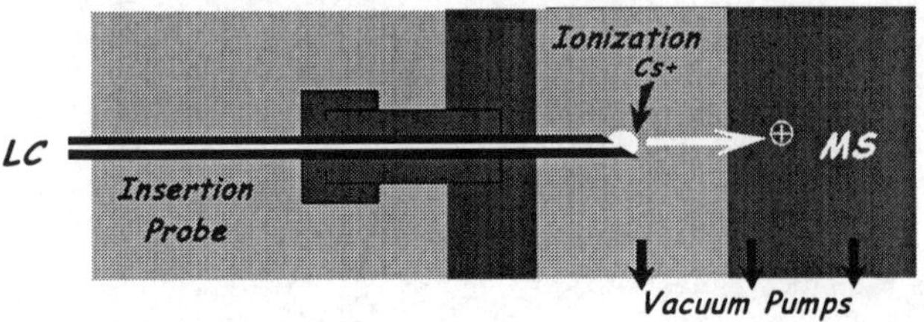

Figure 2.22- The Continuous Flow FAB Interface

Interfacing

Continuous flow fast atom bombardment (CFFAB or flow FAB) is one of the most direct approaches to LC/MS interfacing. The column effluent is directly introduced into the vacuum region ($<10^{-4}$ torr) of a mass spectrometer through a probe at a very low flow rate of 5-10 uL/min.[51] This low flow rate is required because no conventional pumping system could maintain the low pressures required with higher flow rates.

The LC stream is mixed with a matrix material (nitrobenzyl alcohol, thioglycerol or glycerol) to facilitate the ionization process. The mixture of solvent, solute and matrix flows out over the flattened end of the probe to form a very thin layer. Because of the low flow rate, most of the solvent mixture evaporates in the ion source. In some cases, the end of the probe is heated slightly to prevent the solvent mixture from freezing and facilitate solvent evaporation.

Ionization

FAB is a desorption ionization method that occurs when a beam of fast moving atoms or ions (5-8 keV Xe, Ar, Cs) is directed at the surface coated with sample compound dissolved in a matrix material. This is sometimes called liquid-SIMS (secondary ion mass spectrometry). The beam can be made up of atoms or ions depending on the type and pressure of the beam generator. The impact of these atoms or ions on the liquid surface transfers kinetic energy to the sample in a manner that results in the desorption of sample ions into the gas phase. This process is quite gentle with little fragmentation occurring.

The FAB spectra are characterized by ions at every mass, particularly at lower molecular weights (matrix and solvent adducts). These are true spectral interferences and are an artifact of reaction of matrix material under these high energy conditions.

Analytical Utility

CFFAB has shown its greatest utility with implementation on sector instruments. Because sectors are already operating at ultra high voltages, they are more capable of accommodating the high voltages required for the ion gun and optics required for FAB. In addition, the non-linear nature of sectors also minimizes the noise from neutrals. CFFAB will likely not have significant utility on other mass analyzers.

The high ionization efficiency of FAB, makes this technology excellent for trace analysis. Unfortunately, the relatively few stalwart users of CFFAB have had to deal with capillary LC to perform successful separations.

Considerations

The choice of matrix will have dramatic effects on your analytical response. For this reason, CFFAB isn't a technique for the analysis of unknowns. It performs quite well for class analysis where some prediction of response can occur. It has shown excellent results for neuropeptides.

Summary-CFFAB

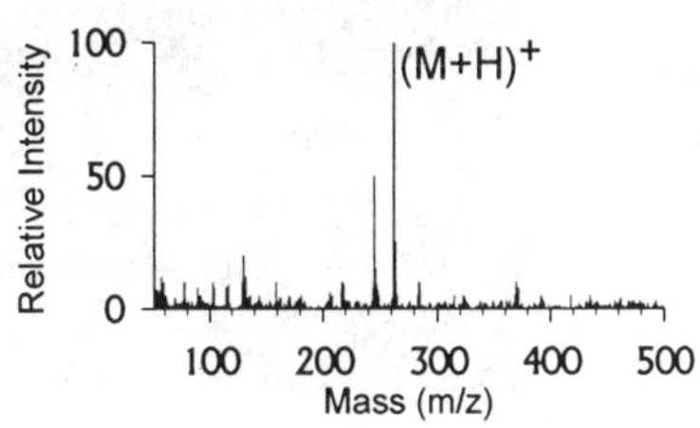

Figure 2.23-CF spectrum of a dipeptide (MW=261).

Information Content
- Molecular Weight
- Structure with MS/MS

Flow Rates
- CFFAB- 5-10 μL/min

Variations
- Primary Ions- Ar, Xe, Cs
- Frit FAB

General Applicability
- Peptides, nucleotides, small polar molecules, carbohydrates

Ionization Modes:
- Positive and Negative

Common Mass Analyzers
- Sectors

Pioneers
- Ito, Takeuchi, Ishii, & Goto(1985)[52] Caprioli, Fan, & Cottrell (1986)[53]

A Good Reference
- Stroh. & Rinehart (1994)[54]

Figure 2.24- The problem solving domain of CFFAB.

Overview: Thermospray (TS)

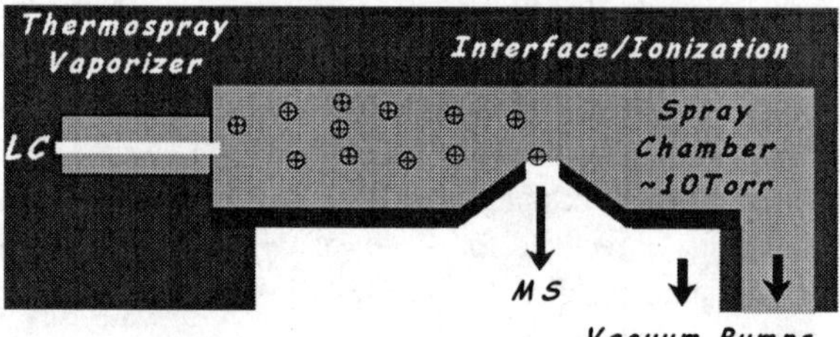

Figure. 2.25- Thermospray Interface

Interfacing

The heart of thermospray LC/MS is the *thermospray vaporizer*. This vaporizer is a heated capillary tube through which the effluent from the LC flows (see Figure 2.22). By precisely controlling the amount of heat delivered to the liquid in the tube, the complete effluent from a conventional LC could be completely evaporated at or near the tip of the vaporizer. The thermospray vaporizer is capable of controlling the evaporation of a wide variety of mobile phases with widely varying heats of vaporization. The heat input can be controlled in real-time as the solvent composition changes across a gradient run.

The vaporized effluent is introduced into a reduced pressure spray chamber (ca. 10 torr). Sample from the spray chamber is introduced into the high vacuum of the mass analyzer through a conical sampling aperture (ca. 250um i.d.).

Ionization

An artifact of the TS vaporization is the production of gas-phase ions, referred to as thermospray ionization. The rapid spray heating of ionic solutions results in the desorption of intact ions from the spray droplets. This mechanism is similar to that observed in ES. The primary ions produced by this process are protonated or ammoniated molecular ions. The fundamentals of this process are discussed in more detail in *Appendix C*.

One of the limitations of thermospray ionization is its restriction to aqueous mobile phases and the requirement of adding buffers to the solution to facilitate ion formation. To broaden the applicability of thermospray to a wider range of LC solvent systems, filaments and discharge electrodes have been added to the spray chamber. These supply electrons and induce chemical ionization processes, the CI reagents being the solvent molecules.

Analytical Utility

Thermospray provides a very versatile and reasonably sensitive approach to LC/MS. It has a huge published applications base with several TS environmental methods making it into the Federal Register. It can operate over a wide range of solvent conditions at flow rates well over 1.0 mL/min.

The analytical utility of TS significantly overlaps that of APCI (which is coupled with ES) and as a consequence may not dominate LC/MS in the future as it has in the past. The chemical space map[55] below shows that TS is suitable for use over a very wide range of polarities.

Considerations

The interface is simple and can easily be fitted to most standard mass analyzers. Successful operation of TS requires that the analyst have some insight into the nature of the analytes in order to select and optimize the ionization conditions.

Summary-TS

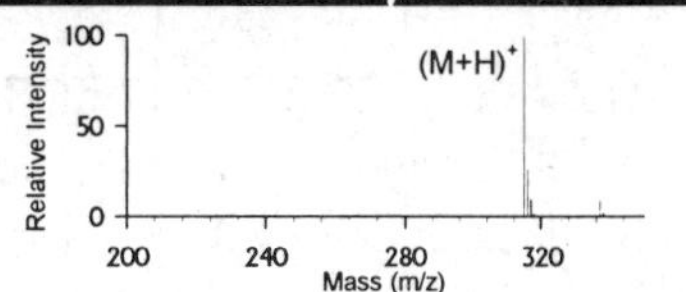

Figure 2.26- TS spectrum of a progesterone (MW=314)

Information Content
- Molecular Weight
- Structural Elucidation: requires MS/MS or discharge CI

Variations
- Thermal Nebulization

General Applicability
- Ionic and polar compounds (<2,000 u)

Flow Rates
- 0.5-2.0 mL/min.

Ionization Modes
- Thermospray Ionization (TSI)
- Filament "on" & Discharge CI
- Polarity- Positive/Negative

Common mass analyzers
- Quads & Sectors

Pioneers
- Blakley,Carmody, &Vestal (1980)[56]

A Good Reference
- Yergey, A.L. et al. (1990).[57]

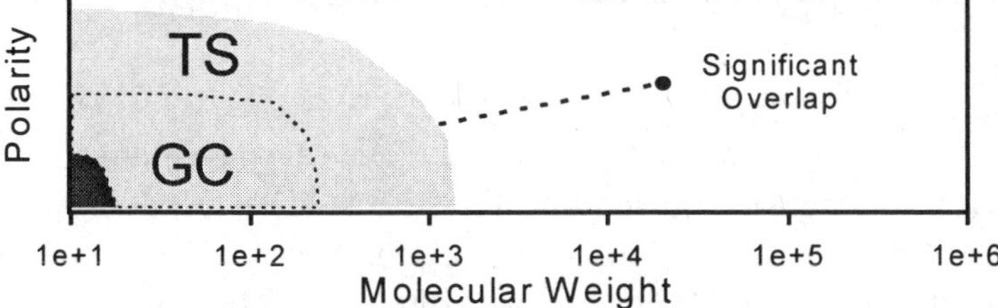

Figure 2.27- The problem solving domain of thermospray.

2.3- Your Separation Alternatives

HPLC & Other Alternatives

One of the most versatile, powerful, and most used analytical techniques ever developed for mixture analysis is High Pressure Liquid Chromatography (HPLC). HPLC can separate compounds with molecular weights into the hundreds of thousands, amounts that range from attograms to grams, aqueous or organic soluble samples, volatile or involatile species, ionic, polar, or neutral compounds; and diasteriomers and racemic mixtures. These separations can be performed quickly and with high reproducibility. If you can dissolve it, you can separate it. All this has come about because of improvements in the design of pumps, controllers and particularly columns.[58] Fortunately, all these capabilities are also available for separation alternatives in LC/MS.

The figure below shows the alternatives available in HPLC. Each of the primary components of an HPLC system has been listed with the options available.

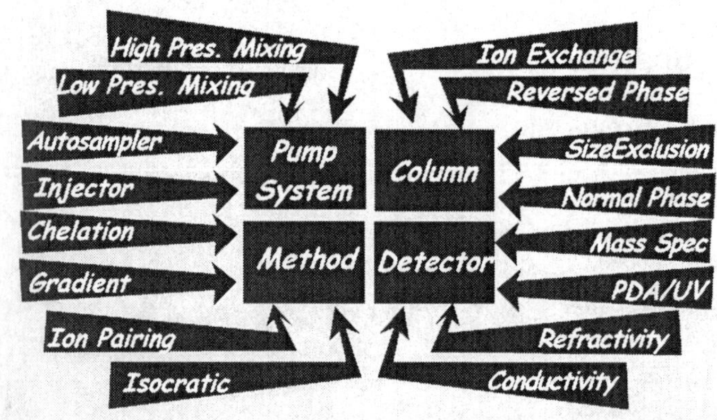

Figure 2.28- A Plethora of Alternatives with HPLC.

At first, this may look formidable. It can even be more confusing when trying to determine the best mix of technology and methods necessary to answer specific problems. But it is really much simpler than it looks. For example, the most basic LC system consists of a high pressure pump, an injector, a column, detector and some sort of data recorder, usually a PC. Injectors have been standardized and autosamplers are readily available. In other words, the hardware has been standardized. Even though the range of methods available is broad, many have been also standardized so the choices are again narrowed. Since this book is limited to interfacing LC with the mass spectrometer, the choice of detectors is also narrowed. In reality, the solution to your problem is not going to be driven by LC hardware and methods. The real problem-solving is done by choosing the separation method most applicable to your need. By looking at LC this way, the number of choices that have to be made is minimized. Another way to think of this is that once it is determined which *mechanism* of separation is best, the hardware will follow. The diagram below shows the common LC separation methods available today.[59] These methods, or combinations of these methods can answer almost any separation problem.

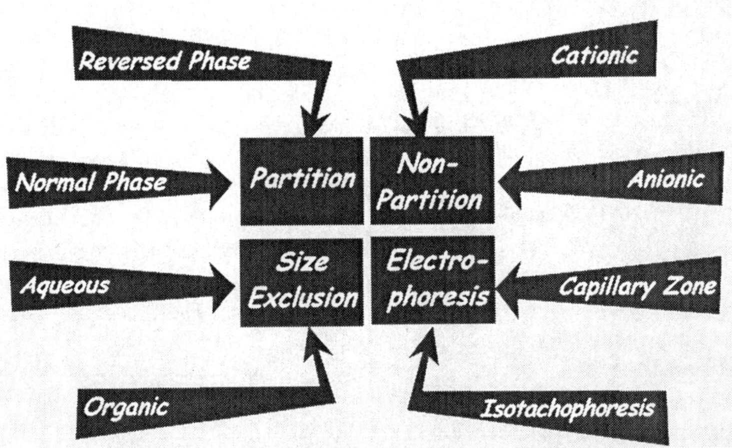

Figure 2.29- Focus on Separation Alternatives for HPLC

This section presents an **Overview** of each separation alternative commonly used with LC/MS. A brief description of the separation mechanism, column types, and analytical utility is provided. Issues relating to the evaluation and selection of various separation technologies are presented in *Chapter Three, Section 3.3*.

Overview: Reversed Phase (RP)

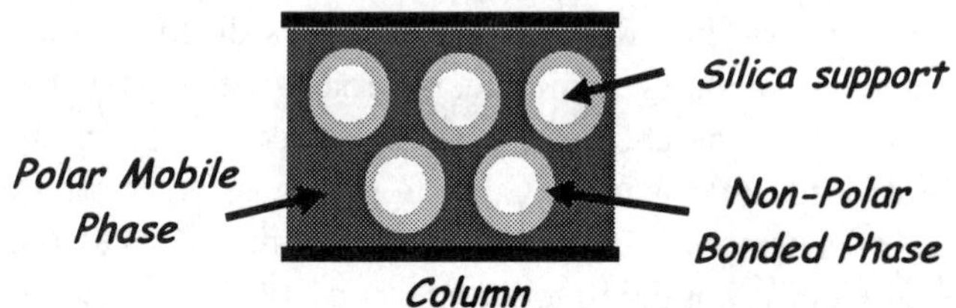

Figure 2.30- Reversed Phase Column

Separation Mechanism

Reversed phase (RP) HPLC is a form of partition chromatography which derives its name from the fact that the analyte must "partition" itself between the (polar) mobile phase and (non-polar) stationary phase of the column. Separations are based on relative polarity between the stationary phase and the mobile phase. A typical RP gradient run consists of loading a sample on a column equilibrated with polar (high aqueous) solvent. The mobile phase polarity is then changed by introducing a different, less-polar, miscible solvent. Since the stationary phase is relatively non-polar, very polar or ionic species are not retained on the column and quickly elute, while nonpolar species are retained on the column until the mobile phase becomes more non-polar. Every compound with differing polarity will spend different proportions of its time on the stationary phase, thereby effecting a separation.

Column Attributes

One of the biggest advances in RP came with the development of stable, bonded phase (BP) packings. This avoided a problem associated with liquid-liquid chromatography; that is, attaching a non-polar liquid phase to a matrix. For that reason, bonded phase LC is referred to as liquid-solid LC. These columns have a BP coating on silica consisting of long chain saturated hydrocarbons or other non-polar functional groups. The most popular packing material is octadecylsilane (ODS) with an 18-carbon aliphatic chain. The others are listed below.

C-18	Non-Polar
Phenyl	
C-8	
C-4	
CN	More Polar

There are no significant limitations on the type of RP columns used with LC/MS interfaces or mass analyzers.

Analytical Utility

There is no question that RP is the most widely used separation technology in liquid chromatography comprising *ca.* 80% of the market.[60] This high percentage of RP will also be observed in LC/MS since the RP is compatible with all interfacing technologies and has been coupled to virtually every mass analyzer.

The reason for the broad applicability of RP to separations problems is found in its ability to handle compounds of widely diverse polarity and molecular weight compared to either gas chromatography or normal phase HPLC. The analytical range of RP is shown in Figure 2.31.

The range of RP is extended further by variations which include ion suppression, ion-pairing and perfusion.[61] Unfortunately, ion-pairing and ion suppression require the use of reagents that may not be compatible with many LC/MS interfaces. Perfusion has shown some advantages for high speed separations.[62]

Summary- RP

Utility:
- Isolating polar analytes
- Desalting your samples
- Removing chemical interferences

Resolution
- High (>1,000 plates/cm)[63]

Mass Range
- <250,000 u

Flow Rates
- 1 uL/min.-2.0 mL/min.

Compatibility with LC/MS:
- All interfaces & mass analyzers

Common LC/MS Methods:
- Isocratic
- Gradient Elution

Column Diameters:
- Capillary: ES, CFFAB
- Microbore: IS, PB
- Analytical: IS, APCI, PB, TS

Considerations:
- Volatile buffers are required
- Analyte charge should be suppressed for retention.
- Response may be affected by mobile phase

A Good Reference
- Smith, R.M. (1995).[64]

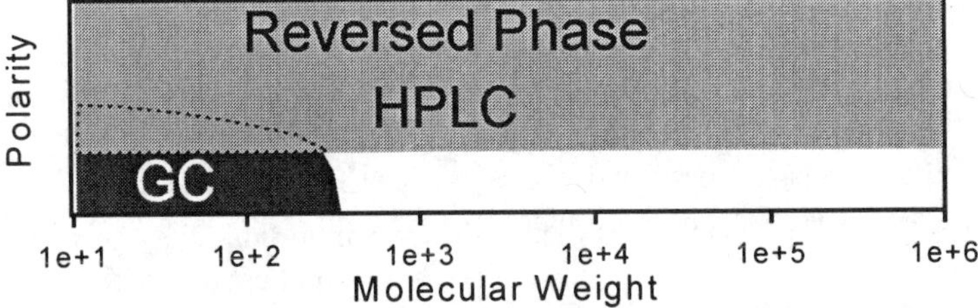

Figure 2.31- The problem solving domain of reversed phase.

Overview: Normal Phase (NP)

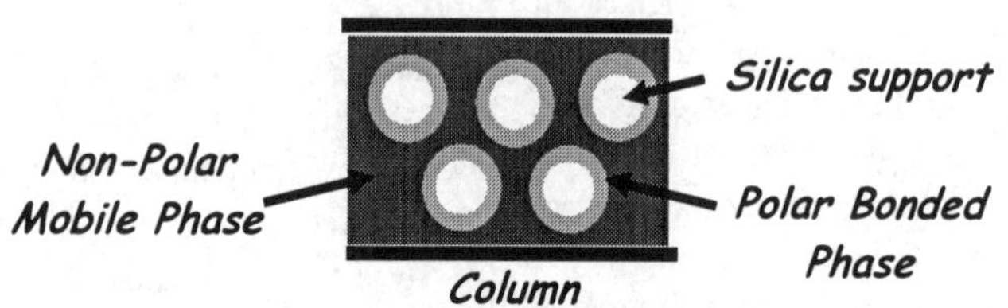

Figure 2.32- Normal Phase Column

Separation Mechanism

The opposite separation method from reversed phase is normal phase (NP). One of the first high pressure packings for HPLC was silica, the same material used for open columns and thin layer chromatography plates. Silica has water bound to it, forming a polar stationary phase.

With normal phase the separation is effected by partitioning analyte between a polar stationary phase and a nonpolar mobile phase. A typical normal phase analytical gradient run consists of loading a sample on a column equilibrated with a non-polar (high organic) solvent. The solvent polarity is then changed by introducing a different, more polar, but miscible solvent. Since the stationary phase is polar, non-polar compounds are not retained and quickly elute. Polar compounds are bound tightly to the stationary phase and elute last as the mobile phase becomes more polar. Figure 2.32 above represents a typical normal phase column.

Column Attributes

As with reversed-phase HPLC, the biggest advance in normal phase LC came with the development of bonded phase packings. These packings made NP much more efficient, reliable and versatile. Bonded phase packings made the columns less susceptible to changes in temperature and solvent composition. Whereas conventional NP could not be used with gradient elution, bonded phases can. The bonded phase consists of polar groups chemically linked to silica particles, or just plain silica. The most common packings are listed in the table below:

Silica	Polar
NH_2	↓
Diol	
CN	Less Polar

There are no limitations of column type for applications with LC/MS in normal phase chromatography.

Analytical Utility

Although normal phase separations only comprise between 10 and 20% of current separations,[65] the applicability of NP to solving problems with LC/MS is limited by a lower demand for separation of lower polarity samples. Much of the analytical domain of NP is covered by gas chromatography; particularly at lower molecular weights. NP has found applications with PAHs, porphorins, lipids, and polymers. Since polar functionalities have far greater relevance to life sciences and biotechnology, the number of RP separations will far outpace NP in applications in LC/MS.

The chemical-space map shown in Figure 2.33 compares the analytical domain of reversed phase, normal phase, and gas chromatography.[66] Note that the analytical range of NP significantly overlaps both RP and GC. NP can be used over a wide MW range but it is limited in its applicability to less polar species. Higher separation efficiencies and better peak shapes make GC and RP the methods of choice.

Summary - NP

Utility:
- Isolating nonpolar analytes
- Removing chemical interferences

Resolution
- Moderate

Mass Range
- >250,000 u

Flow rates
- 1 uL/min - 2.0 mL/min.

Compatibility with LC/MS
- Interfaces- all
- Mass Analyzers- all

Common LC/MS Methods
- Isocratic
- Gradient Elution (BP)

Column Diameters:
- Capillary: ES, NS
- Microbore: IS, PB
- Analytical: APCI, IS, PB, TS

Considerations:
- Need an exchangeable proton on your analyte for APCI and ES. (or redox ionization)
- Interfaces tend to be run cooler than RP due to more volatile mobile phase components.

A Good Reference

Smith, R.M. (1995)[67]

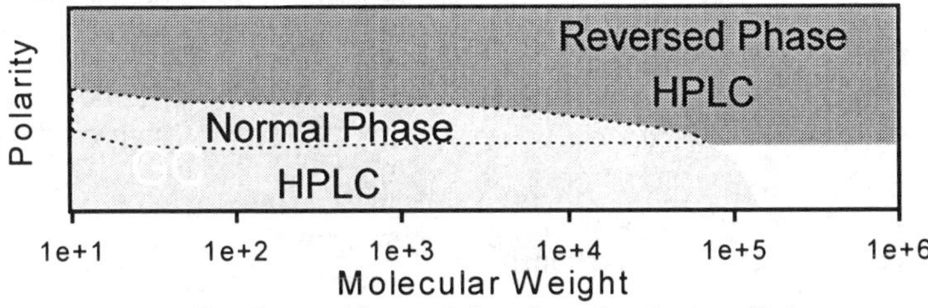

Figure 2.33- The problem solving domain of normal phase.

Overview: Ion Exchange (IE)

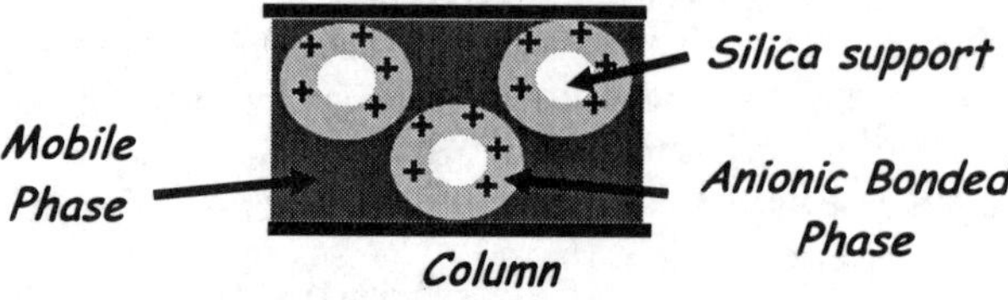

Figure 2.34-Ion Exchange Column

Separation Mechanism

Ion exchange chromatography (IE) depends on the type and degree of ionization of both the column and analyte. The column contains ionogenic groups that attach to opposite charged analyte. Analyte with the same charge as the column support or no charge elutes in the void volume.

A typical ion IE analysis starts by equilibrating the column with low ionic strength solvent. This insures that the sample will stick to the ionogenic groups on the column. After the sample is loaded, the solvent ion strength or pH is changed to either compete with the attached ion for an ionogenic site and displace it, or change the charge of the attached ion or ionogenic group. By changing the pH and/or ion strength, it is possible to control elution. Mass transport limits the rate of these exchange processes-consequently, IE columns are usually run at elevated temperatures.

Column Attributes

IE columns come in two flavors: anionic columns which separate anionic analytes and cationic columns which separate cationic analytes. Generally each type of column has two forms: strong and weak. Strong IE columns have permanent charges that are present irrespective of mobile phase conditions. Weak IE columns have functional groups with charge states induced by the solution equilibrium, usually by changing the pH. IE functional groups and types are:

Group	Type
sulfonic acid	strong acid
carboxylic acid	weak acid
tertiary amine	weak base
quat-ammonia	strong base

The proper use of IE is dependent upon controlling pH and ionic strength. The user must also be aware of analyte pK values which are used to determine separation conditions.

Analytical Utility

The utility of IE is found in its ability to separate ions and ionic species, in contrast to RP where ionization is usually suppressed in order to get effective separations. IE accounts for 12-15% of LC separations done today, split evenly between anionic and cationic applications.[68]

Unfortunately, when interfacing IE for LC/MS, the high ionic strength effluent eluting from the IE column is not always compatible with performance conditions for interfaces or gas-phase ion production. Ionic species tend to be less volatile than electrically neutral species; therefore, many ions decompose at the elevated temperatures required to get them into the gas phase. This limits the utility of IE with APCI, TS, PB, and TS. Even with a technique, such as electrospray that requires your analyte to be ionized, the high ionic strength can suppress ion production.

Figure 2.35 shows most of the range of IE overlapping with RP.[69] Efficiency and selectivity are lower than observed with RP.

Summary- IE

Utility
- Cations and anions
- High MW ions

Resolution
- Low/Moderate

Mass Range
- <1,000,000 u

Flow Rates
- <1.5 mL/min.

Compatibility with LC/MS
- ES

Common Methods
- Step pH and salt gradients
- Continuous salt gradients

Column types
- Silica
- Styrene-divinylbenzene polymer

Considerations
- High ionic strength can suppress ES response
- Mobile phase constituents can corrode system components
- Mobile phase constituents can degrade vacuum pump oil

A Good Reference
- Small, H., (1990)[70]

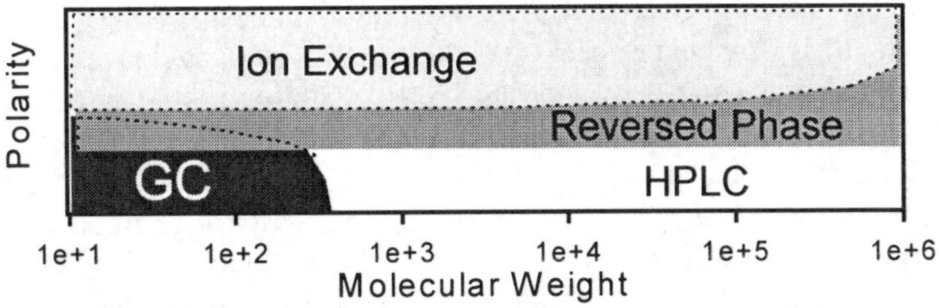

Figure 2.35- The problem solving domain of ion exchange.

Overview: Size Exclusion (SE)

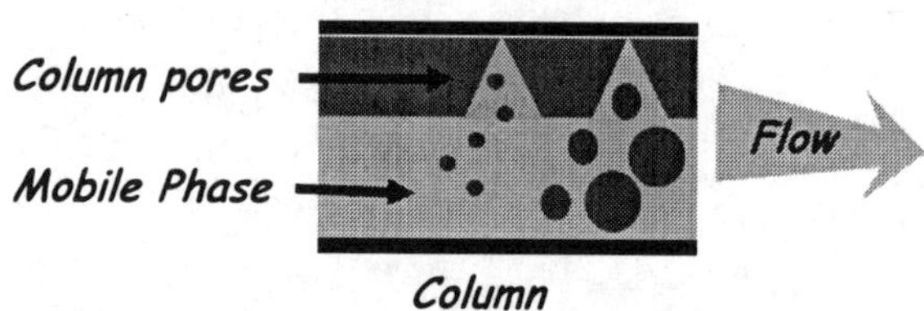

Figure 2.36- Size Exclusion Column

Separation Mechanism

Size exclusion (SE), also referred to as gel permeation, is one of the oldest separation techniques and was the first column type used with a commercial HPLC system. The stationary phase consists of beads with microscopic, tapered pores. These pores fill up with mobile phase that remains relatively stagnant as liquid flows through the column. Separations are based on the amount of time each analyte spends diffusing into and out of these pores. Large molecules are restricted from entering the pores and pass through the column quickly (eluting first). Small molecules diffuse in and out of the pores so they are retained on the column longer and elute later. It is sometimes said that these columns separate by molecular weight, but the separation is related to molecular shape. If a molecule is elongated it can behave as a much larger molecule, when compared to a spherical molecule of the same MW.

Column Attributes

Types of columns are distinguished by their pore size and the MW range. Here, the molecular weight given is the maximum; in general the minimum will be ~5% of that value.

Pore	MW Range
5 nm	5 Kda
20 nm	50 Kda
50 nm	300 Kda
100 nm	800 Kda
1,000 nm	8,000 Kda

SE columns can also display secondary retention effects. Ion exclusion occurs in silica columns because the silanol groups will have a negative charge above pH=3; negatively charged compounds will be excluded from the pores. Conversely, positive charged molecules will behave as if in an ion exchange column. Partitioning can also occur.

Analytical Utility

SE can be used with compounds having very high MW- ca. 8,000Kda (see Figure 2.37).[71] Therefore, the primary use of SE is in determining the MW range (or isolating a particular MW range) of polymers and proteins.

SE columns account for 8% of the separations done today.[72] Depending on the column material, separations can be done using organic or aqueous solvents and therefore can range from very polar biopolymers to very nonpolar industrial polymers.

In LC/MS, only a few applications of SE have been reported; however, there may be special instances where SE will assist the analyst in isolating or purifying a particular analyte. SE is simple and reliable. Do not exclude the possibility of off-line usage of SE for sample cleanup and isolation before analytical separations are performed with RP.

Summary-SE

Utility
- Special cases where high molecular weight sample fractions require isolation
- Could be used in purification of simple protein mixtures

Resolution
- Low

Mass Range
- 1Kda - 8,000Kda

Flow rates
- 1 - 5 mL/min.

Compatibility with LC/MS
- ES (possibly CFFAB)

Column Types
- Styrene divinylbenzene
- Porous silanized silica

Considerations
- Volatile buffers are required
- Salts and high ionic strength can suppress ES response
- Long equilibration times are required (sometimes 24 hours)

A Good Reference
- Wu, C.S. (1995)[73]

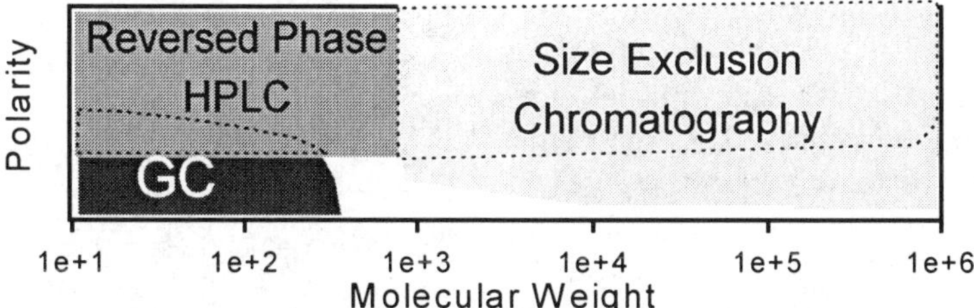

Figure 2.37- The problem solving domain of size exclusion.

Overview: Capillary Electrophoresis (CE)

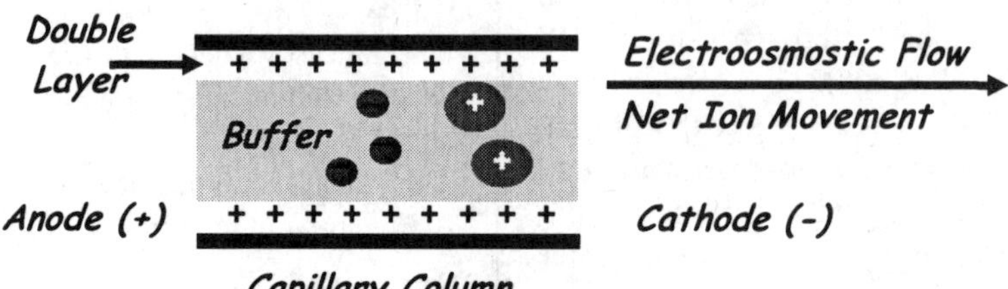

Figure 2.38-Capillary Electrophoresis

Separation Mechanism

Electrophoresis is based on the migration of ions in solution when placed in an electric field; anions move toward the anode and cations toward the cathode. Capillary Electrophoresis (CE) is a type where an electric potential (5-50kV) is applied across the ends of a capillary (0.3-2 meters long x 20-150um I.D.) filled with a buffer solution; causing ions to migrate along the axis. The motion of the ions will cause a net flow of the liquid through the capillary because of viscous drag. This net flow is called electro-osmotic flow (EOF), typically a few nanoliter per minute. The EOF is larger than the electrophoretic mobility of the counterions in solution. Therefore, all sample ions, (anionic and cationic) move to the cathode. Since the cations move faster (are not retarded by their charge) they get to the cathode first. Neutral molecules in the sample are unaffected and are not separated.

Column Attributes

Ion movement is dependent upon ionic strength of the buffer, voltage potential, viscosity, molecular size and charge. By controlling these experimental parameters one can optimize a separation.

The techniques listed below are variations of the CE techniques that have been implemented to overcome some of the limitations of CE. ITP, for example, is a technique that allows the analyst to overcome the sample capacity limitation of CE in order to accumulate a detectable amount of sample on the capillary.

Technique	Mechanism
Isotachophoresis (ITP)	Buffers w/ different mobilities
Capillary Gel (CGE)	Size & Charge
Micellar Electrokinetic (MEKC)	Hydrophobic & Electrostatic
Capillary Electrochromatography (CEC)	Electroosmotic & Reversed Phase

Analytical Utility

The greatest utility of CE is found in its ability to separate complex mixtures with very high separation efficiency, such as, oligonucleotides and peptides. The lower analyte dispersion associated with CE results in significantly higher separation efficiencies and less dilution of the sample on column. Consequently, some of the highest separation efficiencies and sensitivities result. Unfortunately, also associated with CE are lower flow rates (nL/min) and lower sample volume capacity (1-100nL). The low flow rate limits the practical utility of CE to low flow interfacing techniques such as ES and CFFAB. The low sample capacity limits the dynamic range and the utility of CE for trace sample components (<0.1%). At the conventional flow rates required for ES (1-10 uL/min), a makeup flow is required using either sheath flow (concentric capillary flow)[74] or a liquid junction (butted capillaries aligned in a makeup buffer).[75]

Summary-CE

Utility
- High efficiency separations
- Applications where sample size is small

Resolution
- Very High

Mass Range
- <1,000,000 u

Flow Rates
- 0-100 nL/min.

Compatibility with LC/MS
- Direct Interface-ES
- Sheath Flow- ES
- Liquid Junction- ES,CFFAB

Column Types
- Fused silica capillary
- Non-Ionic & Ionic coated capillaries
- Packed capillary(CEC)[76]

Considerations:
- Makeup liquid composition can affect your separations
- Redox chemistry at needle electrode can affect separations
- Involatile buffers are required

A Good Reference
- Altria, K.D. (1996); Landers, J.P. (1994)[77]

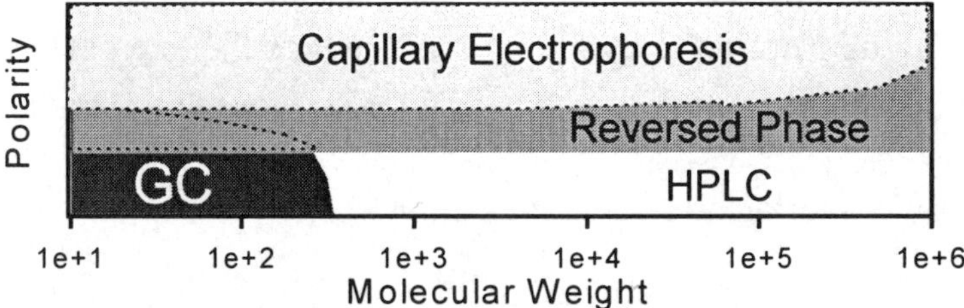

Figure 2.39- The problem solving domain of capillary electrophoresis.

2.4- Narrowing Your Choices

Table 2.1- Summary of Available LC/MS Alternatives.

To determine the interface availability for each separation alternative (*select from left column*) and each mass analysis alternative (*select from top row*).

	Mass Analyzers				
LC	Quads	Traps	TOF	Sectors	FTMS
RP	All	ES, AP	ES, AP	All	ES, AP
NP	All	ES,AP	ES,AP	All	ES,AP
IE	ES	ES	ES	ES (CF)	ES
SE	ES	ES	ES	ES	ES
CE	ES	ES	ES	ES (CF)	ES

RP- reversed phase
NP- normal phase
IE- ion exchange
SE- size exclusion
CE- cap electrophoresis
ES- electrospray
PB- particle beam
CF- continuous flow FAB
TS- thermospray
TOF- time-of-flight
AP- atmospheric pressure CI
FTMS- fourier transform MS
All- All interfacing approaches apply

Worksheet 2.0- Narrowing Your Choices

In *Chapter Two* we have reviewed most of the common approaches to LC/MS in two-page **Overviews** in order to present you with a broad view of the common alternatives available to solve problems with LC/MS. Although some alternatives clearly have a broader analytical utility compared to others, we have attempted to present the information in a uniform and even-handed fashion, giving our best assessment of the compatibility and utility of each. More information and analysis may be required in order to evaluate and select the best technology alternatives to solve your specific problems. We will treat the topic of evaluating your alternatives in *Chapter Three*. For now, let's see if we can narrow your choices based on what we have covered so far. Use your problem definitions from *Chapter One* and the **Overviews** from this chapter to check off the technology alternatives that may possibly meet your problem needs.

1. **Which separation alternatives will you consider for your problems?** *Use the **Overviews** presented in Section 2.4.*

   RP ❐ NP ❐ IE ❐ SE ❐ CE ❐ GC ❐ All ❐

2. **Which mass analyzer alternatives will you consider for your problems?** *Use the **Overviews** pr*

   Quads ❐ Traps ❐ TOF ❐ Sectors ❐ FTMS ❐ All ❐

Note: Copies of this worksheet can be downloaded from **www.LCMS.com**.

Match your selections in question 1 to the Separation Alternatives listed in the left column of Table 2.1. Match your selections in question 2 to the Mass Analysis Alternatives in the top row of Table 2.1. Cross-match your selections for both separation and mass analysis to see the available interfacing alternatives for each cross-match. Write your result.

What are your interface alternatives based on Table 2.1?

Note: Selection of appropriate technologies in LC/MS should involve a thorough and measured approach (i.e. an evaluation). This exercise is intended to give you a broad overview of your alternatives and an idea of the scope of your future evaluation process. Proceed to *Chapter Three*.

Discussion Questions and Exercises

1. Are there any techniques described in these overviews that are clearly not relevant to your problem-solving needs?
2. Would you limit your potential implementation of LC/MS to separations capabilities that already exist in your lab? If so, what are the typical flow rates and mobile phase compositions of your existing separation methods?
3. Are there any separations alternatives that are not currently implemented in your lab but that you may consider for LC/MS (e.g. CE)?
4. Which interfacing techniques can be implemented on the same mass analyzer?
5. Which ionization techniques can be implemented on the same mass analyzer?
6. Make a table matching all commercial LC/MS interfaces and the respective ionization processes. Discuss the general utility of each device.

Notes

[1] It is our view that covering the breadth of your LC/MS alternatives at this point is more relevant to problem-solving than treating each topic with great depth. See *Appendix C* and *D* for a more detailed discussion.

[2] The exception to this are electrospray spectra of polymeric compounds (i.e., biopolymers: proteins, DNA, RNA; industrial polymers: polyethylene and polypropylene glycols). These spectra are characterized by a multiply-charged envelope.

[3] With the advent of ionization techniques that produce multiply charged ions (i.e., electrospray), the observable upper masses have increased significantly for a given upper m/z value of an instrument.

[4] Yost and Enke developed the triple quad mass spectrometers almost two decades ago. By arranging three quads on axis- the first and third used for mass analysis, the second quad used to collisionally fragment sample ions- significant improvements could be made in specificity and selectivity with the wide variety of triple quad operation modes.

Enke, C.X, Yost, R.A., Anal. Chem., 51: 1251A (1979).

[5] Quadrupoles are the "workhorse" of mass spectrometry because they are rugged and reliable. They operate at relatively high pressures compared to other mass analyzers and are tolerant of much more dirt and contamination. They are easy to tune and calibrate with most instruments having computer algorithms to perform many startup operations.

[6] SIM is also referred to as selected ion monitoring- it is convenient that both "selected" and "single" start with the same letter. This technique is very sensitive because dweel times at agiven mass can be 10-100 times longer than when scanning, thus yielding a statistical advantage one to two orders of magnitude higher.

[7] Scanning modes compromise sensitivity because of the very poor duty cycle. For example, the ion may only reside on a single mass value for a fraction of the scan time (e.g. 0.01% of total time compared to SIM at 100%)

[8] This reference is considered a classic. It was recently reprinted- it originally appeared in 1976. Dawson, Peter H., (Ed.), *Quadrupole Mass Spectrometry and its Applications*; American Institute of Physics: Woodbury, 1995.

[9] Miller, P.E., Denton, M.B., "The Quadrupole Mass Filter: Basic Operating Concepts," J. Chem Educ. 7, pages 617-622 (1986)

[10] Lehmann, W.D., "Pictograms for experimental parameters in mass spectrometry," J. Am. Soc. Mass Spectrom. 8, pages 756-759, 1997.

[11] This is referred to as "collisional damping."

[12] It should be remembered that this will change with further development. The number given here is based on information from 1997.

[13] Zoom Scan™ is the trade name on the Finnigan LCQ ion trap operation that acquires a selected mass region at higher resolution and potentially higher mass accuracy. This allows quadrupole ion traps to have some of the capabilities of higher resolution instruments for selected applications. These capabilities may have significant utility for identification and qualitative analysis. Other manufacturers have similar capabilities.

[14] March, R.E., Hughes, R.J., *Quadrupole Storage MS*; John Wiley and Sons: New York, 1989. This is a good general reference that deals with the history of the development of this technology. Other books from the CRC Press deal with methods and applications also.

[15] Lehmann, W.D. *op cit.*

[16] Mamyrin, B.A., Karataev, V.I., Shmikk, D.V., Zagulin, V.A. Zh. Eksp. Teior. Fiz. 64, 82 (1973).

[17] You must scan for structural elucidation problems in order to acquire all the fragmentation information for a given unknown.

[18] Your scan time should be significantly shorter than your chromatographic peak width to accurately record relative ion abundances.

[19] Reflectron and delayed extraction are technique that enhance the performance of TOF. These techniques are discussed further in *Appendix D* on "Fundamentals of Mass Analyzers."

[20] The "reflectron" is a device that reflects the ions traveling in the field-free region of TOF with an ion mirror in order to accommodate energy spread of ions and improve resolution.

[21] MALDI stands for Matrix Assisted Laser Desorption Ionization and is usually coupled with LC/MS in an off-line fashion. We discuss more details of MALDI and its relationship to on-line LC/MS in Appendix C.

[22] Stephens, W.E., Phys. Rev., 69, pages 691-695 (1946). Wiley, W.C., McLaren, I.H., Rev Sci. Instr., 26, pages 1150-1153 (1955).

[23] Cotter, R.J., *Time-of-Flight: Instrumentation and Applications in Biological Research*; ACS: Washington, D.C. (1994). A good general source that includes the historical and theoretical underpinnings of the technology.

[24] Lehmann, W.D. *op cit.*

[25] Magnetic sectors were the first mass analyzers and were developed by J.J. Thompson and F.W. Aston. (Aston, F.W., Proc. Chem. Phil. Soc. 19, 317 (1919)).

[26] Most sectors use magnetic scanning.

[27] Electric sectors are denoted by the letter E, magnetic sectors are denoted by the letter B, quadrupoles are denoted as Q, and time-of-flight are denoted as TOF.

[28] Aston, F.W., Proc. Chem. Phil. Soc. 19, 317-326 (1919). Hipple, J.A., Fox, R.E., Condon, E.E. Phys. Rev. 69, 347 (1946).

[29] Dass, Chhabil, "Chapter 1 Instrumentation and Techniques," IN: *Mass Spectrometry: Clinical and Biomedical Applications*; Desiderio, D.M. (Ed.), Plenum, New York (1994). This reference gives a good, short overview of all sector technologies and provides excellent comparison between the wide variety of sector combinations. Also McLafferty, F.W. (Ed.), *Tandem Mass Spectrometry*; Wiley: New York (1983).

[30] Lehmann, W.D. *op cit.*

[31] Generally uses a superconducting magnet that requires liquid helium, although some small, permanent magnet systems with limited mass range, exist.

[32] Alternatively, infrared lasers are being used for collisional dissociation.

[33] Comisaron, M.B., Marshall, A.G., Chem. Phys. Lett., 25, pages 282-283; and 26, pages 489-490 (1974).

[34] Buchanan, M.V. (Ed.), *Fourier Transform MS*; American Chemical Society: Washington D.C. (1987). A general reference that will give the particulars and peculiarities of FTMS.

[35] Lehmann, W.D. *op cit.*

[36] Although in some cases (as with particle beam) the two processes can be handled as discrete sub-processes, with other cases (as with thermospray) the distinction between interface and ionization becomes less obvious.

[37] Bruins, A.P.,Covey,T.R., and Henion,J.D., Anal. Chem., 59, 2642 (1987). **Ionspray (IS)** is the term used for pneumatically assisted electrospray. This technology permits operation of electrospray at far greater flows than conventional electrospray.

[38] Banks, J.F., Shew,S.,Whitehouse, C.M., and Fenn, J.B., Anal. Chem. 66:406 (1994). **UltraSpray (US)** is another version of aerosol genera-

tion assisted electrospray. In this implementation, high power ultrasonic vibrations are utilized to disrupt the liquid surface. In this specific case very uniform and small droplets were observed and performance at higher flows was improved over conventional electrospray.

[39] Covey, T.R., Lee, E.D., and Henion,J.D, Anal. Chem. 58: 2453 (1986). In addition to pneumatic assisted electrospray, heat was also added to facilitate increased evaporation at the higher flow rates. This technique is known as **Turbospray™ (TS)** and is a trade name of Sciex. The efficient transfer of heat with Turbospray improves sensitivity at the higher flowrates.

[40] Wilm, M.S., and Mann, M. Anal. Chem. 68: 1 (1996). **Nanospray (NS)** is a low flow implementation of electrospray that utilizes an extremely small needle and nL/min flow rates. Significant advantages for some applications occur with nanospray because of the low sample utilization and the high ion collection efficiency. It is an excellent technique for "identity" applications such as protein sequencing. Nanospray has been coupled to capillary scale LC and CE.

[41] Dole,M.,Mack,L.L.,Hines,R.l.,Mobley,R.C.Ferguson,L.D., and Alice,M.B., J Chem. Phys., 49, 2240 (1968).

[42] Yamashita,M. and Fenn, J.B., J. Phys. Chem. 88, pages 4451(1984).

[43] Mann, M. and Fenn, J.B., "Electrospray Mass Spectrometry; Principals and Methods", in Mass Spectrometry; Clinical and Biomedical Applications, Vol. 1, Desiderio, D.M. (ed.) Plenum Press, New York (1992).

[44] Rapid heating is a common method for getting labile molecules into the vapor phase. Cotter, R.J., "Mass spectrometry of nonvolatile compounds, Desorption from extended probes," Anal. Chem. 52, pages 1589A-1606A (1980). Another excellent review of this topic was written by: Vestal, M.L. "Ionization techniques for nonvolatile molecules," Mass Spectrometry Reviews, 2, pages 447-480 (1983).

[45] See Appendix C for further discussion on ion formation with APCI.

[46] Horning, E.C., Carroll, D.I., Dzidic, I., Haegele, K.D., Horning, M.G., Stillwell, R.N., "Atmospheric Pressure Ionization (API) Mass Spectrometry. Solvent Mediated Ionization of Samples Introduced in Solution and in a Liquid Chromatograph Effluent System," J. Chromatogr. Sci. 12, pages 725-729 (1974).

[47] Thompson, B.A.; Danylewch, M.I., "Design and Performance of a New Total Liquid Introduction LC/MS Interface," page 852, Proceedings

of the 31st ASMS Conference on Mass Spectrometry and Allied Topics, Boston, Massacheusetts, May 8-13 (1983).

[48] The momentum separator is similar in principal to a jet separator used with GC/MS interfaces. An excellent reference to gain a greater insight into the enrichment process associated with aerosol beams was written by Israel and Friedlander.

Israel,G.W., and Friedlander, S.K. (1967). J. Colloid Interf. Sci., 24, 330 (1967).

[49] Willoughby, R.C., Browner, R.F., "Monodisperse Aerosol Generation Interface for Coupling Liquid Chromatography with Mass Spectroscopy," Anal. Chem. 56, pages 2626-2631 (1984) (The classic paper on particle beam LC/MS. Note the author!!) and Browner, R.F., Winkler, P.C., Perkins, D.D., Abbey, L.E., "Aerosols as Microsample Introduction Media for Mass Spectrometry," Microchem. J. 34, pages 15-24 (1986).

[50] Bellar, T.A., Behymer, T.D., Budde, W.L., "Investigation of Enhanced Ion Abundances from a Carrier Process in High Performance Liquid Chromatography Particle Beam Mass Spectrometry," J. Am. Soc. Mass Spectrom. 1, pages 92-98 (1990) and Behymer, T.D., Bellar, T.A., Budde, W.L., "Liquid Chromatography/Particle Beam/Mass Spectrometry of Polar Compounds of Environmental Interest," Anal. Chem. 62, pages 1686-1690 (1990).

[51] To achieve the pressure required for ionization and mass analysis, the flow rate of liquid into the mass spectrometer must be limited to 5-to-10 uL per minute. No conventional pumping system could maintain the vacuum with flow rates exceeding these values.

[52] Ito,Y.,Takeuchi,D.,Ishii,D., and Goto,M., J.Chromatogr., 346 (1985).

[53] Caprioli, R.M., Fan, T., and Cottrell, J.S., Anal. Chem., 58, 2949 (1986). Also see: Caprioli, R.M., "Continuous-Flow Fast Atom Bombardment Mass Spectrometry," Anal. Chem. 62, pages 477A-485A (1990); Caprioli, R.M. (Ed.), *Continuous-Flow Fast Atom Bombardment Mass Spectrometry*; Wiley: Chichester (1990); and Caprioli, R.M., "Continuous-Flow Fast Atom Bombardment Mass Spectrometry," Trends Anal. Chem. 7, pages 328-333 (1988).

[54] Stroh, J.G., Rinehart, K.L., "Liquid Chromatography/Fast Atom Bombardment Mass Spectrometry," Top. Mass Spectrom. 1, pages 287-311 (1994).

[55] Mitrovich,S, Willoughby, R.C.,"Mapping Chemical-Space; Application and prediction within LC/MS," Proceeding of the 43rd ASMS Conference on Mass Spectrometry and Allied Topics, page 494, Atlanta,

Georgia, May 21-26, 1995. This work has subsequently been submitted to J. Chem. Education (1998).

[56] Blakley,C.R., Carmody, J.J., and Vestal, M.L., "A New Liquid Chromatograph/Mass Spectrometer Interface Using Crossed-Beam Techniques," Adv. Mass Spectrom. 8B, 1616-1623 (1980).

[57] Yergey, A.L., Edmonds, C.G., Lewis, A.S. and Vestal, M.L., *Liquid Chromatography /Mass Spectrometry*; Plenum Press: New York (1990).

[58] Do you remember large, open columns (usually homemade from Plexiglas tubing) fed by head pressure created by raising or lowering the solvent bottle (the one with the spout on the bottom) filled with calcium phosphate gels, or mixed-bed ion exchangers? If you do, you're giving away your age.

[59] We understand that electrophoretic methods are not technically LC, but they represent a very powerful separation technique that can be interfaced with mass spectrometers.

[60] Siuzdak, G., *Mass Spectrometry for Biotechnology*, page112, Academic Press, New York, (1996).

[61] Perfusion consists of large silica particles (10-50m) that have throughpores which increase the adsorptive surface area to create a higher theoretical plate count for greater resolution without the concomitant problems of slow flow rates and/or high back pressure.

[62] Voyksner, R.D., Keever, J., Enhancing API-MS performance through chromatography: Determination of growth hormone in milk, page 444, Proceedings of the 43rd ASMS Conference on Mass Spectrometry and Allied Topics, Atlanta, Georgia, May 21-26, 1995.

[63] Plates/cm (theoretical plates) are the units that apply to column efficiency as defined by the efficiency factor N. For more detail see Chapter 3 page128.

[64] Smith, R.M. (Ed.), *Retention & Selectivity in Liquid Chromatography*; J. of Chrom. Library 57; Elsevier: Amsterdam (1995).

[65] McMaster, M.C., *HPLC- A Practical Users Guide*, UCH Publishers, New York, page 20 (1994).

[66] Mitrovich, S., *op cit*.

[67] Smith, R.M. (Ed.), *Retention & Selectivity in Liquid Chromatography*; J. of Chrom. Library 57; Elsevier: Amsterdam (1995).

[68] McMaster, M.C., op cit.

[69] Mitrovich, S., *op cit*.

[70] Small, H., *Ion Chromatography*; Plenum: New York, (1990)

[71] Mitrovich, S., *op cit.*

[72] McMaster, M.C., op cit.

[73] Wu, C.S., (Ed.), *Handbook of Size Exclusion Chromatography*; Chromatographic Science Series 69: Dekker (1995). Obviously a mature technique still in great use; hence the "Handbook" in the title.

[74] Smith,R.D., Bartinaga, C.J., and Udseth, H.R., Anal. Chem. 60,1948 (1948).

[75] Lee, E.D., Muck,J.D., Henion,J.D., and Covey, T.R. Biomed.Environ. Mass Spectrom., 18, 844 (1989).

[76] Colon, L.A., Guo,Y., and Fermier, A. "Capillary Electrochromatography," Anal. Chem. 69, 15, pages 461A-467A (1997). The recent review is an excellent overview of the recently developed field of Capillary Electrochromatography (CEC). CEC is a hybrid between capillary LC and capillary electrophoresis. Some would argue that it is the best of both worlds. It may have significant potential for electrospray applications.

[77] Altria, K.D. (Ed.), *Capillary Electrophoresis Guide Book: Principles, Operation and Applications*; Humana Press: Totowa (1996) and Landers, J.P. (Ed.); *Handbook of Capillary Electrophoresis*; CRC Press: Boca Raton (1994).

Chapter Three- Flow Diagram

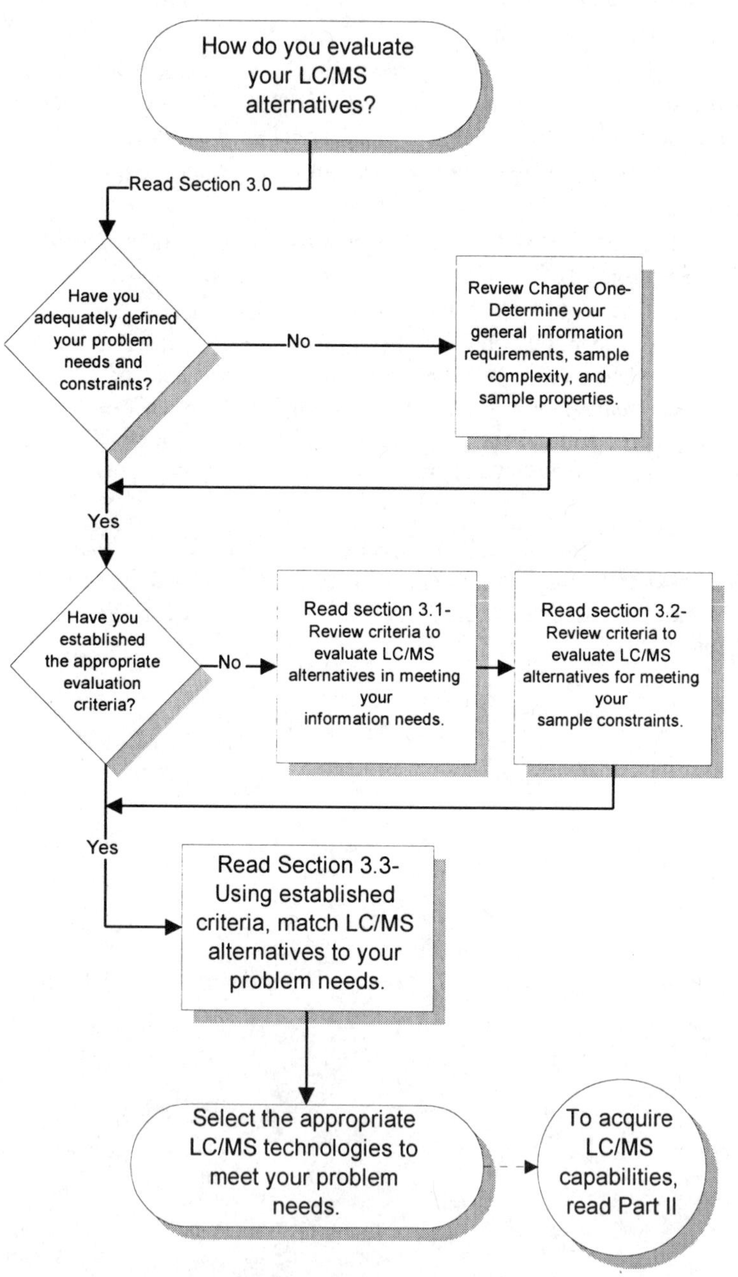

How do you evaluate your LC/MS alternatives?
Read Section 3.0
Have you adequately defined your problem needs and constraints?
No
Review Chapter One- Determine your general information requirements, sample complexity, and sample properties.
Yes
Have you established the appropriate evaluation criteria?
No
Read section 3.1- Review criteria to evaluate LC/MS alternatives in meeting your information needs.
Read section 3.2- Review criteria to evaluate LC/MS alternatives for meeting your sample constraints.
Yes
Read Section 3.3- Using established criteria, match LC/MS alternatives to your problem needs.
Select the appropriate LC/MS technologies to meet your problem needs.
To acquire LC/MS capabilities, read Part II

Chapter Three

Evaluating Your LC/MS Alternatives

(or, Matching Your Needs to Technology)

Problems and solutions are not state functions; the energy getting from one to the other is highly dependent upon the path you choose.

3.0-What criteria should you use?

You cannot adequately answer the question, *Is LC/MS right for you?*, until you first identify your problem needs (*Chapter One*); understand your alternatives to meeting those *needs (Chapter Two*); and finally match those needs to a particular alternative (*Chapter Three*). This process may seem daunting at first but with a systematic evaluation you should be able to reduce the relatively large number of LC/MS alternatives into a manageable number of choices.

The first step in your evaluation should be the selection of the appropriate evaluation criteria. These criteria should be selected on the basis of your problem definition; in particular, your information needs and your sample constraints. Any LC/MS alternative that you choose must meet the information needs and the sample constraints of your problem. *Figure 3.0* shows the connection between *information needs* and *evaluation parameters*. *Figure 3.1* shows the connection between *sample constraints* and *evaluation parameters*. These figures make the link between your problem and the appropriate evaluation parameters needed to select an LC/MS alternative. Using the results from your problem analysis in *Chapter One* (*Worksheets 1.0-1.2*) you can select your appropriate LC/MS evaluation parameters which are reviewed in the next two sections.

3.1-Parameters for Information Needs

The flow diagram in Figure 3.0 illustrates the relationship between the *information needs* of your LC/MS problem and the associated evaluation criteria (parameters). Although most problems in analytical chemistry can be reduced to "identity" or "quantity" problems, the enhanced ability to “identify and quantify” with mass spectrometry results in significantly enhanced problem-solving capabilities with LC/MS. As a consequence, the evaluation criteria relating to your information needs will tend to apply to performance parameters of the mass analyzer alternatives. In other words, selecting the appropriate mass analyzer will go a long way in meeting the information needs of your LC/MS problems. Conversely, selecting an inappropriate mass analyzer can significantly restrict your ability to resolve your information needs.

This section contains a review of each evaluation parameter relating to *information needs* in a two-page summary; a definition of the parameter is presented to the left, and a discussion on the utility of each parameter (point- counterpoint) is presented to the right. At the bottom of the right page, a tabular listing of typical values for each related alternative will be presented.

Note that the evaluation parameters listed in *Figures 3.0* and *3.1* are ones we have deemed important to the process of distinguishing one LC/MS technology from another. You may require other parameters to effectively aid you in your particular evaluation. Additional parameters should be added to your list. Also note, the evaluation discussed in this chapter is intended to address only your LC/MS "technology" needs, not your "broader" problem needs. Broader LC/MS needs (and justification) will be addressed in the next chapter, *Chapter Four*.

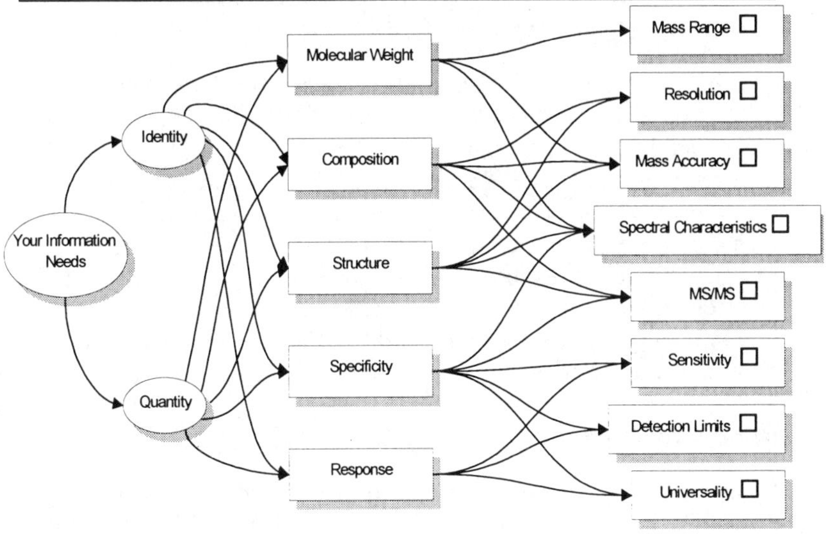

Figure 3.0- Matching your <u>information needs</u> to the appropriate evaluation parameters.

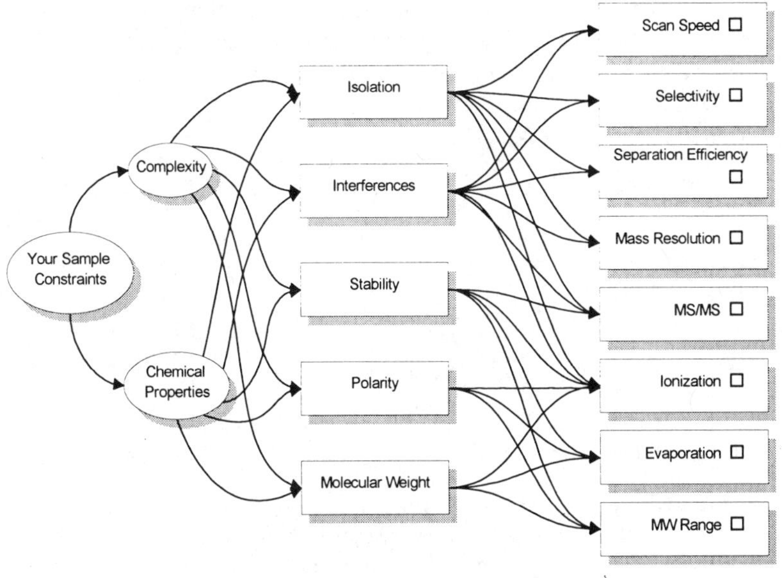

Figure 3.1- Matching your <u>sample constraints</u> to the appropriate evaluation parameters.

Parameter: Mass Resolution

The success of mass spectral analysis and the information derived from an analysis is highly dependent upon being able to resolve ions of one mass from those of another mass. Resolution is the measurement of the mass analyzer's ability to separate one mass from an adjacent mass.[1] The resolution of a mass spectrometer is defined by Equation 3.0, where $\mathbf{R_m}$ is the resolving power, **m** is the measured mass or average mass, and $\mathbf{\Delta m_{resolution}}$ is the difference between two adjacent mass peaks. It is common practice within the instrument industry to use the terms resolution and resolving power interchangeably. The important point is to understand the analytical implications of low versus high resolution.

Resolving Power: $$R_m = m/\Delta m_{resolution} \quad (3.0)$$

Resolving power is determined by two methods shown below. In the case of the Doublet Method, $\mathbf{\Delta m_{resolution}}$ is determined by measuring the mass of two adjacent peaks with a 10% valley; m is the average mass, $\mathbf{\Delta m_{resolution}}$ is the difference. In the case of the more common Singlet Method, $\Delta m_{resolution}$ is the measured full peak width at half maximum (FWHM) and m is the mass at the maximum. In some cases the peak width is measured at 5% of the peak height.[2] The singlet method is simpler and more practical for most applications because it can be applied to virtually any ion in the spectrum.

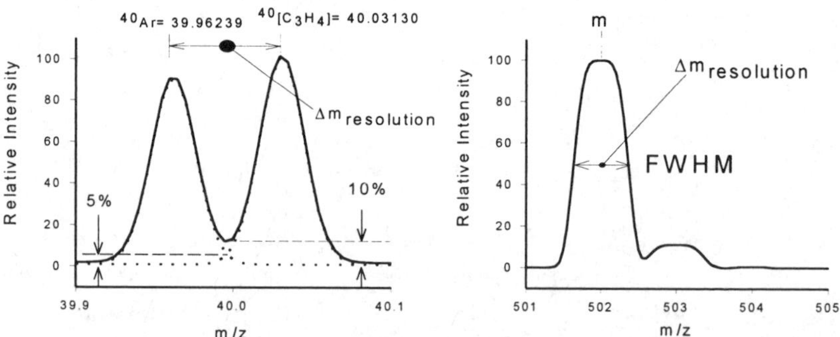

Figure 3.2- Two methods of determining resolution; (left) the Doublet Method where two similar signals from adjacent masses are measured with a 10% valley; and (right) the more practical and commonly used Singlet Method where full width at half maximum (FWHM) is used to determine the distance between two adjacent masses (delta mass).

Low Resolution

A prerequisite to solving most problems in organic analysis is the requirement that the mass analyzer separate masses that differ by at least one mass unit (a proton), known as unit or nominal resolution. The mode of mass analysis that operates with "unit resolution" is by far the most common. Quadrupoles and the ion traps have performed quite effectively at low resolution with GC/MS for the last decade with great analytical utility at affordable instrument prices.[3]

Low resolution mass spectrometry is the ideal mode of operation where on-line coupling to chromatographic techniques are used; particularly, for target quantitation. In this way isobaric and chemical interferences can be removed chromatographically and reliable results obtained.

One important utility of low resolution mass spectrometry has been the matching of acquired spectra to large mass spectral libraries of low resolution reference spectra. These libraries are capable of identifying total unknowns in less than one second.[4]

The advent of triple quadrupole mass spectrometry has also given the analyst higher specificity when coupled to low resolution mass analyzers.

High Resolution

One of the most powerful operational modes of mass spectrometry is high resolution. The combination of high resolution and accurate mass measurement is required for a reliable determination of *elemental composition.* High resolution can be used for *removal of isobaric interferences* or *increased selectivity* in MS/MS experiments. Note, you cannot separate isomers with high resolution. Chromatography usually required for isolation of isomers. High resolution has many applications for accurate quantification where known interferences have to be removed.

With newer ionization techniques and the trend toward analysis of heavier masses, higher resolution is required to isolate and identify analytes. High resolution, therefore, has a greater role in high mass applications. Higher resolution is also used to identify the charge state of multiply-charged ions.

High resolution has also associated with it the requirement of generally higher operator and interpreter skill. High resolution mass spectrometry has historically been the domain of the so-called "mass spectrometrists."

Table 3.0- Typical Mass Resolutions

Analyzer	Resolution	Utility	Type
Quads	1,000-2,000	Nominal mass measurement	Low
Traps	1,000-2,000	Nominal mass measurement	Low
TOF	500-1,000 2,000-10,000	Nominal mass High specificity/accurate mass	Both
Sectors	5,000-100,000	High specificity/accurate mass	High
FTMS	5,000-1,000K	High specificity/accurate mass	High

Parameter: Mass Accuracy

Mass accuracy is the measurement of the closeness of the mass of a given measurement to the true mass of the substance. Here; $\Delta \mathbf{m}_{\mathbf{accuracy}}$ denotes the difference between a measured value ($\mathbf{m}_{\mathbf{measured}}$) and the true mass ($\mathbf{m}_{\mathbf{true}}$) of a substance. "Exact mass" measurement is the measurement of mass accuracy to very high accuracy (ppm). In practice, exact mass measurements require high resolution as well as high mass accuracy in order to ensure that isobaric interferences are not present, which introduce error in the measurement.

Mass Accuracy:

1) $\Delta m_{accuracy} = m_{true} - m_{measured}$ (3.1)

2) $ppm = 10^6 \Delta m_{accuracy} / m_{measured}$ (3.2)

Mass accuracy can be expressed in terms of millimass units (0.001 amu) as in Equation 3.1 or in terms of parts-per-million (ppm) as in Equation 3.2. Today, millimass units is becoming the more acceptable term. Also note the difference between $\Delta \mathbf{m}_{\mathbf{accuracy}}$ and $\Delta \mathbf{m}_{\mathbf{resolution}}$. These two measurements in practice may be similar in value but represent entirely different measurements.

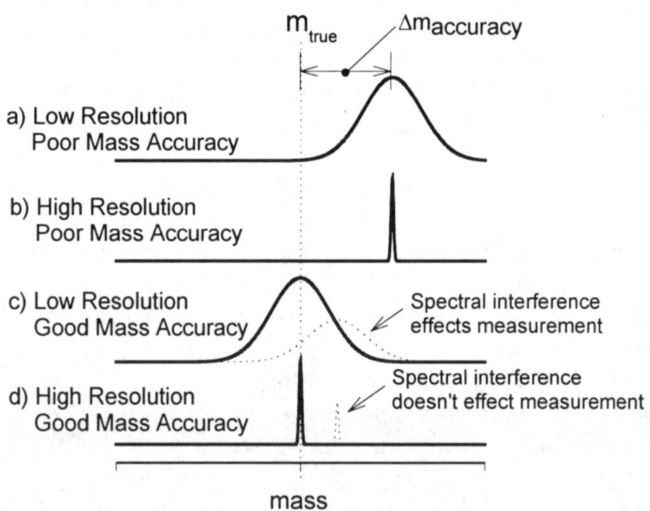

Figure 3.3- Comparison of mass accuracy and resolution: a) low resolution and poor mass accuracy; b) high resolution and poor mass accuracy; c) low resolution and high mass accuracy; and finally d) high resolution and high mass accuracy.

Low Mass Accuracy

Mass accuracy is important to all applications of LC/MS and should be an important consideration when evaluating any mass analyzer. All mass analyzers should be calibrated to the nearest mass unit. The accurate assignment of mass is essential to virtually every aspect of mass analysis.

All instruments should be evaluated to meet a requirement of around 0.1 u drift over a reasonable operating period. Mass changes can come from many sources, including the stability of the instrument's electronics, temperature changes, tuning, and contamination of the mass analyzer, to name just a few.

Mass accuracy must be maintained throughout the entire mass range and throughout an entire experiment. Even small changes in mass accuracy can have a deleterious effect on the interpretation of your spectral results, library matching, and quantitative results.

High Mass Accuracy

High mass accuracy is sometimes referred to as exact mass measurement. When measuring the mass of a particular compound at the ppm level in mass accuracy, one can unequivocally deduce the elemental composition.[5] Knowing the elemental composition of the molecular ion and/or a fragment ion is one of the most powerful analytical tools for identifying and elucidating the structure of unknowns. As shown in Figure 3.3, the ability to perform exact mass measurements is dependent upon resolution. In practice, mass accuracy and high resolution are inextricably linked.

Exact mass spectral results are required for many publications of molecular structure such as JACS[6] and are considered strong experimental evidence of structure in chemical patents.

With the advent of mass spectral data bases, such as PROWL[7] for protein searching, knowledge of the assigned mass to a higher accuracy will narrow the search results significantly. As these data bases get bigger, the utility of accurate mass measurement to confirm identity will only increase.

Table 3.1-Typical Mass Accuracies

Analyzer	Mass Accuracy	Utility	Type
Quads	0.1 u	Identity w/nominal mass frag.	Low
Traps	0.1 u	Identity w/nominal mass frag.	Low
TOF	0.0001 u	Empirical form./elemental comp.	High
Sectors	0.0001 u	Empirical form./elemental comp.	High
FTMS	0.0001 u	Empirical form./elemental comp.	High

Parameter: Mass Range

Mass spectrometers measure mass-to-charge ratio (m/z). Mass range is the difference between the upper and lower limits of that measurement, denoted as $\Delta \mathbf{m}_{range}$. Figure 3.4 shows the difference between $\Delta \mathbf{m}_{resolution}$, $\Delta \mathbf{m}_{accuracy}$, and $\Delta \mathbf{m}_{range}$. All of these measurements are important in determining the analytical capabilities of a mass spectrometer and should not be confused.

Mass range: $\Delta m_{range} = m_{upper\ limit} - m_{lower\ limit}$ (3.3)

The analytical mass range is determined by the instrumental configuration of the various mass analyzers and the ability to generate gas-phase ionic species within that range. Many physical parameters can affect the limits of measurement in a mass analyzer. Each mass analyzer alternative has a different set of parameters that determine the limit of the mass range as described by Equation 3.3.

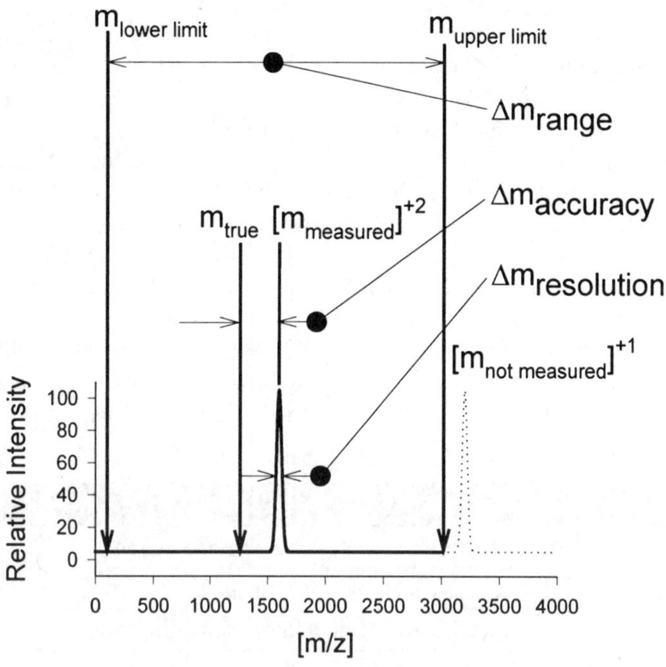

Figure 3.4- Diagram of mass spectrum comparing mass range, resolution, and accuracy. Too many Δms!

High Mass Analysis

Just ten years ago, anything more than 500 u was considered high mass. Now we are analyzing molecules more than one million daltons.[8] The rapid development of the biosciences has pushed the upper mass range envelope. This has resulted from the scientific need to measure and characterize biopolymers. Mass spectrometry has become one of many tools to characterize the complex and dynamic nature of these molecules. High mass measurement is a prerequisite for biomolecule mass spectrometry.

In general, we would define high mass as any analysis more than 2,000u. At this self-imposed threshold you have to begin thinking in very different terms. Tuning is different, your instrumental alternatives may be different, calibration is different, and ionization is usually limited to a few choices- primarily electrospray and MALDI.

The Revolution(s)

Until recently, $m_{upper\ limit}$, defined the upper molecular weight limit of the instrument. The advent of electrospray ionization has revolutionized and redefined the concept of mass range by allowing more than one charge to reside on each molecule.[9] More than one charge on a gas-phase molecule allows the molecule to be measured at a lower m/z. Therefore, the upper limit of mass measurement has now been extended by increasing z instead of extending our instrumental ability to measure m/z.

Another technological revolution in recent years for high mass analysis is the modern-day implementation of time-of-flight (TOF) mass analyzer. TOF effectively has no upper mass limitations. With recent engineering, ionization and electronic development feats, TOF has the resolution, mass accuracy, sensitivity, and scan speed to elevate it to being the preferred tool for high mass analysis.[10]

Table 3.2- Typical Mass Ranges

Analyzer	Upper Limit	Utility for High Mass	Type
Quads	500-3,000	Multicharged species[11]	Low
Traps	500-3,000	Multicharged species[12]	Low
TOF	500->10^6	All masses, no limitation	High
Sector	1,000-15,000	Medium sized molecules	Medium
FTMS	1,000-10,000	Medium sized molecules	Medium

Parameter: Spectral Characteristics

The information derived from any measurement with a mass spectrometer or LC/MS is to a large extent limited by the characteristics of the mass spectrum. There are simple spectra that reveal simply a molecular ion of your analyte material, and there are complex spectra that reveal complex patterns of mass-intensity pairs that can be interpreted and correlated to the specific structural components of your analyte molecules. Although the sources of spectra are many and vary with ionization technique and type of interface, mass analyzer and detection, we can divide spectra into two general types: those with fragmentation and those without. Figure 3.5 shows some examples of spectral characteristics for techniques that yield molecular ion information and those that yield structural information (fragmentation).

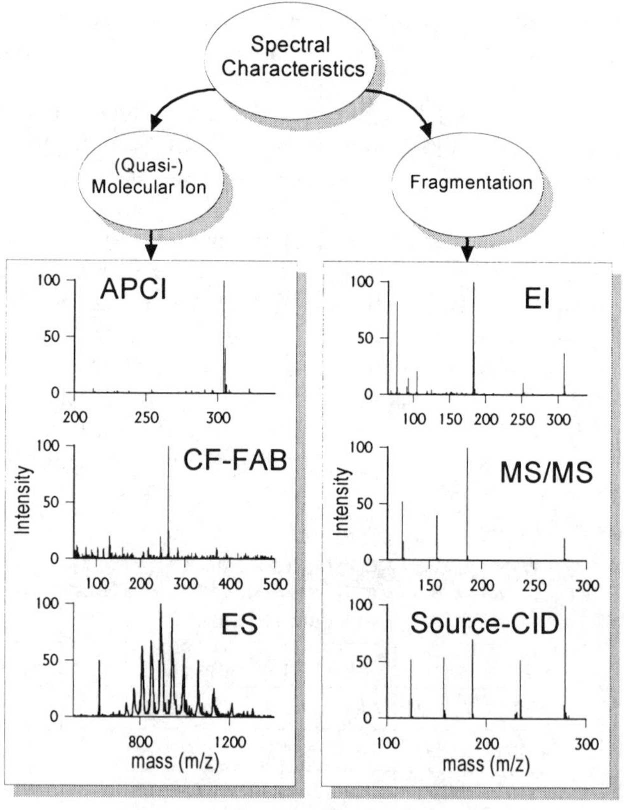

Figure 3.5- Spectral characteristics of LC/MS techniques.

(Quasi-)Molecular Ions

In general, it is preferable that all mass spectra contain a representation of the molecular mass. The molecular mass as seen in its variable forms ($(M+H)^+$, $(M+Na)^+$, $(M-H)^-$, $(M+5H)^{+5}$, etc.) possibly gives us the most important information in the mass spectrum; the physical boundary of the molecule being analyzed.[13] All subsequent fragments, substructures, and compositions must fit within the boundary of the molecular ion.

For many applications in LC/MS, the molecular ion supplies enough information to unambiguously identify specific sample components. In addition, molecular weight information may also provide adequate specificity for many applications in quantitative analysis. (see Table 3.3).

Fragmentation

Unfortunately, in most applications where samples contain unknown components, molecular weight information is not enough. In these applications, some fragmentation spectra are critical to providing compound specific information needed for identity and/or isolation.

Fragmentation spectra are a requirement for many facets of chemical analysis with mass spectrometry, including structural elucidation and specificity for quantification. There are three common approaches to getting fragmentation in LC/MS: 1) as a result of the ionization process (primarily electron ionization, EI), 2) as a result of collisions within the LC/MS interface (Source-CID), and 3) as a result of CID between adjacent mass analyzers (MS/MS).

Table 3.3-Typical Spectral Characteristics

Analyzer	Interface	Ionization	Config.	Spectral Characteristics
Quads	API,TS,PB	ES,APCI, TSCI	Single	Primarily molecular ions
	API,TS,PB	ES,APCI, TSCI	Triple	Fragmentation (MS/MS)
	PB	EI	Single	Fragmentation (EI)
Traps	API	ES,APCI	MS^n	Both MW & fragmentation (MS/MS)
TOF	API	ES,APCI	Std.	Primarily molecular ions
	API	ES,APCI	Hybrids	Fragmentation (MS/MS)
Sectors	API,TS,PB	ES,APCI, TSCI	EB,BE	Primarily molecular ions
	API,TS,PB	ES,APCI, TSCI	EBE,BEB, Hybrids	Fragmentation (MS/MS)
	PB	EI	EB,BE	Fragmentation (EI)
FTMS	API	ES,APCI	MS^n	Both MW & fragmentation (MS/MS)

Parameter: MS/MS

Mass spectrometry/mass spectrometry (MS/MS) or tandem mass spectrometry is the mode of operation that utilizes multiple stages of mass analysis sandwiching a collision (or reaction) region between each mass analyzer.[14] The coupling of multiple stages of mass analysis provides the most useful and structurally dependent information obtainable. A significant percentage of applications of LC/MS today requires operation in the MS/MS mode. Fortunately, virtually all alternatives for mass analysis are equipped with some type of MS/MS capability. If you are considering LC/MS, you most likely should consider LC/MS/MS.

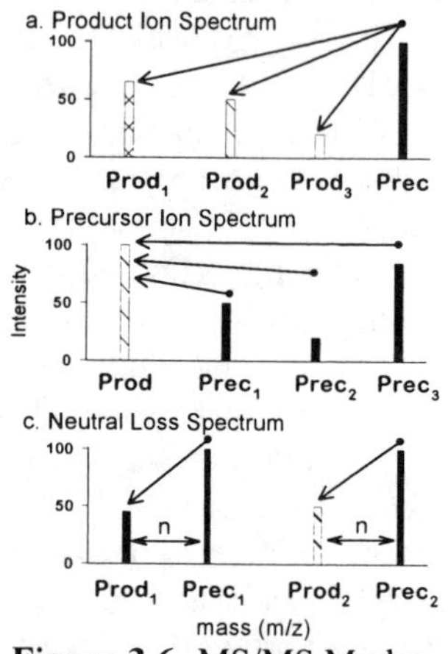

Figure 3.6- MS/MS Modes.

The most common mode of MS/MS is the *product ion mode*. In this mode, the first stage of mass analysis, $(MS)_1$, is used to isolate selected ions into a reaction region. These ions (precursor ions, Prec) are excited in some manner, usually by interaction with a collision gas. This excitation leads to fragmentation processes forming fragment ions (product ions, Prod). The product ions then undergo a second stage of mass analysis, $(MS)_2$. In most cases the product ions are then detected. The three common scanning modes of MS/MS are: *precursor ion* mode, *product ion* mode and *neutral loss* mode (see Figure 3.6).

Non-scanning modes are referred to as Multiple Reaction Monitoring (MRM) or Selected Reaction Monitoring (SRM). In this configuration, selected analyte precursor masses can be continually introduced into the collision region by the first mass analyzer and only selected product ions monitored through the second mass analyzer. This mode of MS/MS yields maximum sensitivity and selectivity for known target analytes when using quadrupole, sector, hybrid mass analyzers. With Ion Traps and FTMS (where only one mass analyzer is used for MS/MS) a second excitation can occur and subsequently a second stage of mass analysis, $(MS)^2$. This continuing sequence of analysis and collision is referred to as MS^n where **n** designates the number of stages of mass analysis in the experiment. Two stages are usually enough but three stages are sometimes used to differentiate very similar compounds that yield indiscernible spectra from the first collision process (such as isomers).

Do you need MS/MS?

Most ionization modes available with LC/MS are considered "soft" ionization techniques, which means they produce predominantly molecular ions, and limited (possibly irreproducible) fragmentation. If your samples limit you to soft ionization (See Section 3.2 on Sample Constraints) and you still need structural information or higher specificity, you need MS/MS.

Your primary alternatives to MS/MS are electron ionization (PB) or interface CID[15]. Choosing these alternatives to MS/MS will compromise your ability to elucidate structure, limit your sensitivity, and/or limit your MW range. The choice of MS/MS vs. EI or CID) will generally be decided based on weighing price and performance issues.

Which one?

Once you decide you need MS/MS, then you are faced with the alphabet soup of alternatives. In general, you can select specific MS/MS alternatives based on your general information needs. All parameters that were important in evaluating MS, such as, including sensitivity, detection limits, resolution, mass accuracy, mass range, scan speed, and component compatibility, should also be considered when evaluating MS/MS. Table 3.4 shows a list of MS/MS alternatives with some of their particular attributes and utilities.

Triple quads are commonly used for high sensitivity target analysis. Sectors are used for high energy collision reactions.[16] FTMS and hybrid TOF (Q-TOF, EB-TOF) are used for high mass accuracy and elemental composition.

Table 3.4- MS/MS capabilities of conventional analyzers

Analyzer	MS/MS	Prod. Ion	Prec. Ion	Neut. Loss	Utility	Resolution	Collision Energy
Quad	QQQ	✓	✓	✓	Sensitive in SRM	Low	Low
	EBQQ[17] BEQQ	✓	✓	✓	High precursor selectivity	Mixed	Both
	Q-TOF	✓			Sens. product ion spectra	High	Low
Traps	MS^n	✓			Versatile/ affordable	Low	Low
TOF	Q-TOF	✓			Sens. product ion spectra	Mixed	Low
Sector	EB	✓				Low	High
	BE	✓				Low	High
	BEB	✓			High perf.	Mixed	High
	EBE	✓			High perf.	Mixed	High
FTMS	MS^n	✓			High perf.	High	Low

Parameter: Sensitivity

Sensitivity in mass spectrometry is a term used to describe the ability of the mass spectrometer to respond to a given amount of sample analyte at a given mass-to-charge ratio (m/z). Sensitivity is usually expressed in terms of coulombs (charge) per microgram of analyte as shown in Equation 3.4, where $\mathbf{A}_{\mathbf{mass\text{-}current}}$ is the area under the time profile or chromatographic peak in coulombs, **G** is the gain of ion detector, and $\mathbf{m}_{\mathbf{sample}}$ the mass of the sample in ug introduced into the mass spectrometer.

Sensitivity: $S_{m/z} = A_{meas\text{-}current} / G\ m_{sample}$ (3.4)

Although we collect ion current at the ion detector in mass spectrometry, we immediately convert the current to voltage with a preamplifier. We can account for current-voltage conversion by including the value of the feedback resistor (R_f) in this expression as shown in Equation 3.5. $\mathbf{A}_{\mathbf{meas\text{-}voltage}}$ is the area under the voltage peak. Feedback resistors generally range from 10^7 to 10^9 ohms.

Sensitivity: $S_{m/z} = A_{meas\text{-}voltage} / R_f\, G\ m_{sample}$ (3.5)

To solve problems in analytical chemistry we really would like to express the sensitivity of the mass spectrometer in terms of analytical response, not sensitivity. Analysts like to see the analytical curve relating measured response with sample concentration. Equation 3.4 and 3.5 can both be rearranged to represent the slope of your quantitative analytical curve (see Equation 3.6).

Response: $A_{meas\text{-}voltage} = (S_{m/z}\ F)\ C_{analyte}$ (3.6)

For flowing systems such as LC/MS we can express the analytical response ($\mathbf{A}_{\mathbf{meas\text{-}voltage}}$) in terms of the product of sensitivity ($\mathbf{S}_{\mathbf{m/z}}$), flow rate **F** (in mL/sec) and $\mathbf{C}_{\mathbf{analyte}}$ (in ug/mL) analyte concentration. This relationship is visualized with the conventional analytical plot shown in Figure 3.7.

To evaluate the sensitivity of an LC/MS system, you should select a known reference compound for each operating mode of your LC/MS alternatives. LC/MS instrument manufacturers provide a list of compounds that they use as a basis for their instrument specifications.

The bottom line for evaluating mass analyzer "sensitivity" is in the ability of a particular mass analyzer to respond to "your" potential analytes at the concentrations required for your analysis. For this reason, detection limit may be a more appropriate criterion for evaluating mass analyzers than the slope sensitivity of the analytical curve (See next parameter).

Collective Efficiencies

The sensitivity of your LC/MS system is directly related to the collective efficiencies of all the processes of the entire system, namely, the LC separation efficiencies, the interface efficiencies, the ionization efficiencies, the mass transmission efficiencies, and the detection efficiencies. A sensitive response requires that each of these components of the LC/MS system operates at its respective optimal efficiency.

In our opinion, an in-depth understanding of the efficiencies of each component in your system is beyond the scope of a reasonable system evaluation.[18] To adequately evaluate sensitivity, one should, however, measure the sensitivity of the system under your expected operating conditions. This means resolution, mass range, gain, ionization, etc. These measurements represent the collective efficiencies of all processes and is appropriate for comparison of one LC/MS system configuration to another.

Duty Cycles

The ideal mass spectrometer system would sample all of the ions, all of the time. Our ability to store and detect ions, particularly with traps and TOF, has allowed us to improve our detection efficiencies and therefore sensitivity. A 100% duty cycle coupled to high collection efficiencies will yield maximum sensitivity. SIM and SRM are high duty cycle operating methods while Q-TOF is a high duty cycle configuration.

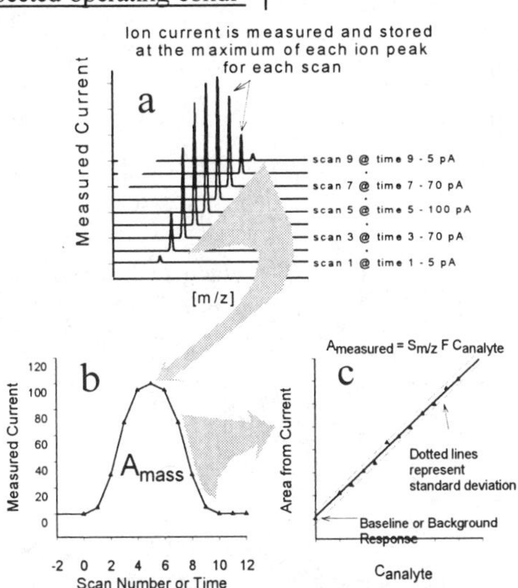

Figure 3.7- Illustration of a) the primary current measurement by mass spectrometry showing consecutive mass scans that are converted to b) area in the time dimension and can be related back to c) a sample concentration. Sensitivity is the slope of the analytical curve.

Parameter: Detection Limits

Detection limit is defined as the smallest amount of material one can measure or detect above the level of the noise. Unfortunately, there are many quantitative descriptions and regulatory definitions for detection limit. There are instrument detection limits, method detection limits, system detection limits, etc. In general, the detection limit is intimately associated with the signal-to-noise ratio (S:N). How you measure the signal and the noise will affect the value of your detection limit.

Detection Limit: $$V_{measured} > V_{baseline} + 3\,\sigma_{baseline} \qquad (3.7)$$

The voltage measured from a given ion signal ($\mathbf{V_{measured}}$) at a given mass should exceed the average baseline signal ($\mathbf{V_{baseline}}$) plus ten[19] times the standard deviation of the baseline. The baseline signal average and standard deviation can be derived from the mass baseline on either side of the ion as shown in Figure 3.8. Conversely, the baseline can be evaluated in the time dimension by acquiring data and using the standard deviation on either side of a chromatographic peak. There are dozens of methods for filtering signals in the presence of noise. Many of these methods are used in mass spectrometry to discern analyte signal from noise. In most cases, you are limited to the signal processing algorithms that reside in your ion detection and processing systems. If you have choices of detection algorithms, learn the relative merits and limitations of each. Table 3.5 lists typical detection limits.

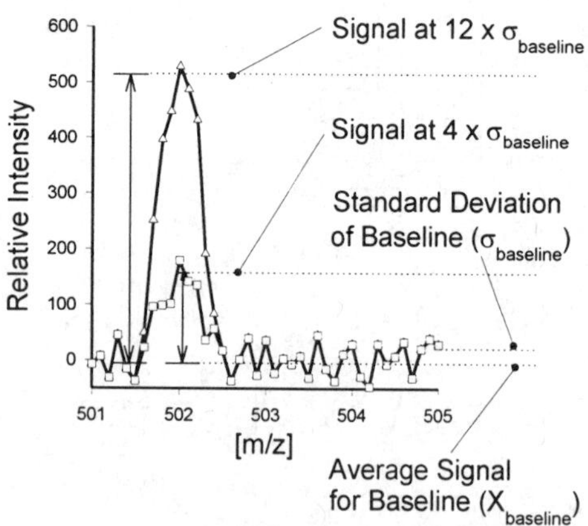

Figure 3.8- Two overlaid ion profiles (or peaks) near the detection limit. The larger ion profile is at ten times the standard deviation of the baseline. The smaller ion profile is at three times the standard deviation of the baseline. Most peak detection algorithms can easily distinguish both of these peaks from background noise.

Minimizing Noise

Since noise is ultimately the primary limitation in terms of detection limits in LC/MS, the system should be operated to minimize any and all components of noise. In most cases, the engineers that designed your instrument have taken every opportunity to engineer noise out of your system. Common sources of noise originate from electronic and mechanical components of the system. You can usually diagnose periodic noise components in your baseline by their frequency (e.g. 60 hz line frequency, turbo-pump frequencies). Other sources of noise are found in chemical background inherent in ionization and the mass analyzer. Examples would be helium metastable neutrals that drift to your detector and ionize background gas after they pass through the analyzer (e.g. conversion dynode). High metastable background noise is a sign that the system is either running at too high a pressure, pumping is inadequate, or the detector optics are poorly designed. Low pressure ion sources such as EI and FAB tend to have more noise than atmospheric pressure sources.

Maximizing Signal

Every LC/MS system under consideration should be tested for detection limits under conditions that are close to "apples-to-apples." To fairly appraise and compare systems, you should run the various systems under the same acquisition conditions. All operating conditions should be at their optimum, such as, tuning and calibration, gain, acquisition parameters, etc.

Many tricks of the trade can maximize the signal at the detection limit. For example, many analysts will increase signal by decreasing the resolution. This is perfectly valid as long as you understand the tradeoffs of decreasing one parameter to improve another. For example, decreasing resolution in MS/MS also decreases selectivity.

Time is our other great ally in terms of signal enhancement. If we can acquire signal for longer periods of time, then we will improve your S:N by the factor $(t)^{1/2}$. Stepping with larger mass increments and integrating specific masses for longer times will greatly enhance signal. You should try to match integration times and step increments when evaluating systems.

Table 3.5-Optimal Detection Limits

Analyzer	Optimal Detection Limits	Utility
Quadrupole	50-500 pg scanning 0.5-5 pg SIM	The ideal mass analyzer for target analysis w/ SIM or SRM.
Ion Trap	1-10 pg	Excels in scanning and MS/MS
TOF	1-10 pg	Excels in scanning mode
Sector	10-100 pg scanning 0.1-1 pg SIM	Excels in SIM mode
FTMS	1-10 pg	Excels in scanning mode

Parameter: Universality

Universality is the ability of a given instrumental system to respond to all components in the sample. Universality is extremely important if you are attempting to characterize unknown components in your samples. Solving problems with LC/MS can be significantly hampered if you cannot guarantee a response from the majority of analytes in your sample. Lack of universality will ultimately lead to missed sample components and false negative results. These gaps in information can lead to either increased time of analyses, primarily because you have to rerun the sample under several detection modes to guarantee a response, or higher uncertainty about results from a given sample.

Electron Ionization (EI) has served as the universal ionization mode for GC/MS for almost 30 years. With GC/MS under EI conditions, each component that elutes from the column yields an information-rich mass spectrum, usually leading directly to compound identification. If a GC-separated compound survives the heated environment of the column, there is a high probability that it will yield a response with EI mass spectrometry. The analytical domain of GC is covered completely by EI. This has distinct benefits when evaluating samples for unknowns. Even if the response is not ideal (e.g. no molecular ion) you still have a response from structurally related fragments at a given retention time. In addition, once you have a response, although, limited, you can always follow up with chemical ionization to confirm molecular weight.

Table 3.6- Universality of Response from LC/MS Interfaces

What is the universality of your interface alternatives?					
Sample Characteristic[20]	ES	APCI	PB	TS	CFF
Ionic	★★★★	★★	★★	★★	★★
Ionizable	★★★★	★★★	★★★	★★★	★★★
Polar Non-ionic	★★	★★★	★★★	★★★	★★★
Nonpolar	★	★★	★★★	★★	★

★★★★ ....denotes universal ability to produce a response

★★★ ....general ability to produce a response

★★ ....intermittent ability to product a response

★ ....limited ability to produce a response

Table 3.7- Aggregate Universal Interface Packages for LC/MS

Interfacing Alternative	Recommended Ionization Modes	Recommended Ion Polarity Modes
Atmospheric Pressure	1. ESI 2. APCI	Positive and Negative Positive and Negative
Particle Beam	1. EI 2. CI (Selective reagents)	Positive Only Positive and Negative
Thermospray	1. TSI 2. CI (Filament/discharge)	Positive and Negative Positive and Negative
Continuous Flow FAB	CI (Selective matrices)	Positive and Negative

Alternatively, LC/MS covers a much broader range of compound size (molecular weight) and polarity compared to GC/MS. Unfortunately, there is no universal ionization technique for LC/MS. The universal response from electron ionization is only commercially available from one interface, particle beam. This, of course, is one of the advantages of the particle beam approach compared to the other LC/MS ionization modes. Unlike GC/MS, the breadth of compounds separated by LC is not completely covered by EI; therefore, the use of EI with particle beam is limited to molecules less than 1,000 u (and for universal response more likely <600 u).

The lack of universal response from LC/MS may mean that each component you observe on a UV trace from your LC/UV/MS may not respond with a given ionization mode on the mass spectrometer. In many cases you may have to run both positive and negative ion modes to get a response from the individual components. Knowledge of sample chemistries is much more important in LC/MS than in GC/MS. Whether you are considering APCI or electrospray, you should develop a feeling for the solution chemistries of your analytes and how they relate to gas-phase ion formation. For example, are you studying acids, bases, cations, anions, or zwitterions? With all LC/MS techniques, solution chemistry will affect gas-phase ion production process. Table 3.6 ranks the various interfaces in terms of response to sample characteristics. Clearly, electrospray (ES) is preferred for samples that tend to be ionic. Particle beam (PB) will tend to favor compounds that are less ionic.

When evaluating your LC/MS alternatives, one should consider one of the aggregate packages in Table 3.7 for at least achieving universal response.

Worksheet 3.0-Evaluating Your LC/MS Alternatives

Selecting Criteria for Information Needs

Part A- Fill out the following checklist to establish your information needs.

What are your information needs?	Yes	No	Comment
–MW information?	❑	❑	________
–Elemental composition?	❑	❑	________
–Structural information?	❑	❑	________
–Compound specificity?	❑	❑	________
–Enhanced response?	❑	❑	________

Part B- Match your evaluation criteria and information needs. (Add any additional parameters that may relate to your particular need)

MW	Composition	Structure	Specificity	Response
Mass Range ❑	Resolution ❑	Spectral Char. ❑	Resolution ❑	Sensitivity ❑
Resolution ❑	Accur. Mass ❑	MS/MS ❑	Spectral Char. ❑	Detect. Limit ❑
Accur. Mass ❑	❑	❑	MS/MS ❑	Universality ❑
Spectral Char. ❑	❑	❑	Sensitivity ❑	❑
❑	❑	❑	❑	❑

Part C- Select values for each evaluation parameter.

Parameters	Selections				Values
Mass Resolution:	High	❑	Low	❑	Required value (resolving power):
Mass Accuracy:	Exact	❑	Low	❑	Required value (ppm or Δmass):
Mass Range:	Upper	❑	Lower	❑	Required value (amu):
Spectral Char.:	MW	❑	Frag.	❑	Required spectral types:
MS/MS:	Hi En.	❑	MS^n	❑	Requirements:
Sensitivity:	High	❑	Low	❑	Required value (Coulombs/ug) :
Detection Limits:	High	❑	Low	❑	Required value (ug/mL):
Universality:	Single	❑	Aggreg.	❑	Required type:

Note: Copies of this worksheet can be downloaded from **www.LCMS.com**.

3.2-Parameters for Sample Constraints

The information that you may obtain from a given experiment in LC/MS will always be constrained in some way by the complexity of your samples and the chemical properties of your analytes. Your need for information to solve problems is limited by the nature of your samples. To properly select the appropriate LC/MS alternatives in LC/MS, you must first select the appropriate evaluation criteria. This section presents evaluation criteria relating to sample constraints as shown in Figure 3.7.

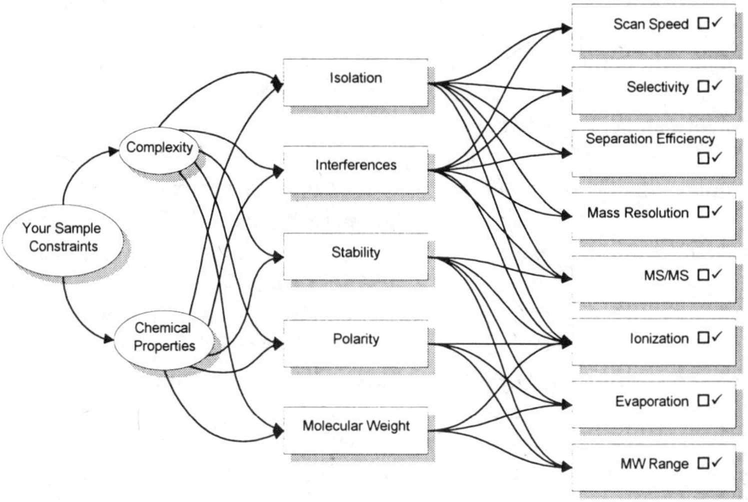

Figure 3.7- Matching your sample constraints to the appropriate evaluation parameters.

The constraints imposed upon your analysis by the nature of your samples will tend to be addressed by the separation and interfacing (ionization) components of an LC/MS system. The complexity of your samples and the properties of your analytes must generally be considered prior to mass analysis. If the separation and interface components of the LC/MS system fail to address these constraints, you will not have detectable analytes or interpretable data.

Picking an interface and separation alternative can be as important (and confusing) as picking a mass analyzer. You have just as many choices and the best choices are not always obvious. As always, you should start your evaluation with your problem needs and the appropriate evaluation criteria. Your evaluation of these system components should then consist of defining where your problems fit into the analytical range, response, and operating attributes of a given alternative.

Parameter: MW Range

We define the molecular weight (MW) range of a given chromatographic or interfacing/ionization technique as the range of compounds bounded by an upper and lower limit of molecular weight. In the previous section we considered mass range as a parameter for evaluating mass analyzers. The molecular weight range of the other LC/MS system components is also of primary importance in evaluating the entire system. All components of the LC/MS system must match the molecular weight properties of your samples. Your samples must fit within the molecular weight range of your separation system, your interface and ionization technique, and finally the mass range of your mass analyzer. Each component of the system can also restrict your ability to measure high molecular weight sample components.

We have shaded the analytical domain of each interfacing and ionization technique on the Figures provided in the **Overviews** presented in *Chapter Two*. These maps show that there is a clear difference between both the separation and interfacing techniques with respect to high mass analysis. Table 3.8 below ranks the various interfaces in terms of their ability to both evaporate and ionize high molecular weight compounds. It is clear that for high mass analysis alone (greater than 2,000 u), electrospray is without competition and should be the technique of choice for high molecular weight analysis.

Conversely, for analysis of compounds less than 2,000 u, there are a wide variety of alternatives and MW range is probably not going to be definitive criteria in your evaluation for interfacing technologies.

Table 3.8- Comparison of molecular weight range of the interfaces relative to sample properties.

What is the upper MW range of your alternatives?					
Chemical Properties[20]	ES	APCI	PB	TS	CF
Ionic	★★★★	★	★	★	★★
Ionizable	★★★★	★★	★★	★★	★★★
Polar Non-ionic	★★★★	★★	★★	★★	★★★
Nonpolar	na	★★	★★	★★	★

★★★★ denotes ability to produce ions above 100,000 u
★★★ above 2,000 u
★★ above 1,000 u
★ above 500 u

MW with Separations

In general, the separation component of LC/MS will not impose a limitation on your experiment and should not be of primary emphasis in evaluating your LC/MS alternatives. As you can see from Table 3.9, all of the conventional separation techniques are capable of separation of high molecular weight species. Your selection of one separation technology over another will likely not be based on MW restriction, but on the chemical properties of your samples and the relative performance characteristics of the specific techniques. As we have stated throughout this text, the majority of separation problems in LC/MS are quite adequately solved with reversed phase. Since reversed phase LC is compatible with all interfacing technologies, you will likely be able to accommodate most conventional separation conditions with any LC/MS system.

MW with Interfaces

The limitations in molecular weight range for LC/MS analysis are brought about by the decomposition of your sample components within the interface or during the ionization process. The transfer of large molecules from solution into the vapor phase is difficult because of the need to overcome the significant intermolecular forces that are responsible for solvating the molecule in solution. The intermolecular forces are particularly large with polar and ionic molecules.

The classical way to overcome these molecular interactions energy is to add heat to your sample by boiling the liquid in the case of thermospray, or heating an aerosol in the case of APCI and particle beam. The addition of excess energy in these interfacing techniques will contribute to thermal decomposition processes of labile molecules, particularly those at high molecular weight.

Table 3.9- Comparison of molecular weight range of the separations relative to sample properties.

What is the upper MW range of your alternatives?					
Chemical Properties[20]	RP	NP	SE	IE	CE
Ionic	★★★★	na	★★★★	★★★★	★★★★
Ionizable	★★★★	★★★★	★★★★	★★★★	★★★★
Polar Nonionic	★★★★	★★★★	★★★★	na	na
Nonpolar	na	★★★★	★★★★	na	na

★★★★ denotes ability to separate analytes above 100,000 u
★★★ above 2,000 u
★★ above 1,000 u
na technique does not apply to this chemical property

Parameter: Ionization

In order to properly evaluate your LC/MS alternatives you must develop an understanding of the types and utility of the various ionization modes (see Table 3.10). Since ionization is a prerequisite for mass analysis, the selection of an ionization mode will restrict the sample components that you can analyze and the type of information that is ultimately derived from a given experiment. Table 3.11 shows the types of ionization modes available for each interface alternative, including mechanism and type of reaction products. This table is included to provide a summary of all the modes of ionization in order that you gain a broad perspective as to the utility and limitations of each mode.

One should select ionization techniques that are applicable to your sample types and that produce the type of spectral characteristics that are required to solve your problem (i.e., molecular weight, fragmentation). Table 3.10 matches a list of selection criteria for LC/MS analysis with the various ionization modes (indicated by ✓). Use this table to narrow your choices of ionization alternatives. After this, match your ionization alternatives to their respective interface. In most cases, you will have more than one mode of ionization available on a given interface. This will enhance your flexibility. It is generally difficult to predict the best mode of ionization or the effects of interferences for a given compound apriori; consequently, having multiple ionization alternatives with a single LC/MS system gives you the highest problem-solving potential. It is highly recommended that you do not limit yourself to one mode of ionization (i.e. APCI and not ES).

Table 3.10-Criteria for Selection of Ionization Mode

Selection Criteria	ESI	CI	EI	TSI	FAB	MALDI
High Mass	✓					✓
MW information	✓	✓	(✓)[21]	✓	✓	✓
Fragment Information			✓			
Requires MS/MS	✓	✓		✓	✓	✓
Universality			✓			
+/- Polarities	✓	✓		✓	✓	✓

Table 3.11- Ionization Processes Found in LC/MS

Ionization	Ionization Mechanism	Ion Production	Product
Electrospray Ionization ESI	$(M+H)^+_{(l)} \ulcorner\!\rightarrow (M+H)^+_{(g)}$	Acid-base equilibrium	Gas-phase quasi-molecular ion of virtually any ion in solution
	$(M\text{-}H)^-_{(l)} \ulcorner\!\rightarrow (M\text{-}H)^-_{(g)}$	Acid-base equilibrium	
	$M^+_{(l)} \ulcorner\!\rightarrow M^+_{(g)}$	Cation-dissolution	
	$M^-_{(l)} \ulcorner\!\rightarrow M^-_{(g)}$	Anion-dissolution	
	$(M+60H)^{+60}_{(sol)} \ulcorner\!\rightarrow (M+60H)^{+60}_{(g)}$	Acid-base equilibrium	
	$(M\text{-}60H)^{-60}_{(sol)} \ulcorner\!\rightarrow (M\text{-}60H)^{-60}_{(g)}$	Acid base equilibrium	
Electron Ionization EI	$M_{(g)} + e^- \rightarrow M^{+\cdot}_{(g)} + 2e^-$	Gas-phase, unimolecular decomposition	Molecular ion radical plus fragmentation from excess internal energy.
Chemical Ionization CI & APCI	$M_{(g)} + (R+H)^+_{(g)} \rightarrow (M+H)^+_{(g)} + R_{(g)}$	Proton Transfer	Protonated molecule
	$M_{(g)} + R^{+\cdot}_{(g)} \rightarrow M^{+\cdot}_{(g)} + R_{(g)}$	Charge Exchange	Molecular cation radical
	$M_{(g)} + e^- \rightarrow M^{-\cdot}_{(g)}$	Electron Capture	Molecular anion radical
	$M_{(g)} + B^-_{(g)} \rightarrow (M\text{-}H)^-_{(g)} + BH_{(g)}$	Proton Transfer	Deprotonated molecule
Thermospray TS	$M_{(sol)} + (NH_4)^+_{(aq)} \ulcorner\!\rightarrow (M+H)^+_{(g)} + NH_{3(g)}$	Proton Transfer	Protonated molecule
	$M_{(g)} + B^-_{(g)} \rightarrow (M\text{-}H)^-_{(g)} + BH_{(g)}$	Proton Transfer	Deprotonated molecule
Fast Atom Bombardment FAB	$M_{(l)} + R_{(g)}^+ \ulcorner\!\rightarrow (M+H)^+_{(g)} + R_{(g)}$	High energy primary beam impacting liquid surface creates ion rich plasma	Protonated molecule
	$M_{(l)} + R_{(g)}^+ \ulcorner\!\rightarrow (M\text{-}H)^-_{(g)} + R_{(g)}$		Deprotonated molecule
Laser Desorption MALDI	$M_{(crystal)} + \lambda \ulcorner\!\rightarrow M^+_{(g)}$ or $(M+H)^+_{(g)}$	Photo desorption and ionization	Cationic or protonated molecule
	$M_{(crystal)} + \lambda \ulcorner\!\rightarrow M^-_{(g)}$ or $(M\text{-}H)^-_{(g)}$		Anionic or deprotonated molecule

$\ulcorner\!\rightarrow$: Desorption process

$\rightarrow$: Gas phase process

Parameter: Evaporation

All LC/MS systems must address the conversion of liquid-phase analyte into the gas-phase (evaporation). We feel that some knowledge of the differences between the evaporation processes for each LC/MS technique may assist you in evaluating and understanding the advantage of one technique over another. Table 3.11 shows a listing of evaporation processes for each interface.

Often ignored, evaporation may be the most significant limitation to obtaining analytical results with many compounds. The most obvious limitation of the evaporation process was discussed already in *Chapter One* with respect to the analytical boundaries of GC/MS. Decomposition and chemical reactions compete with the evaporation processes, particularly at elevated temperatures. The rates of most reactions increase with temperature. The rate of evaporation increases with increased surface area as well as temperature. Most of the LC/MS interfaces use a combination of heat and nebulization to establish conditions that favor evaporation over decomposition. Some of the temperature effects for different nebulizers are discussed in Table 3.12.

Table 3.12- Effects of Nebulizer Temperature on Analyte Decomposition with LC/MS Alternatives

Interface	Nebulization	Temp.^{o}C	Comments
ESI	Electrospray	RT-100	Virtually no thermal degradation because of ion evaporation process
	Ionspray	RT-100	Virtually no thermal degradation because of ion evaporation process
	Turbospray	400-600	Some decomposition is observed but ion evaporation is still favored
APCI	Thermal-pneumatic	400-500	Thermal processes can cause decomposition, some ion evaporation processes may occur with APCI if droplets get charged.
PB	Pneumatic or Thermal-pneumatic	80-100 150-300	Thermal processes can cause decomposition. Micro particles are evaporated on the surface of the ionization source.
TS	Thermal	200-400	Thermal processes can cause decomposition; some ion evaporation processes occur with TS as well.

Aerosol Generation

Most LC/MS interfaces require that the liquid effluent from the LC to be sprayed at some point within the interface (except CFFAB). APCI, ES, PB, and TS all require an aerosol generation step in order to create the high surface area necessary for evaporation. It should be noted that all evaporation is an "interfacial phenomenon" meaning that the process only occurs at a liquid-gas interfaces. It is then obvious that creating more surface will facilitate and enhance the process of evaporation. If your analyte is not at a surface, it will not be transferred into the vapor phase.

The efficiency of aerosol generation will have a profound effect on the production of gas-phase ions. The most efficient aerosol generator used in LC/MS today is electrospray, which generates the highest surface area per volume (or the smallest particles) compared to all other techniques. Unfortunately, the electrospray process is limited to uL/min. (microliter per minute) flow rates. For more details on the fundamentals of aerosol generation, review *Appendix C*.

Desorption

Generally, when we think of evaporation we think of boiling a liquid to the extent of transferring molecules into the gas phase. Adding sufficient amounts of heat (thermal energy) will cause a liquid to evaporate. In recent years, many alternate techniques have been developed to transport molecules into the vapor phase. We designate these processes as "desorption" techniques.

With desorption techniques, other forms of energy are applied to the liquid in a manner that is selective and directed rather than random, as with heat. Examples of desorption techniques are 1) ES, utilizing high electric potential energy, selecting and then ejecting ions in solution; 2) FAB, utilizing high kinetic energy from a primary ion beam, partitioning this energy into the liquid and then ejecting surface molecules (and ions); and 3) MALDI, utilizing intense light from a laser, partitioning this energy into the liquid matrix and then ejecting molecules (and ions).

See *Appendix C* for details on each desorption technique.

Table 3.13- Criteria for Selection of Evaporation Mode

Selection Criteria	ES	APCI	PB	TS	CF	MALDI
Thermal Evaporation		✓	✓	✓		
Desorption	✓				✓	✓
Surface Desorption					✓	✓
Aerosol Desorption	✓			(✓)[22]		
Harshness Rank[23] (1 is softest)	1	4	4	4	3	2

Parameter: Selectivity

Selectivity is defined as the ability to separate or isolate the response of one component in your sample from other components in your sample. This, of course, includes isolation from other analytes, interferences, and matrix components. Selectivity in LC/MS is achieved in a variety of manners, including sample preparation, analytical separations, LC/MS interfacing, ionization, and mass analysis. Obviously, some level of selectivity is required when analyzing samples of higher complexity and variability. Figure 3.8 illustrates and emphasizes the alternatives we have with LC/MS with regard to selectivity.

Mass selectivity allows us to isolate (or select) specific components by virtue of their particular mass specific response, albeit the different components may not be physically separated. In this way, specificity and selectivity are intimately linked.

In contrast, *chromatographic selectivity* allows us to physically separate components in a mixture as they elute through a column and completely isolate the compound response from other components.

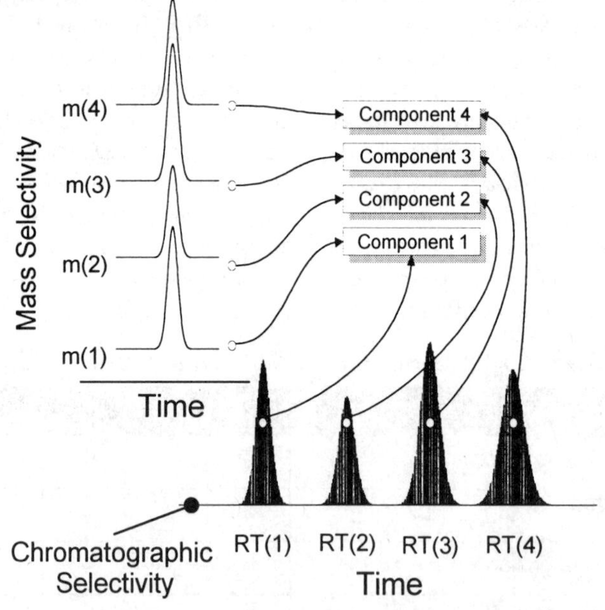

Figure 3.8- Mass selectivity versus chromatographic selectivity in LC/MS. The 3-dimensional aspect of LC/MS data coupled with mass specificity gives the analyst choices as to how to achieve adequate selectivity.

Selectivity with LC

Chromatographic selectivity is a prerequisite for most applications of LC/MS. This is because mass selectivity does not eliminate chemical and physical interferences that may affect the relative response of your analytes. There is generally no better way to deal with chemical interferences in a sample than to physically separate them from your analyte. This is typically done through a combination of extraction (sample preparation) and analytical (LC) separation. As a general rule, the greater the selectivity of your analytical separation, the greater the reliability and ruggedness of your analytical method.

Evaluating LC/MS alternatives should take into account the separation requirements for your sample. What separation technologies are required to isolate your prospective analytes from other sample components? Are these separation technologies compatible with your mass analysis and interfacing requirements? Table 3.14 compared selectivity from LC versus MS.

Selectivity with MS

The mass selectivity of the mass spectrometer gives us significant benefits over other detection modes that lack the compound specificity of MS. Mass selectivity is particularly useful when utilizing MS/MS. A single component can be selected with the first mass analyzer, then dissociated to yield structurally significant fragments that can be analyzed with a second stage of mass analysis.

As stated previously, interpretation requires that the acquired spectrum be traced to a single chemical species. This is readily done with MS/MS. In addition, the mass selectivity can be very useful for target analysis where response is not affected by interferences; or where you have used a standardization technique that accounts for response changes because of interferences (e.g. isotope dilution- see *Chapter 10*).

Selectivity is inherent in all mass analyzers; however, both higher resolution and MS/MS will yield higher selectivity in LC/MS and should be considered for complex and variable samples.

Table 3.14- Selectivity of LC versus MS

	LC	MS
Isolation	Physical separation in time. Can make analysis lengthy.	Separation is due to compound specific response. No physical separation. Not time dependent, fast. May require high resolution or MS/MS.
Interferences	Is capable of complete isolation of analyte from interferences.	Interferences can suppress or enhance response.

Parameter: LC Resolution

The parameters used to describe the degree of separation by which an LC column separates individual components are the *Retention*, *Separation* and *Efficiency* Factors (see Figure 3.9).[24] The first factor, the Retention Factor (k', or *Capacity Factor*), is a measure of the relative retention of each peak on the column. It is the usual starting point for developing an LC method. The major variable controlling k' is solvent polarity, while the values for k' typically range from 1 to 15 for analytical columns. Most high throughput clinical assays attempt to keep k' under 5.

Retention Factor: $k' = (V_A - V_o)/V_o$ (3.8)

The second factor, Separation Factor (α, also referred to as *selectivity factor*), represents the relative separation between two peaks. Components with an α of 1.0 overlap completely. Values for α's are typically less than 2, more than adequate for separation on an analytical column. Temperature, the composition of the mobile-phase, and derivatization of the components (used less often)[25] are used to control α.

Separation Factor: $\alpha = (V_B - V_o)/ (V_A - V_o)$ (3.9)

The third factor, Efficiency Factor (N is also referred to as *Plate Number*), measures peak sharpness. The sharper the peak the better the separation and the higher the efficiency of the column. Values for N may range from hundreds (poor) to hundreds of thousands (good). N is greatly affected by both column effects (e.g.,particle size,column length) and extra-column effects (i.e., tubing size, unions, detector cell volumes).

Efficiency Factor: $N = 16(V_A/W_A)^2$ (3.10)

All of the above parameters combine in the Chromatographic Resolution (R_C) equation, which predicts how each factor will affect the analytical separation. In practice, R_c is determined empirically from the chromatogram.

Resolution: $R_C = \frac{1}{4}(\alpha-1/\alpha)\,(N)^{\frac{1}{2}}\,(k'/1+k')$ (3.11)

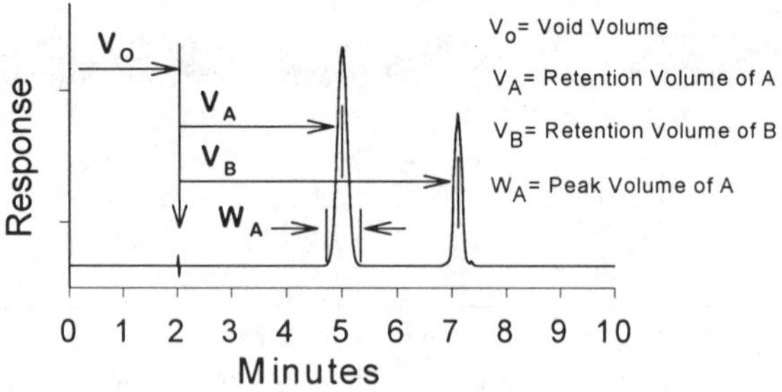

Figure 3.9- Factors that influence separations in LC/MS

LC/MS & Resolution

For most analyses the individual components in a complex sample must be separated by a column with high separation factors. High separation efficiencies are often required to remove interferences, isolate your analytes, and, for trace constituents, concentrate your analytes into narrow chromatographic bands, which improves detection limits. Isolating your sample components from each other is generally a prerequisite for interpretation of mass spectral results, particularly when you do not have MS/MS capabilities.

How important then is resolution for evaluating your LC/MS alternatives? It should be a consideration before you select your interfacing and mass analysis alternatives. The major issue to consider is your ability to have separation capabilities when you need them. Most interfaces allow a wide variety of separation alternatives; check on the compatibilities (see Table 3.15).

Deconvolution

The three-dimensional nature of LC/MS data (time in one dimension and mass in the other) allows one to extract (and display) individual masses, representative of the component. Figure 3.10 shows the deconvolution of three partially resolved peaks (A, B, and C) into individual peaks. This selectivity of mass analysis can overcome "some" of the need for high efficiency separations.

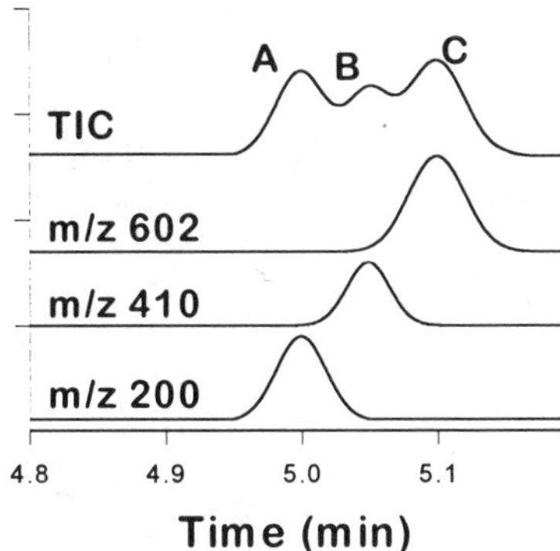

Figure 3.10- Deconvolution of 3 components by their different masses.

Table 3.15- Criteria for Selection of Separation Modes

Selection Criteria	RP	NP	SE	IE	CE
Isocratic Efficiency (plates/m)[26]	10k-100k	10k-100k	poor	100-1k	100k-1M
Gradient Capability	yes	na	na	yes	(yes)[27]
Use in Sample Preparation[28]	SPE, 96-well plates	SPE	Desalting Fractionation	SPE	na
Flow Rates (mL/min)	0.001-1.0	0.2-1.0	0.5-1.5	0.5-1.5	0.0005-0.01

Parameter: Scan Speed

Scan speed (or rate) is an important evaluation criterion in LC/MS because there is a consistent trend in analytical analysis to increase the speed of separations and the sample throughput. Scan speed is an essential acquisition parameter for mass spectrometry because the qualitative information (relative ion peak heights) and the quantitative information can be dramatically affected by operation at inadequate scan speeds (see Figure 3.11).[29] We define scan speed as simply the effective rate that we scan the mass range (dm) in amu per second (dt). (see Equation 3.12)

Scan Rate: $$\phi_{scan} = (dm/dt) \qquad (3.12)$$

Minimum Scan Rate: $$\phi_{10\ points} = 10\ \Delta m_{acquisition} / w \qquad (3.13)$$

The minimum recommended scan rate can be easily calculated with Equation 3.13 by inserting your expected values of acquisition mass range $\Delta m_{acquisition}$ and your narrowest LC peak width, w. This calculated rate will guarantee you at least 10 points across each peak.

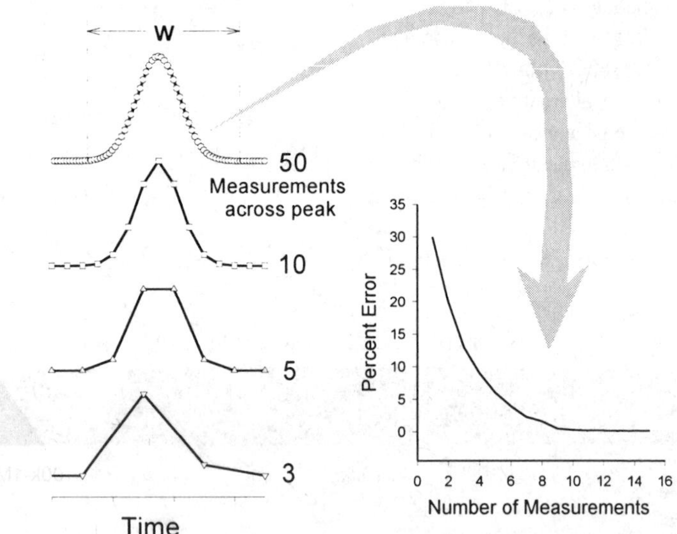

Figure 3.11- The number of measurements across a chromatographic peak will affect the accuracy of peak area determination which is important in quantitative analysis. Significant errors in quantitative results are observed at low sampling rates relative to the chromatographic peak width. Mass scan speed is a critical parameter in determining the number of measurements.

Fast Separations

It is clear from Figure 3.11 that scanning at a slow scan rate relative to the chromatographic peak width will cause significant error in your analytical results. Mass spectrometers are all designed to acquire three-dimensional data at higher rates than conventional detectors. This is the obvious requirement of an information-rich multi-dimensional detector. As we improve the mass spectral capabilities, we also place more pressure on performance characteristics such as scan speed. Time-of-flight is superior for acquisition speeds in the millisecond range. TOF instruments can acquire at scan rates of more than 10,000,000 u/sec (see Table 3.16).

Also consider that a typical 10-or-15-second chromatographic peak at an acquisition mass range of 500 u would only require a scan speed of 500 u/sec. This is easily accomplished with all commercial mass analyzers.

High speed separations can separate hundreds of components in minutes.[30] This places a significant burden on the need for fast scan speeds.

Fast Mass Analysis

The requirement for ultrafast scan rates for mass analysis are found in:

- applications of high mass where each scan takes longer;
- applications with MS/MS where multiple acquisition experiments are carried out within a single acquisition point upon a chromatographic peak.

A common approach to increasing scan speed as mass range increases and peak widths decrease is to skip points on the mass axis. A quadrupole mass analyzer can typically sample 10 points across each mass during an acquisition. Each point has a sampling time before moving to the next mass increment. If we reduce the number of sample points from 10 to 2 points, we can integrate each point five times longer and improve our signal to noise by better than a factor of two.

Recent developments with time-of-flight MS/MS configurations are attempting to perform as many as 100 separate MS/MS experiments per second.[31] This can only be done at the high scan speeds of TOF.

Table 3.16-Typical scan rates for mass analyzers

Analyzer	Upper Scan Rate	Utility
Quad	4,000 u/sec	Scan rate is adequate for conventional LC/MS and LC/MS/MS experiments
Ion Trap	4,000	Scan rate is adequate for conventional LC/MS and LC/MS/MS experiments
TOF	>1,000,000	Ideal for high mass and fast LC
Sector	2,000	Scan rate is adequate for conventional LC/MS and LC/MS/MS experiments
FTMS	10,000	Scan rate is adequate for conventional LC/MS and LC/MS/MS experiments

Parameter: Compatibility

Evaluating your LC/MS alternatives requires that you match your prospective sample properties or constraints to the appropriate LC/MS technologies. Figures 3.12 and 3.13 show a selection chart for compounds of various solubility, charge state, and molecular weight. Using these charts, you will be able to match your sample properties with the appropriate separations and interfacing alternatives. Not all separations or interfacing technologies apply to all sample types; so by using your knowledge of your sample you can narrow your alternatives for LC/MS.

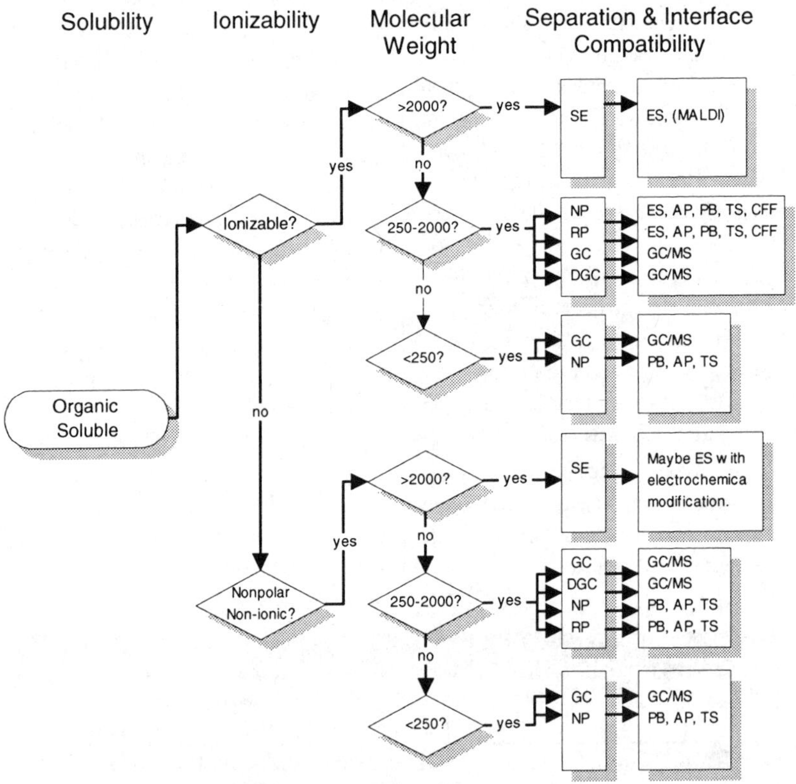

Figure 3.12- Matching separation technologies with interfacing technologies for organic soluble compounds using polarity and molecular weight as selection criteria. (DGC is short for derivatization gas chromatography.)

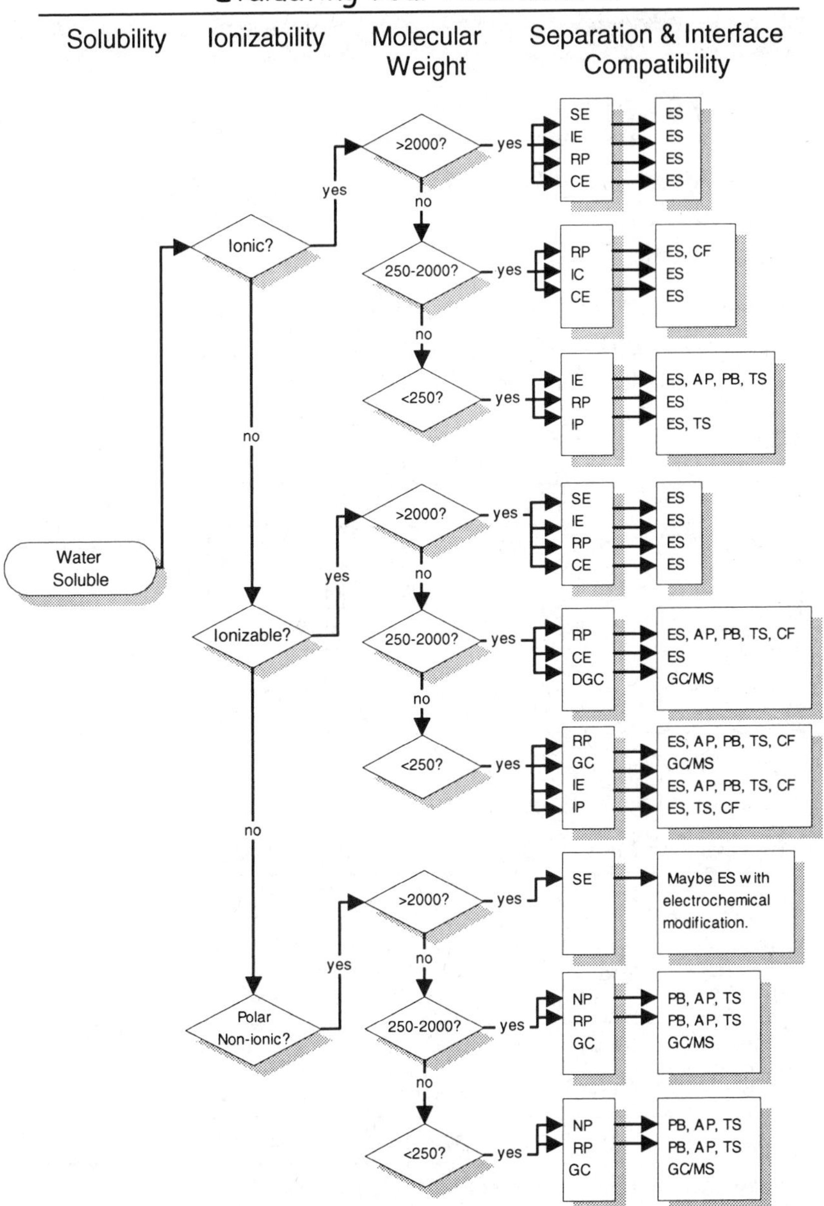

Figure 3.13- Matching separation technologies with interfacing technologies for water soluble compounds using polarity and molecular weight as selection criteria.

Worksheet 3.1 - Evaluating Your LC/MS Alternatives

Selecting Criteria to Address Sample Constraints

Part A- Fill out the following checklist to establish your sample constraints.

What are your information needs?	Yes	No	Comment
– Isolation of analytes	❑	❑	______
– Removal of interferences	❑	❑	______
– Sample components are stable	❑	❑	______
– Sample components polar/ionic	❑	❑	______
– Analytes high molecular weight?	❑	❑	______

Part B- Match your evaluation criteria and sample constraints. (Add additional parameters that may relate to your particular constraint)

Isolation	Interferences	Stability	Polarity	MW
Selectivity ❑	Selectivity ❑	MS/MS ❑	Ionization ❑	Ionization ❑
Separ. Eff. ❑	Separ. Eff. ❑	Ionization ❑	Evap. ❑	Evap. ❑
Resolution ❑	Resolution ❑	Evap. ❑	MW ❑	MW ❑
MS/MS ❑	MS/MS ❑	MW ❑	❑	❑
❑	Ionization ❑	❑	❑	❑

Part C- Select values for each evaluation parameter.

Parameters	Selections				Values
Selectivity:	MS	❑	LC	❑	How selective?
Separation Efficiency:	High	❑	Low	❑	How efficient?
Mass Res:	High	❑	Low	❑	Required resolution:
MS/MS:	Isolate	❑	Identify	❑	MS/MS needs:
Ionization:	Gas Phase	❑	Preformed	❑	Required types:
Evap.:	Heat	❑	Desorp.	❑	Limitations:
MW Range:	Upper	❑	Lower	❑	Required value

Note: Copies of this worksheet can be downloaded from **www.LCMS.com**.

3.3-Selecting Your Alternative

Armed with the proper criteria to evaluate your alternatives in LC/MS, you can now go through the process of selecting the appropriate LC/MS technologies to meet your problem needs. Worksheet 3.2 is included at the end of this section so that you can consolidate all of the information into one exercise and arrive at a reasonable choice.

Always keep in mind the problem definitions that you have developed throughout this chapter. Weigh each parameter according to its ultimate contribution in supporting your information needs and sample constraints of your problem. Figure 3.14 illustrates this process of matching criteria to alternatives to select each link in your analytical chain.

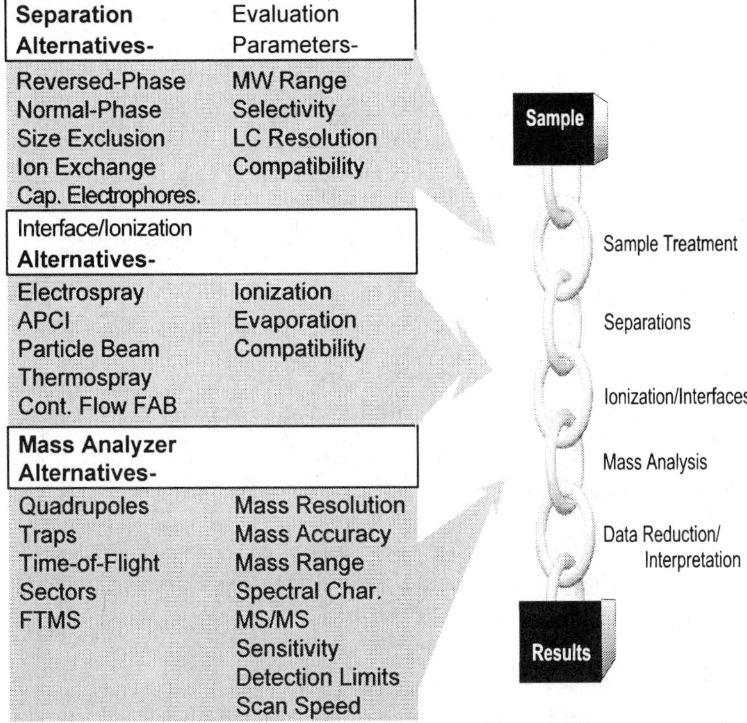

Figure 3.14- Evaluating your alternatives in LC/MS requires 1) knowing your alternatives, 2) matching evaluation criteria to your problem needs and constraints, and 3) using the criteria to select the best alternative.

The actual values found in the various Tables for each parameter discussed in this chapter are only ballpark estimates or ranges. Actual values of any parameter (e.g. resolving power, detection limits, scan speeds) should be acquired from manufacturers' published specifications or measured directly from an instrumental system under evaluation.

Some alternatives will be easy to select while others will be more difficult. For example, electrospray has no competition for LC/MS applications at high molecular weight and for highly ionic species. Don't waste a lot of time considering other techniques. Spend your time concentrating on which mass analyzer in combination with ES is best suited for your information needs.

In contrast, selecting the best interface at lower molecular weights will be much more difficult. A great deal of similarities exist between the performance of APCI, TS, ES, CFF, and PB in the molecular weight range from 100 to 1,000 u. In many cases, all the alternatives will adequately solve your problems. In our view, your most difficult challenge in evaluating LC/MS technologies will be to sort through the large number of mass analyzer alternatives that are available to solve your problems. We have attempted to compartmentalize some of these MS alternatives for an AP-LC/MS system in Figure 3.15.

The following generalities may assist you in making your selections:

- Atmospheric pressure techniques (ES and APCI) when coupled with MS/MS will solve the majority of problems in LC/MS.
- The majority of problem-solvers in LC/MS will need MS/MS capabilities to meet the information and selectivity requirements of their problems.
- The majority of clinical assays in LC/MS are APCI not ES, based methods.
- Reversed-phase is by far the most appropriate separation technology for applications in LC/MS.
- CFF is primarily used with sector mass analyzers.
- TS has essentially been replaced with AP technologies.
- The universal detection and library searching capabilities of electron ionization makes PB the technique of choice for identification of unknowns.

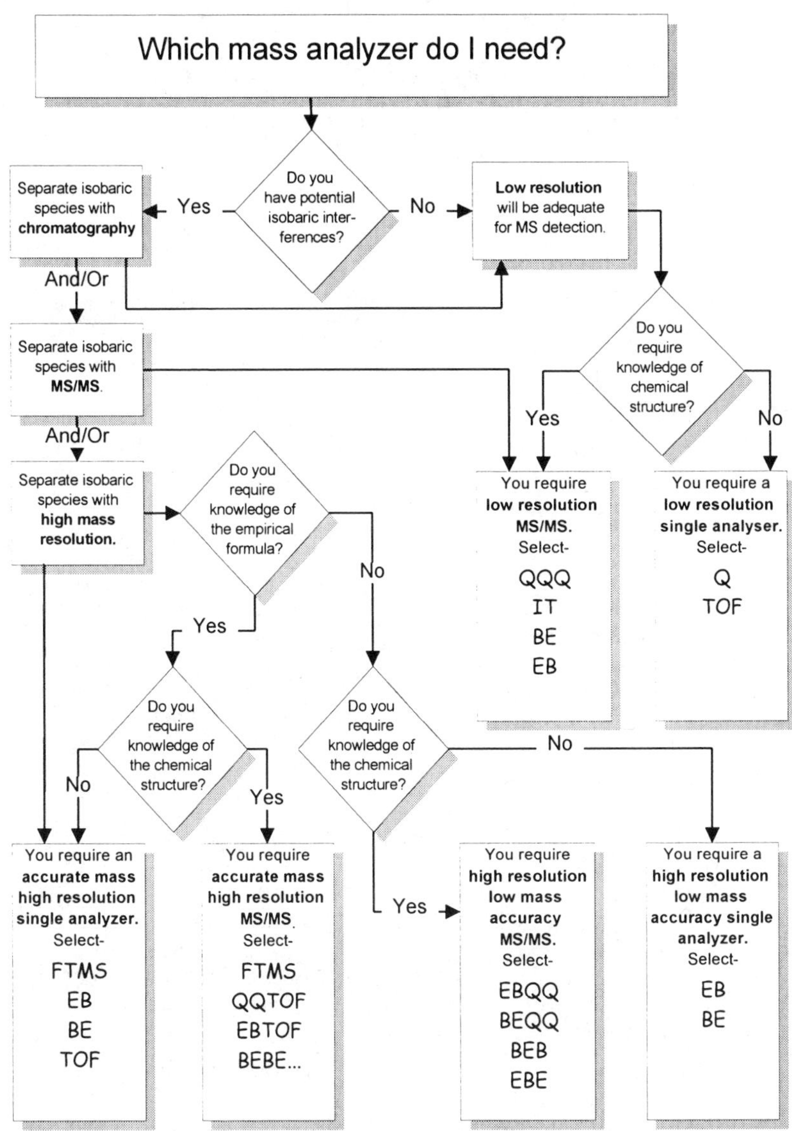

Figure 3.15- Selecting your alternatives for an AP-LC/MS system requires 1) knowing your alternatives, 2) matching evaluation criteria to your problem needs and constraints, and 3) using the criteria to select the best alternative.

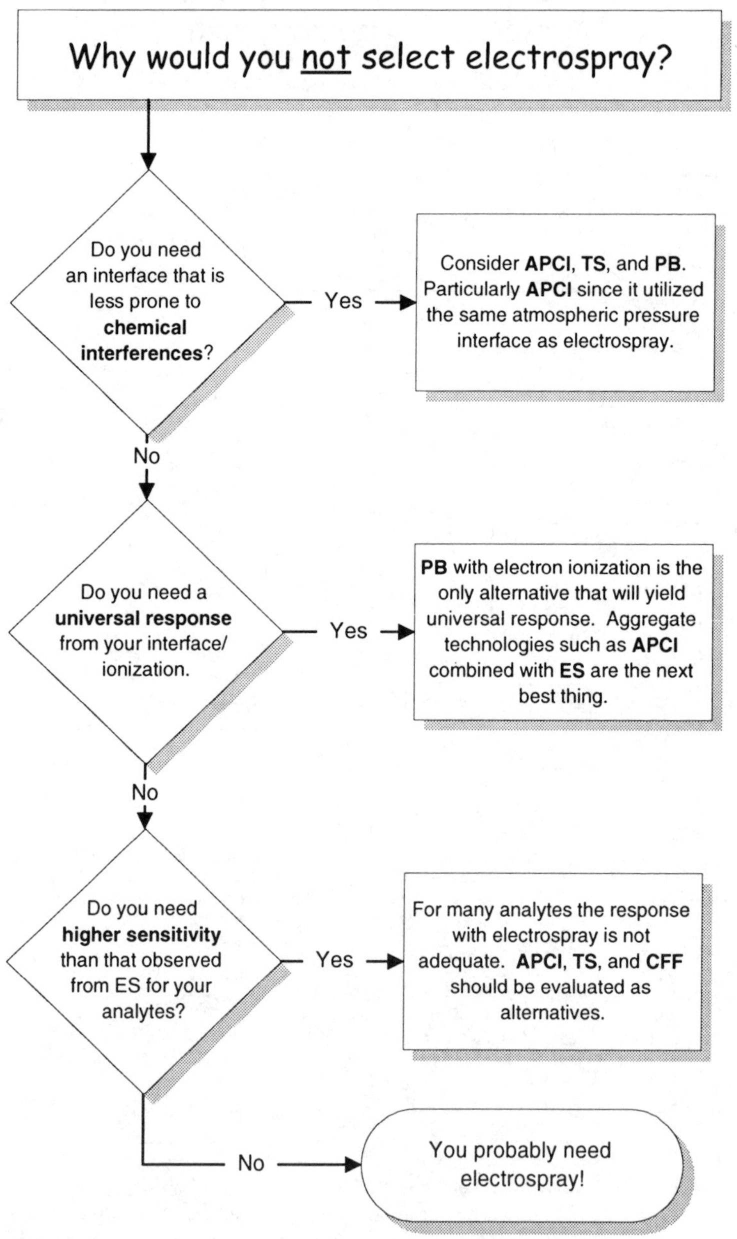

Figure 3.16- Why not select an Electrospray LC/MS system.

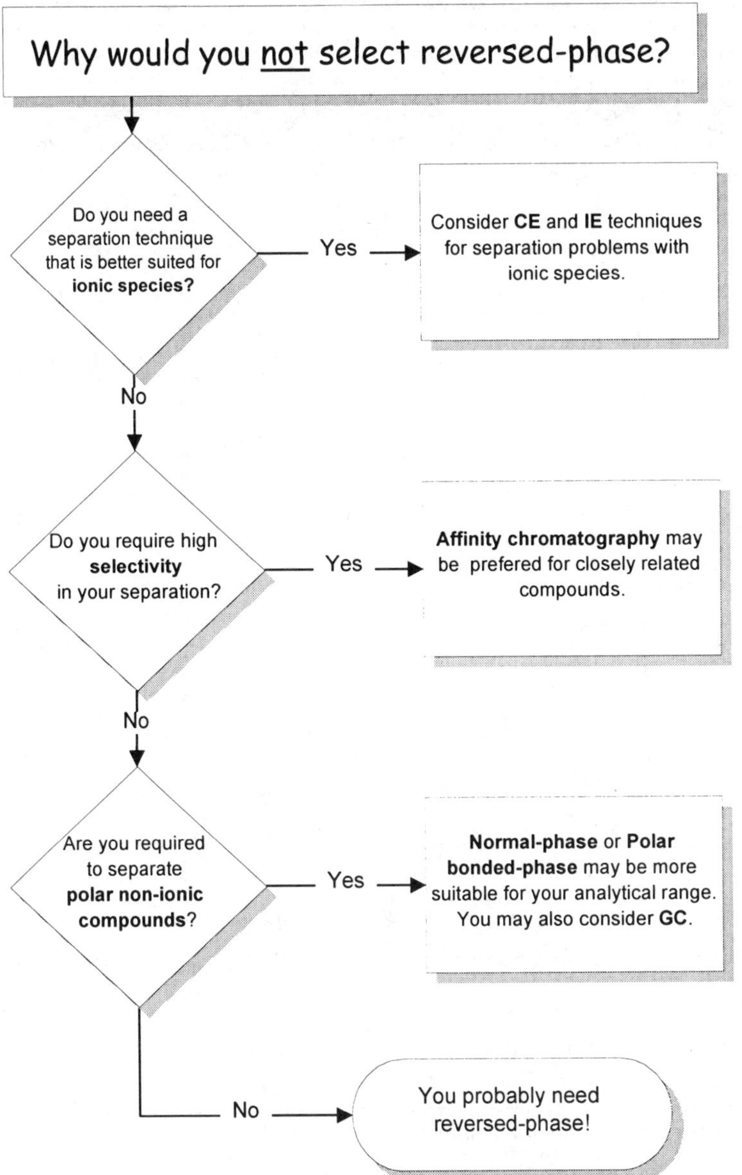

Figure 3.17- Why not select a Reverse-Phase liquid chromatography as a separation means for an LC/MS system.

Worksheet 3.2-Evaluating Your LC/MS Alternatives

Part A-Matching your problem needs to LC/MS alternatives: Match your problem definition to LC/MS alternatives using the criteria you have selected in Worksheets 3.0 and 3.1.

Your Problem Definition ⟷ LC/MS Alternatives

Information Parameter	Constraint Parameter	Param. Value	Mass Analyzer Alternatives: Quad	Trap	TOF	Sector	FTMS
Res. ❑	Res. ❑	______	❑	❑	❑	❑	❑
Mass Ac. ❑		______	❑	❑	❑	❑	❑
Mass R. ❑		______	❑	❑	❑	❑	❑
	Scan Sp. ❑	______	❑	❑	❑	❑	❑
MS/MS ❑	MS/MS ❑	______	❑	❑	❑	❑	❑
Sensit. ❑		______	❑	❑	❑	❑	❑
Det. Lim. ❑		______	❑	❑	❑	❑	❑
Your Mass Analyzer Alternatives			❑	❑	❑	❑	❑

Information Parameter	Constraint Parameter	You Require	Interface Alternatives: ES	AP	PB	TS	CF
Univers. ❑		______	❑	❑	❑	❑	❑
Spectra ❑		______	❑	❑	❑	❑	❑
	Ionizat. ❑	______	❑	❑	❑	❑	❑
	Evap. ❑	______	❑	❑	❑	❑	❑
	MW R. ❑	______	❑	❑	❑	❑	❑
Your Interface Alternatives:			❑	❑	❑	❑	❑

Information Parameter	Constraint Parameter	You Require	Separation Alternatives: RP	NP	IE	SE	CE
Selectivity ❑	Selectivity ❑	______	❑	❑	❑	❑	❑
	Efficiency ❑	______	❑	❑	❑	❑	❑
	Compat. ❑	______	❑	❑	❑	❑	❑
Your Separation Alternatives:			❑	❑	❑	❑	❑

Note: Copies of this worksheet can be downloaded from **www.LCMS.com**.

Worksheet 3.2-Evaluating Your LC/MS Alternatives

Part B- Building a system configuration. From your results in Part A, put together the LC/MS system configurations that best meet your problem-solving needs.

	Mass Analysis	Interface	Separation
LC/MS System One-	______________ /	______________ /	______________
LC/MS System Two	______________ /	______________ /	______________

Part C- Evaluate and compare your system configurations. Using the configurations in Part B, fill out the following checklist to evaluate each system alternative under your problem conditions. In this evaluation consider the entire system performance, not just the individual components. Use the best available values for each parameter. You may insert table values from this chapter or use other sources. This exercise may help you to compare one system configuration to another. Checking (no) would indicate that your evaluation requirement was not met.

LC/MS Systems	Evaluation Parameter	Parameter Value	Acceptability Yes	No	Exceeds
System One	Sensitivity	__________	❑	❑	❑
__________	MS/MS	__________	❑	❑	❑
	Mass Range	__________	❑	❑	❑
	Selectivity	__________	❑	❑	❑
	Universality	__________	❑	❑	❑
Other:		__________	❑	❑	❑
Other:		__________	❑	❑	❑
System Two	Sensitivity	__________	❑	❑	❑
__________	MS/MS	__________	❑	❑	❑
	Mass Range	__________	❑	❑	❑
	Selectivity	__________	❑	❑	❑
	Universality	__________	❑	❑	❑
Other:		__________	❑	❑	❑
Other:		__________	❑	❑	❑

Note: Copies of this worksheet can be downloaded from **www.LCMS.com**.

Discussion Questions and Exercises

1. What are the advantages of operating your mass analyzer in low resolution modes?

2. In your own words, discuss high resolution and high mass accuracy. How are they different? How do they experimentally relate?

3. Name ten instrumental parameters that affect sensitivity.

4. Name three methods of inducing fragmentation in mass spectrometry.

5. List every type of ionization technique, the spectral characteristics that are associated with the technique, and the analytical utility of the technique.

6. What are the advantages and disadvantages of electron ionization?

7. What are the advantages and disadvantages of MS/MS?

8. In your opinion, what is the most universal ionization technique?

9. What is the "softest" ionization technique? Try to rank the ionization techniques in terms of softness (EI, TSI, ESI, CI, APCI, FAB).

Notes

[1] We use mass resolution here to make the distinction between resolving masses in the mass spectrometer and chromatographic resolution, a measurement of the ability to measure two adjacent chromatographic peaks. Since LC/MS addresses these two important audiences with distinct technical jargon, we will make every effort to be specific when there is duplication (or confusion) of terminology.

[2] Chhabil Das "Mass Spectrometry, Instrumentation and Techniques," IN: *Mass Spectrometry, Clinical and Biomedical Applications, 2*, Dominic M. Desiderio, ed. Plenum Press: New York (1994).

[3] Instruments such as the MSD from Hewlett Packard and the ITD from Finnigan allowed a mass market for GC/MS to develop with thousands of instruments being sold each year for under $50,000 into environmental, clinical, and industrial applications.

[4] Mass spectral libraries of electron ionization (EI) mass spectra, which are relevant to hundreds of thousands of compounds at masses typically less than 500 amu, can be purchased. The two primary data bases are Wiley and NIST. See *Appendix A* for details about the availability of these commercial spectral libraries.

[5] McLafferty, F.W. *Interpretation of Mass Spectra,* 3rd. Edition, page 15, University Science Books: Mill Valley (1980).

[6] There is a formal requirement that structural confirmation with high resolution mass spectrometry be submitted with all spectra submitted to the Journal of the American Chemical Society. This general standard is also held in many other journals where new structures are reported.

[7] PROWL (Rockefeller University). Home page: chait_sgi.rockefeller.edu.

[8] Typically, mass spectrometrists refer to the mass of an ion in terms of the atomic mass units (amu) but with the advent of electrospray and the analysis of high molecular weight biopolymers, the term daltons is becoming more prevalent.

[9] US Patent 5,130,538, "Method of producing multiply charged ions for determining molecular weights of molecules by use of the multiply charged ions of molecules," (issued July 14, 1992). Wong, S.F., Meng, C.K., Fenn, J.B., "Multiple charging in electrospray ionization of poly(ethylene glycols)," Phys. Chem. 92, pages 546-550 (1989).

[10] Cotter, R.J., *Time-of-Flight Mass Spectrometry: Instrumentation and Applications in Biological Research*, ACS Professional Reference Books: Washington, D.C. (1997).

[11] It can be difficult to measure low mass offspring of large masses with triple quadrupole mass spectrometry. Also the transmission of quadrupoles falls off significantly for masses more than m/z 1000 and therefore sensitivity is another upper mass limitation.

[12] It can be difficult to measure low mass offspring of large masses with an ion trap. It is also difficult for a trap to capture high mass ions.

[13] McLafferty, F.W., Turecek, F., *Interpretation of Mass Spectra* (4th ed.), University Science Books: Mill Valley (1993).

[14] The exceptions to this "spatial" setup of MS/MS are the trapping mass analyzers, ion traps and FTMS. Here the mass analyzers and the collision region are one. MS/MS is performed in a "temporal" setup. See *Chapter Two* for more details.

[15] "Interface CID" refers to the collisional dissociation that occurs within the intermediate pressure reduction stages of the API interface (ES or APCI). Under conditions of longer mean free path at lower pressures and higher potential between the lense elements within the interface, collisions between an accelerated ion and the surrounding gases can be quite energetic. The increase in internal energy due to these condition can result to significant decomposition.

[16] The relative merits of high energy and low energy collisions have been studied for many years and are discussed in some detail in Gaskell and Reilly. In general, we can conclude that high energy spectra (from sectors) tend to contain fragments that result from relatively fast processes; lower energy single collision spectra (from quads) contain fragments that result from processes that move toward thermodynamic equilibrium. The extreme case is traps (ion traps and FTMS) that fragment with multiple collisions. These spectra tend to be representative of processes that have reached a thermodynamic equilibrium, depending upon how long the ions reside in the trap. (Gaskell, S.J., Reilly, M.H., "The complementary analytical value of high- and low-energy collisional activation in hybrid tandem mass spectrometry," Rapid Commun. Mass Spectrom. 2, pages 139-141 (1968).)

[17] There are a number of variations of magnetic sector-based instruments that are designated by the order in which the magnetic sector (B) and the electric sector (E) are configured. For example, EBQQ stand for the configuration of an ion source connected to an electric sector, then the magnet, followed by an rf only quadrupole (Q), and lastly, an analyzing quadrupole (Q). The exact orientation of each type of sector or analyzer will affect to resolution on the first analysis step or conversely the second. The best summary of the many configurations was published by:

Pesch, R. (of Finnigan MAT), Spectra, Volume 9, Number 4. Spectra is an internally published technical bulletin from Finnigan.

[18] A thorough knowledge and understanding of each component can take a lifetime and most likely requires a physics degree.

[19] The value ten times the standard deviation of the baseline is a safe value for measured detection limits because the ion signal is clearly discernible from the background noise. Other authors recommend a value of three. Others use peak-to-peak values for noise instead of standard deviation because no calculations are required. One or two spikes in the background, however, can have a dramatic effect on detection limit values with this approach.

[20] The compound may be characterized by it polarity and ionizability. For this discussion we have divided sample characteristics into four groups; namely,

1.) **Ionic compounds-** are permanently ionized in solution such as strong acids or bases, or quaternary amines. These compounds cannot be charge neutralized in solution,

2.) **Ionizable compounds-** can exist in either a charged state or neutral state in solution depending on the degree of protonation or deprotonation, etc. (This is where the majority of compounds of interest in LC/MS reside),

3.) **Polar non-ionic compounds-** exist with a permanent dipole, but have not exchangeable protons or other charge generating moieties but may have polar function groups that may undergo hydrogen bonding (e.g. sugars, starch),

4.) **Nonpolar-** are compounds that do not have a significant dipole moment nor do they have polar functional groups (e.g. PNAs).

[21] No molecular ion will be observed in many spectra acquired under the energetic conditions of electron ionization.

[22] As droplets evaporate with Thermospray preformed ions from the solution (buffers) are desorbed into the vapor phase by field induction.

[23] The term "harshness rank" is our attempt to quantify or rank the various interfaces in terms of their potential to destroy labile molecules. Electrospray, which utilizes little to no heat in the desorption process is the least harsh with a value of 1. APCI, PB and TS are the harshest interfaces because that rely heavily on thermal processes to evaporate analyte from droplets and particles. Intermediate are FAB and LD.

[24] McMaster, M.C., *HPLC, A Practical User's Guide*, VCH Publishers: New York (1994).

[25] Usually most components can be separated without derivatization. Derivatization is more commonly used to change a mixture's solubility or increase detection sensitivity (McMaster M.C., *HPLC, A Practical User's Guide*, VCH Publishers: New York (1994)).
[26] A measured value of the efficiency factor N for a given column can be normalized to the column length (L) by dividing by L to yield "plates/meter." A typical 25 cm reversed-phase column will have 25,000 plates if packed to an efficiency of 100,000 plates/meter. This, of course, depends on the particle diameter, quality of packing, and plumbing.
[27] The hybrid CEC technologies allow the use of gradients with CE.
[28] Separation technologies are used for both course separations or isolation at the sample preparation stage of analysis or the fine separation that occurs on the analytical column. These technologies are used at for both applications. In some cases, there is a gray area where sample prep and separations are affectively the same.
[29] Banks, F.J., Gulcicek, E.E., "Rapid peptide mapping by reversed-phase liquid chromatography on nonporous silica with on-line electrospray time-of-flight mass spectrometry," Anal. Chem. 68, pages 3973-3978 (1997).
[30] Banks, F.J., Gulcicek, E.E., *op cit.*
[31] Criag Whitehouse, Analytica of Branford, private communication.

Part II

How do you acquire LC/MS capabilities?

1. Intelligence

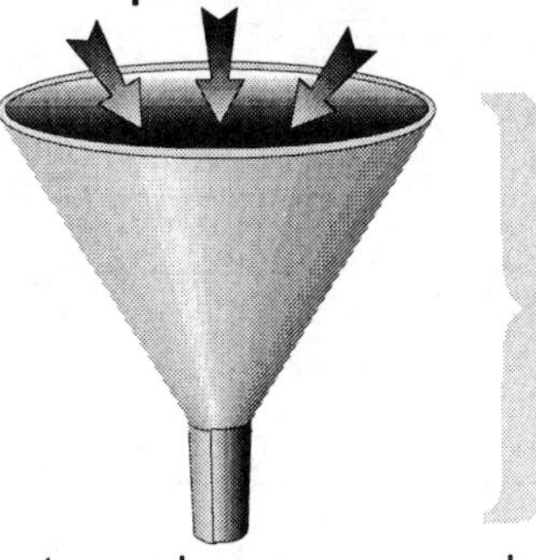

Chapter Four- Justifying Your LC/MS Acquisition

Send out, purchase, or upgrade

2. Design

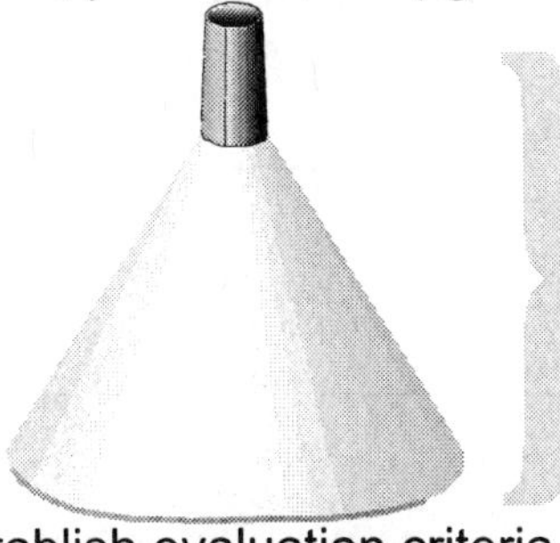

Chapter Five- Locating Your LC/MS Alternatives

Establish evaluation criteria

3. Choice

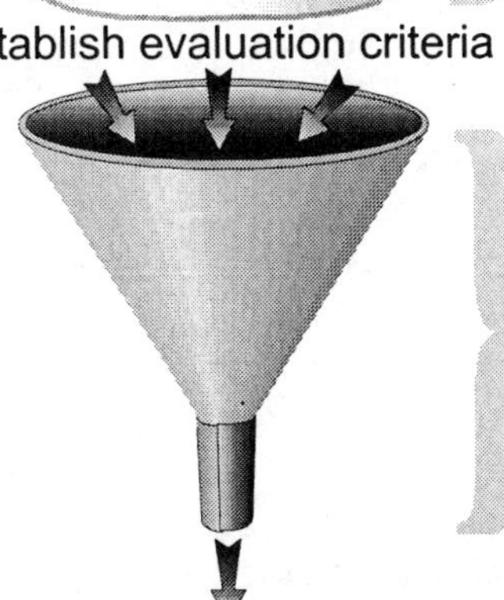

Chapter Six- Choosing Your LC/MS Alternatives

Chapter Seven- Implementing Your LC/MS Lab

Implement your LC/MS capability

Chapter Four- Flow Diagram

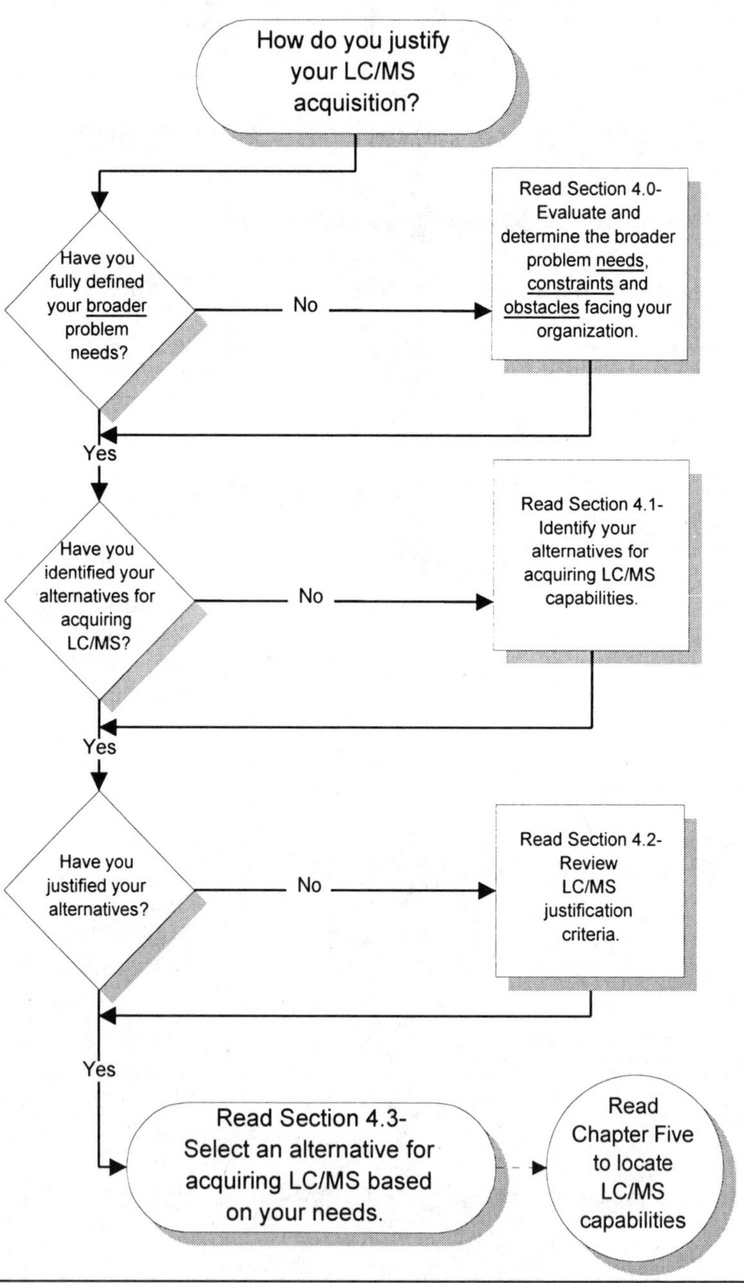

How do you justify your LC/MS acquisition?
Have you fully defined your broader problem needs?
No
Read Section 4.0- Evaluate and determine the broader problem needs, constraints and obstacles facing your organization.
Yes
Have you identified your alternatives for acquiring LC/MS?
No
Read Section 4.1- Identify your alternatives for acquiring LC/MS capabilities.
Yes
Have you justified your alternatives?
No
Read Section 4.2- Review LC/MS justification criteria.
Yes
Read Section 4.3- Select an alternative for acquiring LC/MS based on your needs.
Read Chapter Five to locate LC/MS capabilities

Chapter Four

Justifying Your LC/MS Acquisition

(or, LC/MS is More Than Technology)

If you tell what you want, you're quite likely to get it.
If you don't tell what you want, you're quite unlikely to get it.[1]

Donald Gause and Gerald Weinbert

4.0-Defining Your Broader Needs

To this point we have considered LC/MS from a "technology" point of view. Unfortunately, matching your problems with the appropriate technology is only half the battle. A more thorough solution to your problem will involve sorting out all of the other (so-called "broader") issues related to implementing a problem solution, such as: How do you acquire the technologies? Where do you find them? Can you afford them (budgeting, $)? How should you implement them? These are all issues that have less to do with sample chemistries or information requirements for a given analysis, and more to do with your current organization's needs, priorities, structure, and budgetary constraints.

This chapter identifies these broader issues that are critical considerations in your process of acquiring LC/MS capabilities. Should you run out to your nearest instrument company and purchase an instrument; or consider other alternatives to acquiring LC/MS capabilities? Are sending samples to a contract lab or upgrading existing instrumentation possible alternatives? Does the cost of LC/MS technologies justify the benefit of the information obtained from these technologies?

Defining Your Analysis Needs

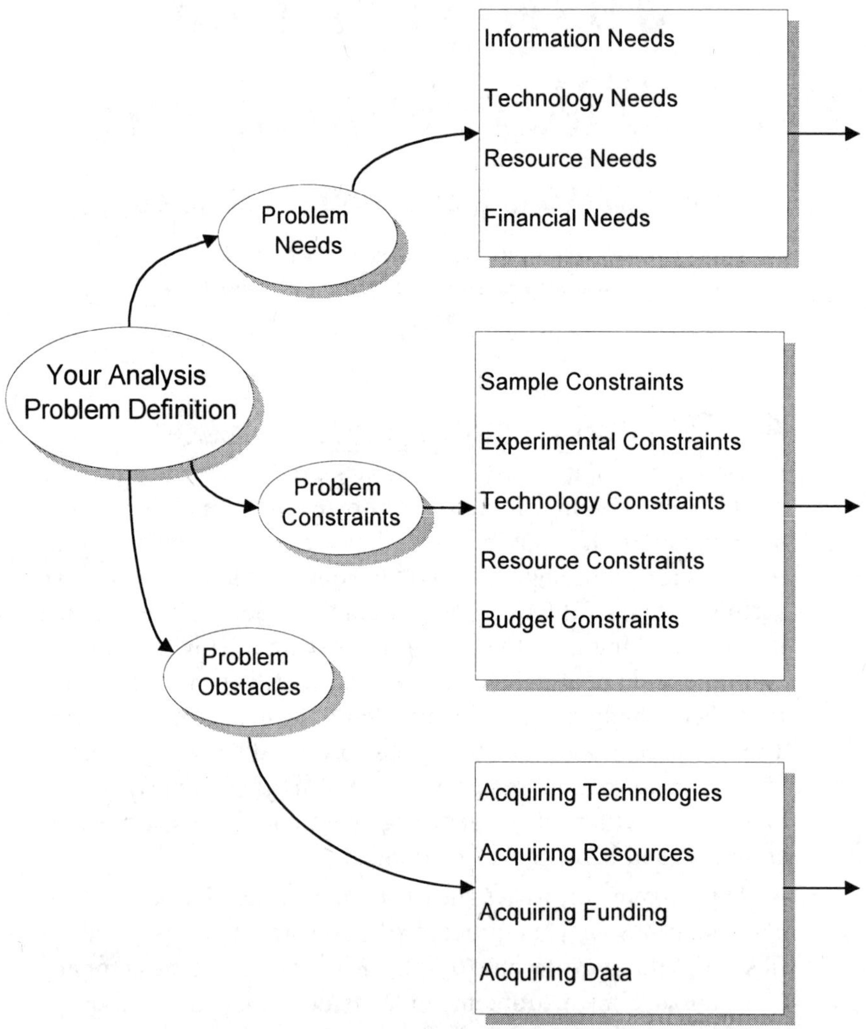

Figure 4.0- Fully defining problems addressed by LC/MS requires that you adequately define your problem needs, constraints, and obstacles (both pages).

Matching Your Needs to Alternatives

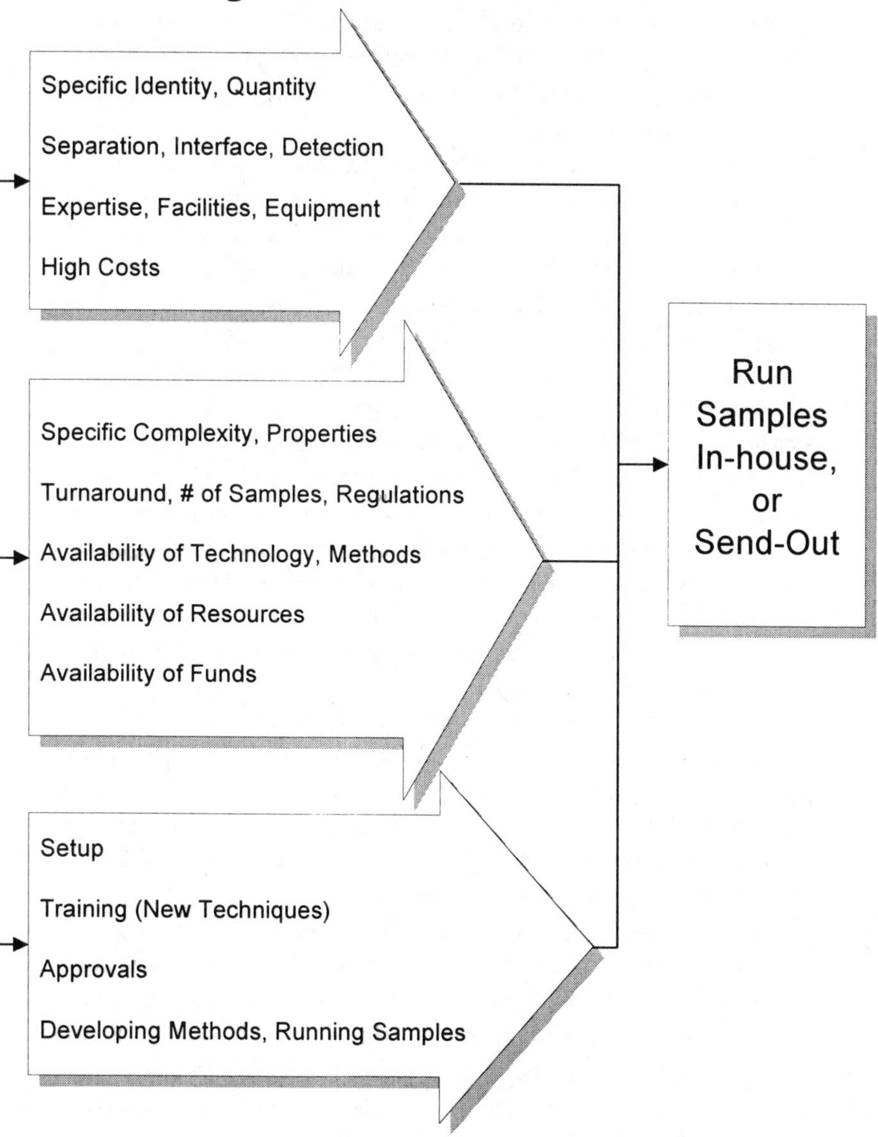
Specific Identity, Quantity
Separation, Interface, Detection
Expertise, Facilities, Equipment
High Costs
Specific Complexity, Properties
Turnaround, # of Samples, Regulations
Availability of Technology, Methods
Availability of Resources
Availability of Funds
Setup
Training (New Techniques)
Approvals
Developing Methods, Running Samples
Run Samples In-house, or Send-Out

Problem definition is the process of:

1.) defining what is needed to solve your problem (i.e. The goal),
2.) defining the constraints imposed upon you in the process of solving the problem (i.e. The boundaries), and
3.) identifying obstacles that you will encounter along the way (i.e. The hurdles).

Defining problems relating to acquiring LC/MS capabilities requires these three aspects of problem definition to be addressed. Figure 4.0 itemizes problem *needs*, *constraints*, and *obstacles* as they relate to problem definition in LC/MS. Identifying the specific items that compose your problem definition is a prerequisite to justifying which alternative you should consider in LC/MS.

Problem *needs* in LC/MS are the technology, resource, and financial requirements that are necessary to solve your LC/MS problems. Which technologies are required? What expertise? What is the cost? Problem *constraints* in LC/MS are all the limitations that you face when trying to meet the problem needs. Do you have the expertise? Do you have the equipment, facilities, and technologies required to solve the problem? Are they available? Problem *obstacles* in LC/MS are the barriers that you may face in order to solve your problems. These might include buying instrumentation, hiring qualified people, training existing people, or having to develop methods.

Worksheet 4.0 is provided to assist you in the process of defining your specific *needs* (Part A), *constraints* (Part B), and *obstacles* (Part C). Your actual problem definition may require a higher level of definition beyond this two-page worksheet; however, this broad overview may facilitate a more complete definition and force you to consider the "big picture" at the outset of your justification process. Fill out each part of the worksheet to accurately reflect your problem.

Also keep in mind that problem-solving is a dynamic process; consequently, the problem definition that you characterize today may change in the future as you gather additional information and weigh the merits of your various alternatives. Always modify and adapt your problem definition to your changing demands. You can get additional copies of this worksheet (as well as others) on the Internet at `www.lcms.com`.

Worksheet 4.0- Defining Your LC/MS Needs

Part A- Defining your broader problem **needs**. Check off your specific problem requirements and comment on each item.

What are your information needs?	Yes	No	Comments
Do you require:			
– MW information?	❑	❑	______
– Elemental composition?	❑	❑	______
– Structural information?	❑	❑	______
– Compound specificity?	❑	❑	______
– Enhanced response?	❑	❑	______

What are your technology needs?			Comments
Quads- ❑	ES- ❑	RP- ❑	______
Traps- ❑	AP- ❑	NP- ❑	______
TOF- ❑	PB- ❑	SE- ❑	______
Sectors- ❑	TS- ❑	IE- ❑	______
FTMS- ❑	CF- ❑	CE- ❑	______

What are your resource needs?	Yes	No	Comments
Do you require:			
– specialized expertise?	❑	❑	______
– an increased headcount?	❑	❑	______
– special facilities?	❑	❑	______
– special equipment?	❑	❑	______

What are your financial needs?	Yes	No	Comments
Do you require:			
– equipment purchase?	❑	❑	______
– hiring personnel?	❑	❑	______
– additional funding?	❑	❑	______

Part B- Defining your broader problem **constraints**. Check off your specific problem requirements and comment on each item.

What are your sample constraints?	Yes	No	Comments
Do you require:			
– isolation of sample components?	❑	❑	______
– removal of interferences?	❑	❑	______

Note: Copies of this worksheet can be downloaded from **www.LCMS.com**.

Worksheet 4.0- Defining Your LC/MS Needs

Do you have:

– unstable sample components?	❑	❑	______
– polar sample components?	❑	❑	______
– high MW species?	❑	❑	______

What are your experimental constraints?	Yes	No
Do you have:		
– specific time constraints?	❑	❑
– equipment constraints?	❑	❑
– specific sample constraints?	❑	❑
– specific regulatory constraints?	❑	❑
– specific legal constraints?	❑	❑
What are your technology constraints?	**Yes**	**No**
Is required LC/MS technology available?	❑	❑
Are methods available?	❑	❑
What are your resource constraints?	**Yes**	**No**
Is required expertise available?	❑	❑
Are facilities available?	❑	❑
Is required LC/MS equipment available?	❑	❑
What are your budget needs? Do you require:	**Yes**	**No**
Equipment purchase?	❑	❑
Hiring personnel?	❑	❑
Additional funding?	❑	❑

Part C- Defining your broader problem **obstacles**. Check off your specific problem requirements.

What are your problem obstacles? Will you:	Yes	No
Acquire specific technologies?	❑	❑
Acquire specific resources?	❑	❑
Need specific funding?	❑	❑
Need to develop methods?	❑	❑

Notes: ____

____

____

Note: Copies of this worksheet can be downloaded from `www.LCMS.com`.

4.1-Identify Your Acquisition Alternatives

Fortunately, the many newcomers to LC/MS are presented with several alternatives to allow them to acquire LC/MS capabilities. For the laboratory with a limited number of samples, the samples can be sent to contract labs or universities for several hundred dollars and up per sample. Acquiring LC/MS capabilities does not mean that you necessarily have to acquire LC/MS equipment; sending samples “out” is more practical for many laboratories. For the laboratory with a more substantial sample demand, a wide variety of in-house alternatives exist to acquire LC/MS capabilities. In-house analysis may include purchasing (or leasing) new or used equipment or upgrading existing equipment. Figure 4.1 shows a diagram of your alternatives for acquiring LC/MS.

In general, you have two choices when sending samples out, contract labs and universities. Universities are typically endowed with some of the most sophisticated technologies and can provide some of the highest performance instrumental techniques (e.g. FTMS or high resolution sectors). Contract labs are generally (but not always) associated with higher volume production applications; quadrupole technologies tend to be most common. A list of contract labs, universities, and institutes providing LC/MS capabilities is presented in *Chapter Five, Section 5.1*. Specific contact information are listed in *Appendix A*.

Figure 4.1 indicates that in-house alternatives for LC/MS requires considerably more justification than sending samples out. Obviously, the investment of hundreds of thousands of dollars for purchasing LC/MS equipment versus hundreds of dollars for sending out requires a higher level of justification. To justify the purchase of LC/MS equipment requires that the organization place a substantial value on the information obtained from the equipment.

The next two sections will discuss some aspects of justifying the acquisition of LC/MS technologies. In *Chapter Five*, *Sections 5.2* and *5.3*, are several lists of the various providers of commercial LC/MS systems and upgrades. *Appendix A* also provides contact information for manufacturers, leasing and upgrades. Updated lists can be obtained on the Internet at `www.lcms.com`.

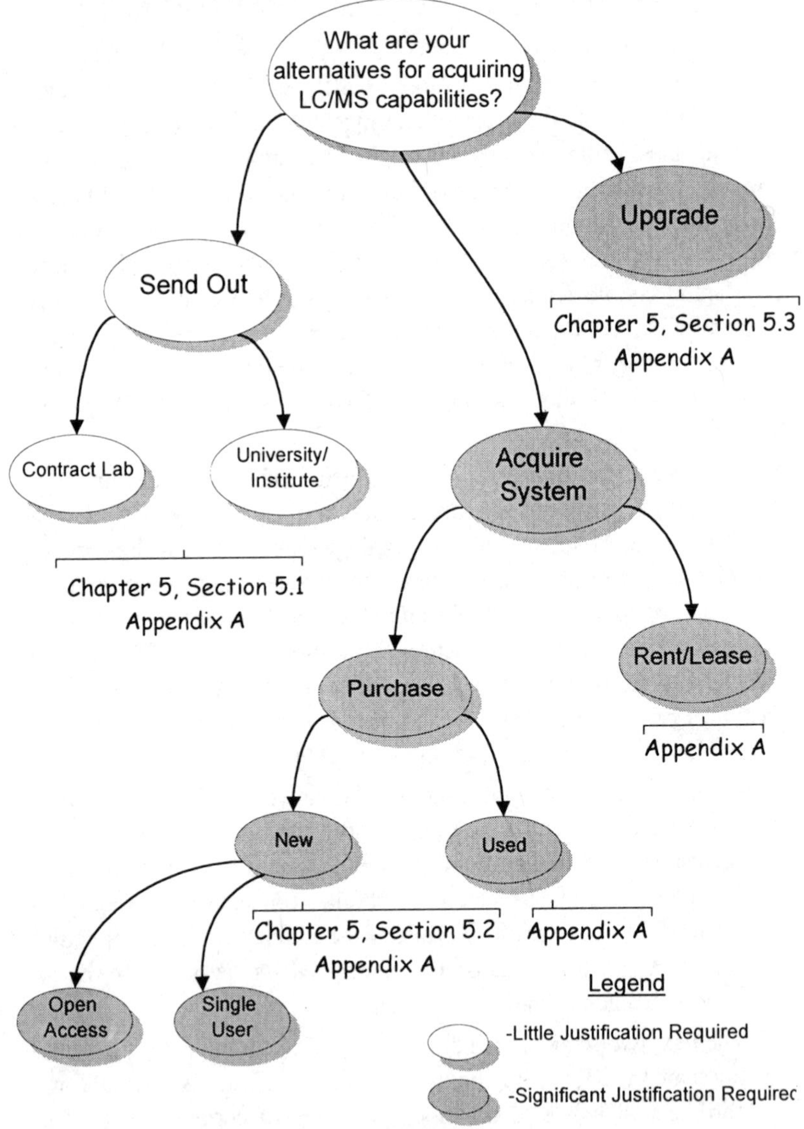

Figure 4.1- There are many alternatives to acquiring LC/MS capabilities (shown here). Acquiring or upgrading alternatives will require significant justification.

4.2-Justifying Your Alternatives

Justifying the purchase of new and expensive technologies (like LC/MS) by any organization can be a very complex process. This process is difficult to characterize because each organization has a unique structure, unique needs, and unique people.[2] The interplay between all these variables makes any single model or approach to justification difficult. We will endeavor to present a problem-solving framework within which you may justify your alternatives. Even if this approach does not meet the specific requirements of your organization, it should provide useful input to your justification process.

This framework is established by first defining your problems (*Section 4.0*), then searching for and identifying your alternatives (*Section 4.1*), and finally choosing the appropriate alternative (*Section 4.3*). This Section will discuss some of the justification criteria that we feel may be relevant to your problem needs. In each case we will attempt to weigh the given criterion against your alternatives to *send-out*, *purchase*, or *upgrade*. Table 4.0 summarizes the criteria that are discussed in this section.

Table 4.0- Justification Criteria for LC/MS Alternatives

Justification Criteria	Considerations
Sample Demand- Number of Samples	• Too many • Too few • Variability
Sample Demand- Sample Loading	• Routine • Batch vs. Level Loaded • Capacity
Sample Control	• Intellectual Property • Safety • Regulations
Efficient Information Flow	• Efficient Interaction • Turnaround • Speed of Decisions

Alternatives
Send-Out
Purchase
Upgrade

Justification: Number Of Samples

You may need LC/MS technologies, but do you have the sample volume (e.g., number of samples per unit time) to justify the purchase of an instrument? This question may depend on which LC/MS technologies you require as well as the number of samples your lab generates. If your lab generates too few samples, there is no way to justify the purchase of LC/MS from a "per sample" perspective. A recent survey presented in LC-GC cited that 67% of respondents had fewer than 50 samples per week per instrument (30% had less than 20 samples per week per instrument).[3] At this level, the cost-per-sample arguments can go either way in a send-out versus purchase analysis.

Figures 4.2 and 4.3 plot the unit sample cost as a function of the number of samples run/week for two different instrument scenarios.[4] Obviously, as the number of samples increase per week, the cost per sample will decrease.

When to Send-Out

If your lab doesn't do more than 10 samples per week you should consider sending samples out. The price per sample from contract labs will vary widely from $100-$1,000 per sample. Contract labs will charge on a per sample or a per hour basis, depending on the nature of the analysis. Methods development can also be quite expensive on a per sample basis.

Sending out when you have a relatively low volume of samples will allow you flexibility to choose techniques and also avoid hiring expertise in-house. You may even want to send your samples out first to establish a track record and a run rate, then bring the analysis in-house.

Sending out occasional samples to contract labs and universities should become a common practice for labs requiring identification of unknowns and confirmation of impurities.

Table 4.1- Cost scenarios for Figures 4.2 and 4.3[5]

		Scenario 1		Scenario 2
Instrument Type:		High Perf.		Mass Detector
Analyzer:		MS/MS		Single Analyzer
Est. Instrum. Price:		$400,000		$200,000
Est. Ann. Deprec.:		$80,000		$40,000
Est. Ann. Sal/Ben.	Ph.D.	$80,000	B.S.	$50,000
Overhead Rate:	100%	$80,000	10%	$50,000
Ann. Cost/Instrum:	Total-	$240,000		$140,000

When to Purchase

If the demand for LC/MS analysis exceeds about 20 samples per week, you can justify purchasing instrumentation relative to sending out samples on a cost-per-sample basis. This threshold seems to hold for both high performance and detector products in LC/MS. Of course, as the demand increases, the cost per sample goes down with in-house analysis. This is also true to a lesser extent when samples are sent out.

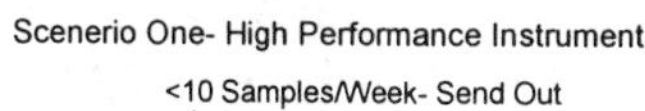

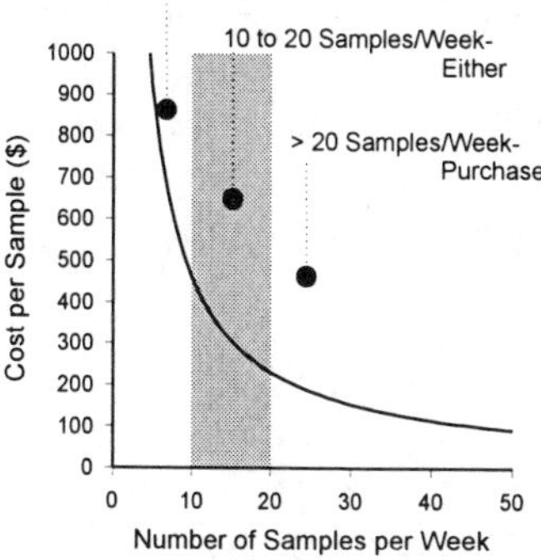

Figure 4.2- Cost-per-sample versus samples per week for a high performance MS/MS.

When to Upgrade

Sample demand can also influence your decision to upgrade an existing mass spectrometer to LC/MS capabilities. For the occasional sample, it still may be easier to send your sample out rather than spend significant time (and effort) in reconfiguring systems. The conversion from one configuration to another can take days in some instances, and infrequent use can result in higher occurrence of operator error.

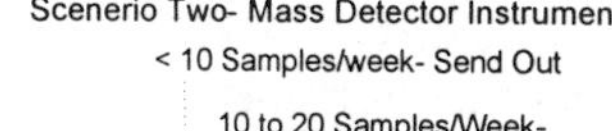

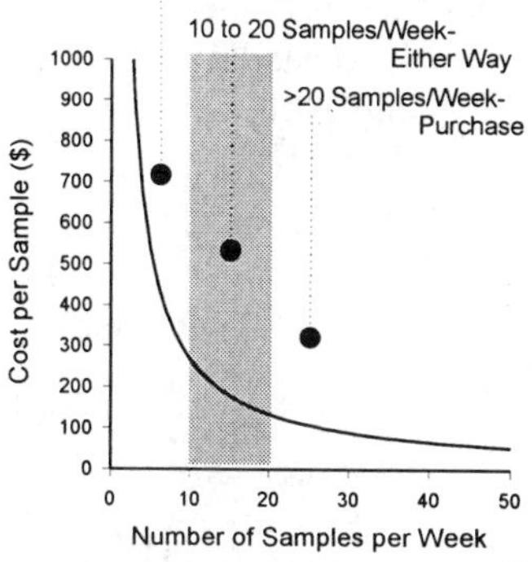

Figure 4.3- Cost-per-sample vs samples per week for a single mass analyzer instrument.

Table 4.2- Justification based on your sample demand

Criterion	Send-Out	Purchase	Upgrade
Sample Demand Number of samples per unit time	At less than 10 samples/week, sending out is more cost-effective than purchasing new equipment.	At more than 20 samples per week, consider purchasing a system. If you can't justify new equipment, consider used.	A few samples/week can justify the investment in upgraded equipment.

Justification: Demand-Sample Loading

Another important consideration when acquiring LC/MS capabilities is to consider the variability of sample demand on your lab. In general, problems don't come along in a uniform manner (40 samples per week); rather, they come in batches. Often the batches are time-dependent.

Clinical trials are a good example. If you get 2,000 samples into your lab, generally there is urgency to turn the samples around inside of a week. Even if you have an instrument, you may not be able to handle such a large lot of samples in a short time frame.

If you can estimate your average single analysis time, use Fig. 4.4 to estimate weekly capacity.

When to Send-Out

If the short-term demand placed on your lab exceeds your capacity, you should consider sending samples to outside contract labs. It is becoming quite common for many labs to send their high volume batch sample loads to contract labs. In many cases, they are much better equipped to handle production LC/MS analysis than a research environment. Many labs with LC/MS capabilities level-load their internal instrumentation with non-routine or critical internal research problems, and off-load the excess to contract labs.

Contract labs are listed in *Chapter Five* and *Appendix A.*

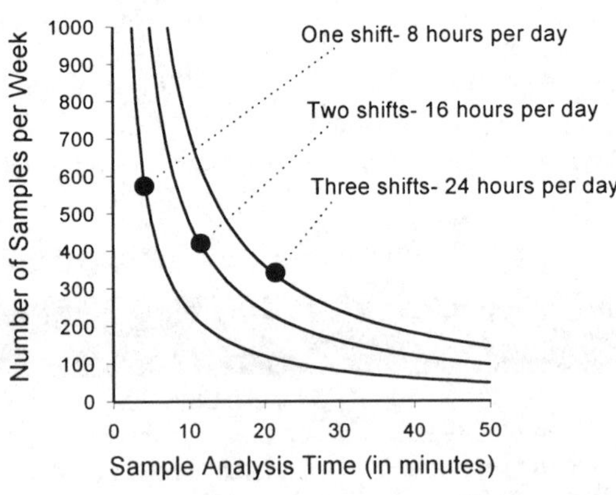

Figure 4.4- Number of samples per week as a function of individual sample analysis time. Estimate the weekly sample capacity of your lab for an average single analysis time.

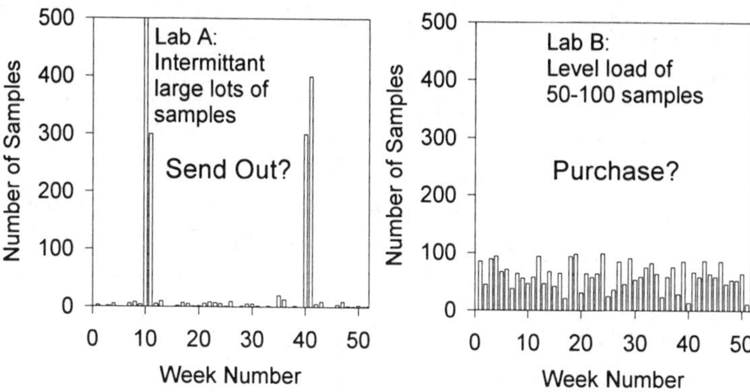

Figure 4.5- Sample loading is an important consideration. Here are two labs with different weekly sample loading, Lab A with intermittent large batches, and Lab B with between 50-100 samples per week.

When to Purchase

Purchasing alternatives should be considered when your sample demand exceeds about 10 samples per week and can be level-loaded. Note Figure 4.5.

Production labs are capable of exceeding 1,000 samples per week.[6] Pharmaceutical industry goals are currently targeting greater than 400 samples per day.[7] Another recent trend is the use of "open access" (*Chapter Six)* systems to assist an organization in increasing its sample load on a given instrument by increasing the access of the instrument to the entire organization.[8]

When to Upgrade

Upgrading existing instrumentation to LC/MS capabilities is a cost-effective way to address any type of potential sample demand, whether level-loaded or batch. Upgrades are a good alternative for labs that have sample demand placed upon them in batches. An existing instrument (e.g. GC/MS) can be more easily reconfigured for a given batch of samples than for intermittent samples on a daily basis.

Table 4.3- Justification based on your sample loading

Criterion	Send-Out	Purchase	Upgrade
Sample Demand- Loading of samples on your laboratory	Consider sending out when sample demand exceeds lab capacity or with large batches of single method samples with rapid turnaround demand.	Consider purchasing when the sample load on your lab can be distributed over time within your lab sample capacity.	Consider upgrading when the sample load can fit within your existing equipment utilization.

Justification: Control of Samples

There are many reasons today why corporations and institutions want to control access to both samples and the information derived from the analysis of samples. These include:

- the need to control access to *intellectual property*,
- the need to limit the transport of materials that present *safety*, *liability*, and *regulatory concerns*,
- the need to control handling of materials that present *safety*, *liability*, and *regulatory concerns*,
- and the need to control the access to information derived from a given sample analysis that may contribute to *decreased competitive positions* or *increased liability*.

Since LC/MS is one of the primary tools used in the identity of new chemical or biochemical species discovered or synthesized in research laboratories, there is significant incentive to keep the analysis by LC/MS as an in-house activity. This is reflected in the fact that most companies that rely on LC/MS for identity of product and process components, tend to buy their own instruments.

When to Send-Out

Some companies are quite comfortable sending proprietary samples outside of the organization for analysis, particularly to contract labs that are equipped to deal with proprietary information.

There are two approaches to safeguard proprietary information when sending samples out of the lab. These may be worth considering if you would otherwise discount sending out as an alternative for your analysis.

The first approach to safeguarding your sample information is to address the issues of access and disclosure of information contractually through a confidentiality agreement.[9]

The second approach to safeguard samples is to limit the outside lab to information about the samples that would not lead to substantive interpretation. In this situation the deliverables of the contract lab would simply be raw data supplied back to your lab for subsequent interpretation. It is wise to realize that the information derived from a mass spectrometer can be very revealing to a trained analyst without any other input.

When to Purchase

Many organizations, particularly those in pharmaceutical and biotechnology industries, do not want wide access to their intellectual property. LC/MS is an integral asset to their discovery and synthesis research.

In early discovery (pre-patent) processes, the goal is to prevent access at all cost to potential drug candidates. Even in product development and formulation (post-patent filing), significant effort is made to limit access to information about either the products or processes. The best way to control access is to analyze the samples yourself. This, of course, means purchasing an instrument.

Whenever the information derived from a given sample analysis has significant economic value, or significant economic liability, purchasing becomes a highly preferred alternative.

When to Upgrade

Clearly, any organization that has an incentive to control and run samples in-house should strongly consider upgrading its existing instrumentation. If the LC/MS technology required to solve the particular problems is available commercially, upgrading is a cost-effective and efficient method of acquiring this capability.

Table 4.4- Justification based on control of samples

Criterion	Send-Out	Purchase	Upgrade
Control of Samples	If control of, or information derived from, your samples is essential to your organization, samples should be sent out only under secure and well-established contractual relationships.	Consider purchasing when control of samples is required for intellectual property, regulatory, safety, or other non-financial reasons.	Consider upgrading existing equip. when control of the samples is required for intellectual property, regulatory, safety, or other non-financial reasons.

Justification: Efficient Information Flow

Why is efficient information flow important when selecting an LC/MS alternative? Most chemical analysis is an interactive process; that is, more than one individual or group is involved. The efficiency of the interaction between various groups will affect the speed with which decisions are made, the utilization of resources, the rate of development, and ultimately the profitability of the organization.

Efficient information flow may outweigh any other criteria for justifying your acquisition. One timely piece of information can sometimes justify the expense of even the most expensive piece of equipment.

An example where efficient information flow is important would be found with the identification of impurities in raw materials feeding a manufacturing process. Rapid identification of impurities may prevent costly downtime or out-of-spec. finished products. In general, the closer the measurement is to the process, the more timely an action to the measurement can take place.

Efficient information flow is affected by sample turn-around and the knowledge of the chemistries by the participants involved (or participating) in the analysis.

When to Send-Out

Lack of efficiency is an obvious concern when sending samples out. Questions that you must consider are: Can an outside lab meet the sample turnaround requirements of your organization? Can it participate in the interpretation and information feedback process to a level adequate to meet the needs of your organization? These questions may only be answered by evaluating the specific outside lab.

With respect to turnaround, some labs can turn samples around in a short period (several days), although you may have to pay a premium. Other labs may not be equipped to handle "the one of" type samples at all.

In general, in-house analysts are more familiar with the specific problems, chemistries, and products that you may be analyzing. In contrast, contract lab analysts may have greater expertise in terms of mass spectral interpretation and LC/MS sample analysis, but little knowledge of the chemistries of your particular sample. The balance of product expertise versus LC/MS expertise may be an important determinant when selecting between the choices of in-house analysis versus sending out.

When to Purchase

One of the most compelling justifications for purchasing LC/MS equipment is the need for efficient information flow between the various members of your organization. Measurements can be made in the same lab where the compounds are synthesized. Impurities can be identified within minutes or hours of becoming apparent. Decision making and information flow is efficient.

Research and development is not a linear process; rather it is a cyclical process where the flow of information to and from each part is key to the success of the overall process. Any impediment to efficient information flow (such as taking the time to have samples analyzed by an outside lab) may have a deleterious effect on the entire process.

When to Upgrade

For most organizations that have existing mass spectrometry capabilities, upgrades are easily justified on the grounds of efficiency when the new technique will provide needed information that is not available by other means. In general, upgrading by definition is placing a new capability into an existing information flow. The sample formats, communication protocols, and reporting protocols are usually already established. Therefore, training of all users in the new technique becomes less burdensome when upgrading.

If the upgrade requires special configurations and a time-consuming changeover, the efficiency benefits of upgrading may be lost.

Table 4.5- Justification Based on Information Flow

Criteria	Send-Out	Purchase	Upgrade
Efficiency Information Flow	Consider sending out if rapid turnaround and efficient interaction can be established between an outside lab and in-house problem source (sample source).	Consider purchasing when a high level of interaction and interpretation is required between LC/MS analysts and the research scientists.	Consider upgrading when a high level of interaction and interpretation is required between LC/MS analysts and the research scientists.

4.3-Selecting Your Alternative

You need LC/MS. Your organization needs LC/MS. So go get it. But, select the right alternative. Do your homework and planning upfront and you will be glad in the end. Selecting the appropriate alternatives for acquiring LC/MS capabilities will more likely depend on your organizational needs than your information needs. In terms of getting information, you generally get similar information irrespective of where the samples are run. You can purchase almost any capability (with the exception of state-of-the-art technology that has yet to be engineered into a commercial package) and you can send out to have virtually any analytical sample analyzed. LC/MS is available to all organizations, whether or not they have the expertise or financial means to purchase a system of their own. Information needs alone do not justify the purchase of a system; rather, questions about sample type, control, and information flow all must be weighed in your decision.

To assist you in your decision making process we provide two worksheets to prompt your evaluation. Worksheet 4.0 can assist you in assessing your broader problem needs. Worksheet 4.1 can assist you in matching your problem needs to an appropriate alternative. Worksheet 4.1 has three parts:

1.) *Part A* is included to address the justification criteria to distinguish your needs for acquiring a <u>system versus sending out</u> the work.

2.) *Part B* provides criteria to evaluate two alternatives; <u>purchase versus upgrade</u>.

3.) *Part C* presents issues relating to purchasing <u>new versus used</u> equipment.

Only you are capable of weighing the importance of each criterion. Your selection of a specific alternative, whether purchase or send-out, is only the beginning of the process. If you justified sending out the work, you must now locate prospective outside labs, evaluate their capabilities, and select one that meets your problem needs. If you justified purchasing an instrument, you must locate prospective manufacturers, evaluate their products, and select one as well. The processes of locating, evaluating, and selecting a specific LC/MS alternative are addressed in *Chapters Five* and *Six*.

Worksheet 4.1-Justifying Your Acquisition

Part A- Selection criteria for the justification of acquiring an LC/MS system versus sending out samples. Check off the appropriate selection.

Acquire versus **Send-Out**	Comments	Acquire	Send-Out
– How many samples/week (avg.)?	________	❑	❑
– What is your sample loading?	________	❑	❑
– Do you require a high level of sample control?	________	❑	❑
– Do you require interaction between analyst and researchers?	________	❑	❑
– Do your analytical requirements vary considerably?	________	❑	❑
– Other criteria:	________	❑	❑

Part B- Selection criteria for the justification of purchasing an LC/MS system versus upgrading an existing system. Check off the appropriate selection.

Purchase versus **Upgrade**	Comments	Purchase	Upgrade
– Do you have existing equipment?	________	❑	❑
– Are upgrades available?	________	❑	❑
– Does an upgrade meet your performance requirements?	________	❑	❑
– Does an upgrade meet your sample demand requirements?	________	❑	❑
– Does an upgrade meet your performance reliability?	________	❑	❑
– Other criteria:	________	❑	❑

Part C- Selection criteria for the justification of acquiring a new versus used LC/MS system. Check off the appropriate selection.

Purchase New versus **Used**	Comments	New	Used
Can you afford new?	________	❑	❑
Will used meet your performance demands?	________	❑	❑
Will used meet your sample demands?	________	❑	❑
Other criteria:	________	❑	❑

Note: Copies of this worksheet can be downloaded from `www.LCMS.com`.

Discussion Questions and Exercises

1. Discuss the differences between "technology" and "non-technology" aspects of problem-solving in LC/MS.
2. What are your alternatives if you can't afford to purchase an instrument?
3. If you have decided to purchase an instrument, what are your alternatives if you can't afford to purchase a new instrument?
4. Estimate your cost per sample if your lab averaged between 30 and 40 LC/MS samples per week. For an APCI/LC/MS system? For an APCI/LC/MS/MS (triple quadrupole) system? For an APCI/LC/MS^n (ion trap) system?
5. Give five reasons for purchasing an instrument for reasons other than a large number of samples per unit time.
6. Give five reasons for sending samples to a contract lab instead of purchasing an instrument.

Notes

[1] Gause, D.C., Weinberg, G.M., *Exploring Requirements: Quality before Design*, Dorset House Publishing: New York (1989).

[2] Webster, F. and Wind, Y., *Organizational Buying Behavior*, Prentice-Hall: New Jersey, (1972).

[3] Majors, R.E., "Trends in Sample Preparation," LC-GC, 14, pages 754-766 (1996).

[4] With this evaluation we simply indicate the sample costs for two instrument purchase scenarios, compared to sending samples to an outside contract lab. The first scenario evaluates the analysis of samples with a high performance MS/MS instrument operated by a Ph.D.-level scientist. The second scenario evaluates the analysis of samples with a "lower priced" single analyzer instrument operated by a B.S.-level scientist.

[5] The values for Scenarios One and Two are rough estimates of the expenses incurred for each type of analyzer. These estimates are intended to yield the approximate expenses in order to arrive at an estimated cost per sample value. Tax implications and inflation were not considered in

this estimate. Depreciation was estimated using a straight-line method over a period of five years.

[6] Phoenix International Life Sciences Inc. (2350 Cohen St., Montreal, PQ, Canada, H4R 2N6). Check out its web-site: www.pils.com/services/bio_services/lcms.html.

[7] Pleasants, S., Biddlecombe, R.A., "Automating 96 well format SPE for bioanalysis using LC/MS/MS," Proceedings of the 14th (Montreux) Symposium on Liquid Chromatography/Mass Spectrometry, Cornell University, Ithaca, New York, July 23-25, 1997.

[8] Open Access. Spreen, R.C., Schaffter, L.M., "Open Access MS: A walk-up MS service," Anal. Chem. 68, pages 414A-419A (1996).

[9] It is advisable to seek legal counsel for situations where you are exposing intellectual property to outside observers. As access to information increases, so does risk; consequently, technical, legal, and financial considerations should always be thoroughly addressed through the use of qualified professionals.

Chapter Five- Flow Diagram

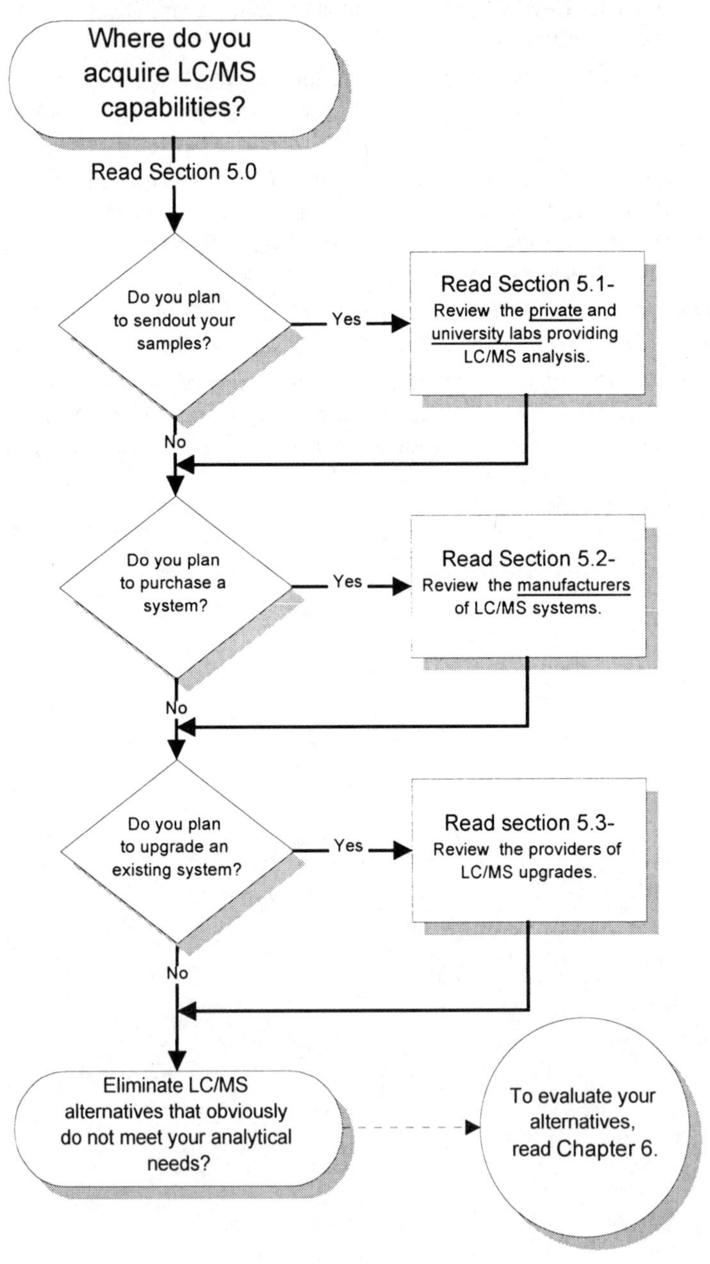

Chapter Five

Locating Your LC/MS Alternatives

(or, Where to Find the Stuff)

A bird and a fish can fall in love, but where do they build their nest?

Whoopie Goldberg
in the movie ***Corina Corina***

5.0- Making Contact

Regardless of which alternative you choose, the process of acquiring LC/MS capabilities involves making contact with either contract labs, universities, or instrument companies. The earlier, the better. The more, the better. Acquiring capabilities requires that you make a decision; the quality of your decisions will depend on the quality and completeness of the information you use to evaluate your decisions. Going directly to the source is the best way to ensure that your information is accurate and complete. Even if you have a history, and therefore a bias, bias with one lab or one instrument company, you should always contact a variety of competitors to validate your preferred source.

Contacting instrument manufacturers has become easier in recent years as the entire instrument market has undergone significant changes primarily because of market contraction; this leads to four major suppliers of LC/MS equipment with only a few second tier suppliers remaining.[1] Fortunately, in our view, this contraction has brought the mass spectrometer manufacturers and the liquid chromatography manufacturers together, the result being better integration of the entire LC/MS system and a steady decrease in entry level product prices. This chapter includes information on locating and contacting the LC/MS alternatives.

5.1-Locating Contract Labs

The number of laboratories offering LC/MS services has significantly increased in recent years and continues to grow, both private and public (universities, colleges and institutes) laboratories. Many laboratories that were originally GC/MS-based contract labs are beginning to diversify into LC/MS technologies. Obviously, local labs make transport of samples more convenient and should be considered first.

Private Contract Labs

We have segregated the following tables by technology (e.g. atmospheric pressure interface technologies, particle beam, thermospray, etc.).

Table 5.1- Contract Laboratories- API/LC/MS

Company	Interface	Analyzer	Comment
ABC Laboratories	ES, APCI	na	www.abclabs.com
Advanced Bioanalytical Services	ES, APCI	Q, QQQ	Jack Henion's lab. Quantitative & qualitative analyses for the pharmaceutical industry; screening; identification. www.abs-cms.com
Alta Analytical Laboratories	ES, APCI	QQQ	Bob Bethem's lab. Quantitative method development, validation and analysis for the pharmaceutical industry. www.altalab.com
ANAPHARM	ES, APCI	QQQ	Phase I CRO with bioanalytical services. Other techniques: HPLC, GC & GC/MS. www.anapharm.com
Appollin	na	na	banchan@erols.com

Abbreviations are summarized at the end of this chapter (see Notes).[2] Addresses and contact information are contained in *Appendix A, Sources of LC/MS Information.* Updated information on existing or new contract labs can be found at the Internet site: `www.lcms.com`.

Table 5.1- Contract Labs- API/LC/MS (cont.)

Company	Interface	Analyzer	Comment
Axelson & Kwok Research Associates	ES, APCI	na	www.asi.bc.ca/bcba
Cedra	ES, APCI	na	www.cedracorp.com
Harris Lab	ES, APCI	na	www.dataedge.com
Kansas City Analytical Services	ES	QQQ	na
Keystone Analytical Laboratories	ES	QQQ	na
LAB International	ES, APCI	QQQ	Other techniques: MS/MS; GC/MS. champagn@lab-cdn.mds.Compuserve.com
Magellan Laboratories	ES, APCI	Q	Develops analytical solutions for pharmaceutical clients. Other techniques: FAB; DP; GC/MS. www.magellanlabs.com
Maxxam Analytics	ES, APCI	na	Service for the pharmaceutical and environmental industries. Other techniques: MS/MS; GC/MS; ICP. www.maxxam.ca
Midwest Research Institute	ES, APCI	B, IT, Q, QQQ	Other techniques: MS/MS; DP; CE/MS; ICP; GC/MS. www.mriresearch.org

Abbreviations are summarized at the end of this chapter (see Notes).[2] Addresses and contact information are contained in *Appendix A*, *Sources of LC/MS Information*. Updated information on existing or new contract labs can be found at the Internet site: `www.lcms.com`.

Table 5.1- Contract Labs- API/LC/MS (cont.)

Company	Interface	Analyzer	Comments
M-Scan	ES	B, Q	Drug & drug metabolites; biomolecule characterization. Other techniques: MS/MS; MALDI-TOF; CI; EI; GC/MS. www.m-scan.com
Northwest Bioanalytical	ES, APCI	Q, QQQ	Quantitative determination of drugs and metabolites in clinical & preclinical samples. Other techniques: GC/MS & GC/MS/MS. ww.xmission.com/~nwt/bio.htm
Oneida Research Services	ES, APCI	Q, QQQ	Drug metabolite & impurity identification & screening; pharmacokinetic & toxicology studies; data analyses. Other tech.: MS/MS; DP; FAB; GC/MS. dfwelpe@borg.com
Panlabs	ES, APCI	na	www.panlabs.com
PPD Pharmaco	ES, APCI	Q, QQQ	Quantitative analyses; drugs & drug metabolites. Other tech.: MS/MS; GC/MS. www.ppdpharmaco.com
Pharma-Kinetics Laboratories	ES, APCI	na	www.pharmakinetics.com
Phoenix International Life Science	ES, APCI	QQQ	Specializes in pharmaceutical applications. www.pils.com

Abbreviations are summarized at the end of this chapter (see Notes).[2] Addresses and contact information are contained in *Appendix A*, *Sources of LC/MS Information*. Updated information on existing or new contract labs can be found at the Internet site: `www.lcms.com`.

Table 5.1- Contract Labs- API/LC/MS (cont.)

Company	Interface	Analyzer	Comments
Quanterra	ES, APCI	QQQ	Paul Winkler directs the LC/MS facilities. Method development and validation for the pharmaceutical, biotech, and environmental industries. GLP/GMP compliant. Other tech.: GC/MS.
Quintiles	ES, API	Q, QQQ	Pharmacology; toxicology; pharmaceutical sciences, & clinical analysis. Other techniques: MS/MS; GC. www.quintiles.com
R&W Bio Research	ES, APCI	na	www.rwbio.nl
Research Triangle Institute	ES, APCI	IT, Q, QQQ	Bob Voyksner's lab. Broad capabilities for qualitative and quantitative method development, validation and analysis. www.rti.org
Ricera	ES, APCI	Q, QQQ	Quantitative & qualitative analyses for the pharmaceutical industry- identification & confirmation. Other techniques: MS/MS; GC/MS; EI. www.ricerca.com
Stanford Research Institute	ES, APCI	na	www.sri.com
Taylor Technology	ES, APCI	QQQ	Paul Taylor's lab. sag@taytech.com
TexMS	ES, APCI	QQQ	Dan Gartiez's lab. www.ghg.net/texms

Abbreviations are summarized at the end of this chapter (see Notes).[2] Addresses and contact information are contained in *Appendix A*, *Sources of LC/MS Information*. Updated information on existing or new contract labs can be found at the Internet site: `www.lcms.com`.

Table 5.1- Contract Labs- API/LC/MS (cont.)

Company	Interface	Analyzer	Comments
TNO Pharma	ES, NS, APCI	QQQ, IT, B	Quantitative, qualitative, drug residue & doping analyses; biomolecule characterization; and contract research for the pharmaceutical industry. www.tno.nl/instit/pharma
Triangle Laboratories	ES; APCI	QQQ	Other techniques: GC/MS (high & low resolution; negative CI), method development, validation and analysis. harvan@compuserve.com
Xenobiotic Laboratories	ES, APCI	QQQ	Quantitative and qualitative, analysis; and contract research for the pharmaceutical industry. www.xbl.com

Table 5.2- Contract Labs- PB/LC/MS

Company	Interface	Analyzer	Comment
Midwest Research Institute	PB	B, IT, Q, QQQ	Other techniques: MS/MS; DP; CE/MS; ICP; GC/MS. www.mriresearch.org
TNO Pharma	PB	QQQ, IT, B	Quantitative, qualitative, drug residue & doping analyses; biomolecule characterization and contract research. www.tno.nl/instit/pharma

Abbreviations are summarized at the end of this chapter (see Notes).[2] Addresses and contact information are contained in *Appendix A*, *Sources of LC/MS Information*. Updated information on existing or new contract labs can be found at the Internet site: `www.lcms.com`.

Table 5.3- Contract Labs- TS/LC/MS

Company	Interface	Analyzer	Comment
LAB International	D, F	QQQ	Other techniques: MS/MS; GC/MS. champagn@lab-cdn.mds.Compuserve.com
Midwest Research Institute	D, F	B, IT, Q, QQQ	Other techniques: MS/MS; DP; CE/MS; ICP; GC/MS. www.mriresearch.org
Oneida Research Services	D, F	Q, QQQ	Drug metabolite and impurity identification and screening; pharmacokinetic and toxicology studies; and data analyses. Other tech.: MS/MS; DP; FAB; GC/MS. dfwelpe@borg.com
TNO Pharma	TSP	QQQ, IT, B	Quantitative, qualitative, drug residue & doping analyses; biomolecule characterization; and contract research. www.tno.nl/instit/pharma

Table 5.3- Contract Labs- CF-FAB/LC/MS

Company	Interface	Analyzer	Comment
M-Scan	CFF	B	Drug & drug metabolites; biomolecule characterization. Other techniques: MS/MS; MALDI-TOF; CI; EI; GC/MS. www.m-scan.com
Oneida Research Services	CFF	Q, QQQ	Drug metabolite and impurity identification and screening; pharmacokinetic and toxicology studies; and data analyses. Other tech.: MS/MS; DP; FAB; GC/MS. dfwelpe@borg.com
TNO Pharma	TSP	QQQ, IT, B	Quantitative, qualitative, drug residue & doping analyses; biomolecule characterization; and contract research. www.tno.nl/instit/pharma

Abbreviations are summarized at the end of this chapter (see Notes).[2]

Public Contract Labs

An important resource for LC/MS sample analysis is found at many universities, institutes, and colleges. Check your local universities and colleges first to see if they have LC/MS capabilities. Alternatively, use the list provided in Table 5.4. Detailed contact information for each lab is contained in *Appendix A*. In addition, an up-to-date listing of labs offering LC/MS analysis is made available on the "LC/MS Home Page" (`www.lcms.com`).

Table 5.4- Public Labs for LC/MS Analysis

USA			
Location	**Institution**		
Arizona	Univ. of Arizona		
California	U.C. Berkeley	U.C. Davis	U.C. Los Angeles
	U.C. Riverside	U.C. San Fran.	Scripps Res. Instit.
Wash, D.C.	Naval Res. Lab.		
Florida	Univ. of Florida	Florida St. Univ.	
Georgia	Emory Univ.	Georgia Tech.	Univ. of Georgia
Iowa	Iowa State Univ.		
Illinois	Univ. of Illinois		
Indiana	Notre Dame		
Louisiana	Louisiana State Univ.		
Maryland	Johns Hopkins	NIH	Univ. of Maryland
Mass.	Harvard Univ.	MIT.	Northeastern Univ.
Missouri	Washington Univ.		

Internet addresses are contained in *Appendix A, Sources of LC/MS Information*. Updated information on existing or new labs can be found at the Internet site: `www.lcms.com`.

Table 5.4- Public Labs for LC/MS Analysis (cont.)

USA			
Location	**Institution**		
Nebraska	Univ. of Nebraska		
New Jersey	Rutgers Univ.		
New York	Albert Einstein Col. of Medicine	Cornell Univ.	Rockefeller Univ.
	SUNY-Buffalo		
North Carolina	Nat. Instit. of Environmental Health Sciences	North Carolina State	
Penna.	Carnegie Mellon Univ.	Univ. of Penna.	
South Carolina	Medical Univ. of South Carolina	Univ. of South Carolina	
Texas	Univ. of Texas		
Virginia	Virginia Commonwealth Univ.		
Washington	Pacific Northwest Nat. Labs.	Univ. of Washington	
Canada			
Location	**Institution**		
Alberta	Univ. of Alberta		
Ontario	McMaster Univ.	Ottawa-Carleton Univ.	
Asia			
Location	**Institution**		
Korea	Korea Basic Science Instit.		

Internet addresses are contained in *Appendix A*, *Sources of LC/MS Information*. Updated information on existing or new labs can be found at the Internet site: `www.lcms.com`.

Table 5.4- Public Labs for LC/MS Analysis (cont.)

Australia			
Location	**Institution**		
Melbourne	Assoc. of Biomolecular Resource Facilities	Royal Melbourne Instit. of Tech.	
Sydney	Univ. of New South Wales	Univ. of Sydney	
Europe			
Location	**Institution**		
Belgium	Univ. of Antwerp	Univ. of Gent	Univ. of Leuven
France	Univ. of Liège		
Germany	European Molecular Biology Lab.	Fraunhofer Instit.of Food Tech. and Packaging	Univ. Konstanz
	Max-Planck Inst. in Mulheim		
Netherlands	Utrecht Univ.		
Switzerland	Swiss Fed. Instit. fir Environmental Science and Technology		
United Kingdom	Cambridge Univ.	Ludwig Instit. for Cancer Research and Univ. College London	Univ. of Manchester
	Nottingham Trent Univ.		
South America			
Location	**Institution**		
Brasil	Univ. of Brasília		

Internet addresses are contained in *Appendix A, Sources of LC/MS Information*. Updated information on existing or new labs can be found at the Internet site: `www.lcms.com`.

5.2-Locating Manufacturers

A list of manufacturers[3] and the instruments they offer are provided in Tables 5.5 thru. 5.7 (for specific information pertaining to each company, e.g., address, phone number, home page,...; see *Appendix A: Sourses of LC/MS Information*). The tables are organized by LC/MS interfaces.

Table 5.5- Manufacturers-AP/LC/MS

Company	Interface	Model	Analyzer	Comment
Analytica of Branford	ES, US, APCI	Enterprise	TOF	Introduced benchtop LC/MS System at Pittcon'97.
	ES, US, APCI	na	na	Electrospray and APCI LC/MS interface..
Bruker, Daltonics	ES, IS, APCI	Esquire-LC/MS^n	IT	Joint marketing with Hewlett Packard of LC/MS/MS system.
	ES, IS, NS, APCI	APEX II	FT	Multiple inlet system (e.g., MALDI).
	ES, IS, NS, APCI	BioAPEX	FT	Multiple inlet system (e.g., MALDI).
Finnigan	ES, IS, APCI	Navigator	Q	Benchtop LC/MS detector.
	ES, IS, APCI	LCQ	IT	Benchtop LC/MS detector.
	ES, IS, APCI	SSQ, TSQ	Q, QQQ	Multiple inlet MS & MS/MS systems (e.g., GC).
	ES, IS, APCI	NewStar-2000	FT	Multiple inlet system (e.g., MALDI).
	ES, IS, APCI	MAT 95's; 900's	B	Multiple inlet MS & MS/MS systems

Abbreviations are summarized at the end of this chapter (see Notes).[2] Addresses and contact information are contained in *Appendix A*, *Sources of LC/MS Information.* Updated information on individual instrument manufacturers can be found at the Internet site: `www.lcms.com`.

Table 5.5- Manufacturers- AP/LC/MS (cont.)

Company	Interface	Model	Analyzer	Comment
Hewlett Packard	ES, IS, APCI	1100- LC/MSD	Q	Introduced benchtop LC/MS detector at Montreux 1996
	ES, IS, APCI	Esquire- LC/MS^n	IT	Joint marketing with Bruker of a LC/MS/MS system.
	ES, IS, APCI	Engine	Q	Multiple inlet MS system (e.g., GC).
Hitachi	ES, IS, APCI	M-1200H	Q	Benchtop LC/MS detector.
IonSpec	ES, IS, APCI	HiResESI	FT	Multiple inlet system (e.g., MALDI).
JEOL	ES, APCI	Automass	Q	Benchtop LC/MS system.
	ES, APCI	AX505	B	Small forward geometry, multiple inlet MS and MS/MS system.
	ES, APCI	RSVP	B	Low cost full feature sector, multiple inlet system.
	ES, APCI	MStation	B	Double focusing sector, multiple inlet system.
	ES, APCI	HX110A	B	Multiple inlet MS and MS/MS systems.
Micromass[4]	ES, IS, APCI	Platform-LC	Q	Multiple inlet LC/MS system (e.g., GC).
	ES, IS, APCI	Quatro-LC	QQQ	Multiple inlet LC/MS system (e.g., GC).
	ES, IS, APCI	AutoSpec's ProSpec's	B	Multiple inlet MS & MS/MS systems (GC).
	ES, IS, NS, APCI	LCT	TOF	Benchtop LC/MS system introduced in mid-1997.
	ES, IS, NS, APCI	Q-TOF	Q-TOF	LC/MS/MS system introduced in mid-1996.

Abbreviations are summarized at the end of this chapter (see Notes).[2] Addresses and contact information are contained in *Appendix A, Sources of LC/MS Information*. Updated information on individual instrument manufacturers can be found at the Internet site: `www.lcms.com`.

Table 5.5- Manufacturers- AP/LC/MS (cont.)

Company	Interface	Model	nalyzer	Comment
New Objective	NS	na	na	Nanospray source for various electrospray LC/MS instruments.
Protana	NS	na	na	Nanospray source for various electrospray LC/MS instruments.
Perkin-Elmer	ES, IS, APCI	API 150 MCA	Q	Introduced benchtop LC/MS detector at Montreux 1996.
	ES, IS, APCI	API 165 LC/MS	Q	Benchtop LC/MS detector.
	ES, IS, APCI	API 365 LC/MS/MS	QQQ	Benchtop LC/MS/MS system.
PerSeptive Biosystems[5]	ES, IS, APCI	Mariner	TOF	Introduced benchtop LC/MS system at Pittcon'97.
Sensar	ES, APCI	Jaguar	TOF	Introduced benchtop LC/MS system at Pittcon'97.
Shimadzu	ES, APCI	na	Q	Previewed benchtop LC/MS detector at Pittcon'97.
Waters	ES, IS, APCI	Platform-LC	Q	Alliance LC/MS system featuring the Micromass Platform-LC detector.
World Precision Instruments	NS	na	na	Nanospray source for various electrospray LC/MS instruments.

Abbreviations are summarized at the end of this chapter (see Notes).[2] Addresses and contact information are contained in *Appendix A*, *Sources of LC/MS Information*. Updated information on individual instrument manufacturers can be found at the Internet site: `www.lcms.com`.

Table 5.6- Manufacturers- PB/LC/MS

Company	Interface	Model	Analyzer	Comment
Finnigan	PPB	SSQ; TSQ	Q, QQQ	Multiple inlet MS & MS/MS systems (e.g., GC); EI & CI.
	PPB	MAT 95's; 900's	B	Multiple inlet MS & MS/MS systems (e.g., GC); EI & CI.
Hewlett Packard	PPB	Engine	Q	Multiple inlet MS system (e.g., GC); EI & CI ionization.. HP licenses the (MAGIC) PB interface from Georgia Tech.
JEOL	PB, UPB	AX505	B	Small forward geometry, multiple inlet MS and MS/MS system; EI & CI.
	PB, UPB	RSVP	B	Low cost full feature sector, multiple inlet system; EI & CI.
	PB, UPB	MStation	B	Double focusing sector, multiple inlet system; EI & CI.
	PB, UPB	HX110A	B	High performance forward geometry, multiple inlet MS and MS/MS systems; EI & CI
Micromass[5]	PPB	Quatro LC	QQQ	Multiple inlet LC/MS/MS system (e.g., GC); EI & CI.
	PPB	ZabSpec, AutoSpec; ProSpec Series	B	Multiple inlet MS and MS/MS systems (e.g., GC); EI & CI.
Waters	TPB	Alliance-Integrity	Q	Benchtop LC/MS detector; EI.

Abbreviations are summarized at the end of this chapter (see Notes).[2] Addresses and contact information are contained in *Appendix A*, *Sources of LC/MS Information.* Updated information on individual instrument manufacturers can be found at the Internet site: `www.lcms.com`.

Table 5.7- Manufacturers- TS/LC/MS

Company	Interface	Model	Analyzer	Comment
Finnigan	TS, D, F	SSQ, TSQ	Q, QQQ	Multiple inlet MS & MS/MS systems (e.g., GC).
	TS, D, F	MAT 95's; MAT 900's	B	Multiple inlet MS & MS/MS systems (e.g., GC).
Hewlett Packard	TS, D, F	Engine	Q	Multiple inlet MS system (e.g., GC).
JEOL	TS, D, F	AX505	B	Small forward geometry, multiple inlet MS and MS/MS system.
	TS, D, F	RSVP	B	Low cost full feature sector, multiple inlet system.
	TS, D, F	MStation	B	Double focusing sector, multiple inlet system.
	TS, D, F	HX110A	B	High performance forward geometry, multiple inlet MS and MS/MS systems.
Micromass[5]	TS, D, F	Quatro LC	QQQ	Multiple inlet LC/MS/MS system (e.g., GC).
	TS, D, F	ZabSpec, AutoSpec; ProSpec Series	B	Multiple inlet MS and MS/MS systems (e.g., GC).

Abbreviations are summarized at the end of this chapter (see Notes).[2] Addresses and contact information are contained in *Appendix A, Sources of LC/MS Information*. Updated information on individual instrument manufacturers can be found at the Internet site: `www.lcms.com`.

Table 5.8- Manufacturers- CF/LC/MS

Company	Interface	Model	Analyzer	Comments
Bruker, Daltonics	CFF	Esquire	IT	Multiple inlet MS/MS system (e.g., LC and GC).
	CFF	APEX II, BioAPEX	FT	Multiple inlet systems (e.g., MALDI).
Finnigan	CFF	SSQ, TSQ	Q, QQQ	Multiple inlet MS & MS/MS systems (e.g., LC and GC).
	CFF	MAT 95's; MAT 900's	B	Multiple inlet MS & MS/MS systems (e.g., LC and GC).
JEOL	CFF	AX505	B	Small forward geometry MS and MS/MS.
	CFF	RSVP	B	Low cost full feature sector.
	CFF	MStation	B	Double focusing sector.
	CFF	HX110A	B	High performance forward geometry MS and MS/MS.
Micromass[5]	CFF	Quatro II	QQQ	Multiple inlet LC/MS/MS system (e.g., LC and GC).
	CFF	ZabSpec, AutoSpec; ProSpec Series	B	Multiple inlet MS and MS/MS systems (e.g., LC and GC).

Abbreviations are summarized at the end of this chapter (see Notes).[2] Addresses and contact information are contained in *Appendix A*, *Sources of LC/MS Information.* Updated information on individual instrument manufacturers can be found at the Internet site: `www.lcms.com`.

5.3-Locating Upgrade Alternatives

If you plan to upgrade an existing system, the first place to inquire is the original manufacturer (see *Appendix A*). The manufacturer is most familiar with the needed requirements or the possibility of upgrading a specific system. Alternatively, there are several companies that sell after-market LC/MS interfaces for various instruments (see Table 5.9 and *Appendix A, Sources of LC/MS Information*).

Table 5.9- Suppliers of LC/MS Upgrades

Company	ES	APCI	PB	TS	CF
Atlanta Research Laboratory Supplies	✓				
AMD Intectra	✓				✓
Analytica of Branford	✓	✓			
ASN Instruments					✓
BCP Instruments					✓
Hiden Analytical					✓
Kratos Analytical					✓
Mass Evolution					✓
Mass Spectrometry Intern.	✓		✓	✓	
M-Scan	✓				✓
New Objective	✓				
PerSeptive Biosystems				✓	
Phrasor Scientific					✓
Premier American Tecnologies	✓		✓	✓	✓
The Protein Analysis Company	✓				
Scientific Instrument Services					✓
World Precision Instruments	✓				

Abbreviations are summarized at the end of this chapter (see Notes).[2] Addresses and contact information are contained in *Appendix A, Sources of LC/MS Information*. Updated information on individual instrument manufacturers can be found at the Internet site: `www.lcms.com`.

5.4- Narrowing Your Choices

You may save some time by narrowing the number of contacts to those with the technologies that you need. Table 5.10 matches your interface and mass analyzer needs to instrument manufacturers that supply that technology. Use this table and *Worksheet 5.0* to record your prospective alternatives. This information may assist you in developing a shorter, more manageable evaluation list before going on to *Chapter Six.*

Table 5.10- Availability of Commercial LC/MS

Interface Alternatives	Mass Analysis Alternatives: Quads	Traps	TOF	Sectors	FTMS
AP	FI, HP, HI, JE, MI, PE, SH, WA	BR, FI, HP	AN, IS, PS, SE	FI, JE, MI	BR, FI
PB	FI, HP, MI, WA			FI, JE, MI	
TS	FI, HP, MI			FI, JE, MI	
CF	BR, MI	BR		BR, JE, MI	BR

AN- Analytica of Branford
AP- APCI & ES
BR- Bruker
CF- Continuous Flow FAB
FI- Finnigan
FTMS- Fourier Transform MS
HI- Hitachi
HP- Hewlett Packard
IS- IonSpec
JE- JEOL
MI- Micromass
PB- Particle beam
PE- Perkin-Elmer
PS- PerSeptive Biosystems
SE- Sensar
SH- Shimadzu
TOF-Time-of-flight
TS-Thermospray
WA- Waters

Worksheet 5.0 - LC/MS Contact Sheet

Part A- Prospective Contract Labs- Fill out the following checklist using the tables in Section 5.1 or other sources.

Are there contract labs in your local area?	Do their services meet your needs?	
1. Name ________________ # _________	Yes ❐	No ❐
2. Name ________________ # _________	Yes ❐	No ❐
Are there contract labs outside your local area?		
1. Name ________________ # _________	Yes ❐	No ❐
2. Name ________________ # _________	Yes ❐	No ❐
3. Name ________________ # _________	Yes ❐	No ❐

Part B- Prospective Manufacturers- Fill out the following checklist using the tables in Section 5.2 or other sources.

What companies sell your required LC/MS technologies?	Have you requested information?	
1. Name ________________ # _________	Yes ❐	No ❐
2. Name ________________ # _________	Yes ❐	No ❐
3. Name ________________ # _________	Yes ❐	No ❐
4. Name ________________ # _________	Yes ❐	No ❐
5. Name ________________ # _________	Yes ❐	No ❐

Part C- Prospective Upgrades-Fill out the following checklist using the tables in Section 5.3 or other sources.

What instruments in your lab can be upgraded?	Do they supply an upgrade pathway?	
1. Company __________ Model# __________	Yes ❐	No ❐
2. Company __________ Model# __________	Yes ❐	No ❐
Is there a third-party supplier of upgrades?		
a. Company __________ Model# __________	Yes ❐	No ❐
b. Company __________ Model# __________	Yes ❐	No ❐

Note: Copies of this worksheet can be downloaded from `www.LCMS.com`.

Discussion Questions and Exercises

1. What specific information should you exchange with a prospective contract lab?
2. What specific information should you exchange with a prospective instrument manufacturer?
3. What specific information should you exchange with a prospective upgrade supplier?
4. Make your contact list (Worksheet 5.0) and phone, e-mail, or fax for information about their products and services.
5. Browse the Internet to obtain information about your prospective suppliers of LC/MS capabilities. Begin at `www.lcms.com`

Notes

[1] The LC/MS instrument market has contracted from six or seven major suppliers to four in recent years: Finnigan (ThermoQuest), Hewlett-Packard, Perkin-Elmer, and Waters. In recent years, Thermo has bought Finnigan, Fisons, Extrel-FTMS; HP is co-marketing instruments with Bruker; PE has a joint venture with Sciex, and has bought PerSeptive Biosystems; and Waters has bought Extrel and Micromass.

[2] **Abbreviations:**

API	Atmospheric pressure interface for; APCI & ES
APCI	Atmospheric pressure chemical ionization LC/MS
B	Magnetic sector mass analyzer
CE/MS	Capillary electrophoresis (electrospray)/mass spec
CF	Continuous flow FAB
CI	Chemical ionization
CTS	Corona discharge thermospray LC/MS
D	Discharge thermospray
DP	Desorption probe for EI and CI
EI	Electron ionization
ES	Electrospray LC/MS (may also include IS & NS)
F	Filament thermospray

FAB	Fast atom bombardment
FT	Fourier transform mass analyzer
GC	Gas chromatography
GC/MS	Gas chromatography/mass spectrometry
ICP	Inductively coupled plasma
IS	Ionspray or pneumatically assisted electrospray
IT	Ion trap mass analyzer
LC	Liquid chromatography
MAGIC	Monodisperse aerosol generator interface for chromatography
MALDI	Matrix assisted laser desorption ionization
MS	Mass spectrometry
MS/MS	Tandem mass spectrometry; mass spectrometry/mass spectrometry
na	Not available
NS	Nanospray (electrospray) LC/MS
PB	Particle beam LC/MS
PPB	Pneumatic nebulization particle beam LC/MS
Q	Quadrupole mass analyzer
QQQ	Triple quadrupole mass analyzer
TES	Thermally assisted electrospray
TOF	Time of flight mass analyzer
TPB	Thermal nebulization particle beam
TS	Thermospray LC/MS
UPB	Ultrasonically nebulization particle beam LC/MS
US	Ultraspray or ultrasonically assisted electrospray

[3] Lammert, S.A., "1997 directory of mass spectrometry manufacturers and suppliers," Rapid Commun. Mass Spectrom. 11, pages 821-845 (1997). This is an annual directory of companies that supply mass spectrometers systems (GC/MS, LC/MS, MS, MS/MS), MS components (electrical supplies, vacuum chambers, interfaces and add-ons, custom modifications), services (repair, cleaning, maintenance, software, analyses), and non-conventional mass analyzer (e.g., QQ-IT, BE-TOF) and LC/MS configurations (e.g., CF-IT).

[4] Waters Corporation purchased Micromass, Ltd. in late 1997.

[5] PerSeptive Biosystems merged with Perkin-Elmer Corp. in 1997. PerSeptive will operate as a division and subsidiary of Perkin-Elmer Corporation.

Chapter Six- Flow Diagram

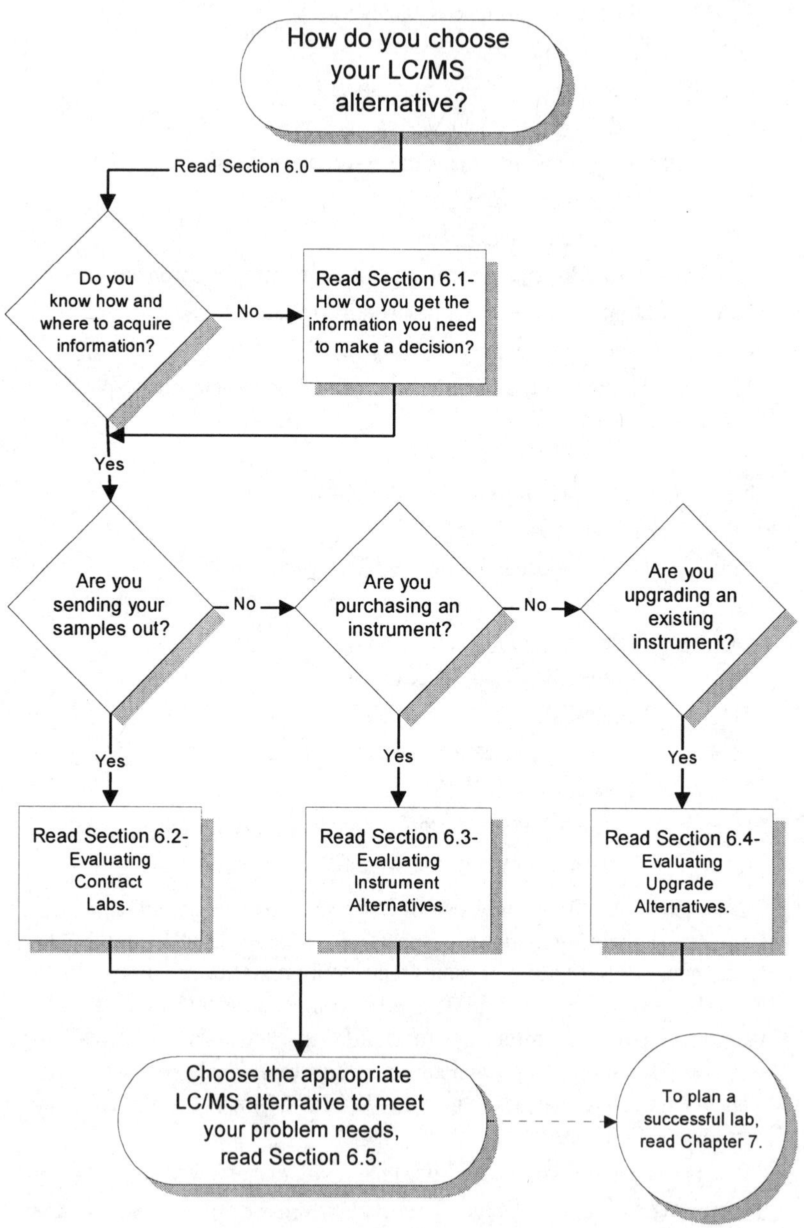

How do you choose your LC/MS alternative?
Read Section 6.0
Do you know how and where to acquire information?
No
Read Section 6.1- How do you get the information you need to make a decision?
Yes
Are you sending your samples out?
No
Are you purchasing an instrument?
No
Are you upgrading an existing instrument?
Yes
Yes
Yes
Read Section 6.2- Evaluating Contract Labs.
Read Section 6.3- Evaluating Instrument Alternatives.
Read Section 6.4- Evaluating Upgrade Alternatives.
Choose the appropriate LC/MS alternative to meet your problem needs, read Section 6.5.
To plan a successful lab, read Chapter 7.

Chapter Six

Choosing Your LC/MS Alternatives

(or, Selecting your best alternative)

Having a compelling need to change is a better predictor of success than any given methodology chosen to drive the change.[1]

Barry Sheehy, Hyler Bracey, and Rick Frazier

6.0-What Are The Steps?

Acquiring LC/MS capabilities requires more than simply justifying your needs. Whether you plan to acquire in-house capabilities or send your samples out, you eventually have to decide on a specific instrument or a specific contract lab. The major steps of this process are outlined in Figure 6.0. We divide this process into four steps:

1.) a *justification* step,
2.) a *search* step,
3.) an *evaluation* step, and
4.) an *implementation* step.

Each step is separated by one or more important decisions. This process is the same as any other problem-solving process in which your ultimate success is dependent upon how well you define your problems and needs.[2]

We have already discussed the justification step in *Chapter Four*. In *Chapter Five* we addressed the directions that you should pursue in your search for specific alternatives in LC/MS. The last two chapters of Part II of this book deal with the evaluation and selection of your specific alternatives (*Chapter Six*) and the implementation (*Chapter 7*) of your choices. These two chapters may be the most important of the book in the sense that the actions you take while acquiring your LC/MS capabilities will have a permanent effect on your ability to solve each specific analytical problem that faces your lab in the future.

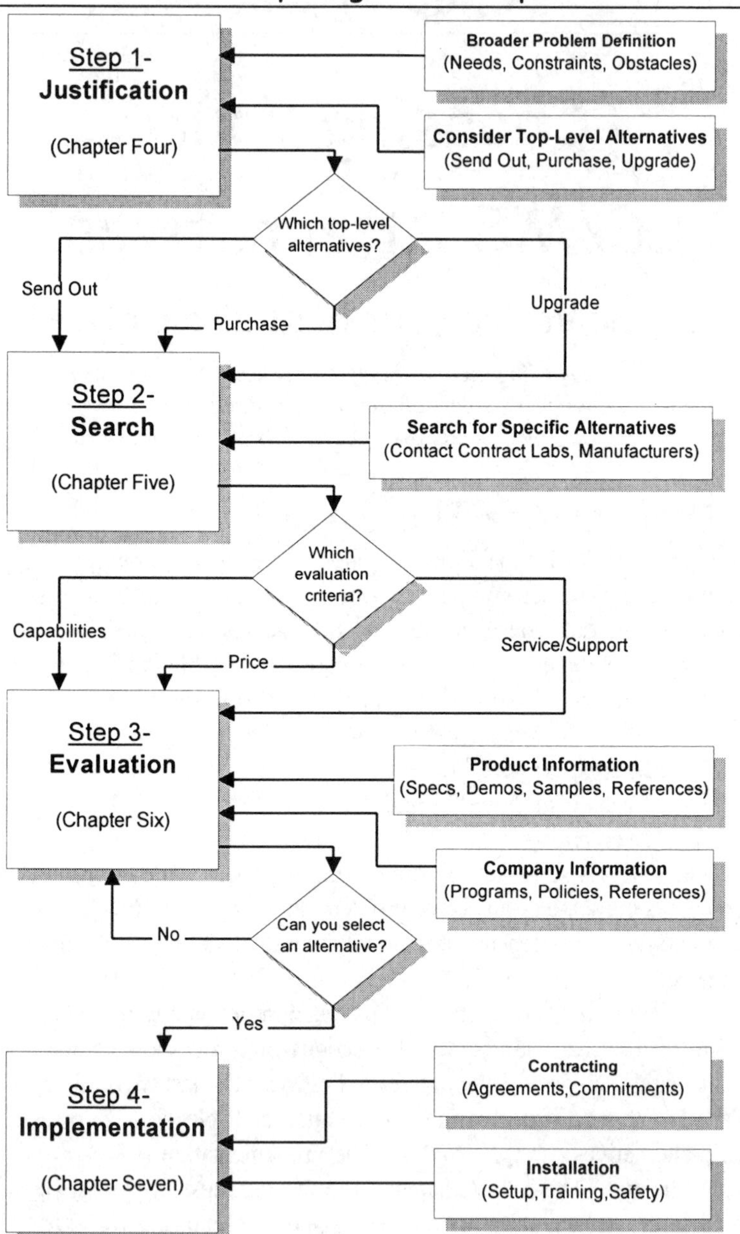

Figure 6.0 The steps to acquiring LC/MS capabilities.

6.1-How Do You Get Enough Information?

The evaluation of any alternative for acquiring LC/MS capabilities should always begin with the gathering of information about the specific products or services that fall within the scope of your problem definition (*Worksheet 4.0*). This can be done by contacting any of the LC/MS resources listed in *Chapter Five* and *Appendix A*. Complete and accurate information is your key to conducting a thorough and effective evaluation. During your process of searching for information you should be careful not to select a specific alternative until you have investigated and considered all practical alternatives. You can, however, eliminate particular alternatives at the outset if they are deficient in some critical parameter (e.g. they lack a critical capability, they are not compatible with your existing equipment, etc.).

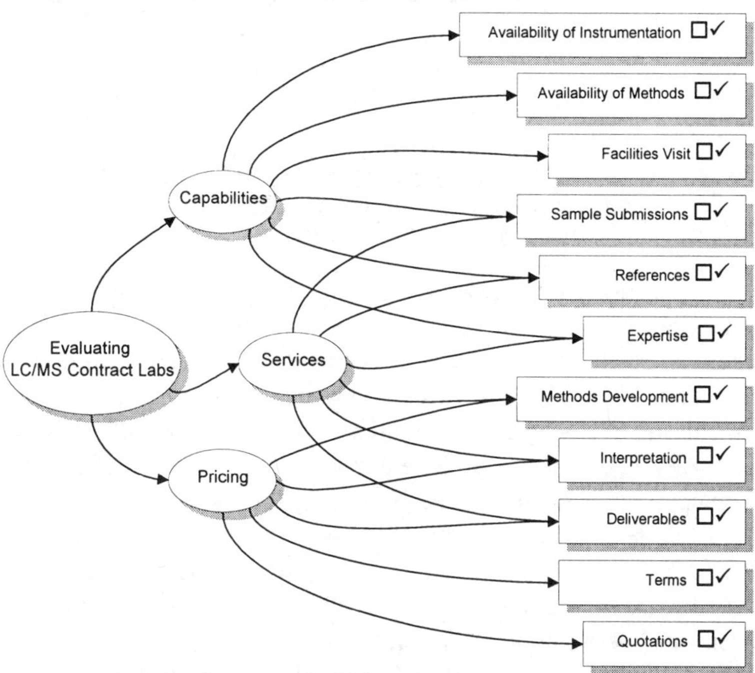

Figure 6.1- Evaluating LC/MS contract labs requires that you weigh the capabilities, ancillary services, and pricing for each lab under consideration.

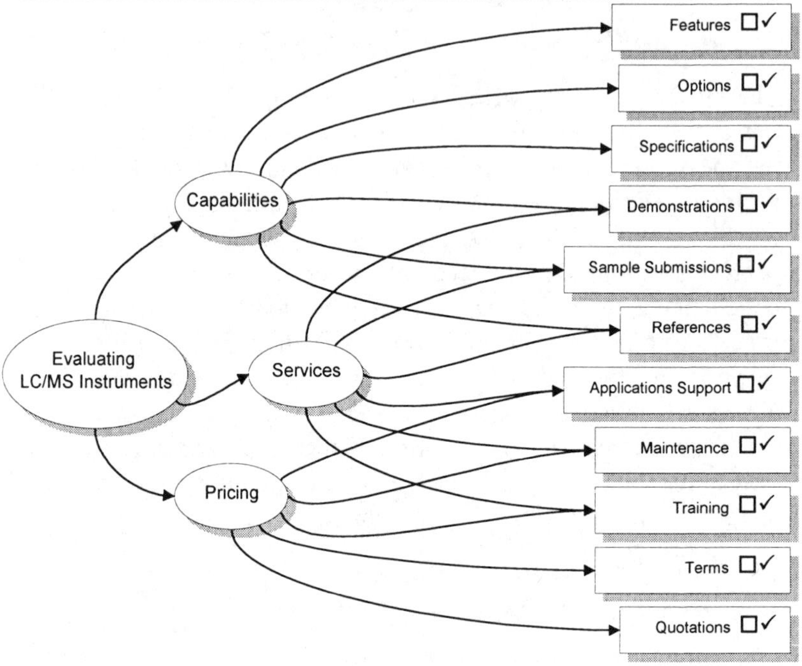

Evaluating LC/MS Instruments
Capabilities
Services
Pricing
Features
Options
Specifications
Demonstrations
Sample Submissions
References
Applications Support
Maintenance
Training
Terms
Quotations

Figure 6.2- Evaluating LC/MS products requires that you weigh the product capabilities, ancillary services, and pricing for each alternative under consideration.

In general, each alternative for LC/MS can be evaluated by gathering information about the *capabilities*, *services*, and/or *prices* as related to your needs. Some of the sources of information are listed in Figure 6.1 (for Contract Labs) and Figure 6.2 (for Purchasing LC/MS Instrumentation). Most of the information you need to evaluate a given alternative will come directly from the supplying company itself. Requesting specs, quotations, standard terms, delivery schedules, company information, reference lists, applications notes, etc., is readily obtainable by contacting the company and/or meeting with a sales representative. You may even need some information from several alternatives early in your *justification* process. Instrument demonstration, visits, evaluation sample analysis, contacting references, are generally performed later in the evaluation process.

6.2-Evaluating Contract Labs

You have justified your need for outside LC/MS sample analysis. It may be because of your in-house capacity or demand. It may be because of your in-house expertise, or lack of expertise. Now you want to select one or more contract labs to meet the needs of your organization. The time and effort you spend on evaluating contracts labs should be commensurate with the number of samples you intend on analyzing. If you have a few samples, finding a local lab or university with the capability to run them is probably quite adequate to meet your needs. As your sample load increases, your risk goes up and therefore your evaluation effort should also increase. Table 6.0 reviews some of the pros and cons of sending out. These considerations should be weighed during your evaluation of each lab under consideration.

Table 6.0- The Pros and Cons of Sending Out

Consideration	Pro	Con
Availability	• Almost any LC/MS technology is available through a contract lab. • You can access very diverse and complex technologies without having to acquire them.	• Some technologies may not be available for rapid sample turnaround. • You may have to send your samples a considerable distance to get the type of information required to solve your problem. • Communication and interaction is usually less effective when dealing with remote locations.
Familiarity	• You don't have to be an expert in LC/MS; you can rely on the expertise of the contract lab personnel.	• Knowledge and effort is required to ensure yourself that the results you obtain from the contract lab meet your problem needs.
Performance	• Many experts are available for solving your problems within the contract laboratory community.	• You can't assume everyone in the contract lab business is an expert.
Price	• Sending out is the least expensive way to perform an analysis on a single sample.	• Sending out may be the most expensive way to analyze a large number of samples.

Evaluating Contract Lab Capabilities

What instrumentation is available at each lab under consideration? What techniques are available? What methods? What expertise? What sample volume? What turnaround times? Acquiring information about the capabilities of each lab is essential to a thorough evaluation. The necessary information to assess the capabilities of a given lab can be acquired in a number of ways. We recommend the following:

- **Request Information**

  Contact each contract lab and request all available information concerning capabilities and services (*Chapter Five* and *Appendix A*). Most labs have brochures that detail their instrumental and experimental capabilities. Some labs have home pages on the Internet. Larger labs have sales and marketing departments that can provide you with a wide variety of information.

  A formal SOQ (Statement of Qualifications) is available from many labs, particularly those with environmental backgrounds. This is a document that lists equipment, services, and personnel qualifications.[3]

  Also request a list of SOPs (Standard Operating Procedures) and a list of methods with validation information (if available). SOPs will give you an overview of the labs GLP and GMP compliance. A list of methods will give you information about the lab's experiences.

  General information provided in brochures, SOQs/SOPs, or from the Internet is adequate for screening potential labs. If a lab lacks the specific technologies required to address your problem needs you don't have to waste time evaluating its expertise. This information, however, may not be adequate to make your final choice.

  Also note that most industries that routinely send samples to contract labs do not typically use a single lab; rather, they will deal with three or four different contract labs. In this manner they can have some level of redundancy in their capabilities and increase their likelihood of a rapid turnaround. While one lab may be busy, another qualified lab may be available. Consider this when gathering information about labs. You may choose to evaluate several labs to meet your needs. Your evaluation process may not end with one lab being selected and every other lab being rejected, rather, you may end with a number of qualified labs that you can use as demands warrant.

- ## Visit Their Lab

A picture is worth a thousand words; seeing a lab can tell you a great deal about the quality and proficiency of the organization. Deficiencies in basic operating processes are easy to spot if you have the opportunity to see the lab, particularly in a production environment. Production labs have to be well-managed and well-organized to handle high throughput and a continuous stream of deliverables. Every aspect of analysis should be documented with clear procedures or protocols. Safety, security, chain-of-custody, and expertise of lab personnel are all issues that are better addressed in person.

By visiting a lab you have the opportunity to interview the managers and analysts that will be responsible for your samples. These interviews are an opportunity for you to establish a relationship with the lab organization. The quality of your relationship with these individuals may have a large bearing on your ability to solve problems using their services. Do you communicate effectively with these people? Do they seem flexible in working with you? Do they seem responsive to your needs?

It should also be noted that some labs (e.g. university labs) are not strictly a production environment, but tend to focus on qualitative analysis and interpretation. Many of these labs may not have the same emphasis on procedure and protocol as production labs; however, they are still capable of excellent problem-solving. Note, however, that your problems may not alway be related to high throughput. Non-production labs may be better equipped to solve your qualitative problems. You should always consider how the specific lab matches your specific needs.

- **Contact References**

  An excellent method for gaining information about the capabilities of a lab is to contact its existing customers. Most labs should be willing to provide you with several references. You should request references with problems and sample types similar to yours (e.g. clinical, environmental, industrial).

  Make contact with one or more references. You should review with each reference any critical issues relating to your specialized needs. Ask specific, not general questions (Have they met all their deliverables? Have they required any rework? Are their deliverables accurate and complete?) Try to construct your questions to coincide with your problem and sample requirements. When interviewing more than one reference, try to ask the same questions? Are their responses consistent? Are there important areas of inconsistency?

- **Request Credentials**

  The credentials of the personnel in the lab will have a significant bearing on your analytical results. Request the credentials of each relevant analyst in the lab, particularly the staff that will be directly involved with your samples. It is reasonable to expect a brief biography of each member of the lab staff with relevant experience in LC/MS and sample analysis. It is also reasonable to request a list of publications and presentations of the members of the lab. This information may be more informative in certain situations compared to others.

  It is prudent to match the skills and background of the members of the lab to your specific problems. Remember, problems are solved with people; even if the lab is adequately equipped to solve your problems, the need for qualified personnel is equally important to problem-solving.

  In addition to personnel, laboratory credentials are also a prerequisite to properly solve many problems. Depending on your particular industry, there may be a wide variety of national, state, local, and industry certifying and accrediting organizations. Table 6.1 lists some examples of these organizations. Information on many of these accrediting organizations and agencies are in *Appendix A, Sources of LC/MS Information*.

In addition, make sure the lab facility has the certifications and accreditations required to meet your analytical needs. Today, more and more labs are operating under the protocols of GLP and GMP (Good Laboratory Practices and Good Manufacturing Practices).

Table 6.1- Examples of Certifying Organizations

Certification/Accreditations and Proficiency Testing	
International	• ISO/IEC Guide 25-1990
National	• National Laboratory Certification Program (Dept. of Health and Human Services) • U.S. Drug Enforcement Admin. License • EPA • FDA • NIDA • OSHA • Department of Transportation • Permit to Receive Soil (Department of Agriculture) • Certifications for Specific EPA Methods • GLP Guidelines (FDA, EPA and Toxic Controlled Substance) • Medicare Provider
State	• Department of Health • Department of Human Services • EPA
Professional	• College of American Pathologist/American Association for Clinical Chemistry (CAP/AACC) • CAP/AACC Forensic Urine Drug Testing • CAP, Toxicology Proficiency • CAP, Therapeutic Drug Monitoring Proficiency

This list was compiled in part from MEDTOX Scientific Inc. (St. Paul Minnesota; home page: `www.medtox.com`).

- **Request Evidence of Related Experience**

  If you are evaluating a lab's ability to perform a given task, you would like to know that it has some experience in performing that task. A lab should be able to demonstrate a history of running samples of your type, particularly if the number of samples you are sending happens to be large. Even with a new lab, the personnel running the samples should have some experience with your sample types. Many labs will publish lists of methods that they have developed and validated. Tables 6.2 and 6.3 are examples of the type of information labs provide to their customers.

Table 6.2- LC/MS/MS Validated Methods

Analyte	Method	LLQ (ng/mL)	ULQ (ng/mL)	Matrix
Morphine	ES, QQQ	0.500	50.0	Human plasma
Morphine-3-glucuronide	ES, QQQ	2.00	200.	Human plasma
Morphine-6-glucuronide	ES, QQQ	2.00	200.	Human plasma
Perphenazine	APCI, QQQ	0.050	3.0	Human plasma

Table 6.3- Method Summaries

Item	Criteria
Analyte	• Name of compound
Quantitation	• LLQ, ULQ • Correlation coefficient • Imprecision and Inaccuracy
Detection	• LLD
Matrix	• Matrix type (plasma, serum, tissues; plant, animal) • Verified free of spectral and chemical interferences • Volume
Sample Storage Preparation Analysis	 • Conditions, type of freezers • Solid-phase extraction • APCI LC/MS/MS
Sample Stability	• Room and storage temperature • In solvent • In the matrix • Freeze/thaw cycle

LLQ: lowest level of quantitation; ULQ: upper level of quantitation; LLD: lowest level of detection; APCI: atmospheric pressure chemical ionization; MS/MS: tandem mass spectrometry. Adapted from Maxxam Analytics, Inc. (May 1997).

- **Send Evaluation Samples**

  Demonstrating the ability to run a few samples may indicate a lab's ability to run hundreds of samples. An effective method of evaluation is to send *evaluation samples* to the lab for analysis. You should send the same samples to all labs under consideration to see how each performs. You should be willing to pay for these analyses.

  An effective way to evaluate the routine performance of a lab is to send the samples and instructions as to how they should be run (e.g. storage, methods, deliverables). Do not inform the lab that they are evaluation samples. In this manner you will get a true representation of the lab performance. If the samples are run under the expectation of a larger account, a lab may focus extraordinary effort on the samples just to gain your business.

Evaluating Contract Lab Services

Sending out samples to a contract lab may mean much more than simply running your samples. Many other activities may be required in order to solve your problems.

Many contract labs are set up with the specific intent of helping you solve analytical problems. This generally means that they provide higher levels of service and support than just running your samples. Others labs are set up to run a specific technique irrespective of the sample characteristics (This is fine for some problems!). The level of service and support may be as important to you in your evaluation as their analytical capabilities.

- **Will the lab adapt or develop methods?**

  For most samples, particularly when quantification is required, a valid method is a prerequisite to running your samples. If you intend to send your samples to a contract lab, the lab may find it necessary to develop and validate a method for your target analyte and matrices before your samples can be run. Does it provide these services?

  In some cases, a contract lab may be required to adapt an existing published method to your particular samples. At a minimum, it may have to validate the method under its experimental conditions.

These activities can be time-consuming, labor intensive and expensive.

- **What part of the analytical chain will they support?**
You should not only evaluate the instrument technologies and the expertise of the staff but also their complete analytical services. Some contract labs are equipped to handle all aspects of sample analysis, from specialized storage, to all aspects of sample preparation and pretreatment, to interpretation and report creation. The ability of a lab to provide these extended services will generally improve the quality of your results.

Some of the larger contract labs supporting the pharmaceutical industry provide complete Phase I through Phase III clinical trial - including, patient recruitment, method development, and sample analysis.

Table 6.4- Examples of Pricing from Contract Labs

Technique	Analysis	Fee/Hr
Probe	FAB	
	Low Resolution	$125
	Peak Match	150
	EI	
	Low Resolution	125
	High Resolution	150
	Peak Match	150
	CI	
	Low Resolution	150
	High Resolution	150
	MALDI	125
Flow Injection	ESI	
	Low Resolution	100
	High Resolution	150
LC/MS	ESI	150
	(separation method provided by user)	
MS/MS	(& other specialized experiments)	125

Information compiled in-part from various sourses (e.g., Washington University Center for Biomedical and Bioorganic Mass Spectrometry home page (`wunmr.wustl.edu/~msf`); Carnegie Mellon University's Center for Molecular Analysis (`www.chem.cmu.edu/cma`).

Evaluating Contract Lab Pricing

What is the pricing for each contract lab under consideration? Each contract laboratory may have very different pricing schemes depending on the nature and specialties of its lab. Most labs are willing to price the work in a variety of ways: per sample, per hour, per service and per project; depending on your needs. When comparing labs, you should make sure that you are comparing *apples-to-apples*.

One lab may have a very low price per sample but require a minimum number of samples (e.g. 1000). Other labs may have graduated pricing scales that depend on the number of samples. A good rule of thumb is that a contract lab would like to make $2,000 to $4,000 per day per instrument.[4] If your one sample takes all day, expect to pay several thousands of dollars. If you have 20 samples with fifteen-minute run-times, expect to pay for an entire days worth of instrument time.

- **What is their pricing per sample?**

  Many standard LC/MS techniques can be easily priced on a per-sample basis. Many labs, including university labs, will have pricing schedules for a specific type of analysis. Table 6.4 shows an example of a pricing schedule for a university lab. These schedules usually apply to samples that are routine and require limited sample pretreatment.

  The for-profit LC/MS contract service market is very price competitive. Prices can run from $25 to $100/sample for a flow injection LC/MS analysis of a single (fairly pure) sample, to $1,000 or more for advanced work, such as protein digestion, isolation and mass spectral identification.[5]

  For method development and subsequent LC/MS analysis, the price may be quoted on a per sample basis for a specific lot of samples, such as $125 to $250/sample for 1,000-2,000 samples and require 2-4 weeks to develop a method.[6] Method development may even be an extra charge, say $10,000 to $20,000.

- **What is their pricing per hour for specialized services?**
  Non-routine samples and methods-development are typically charged on an hourly basis. Hourly rates also apply to problems that have a degree of uncertainty. Typical hourly fees run from $50 to $250 per hour for various types of service and support. It is prudent to set a limit on the number of hours that a lab applies to a given problem and to define interim review points along the way to ensure some progress toward a solution. Otherwise, you could invest significant dollars toward a problem and end up with no results. You should have a high confidence level in the ability of the lab to solve a problem (or a track record) before committing significant dollars to hourly rates.

- **Will they bid on a project basis?**
  Some projects involving methods-development and sample analysis are quoted on a project basis. This requires that the lab have a clear idea of the total project effort: methods development, validation, sample storage, analysis, interpretation, and reporting. This may be a simpler method for both the lab and you, the sample provider.

  You and the lab can both benefit from this approach if the project definition and specifications that you provide are clear and realistic, and the estimations of the lab are accurate. But be cautious, these two conditions are rarely easy to accomplish.

- **Do they itemize pricing for each specific task?**
  Some labs may have a low price per sample rate, but charge itemized fees for every service provided from shipping to photocopying. This approach is an attempt to capture all expenses in an itemized manner. This is not unreasonable as long as you are aware of their pricing structure and invoicing policies upfront. Realizing this, you can make a fair comparison with other companies who bundle many of their expenses into a single fee.

Evaluating Contract Labs

Use *Worksheet 6.0* to evaluate the individual labs. Use additional sheets if required.

Worksheet 6.0-Evaluating Contract Labs

Name of Contract Lab or University:

____

Part A-Fill out the following to evaluate the capabilities of the above lab. Does the lab meet or exceed the requirements of your analysis?

Lab Capabilities	Comments	Yes	No	Exceeds
Does the lab have the instrumentation?	____________	❑	❑	❑
Is the lab properly equipped & operated?	____________	❑	❑	❑
Is the lab capable of running your samples?	____________	❑	❑	❑
Will the lab provide you with any referrals?	____________	❑	❑	❑
Does the lab have the required expertise?	____________	❑	❑	❑
Does the lab have the required certifications?	____________	❑	❑	❑

Part B- Fill out the following to evaluate the services of the above lab.

Lab Services	Comments	Yes	No
Does the lab develop methods?	____________	❑	❑
Does the lab interpret results?	____________	❑	❑
Does the lab provide additional services?	____________	❑	❑

Part C- Fill out the following to evaluate the pricing of the above lab.

Lab Pricing	Amount	Comments
Price per sample?	________	____________
Price per hour per service?	________	____________
Price for special services?	________	____________
Are there additional or hidden charges?	________	____________

Note: Copies of this worksheet can be downloaded from `www.LCMS.com`

6.3-Evaluating Instrument Purchases

You have justified your need to purchase an LC/MS instrument. There is a clear need to perform the analysis in-house. You have both the funding and the resources to support your acquisition. Now you must consider which alternatives are best-suited for your specific needs. The most common method of purchasing LC/MS is to acquire new instrumentation from the manufacturers listed in *Chapter Five*; however, there is a considerable after-market for mass spectrometers. Some after-market sources are provided in *Appendix A*. Table 6.5 reviews some of the pros and cons of purchasing. These should be considered when evaluating your purchasing alternatives.

Table 6.5- The Pros and Cons of Purchasing

Consideration	Pro	Con
Availability	• Purchasing an instrument gives you the opportunity to acquire the highest performance and newest technologies.	• Purchasing can have long lead times.
Familiarity	• The trend with software is toward more intuitive interaction, requiring less training. • The trend with hardware is to provide simpler interaction, eliminating tools and adjustments.	• You will require training on all new instrumentation.
Performance	• Generally, new equipment will have the highest performance.	• The level of performance will be highly dependent upon your expertise.
Price	• Purchasing is the most cost effective alternative for high sample loads; the price performance trend is downward.	• The starting point for new LC/MS instrumentation is $100,000; this may be prohibitive for many LC labs with much lower capital budgets.[7]

Evaluating Instrument Capabilities

What technologies are available commercially? Which technologies match your problem needs? When contacting a prospective instrument company you should acquire as much information about both the company and its instrumentation as possible. Get to know your local sales representative.[8] Most manufacturers have home pages on the Internet that provide contact and product information (See *Appendix A*).

We recommend the following evaluation activities when evaluating the purchase of one instrument relative to another:

- **Request Information**

  Contact each instrument company and request all available information concerning its products and services *(Chapter Five* and *Appendix A*). Most companies will provide detailed product descriptions and specifications. Most will also provide a wealth of published reference and applications data to support various applications of their instrumentation for real problem-solving.

  You should also request a quotation for the system(s) that are under consideration. Have your sales representative line-item all options and accessories that relate to your needs. This quotation will rarely be the final version, but is a good starting point for comparison. In some cases, your purchasing department may be making this request; if they are, obtaining a copy of the information on the quotation is useful for your evaluation as well.

  Try to sort out all the product information in a manner that is understandable to you. Sometimes features are confused with specifications. Unlike actual performance specifications, features are often artifacts of marketing.[9] A feature is an innovative design or special function that may or may not enhance the analytical performance or operational ease of a particular instrument. In some cases a particular feature may assist you in your problem-solving; in other cases you may need a certain performance characteristic (as defined by the specifications) to address your needs. Learn what specifications are of value to you and the difference between what is a feature and what is a specification.

- **Send Evaluation Samples**

It is customary to send samples to prospective instrument companies to evaluate the performance of their instrument compared to competitive systems. Samples are sent to instrument companies for a variety of reasons. First, some data may be required for justification of an instrument for a given problem or application. In these situations, the sample data may be used in an internal proposal or funding agency proposal. Other samples are sent to evaluate the performance of one instrument relative to another. In rare cases, samples are sent to instrument companies to pass a performance requirement for shipping a new instrument to the customer. Irrespective of the reason the samples are sent, they should always have the intended purpose of measuring performance of the instrument for an intended application or use.

When sending samples to different manufacturers in a competitive evaluation, detailed and specific information about the experimental and operational conditions should be communicated as well. In some cases it is recommended to send closely matched columns to each manufacturer to ensure that the separation conditions are identical. If you rely on manufacturer's columns, column-to-column variability may enter an added degree of uncertainty to the evaluation.

- **Demo Their Equipment**

An important part of the evaluation process for instrument purchases is to personally observe a demonstration ("a demo") of the instrument before making your decision. The demo can provide you with information about the usability and performance of an instrument. The demo is also an opportunity for you to assess the company that will supply your instrument, particularly if you have an opportunity to visit the manufacturing facility of the instrument company.

At a demo, you can meet the management and staff responsible for applications support, service and repair, training, and installation. Having an opportunity to assess the entire support organization, as well as the product, is important in your purchase evaluation process.

During a demonstration you should evaluate the entire analytical system under your intended operating conditions. Tables 6.6 and 6.7 list some of the operational and performance areas that should be considered during the demo. In some cases you can specify (within reason) the samples and conditions of the evaluation. It is customary to send selected samples to the demo lab to evaluate the performance of the instrument during your visit. You should avoid research samples that are associated with a high degree of uncertainty or samples that have "unknown" composition. During the demo you should discuss your needs (*Worksheet 4.0*) and question the application chemist or sales representative on how the instrument will meet them.[10]

You should also avoid problems (samples) that test the skill of the lab's application chemist rather than the performance of the instrument. It is in your best interest to assess the performance of the instrument during the demonstration, rather than the chemist's skill. Save your research problems until after you have acquired your instrument.

In general, picking a well-characterized problem in your field in order to challenge the instrumentation performance is reasonable at a demonstration. For example, in a peptide or protein application, you may select a known high molecular weight protein and an enzymatic digest of a known protein. Evaluate the performance of the instrument based on accepted standards and parameters such as sensitivity, resolution, mass range, scan speed, etc. (Review *Chapter Three*).

It is good to remember that knowing what an instrument cannot do is often as important as knowing what it can do.[11] Avoid making negative comments about its products or its competitors' products; you are a guest at their company and should treat the staff with respect. If you have specific evaluation needs, try to be constructive in discussing how the product meets (or fails to meet) your needs.[12]

Table 6.6- Operational Items

System Component	Items to Evaluate at Demos Relating to Operation
Sample Treatment	• Compatibility with autosamplers • Compatibility with automated sample preparation • Check maximum sample rates • Check maximum and minimum sample sizes • Check on expected dead volumes • Check on fault recognition
Separations	• Compatibility with HPLC solvent, flows, etc. • Compatibility with CE (if applicable) • Check on dispersion and couplings • Check on use with in-line ancillary detectors
Interface/Ionization	• Review types of interfaces, nebulizers, options • Review types of ionization modes and polarities • Review how to apply these parameters to your problems
Mass Analyzer	• Review type(s) of mass analyzer • Review and test performance • Evaluate sensitivity and detection limits • Evaluate mass range, scan rates, resolution, mass accuracy
Data Collection, Data Reduction, and Instrument Control	• Review instrument setup- manual/automated tuning and calibration, optimization, suitability testing • Review data collection modes- full profile, stepping • Review acquisition modes- batch, method, loops, timed windows, timed events • Review quantification modes- internal, external, isotope dilution, standard addition (see *Chapter Ten*) • Review real-time display- range, # windows, tracing • Review multitasking capabilities • Review data reduction- algorithms, raw data access, reduced data access • Review libraries- public, user built • Review file structures- formats, organization, transfer • Review multi-channel display- UV, LC/MS pressures • Review data review and integrity- access, audit trails, time-stamps, reintegration
Reporting	• Review report formats- standard, customized • Review data export- compatibility, formats • Review graphics export- compatibility, formats • Review data

Table 6.7- Maintenance and Repair Items

System Component	Items to Evaluate at Demos Relating to Maintenance/Repair
Sample Treatment	• Check on needle maintenance and reliability • Check on carryover • Check on seals, septa, and general maintenance • Review materials and solvent computability
Separations	• Review access to LC • Review plumbing • Review calibration procedures • Review maintenance items and schedule • Review spare parts- heads, seals, valves, unions
Interface/ Ionization	• Review changeover- procedures, time, complexity • Review cleanliness and cleaning procedures • Review maintenance items and schedule • Review filament repair or replacement
Mass Analyzer	• Review access • Review maintenance items and schedule • Review detector replacement and longevity • Review vacuum system access and maintenance • Review trap cleaning • Review gas connection and installation
Data Collection, Data Reduction, Instrument Control	• Review diagnostic software • Review general troubleshooting of instrument • Review data backup, storage, and retrieval • Review all system testing procedures • Review computer maintenance
Reporting	• Review printer options and maintenance • Review diagnostic reports • Review system suitability reports

When should you skip the demo? In some cases people who have a clear knowledge of their product needs do not have to have a demonstration of the instrument. Established products with repeat orders are an example. You don't have to spend the time and resources to evaluate something that you have already established as a choice in your lab. The risk is that you miss changes in the marketplace and technology. In a rapidly changing field like LC/MS you should at a minimum contact alternative manufacturers and inquire about the capabilities (and pricing) of their instruments in respect to your analytical and budgeting needs.

- **Contact References**

  Obtain a list of current customers. Review the list and call a couple of customers who you believe have LC/MS needs similar to your own. Use the telephone conversation to get a feel for the instrument, support services provided, potential problem areas, and likes or dislikes.[13] Remember, everyone has unique prejudices and work environments. Each instrument manufacturer has its fans and its critics. The manufacturer's list may contain only "qualified references."[14]

  If you know of other labs in your industry with a given instrument, contact them as well.[15] Try to gain a balanced point-of-view from references.

Evaluating Instrument Company Support

When evaluating an instrument purchase, the organization supplying the instrument will have a great deal to do with your ultimate ability to solve problems with its instrument. Note that support is not just fixing a problem; it's answering a phone and giving advice, its training operators to understand and maintain their own equipment and solve problems.[16] Services that are critical are maintenance and repair, application's support, and training services. Most of the suppliers of instrumentation will provide you with written service and support policies, as well as training programs. Every company has a vested interest in providing you with a wide variety of programs, from start-up to advanced.[17] In general, the larger the company, the more variety you will find in these programs.

Evaluating Instrument Pricing

Instrument price, for some, is the only purchase criteria. But this is not necessarily correct. As suggested in the *Analytical Consumer*, the price of an instrument should take a secondary role in deciding which instrument to buy.[18] Other dynamics come into play: service support, applications support, ease of software, instrument size, lab politics, etc.

Unfortunately, a finite sum of money is typically budgeted for the purchase; therefore, it is necessary to weigh price against performance and versatility. The manufacturer's price lists are useful for budgeting and determining the cost of accessories. Table 6.8 lists the price ranges for various LC/MS instruments.[19]

Table 6.8- Typical price ranges for LC/MS(/MS)

Analyzer	Price	Comments
Quadrupole	$100K-125K	• Fully integrated benchtop "Detector" systems.
	150K-200K	• Floor model & benchtop-style systems.
	300K-400K	• LC/MS/MS systems.
Ion Trap	150K-250K	• Benchtop MS/MS systems.
TOF	150K-250K	• LC/MS benchtop system.
	300K-500K	• LC/MS/MS.
Sector	250K-400K	• LC/MS/MS systems.
	400K-500K	• LC/MS/MS, with various configurations.
FTMS	500K-1,000K	• LC/MS (MS/MS) systems.

Sales Agreement

The last step in the purchase of an LC/MS instrument is negotiating the final price and sales contract. It is here that all discussions about technology vanish, and discussions concerning business and legal matters take over. We will not concern ourselves in this text with the details of contract negotiating and the purchasing of capital equipment. We will leave this to the financial, purchasing and legal experts of both groups (your company and the instrument company). For general information concerning laboratory management and financial matters see *Appendix A, Sources of LC/MS Information.*

We suggest that during contract negotiations is when you should voice your desires about accessories, spare hardware, upgrades, delivery times, payment and training. The bottom line is to get the instrument (and company) that best suits your needs for the best price.[20]

Leasing

An alternative to purchasing an instrument is leasing. Today, more companies are choosing the option of leasing computers and instrumentation than purchasing the systems outright. The particular combination of reasons that make a firm decide to lease or purchase has grounding in basic economics (see Table 6.9).[21] For some, the reason is a lack of cash; there is no other choice, it is either lease or do without. Unfortunately, this is often the situation for smaller firms, and the decision then is to find the best leasing arrangement, tailored to meet the specific needs of the lessee. Larger firms are currently weighing quantifiable economic parameters as well as those which are not so easy to label with a dollar value (technical or market obsolescence).[22]

Table 6.9- Economic Reasons to Lease

Why Lease?	
• Offers potential savings compared to purchasing. • Provides an alternative source of capital. • Provides constant-cost financing. • Extends length of financing. • Allows flexible budgeting. • Conserves existing credit.	• Provides total financing. • May provide financing for acquisition plus related costs. • Provides a hedge against inflation. • Provides fast, flexible financing. • Simplifies bookkeeping. • Reduces the risk of obsolescence. • Provides a trial use period.

Evaluating Instrument Purchases

Use *Worksheet 6.1* to evaluate the individual instrument manufacturers. Use additional sheets if required.

Worksheet 6.1- Evaluating Purchases

Name of provider of LC/MS instrumentation (and instrument):

________________________(________________________)

Part A-Fill out the following to evaluate the capabilities of the above instrument. Does this instrument meet the requirements of your lab?

Evaluating Capabilities	Comments	Yes	No
Does the instrument have the necessary features and options?	____________	❑	❑
Does the instrument meet your required specifications?	____________	❑	❑
Have you sent evaluation samples for analysis?	____________	❑	❑
Have you had a demonstration of the instrument?	____________	❑	❑
Have you contacted third-party references?	____________	❑	❑
Are there any deficiencies of the instrument?	____________	❑	❑

Part B-Fill out the following to evaluate services.

Evaluating Services	Comments	Yes	No
Is training available? In-house? On-site?	____________	❑	❑
Is service and support readily available?	____________	❑	❑

Part C-Fill out the following to evaluate the pricing of the instrument provider.

Evaluating Purchase Pricing	Amount	Comments
What is the quoted price of the instrument?	____________	____________
Is training included?	____________	____________
Is installation included?	____________	____________
Are there additional or hidden charges?	____________	____________

Note: Copies of this worksheet can be downloaded from `www.LCMS.com`.

6.4-Evaluating Upgrade Pathways

You have justified your need for in-house LC/MS capabilities. You have an existing mass analyzer in your lab or have access to one. Now you must evaluate your alternatives for upgrading your existing equipment in comparison to purchasing an instrument. Table 6.10 itemizes some of the pros and cons of upgrading an existing instrument with LC/MS capabilities.

Table 6.10- The Pros and Cons of Upgrading

Consideration	Pro	Con
Availability	• You already have the mass analyzer. • An upgrade path is available from the original manufacturer or after-market supplier.	• It is sometimes difficult or impractical to upgrade some models. • Upgrades are not always available.
Familiarity	• Most of the hardware mechanics are familiar. • Most of the software mechanics are familiar.	• The personnel required to operate your LC/MS may not be the same as your existing staff.
Compatibility	• Upgrades generally require simply a bolt-on accessory.	• In some cases hardware upgrades may be required, such as, a vacuum system upgrade. • In some cases software upgrades may be required.[23]
Performance	• You can take advantage of all your existing performance capabilities (e.g. upgrading a high resolution magnetic sector analyzer)	• Older systems may be compromised in both analytical performance and capabilities of the software.
Price	• Upgrading an existing system is generally cheaper than purchasing a new one.[24]	• The bundled price for some system upgrades may make it prohibitive.[25]
Reliability	• You are dealing with a known commodity when upgrading an existing system.	• Older systems tend to have higher and more costly service requirements.
Referrals	• You have access to a large installed userbase.	• For some models, few upgrades may exist.

Evaluating Upgrade Capabilities

Upgrading an existing instrument can be as simple as ordering a bolt-on accessory for a fixed price or as complex as reconfiguring your entire instrument with new vacuum, electronics, inlets, sources, detectors, and software. Sometimes, even a new computer is required. The specific type of system that you currently own will determine where in this range you may lie. Your original manufacturer (if it is still around) is the best place to start. If upgrading is a practical alternative for your model, the manufacturer may offer upgrade packages. If upgrading is too difficult or impractical, the manufacturer should be able to give you an itemized list of the difficulties. If they are not doing it there are probably many good reasons why.[26]

Third-party upgrades are an alternative that should not be overlooked (see *Chapter Five* and *Appendix A*). For example, Analytica of Branford supplies electrospray and APCI upgrades for many models of mass spectrometers. For a straightforward upgrade, this approach may be very practical. As the number of system modifications increase and the lines of responsibility between the third-party supplier and your original manufacturer become fuzzy you should carefully weigh risks against rewards.

Evaluating the performance capabilities of a given upgrade can get confusing if you are undertaking a significant and extended upgrade to the mass spectrometer system. As a general rule, assume that you will get no better "raw" performance than you are currently getting from your current system. Your manufacturer may provide performance data on systems similar to yours and in some cases may be willing to commit to a performance specification on a given test compound. You may also be able to arrange a demonstration at your manufacturer's Applications Lab or at another laboratory with similar equipment.

Some upgrade providers require that you provide benchmark performance capabilities on your instrument before they will install their interface. This may be as simple as acquiring an electron impact spectrum of a prescribed test compound on your existing instrument to provide a baseline of performance for your system.

When considering an upgrade, make sure you have a clear understanding of the final capabilities associated with the upgrade. Important questions are: How will the interface and ion source be controlled? Will the LC/MS upgrade be software or manually controlled? Can the experimental parameters be changed within an acquisition method? Is the upgraded system fully automated?

Evaluating Available Services When Upgrading

Some attention to the service and support of your upgrade provider is essential. You likely have a track record with your existing instrument supplier that you can draw on to evaluate the service and support availability for their potential upgrade. If you have received good service and support throughout the lifetime of your existing instrument, it is reasonable to expect the same high quality of support on an upgrade. Conversely, if your support has been less than satisfactory, you may want to avoid upgrading or else consider a third-party supplier of LC/MS interfaces..

Training is also relevant when upgrading an existing system. For example, Analytica of Branford has a policy of offering both on-site training and training at its facilities.[27] Aanlytica of Branford has built its upgrade business on their ability to pull all the necessary system components and suppliers together to produce an integrated system. This takes extremely broad industry knowledge and technical expertise. All members of its installation staff are Ph.D.-level scientists with significant experience in the field of LC/MS.[28]

Evaluating Pricing When Upgrading

The price of an upgrade is highly system-dependent. Ballpark pricing for “bolt-on” LC/MS interfaces will run in the range of $30,000. Add-on accessories such as high energy conversion dynodes or software upgrades can be an additional $10,000 to $50,000.

Evaluating Upgrade Providers

Use *Worksheet 6.2* to evaluate the individual providers. Use additional sheets if required.

Worksheet 6.2- Evaluating Upgrades

Name of provider of upgrade:

____

Part A-Fill out the following to evaluate the capabilities of the upgrade provider. Does the upgrade and provider meet the requirements of your lab?

Upgrade Capabilities	Comments	Yes	No
Does the provider upgrade your existing instrumentation?	____________	❑	❑
Will the proposed upgrade meet your analytical needs?	____________	❑	❑
Will the provider provide you with referrals from previous upgrades?	____________	❑	❑
Does the upgrade require special hardware modifications to your existing equipment?	____________	❑	❑
Does the upgrade require special software modifications to your existing equipment?	____________	❑	❑
Will the proposed upgrade degrade the utility or access of your existing equipment?	____________	❑	❑

Part B- Fill out the following to evaluate the services of the upgrade provider.

Upgrade Services	Comments	Yes	No
Does the upgrade require special training? Is it available?	____________	❑	❑
Does the upgrade provider supply service and support?	____________	❑	❑

Part C- Fill out the following to evaluate the pricing of the upgrade provider.

Lab Pricing	Amount	Comments
What is your upgrade price?	________	________
Is training included? On-site?	________	________
Is installation included?	________	________
Are there additional or hidden charges?	________	________

Note: Copies of this worksheet can be downloaded from `www.LCMS.com`.

6.5-Choosing Your Alternative

The time comes when you have to make a decision. Unfortunately, the decision is yours. We have emphasized the importance of the process, the importance of having a process, and the importance of weighing capabilities against your specific problem needs; but we cannot make the decision for you. We don't have enough information for that. You are the one who has to fill in the blanks.

Here are some rules-of-thumb that you may want to consider in your final decision making process:

- The operators should have a significant voice in the final decision to purchase an instrument; after all, they will be the ones responsible for solving problems with the instrument.
- The stability and longevity of the manufacturing organization is and should be an issue when evaluating a prospective product.
- In general, as the number of installed units increases, so does the reliability of that product. The first 25 to 50 units of any new product will likely have a somewhat higher level of engineering and software defects.
- When considering analytical results from evaluation samples submitted to instrument manufacturers, remember that the results may be more indicative of the expertise of the applications chemist than the performance of the instrument. Remember, you are buying an instrument, not a chemist!
- When considering analytical results from evaluation samples submitted to contract labs, remember that the results are indicative of the expertise of the application's chemist who will be solving your sample problems. Remember, you are buying a chemist, not an instrument!
- If given a choice between added options or training, pick training. We think training is one of the most important line items on a sales agreement.
- Third-party referrals can provide the most realistic appraisal of both contract labs and instrument companies. One should be wary of any company that avoids providing you with referrals.

Discussion Questions and Exercises

1. What is the estimated time required for your organization to complete the four steps (Figure 6.0) involved with purchasing an instrument? ...with sending out samples for analysis?
2. Discuss the differences between *Justifying* and *Evaluating* the purchase of an instrument.
3. What are the criteria that you would use to evaluate a contract lab for a 10,000-sample clinical trial?.....for a low-level forensic sample?
4. What are four *advantages* of upgrading an existing instrument over purchasing a new instrument?
5. What are four *disadvantages* of upgrading an existing instrument over purchasing a new instrument?
6. What attributes of an instrument company should be evaluated at an instrument *demonstration*?
7. What attributes of a contract lab should be evaluated during a *facilities visit*?

Notes

[1] Sheehy, B., Bracey, H., and Frazier, R., Winning the Race for Value-Strategies to Create Competitive Advantage in the Emerging Age of Abundance, Amacom: New York (1996).

[2] Webster, Jr., F.E., Wind, Y., Organizational Buying Behavior, Prentice-Hall: New Jersey (1972).

[3] Paul Winkler, personal communication.

[4] Paul Winkler, ibid.

[5] Prices can also vary based on your tax status (for profit or not for profit) and whether you are associated with the lab or institution. For example, see the internet page (`pdtc.rockefeller.edu`) for Rockefeller University's Price List for Protein/DNA Analysis.

[6] Quotes for method development and subsequent analysis are typically worded two ways. A comprehensive dollar amount based on the number of samples, such as $150/sample for 1,500 samples. This will include report generation, sample storage, phone support, etc. Or, segmented pricing, a separate price per item; such as, a quoted price for methods development, another based on the number of samples, another for gen-

erating a report, another for sample storage, etc. To fairly compare the two pricing structures, all services must be included in the final analysis.

[7] A recent survey indicates that industrial chromatography labs have an average capital budget of just over $50,000 per year. While academic labs have an average capital budget ~ $10,000 per year. Under these current budget constraints, purchasing LC/MS may require a major shift in capital funding in the chromatography community. Survey from: R&D Magazine 38, pages 48-51 (1996).

[8] Alternatively, your first contact may be at a scientific show, such as the Pittsburgh Conference (USA), Analytica (Europe), or AnalabAsia (Asia) (see Appendix A, Sources of LC/MS Information).

[9] Jordon, J.R., "The dynamics of buying an instrument," Inside Laboratory Management 1, pages 10-13 (1997), *op cit.* Do not confuse features for performance items.

[10] Jordon, J.R., *op cit.*

[11] Jordon, J.R., *op cit*

[12] Jordon, J.R., *op cit.*

[13] Jordon, J.R., *op cit.*

[14] It makes perfect sense that an instrument manufacturer is not going to give you a list of references containing unsatisfied customers. They will likely give names of key accounts that have a good operating record. In addition, they may be barred by privacy laws from giving you every customers name.

[15] A good place to find labs using a particular instrument is in the analytical literature, such as Journal of the American Society of Mass Spectrometry, LC/MS Update, etc. (see Appendix A, Sources of LC/MS Information). You may want to contact labs that have published results using a particular instrument.

[16] Jordon, J.R., *op cit.* It is both "fixing" the problem and customer training.

[17] Service and support (application and phone support, troubleshooting and repair) are an important and profitable part of the instrument business. In many companies, more than 20% of an instrument manufacturer's revenue comes from service and support activities.

[18] Jordon, J.R., *op cit.*

[19] You always run the risk of controversy when you tabulate and compare the price of instruments. These values represent the range of advertised and quoted prices, not absolute values. Clearly, if you are considering the purchase of an instrument, the price should be known before getting too far along in the decision process.

[20] Jordon, J.R., *op cit.*

[21] Pritchard, R.E., Hindelang, T.J., The Lease/Buy Decision, Amacon: New York (1980).

[22] Steve Ceppa, personal communication. Today large corporations are considering leasing an instrument, instead of purchasing. Orginally, the typical working life expectancy for a mass spectrometer was ~10 years. But in the past two to three years the working life of an LC/MS instrument has been reduced to five to seven years- a result of new LC/MS technology and instrumentation. The effect of this is that corporations are now considering leasing as an economical alternative. In-house personnel still do the evaluation and make the final decision of which instrument to purchase, but the instruments are purchased by an outside leasing company who in turn leases the instrument to the corporations.

[23] ....installing the LC/MS interfaces is the easy part. The hard part is updating the mass analyzer in terms of hardware and updating the software to current revisions, if possible. (Whitehouse, C., personal communication.)

[24] The price of LC/MS upgrades typically ranges from $25,000 to $100,000 depending on the state of the original system.

[25] The additional price of updating your hardware (e.g. vacuum pumps), electronics (e.g. high energy conversion dynode, negative ion) and new versions of software, may be many times the price of the LC/MS interface alone. In cases where the original in-house system requires extensive system upgrades, you may be better off purchasing an entire system instead of upgrading an existing system.

[26] For many instruments that were not originally designed specifically for LC/MS, upgrades can be very complicated and labor intensive. Many instrument companies find it too risky and difficult to make a reasonable profit on upgrading existing hardware in the field.

[27] Craig Whitehouse, personal communication. In upgrade cases where third-party system upgrades require several component suppliers to work together, Analytica will coordinate the activities of each party in order to produce a smooth process and integrated upgrade.

[28] James Boyle, personal communication.

Chapter Seven- Flow Diagram

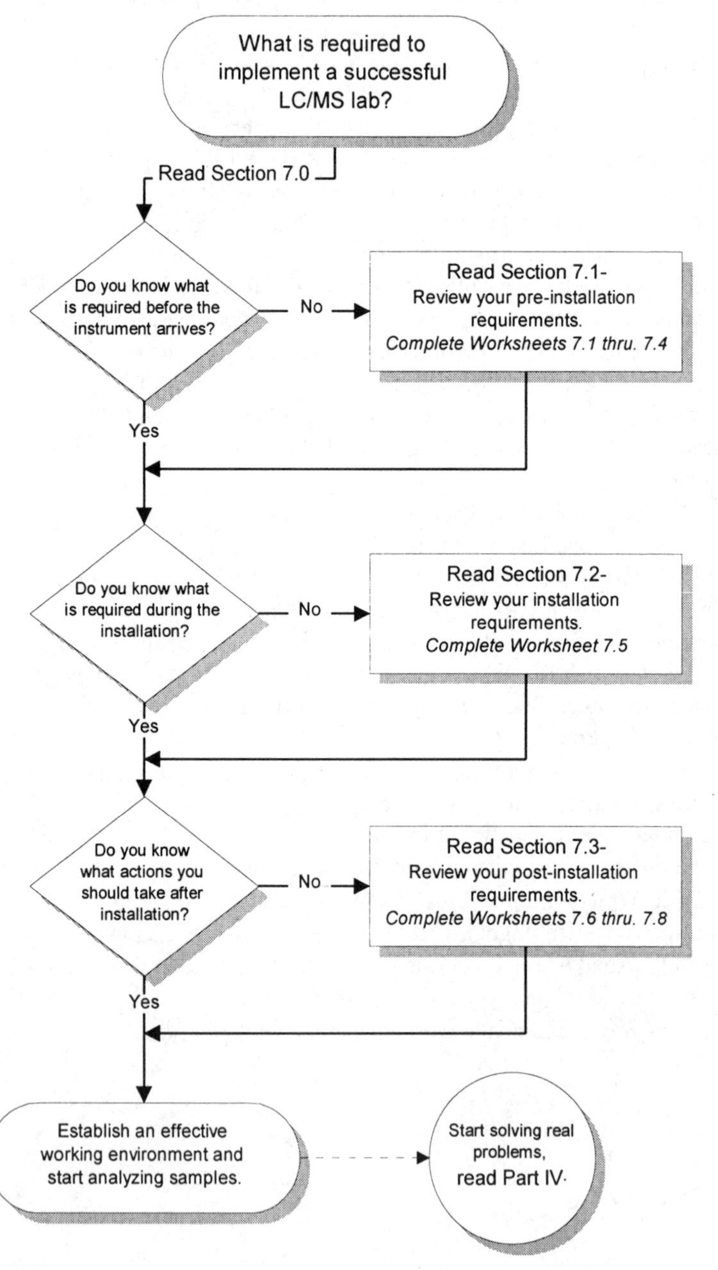

What is required to implement a successful LC/MS lab?
Read Section 7.0
Do you know what is required before the instrument arrives?
No
Read Section 7.1-
Review your pre-installation requirements.
Complete Worksheets 7.1 thru. 7.4
Yes
Do you know what is required during the installation?
No
Read Section 7.2-
Review your installation requirements.
Complete Worksheet 7.5
Yes
Do you know what actions you should take after installation?
No
Read Section 7.3-
Review your post-installation requirements.
Complete Worksheets 7.6 thru. 7.8
Yes
Establish an effective working environment and start analyzing samples.
Start solving real problems, read Part IV.

Chapter Seven

Implementing Your LC/MS Lab

(or, How to Succeed at LC/MS)

The plan is nothing, the planning is everything.
Dwight D. Eisenhower[1]

7.0- Planning a Successful Lab

The planning and maintenance of your LC/MS lab may have more to do with your ultimate success in problem-solving than which instrument you purchase. If you have made a commitment to acquire LC/MS capabilities, you should start planning the implementation of your lab as early into the process as possible. This chapter divides the planning process into three stages; pre-, during-, and post-installation (Figure 7.0). Although much of the information relating to installation will be supplied by your instrument company, companies tend to focus more on instrument specific aspects of the setup, and in some cases neglect the broader issues of LC/MS laboratory setup.

In addition to this chapter, general information and training on setting up and maintaining a lab is available from professional societies such as the American Chemical Society, American Council of Independent Laboratories, American Organization of Analytical Chemist International or the Analytical Laboratory Managers Association. The American Society of Mass Spectrometry also conducts a laboratory managers workshop. These professional groups offer pamphlets, books, video tapes and sponsor interest groups that will help you in setting up and maintaining a lab facility (see *Appendix A, Sources of LC/MS Information*).

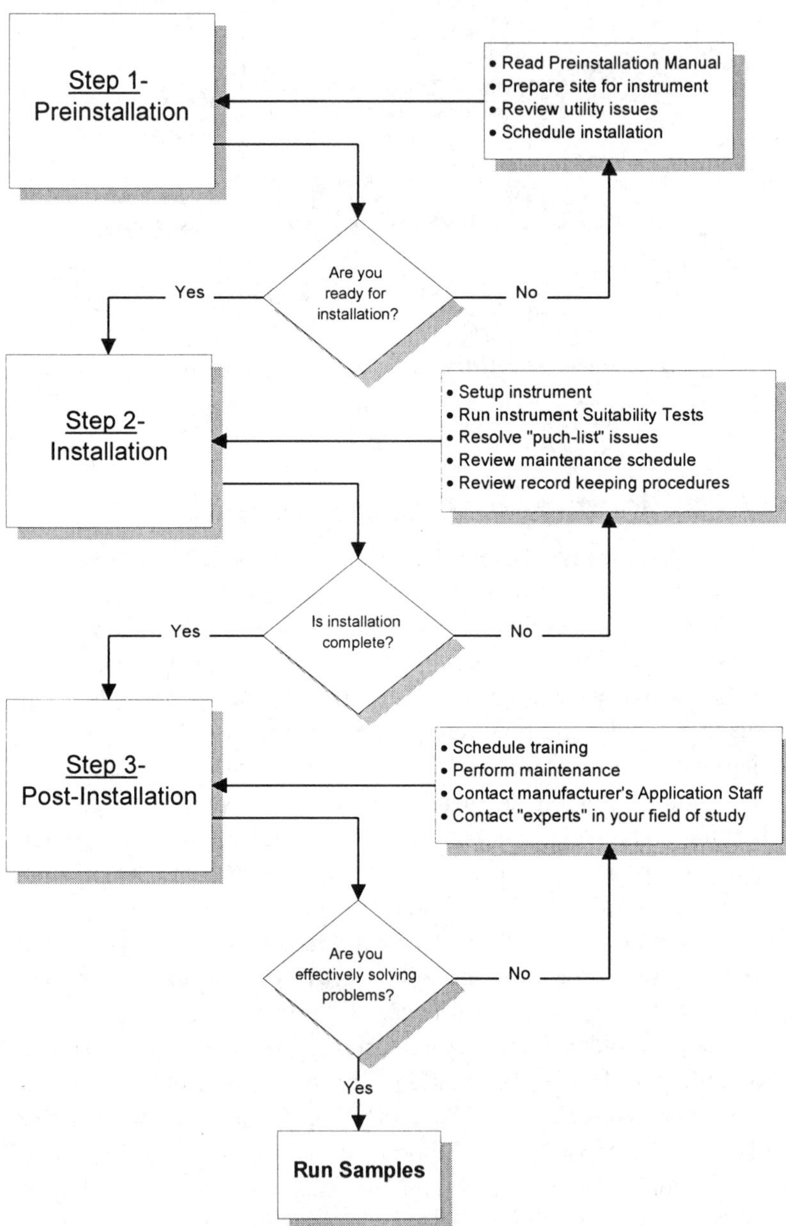

Figure 7.0- The steps for implementing an LC/MS lab.

Setting up and maintaining a lab starts with a well-thought out plan. You should have a clear idea of your laboratory needs before acquiring your instrument. Overall laboratory issues should be considered during the time period when you are evaluating your instrumental alternatives, not after you are finished. During your evaluation you should consider the impact of any given instrument on your lab. Table 7.0 outlines the main areas that should be considered in your laboratory plan.

Table 7.0- Considerations for your LC/MS lab

Stages	Considerations
Before Installation	• What will the function of the LC/MS lab be? • How much space is needed? • What are the electrical, gas, water and cooling requirements • Are there any environmental hazards to consider? • What safety precautions (if any) will have to be taken into consideration? • Do we have SOP's? • Do we need training before the instrument arrives?
During Installation	• Who will set up the instrument? • What components make up the instrument? • What plans do I need to make for maintenance & record keeping? • Who do I contact for help? • Will training occur at installation?
After Installation	• What is the maintenance schedule? • What safety measures do we need to institute? • What records, archiving, and backup policies should we maintain? • What post-installation training is required?

7.1-Before Installing Your LC/MS

You should have a clear picture of how your instrument will be used well before you close on any instrument purchase order. If you cannot qualitatively describe how the instrument will be used, you are likely not prepared to purchase an instrument in the first place. Key questions to answer, of course, are, "Who will be using the instrument (single vs. multiple operators)?" "Who will be preparing the samples (in-lab versus remote)?", and "Who will be developing your methods?". The interaction between all participants in the laboratory will dramatically affect the setup and implementation of your LC/MS lab.

The lab you finally implement should be expected to do more than hold your instrument. It should also facilitate the use of the instrument and all ancillary activities in your problem-solving endeavors.

Pre-Installation Manual

All instrument manufacturers provide a "Pre-Installation Manual" to inform their customers about requirements for installation. You should request a copy of the pre-installation manual before you close on your purchase. In this way you can be aware of the impact and requirements of a new instrument before you purchase it. For example, you can consider the extra-instrumental budgets and activities that are required to support a successful lab.

Some common planning failures of new instrument purchasers are lack of adequate power, an approved exhaust system, adequate space for the system or access space to the lab. It is not unheard of for installers to arrive at the customer's site only to find the instrument could not fit through the doorway to the lab. These encounters can all be avoided by proper planning. In general, the larger and higher the complexity of the instrument (such as an FTMS LC/MS instrument), the higher the preinstallation requirements.

Important: Don't wait until installation day to prepare!

Surveying Your Facilities

Whether you plan to use an existing lab or remodel, care should be taken to ensure that the lab facilities are appropriate for your specific needs. A proper working environment for an LC/MS instrument requires space, proper utilities, controlled temperatures, a safe lab environment, areas for particular functions (such as, sample prep, storage), proper record-keeping, and supplies (see Table 7.1).

Floor Plan

The better the layout, the easier it is for personnel to do their jobs. Arriving at the right layout involves more than just moving things around to a location where they will fit; rather, it involves employees planning the use of the space for maximum benefit. The objective of a good floor plan is to conserve time and motion in all activities relating to the instrument.

Table 7.1- Requirements for an LC/MS lab facility

Facilities	Typical Requirements
Space	• Ample floor space for equipment, service access and accessory storage. • Sufficient clearance through hallways, doorways and elevators for adequate access to the installation site. • Clean, dust free work space. • Fume or exhaust hood for maintenance operations of disassembly and cleaning. • Sample preparation area. • Area for administration tasks: analyzing, reporting. • Storage areas for supplies, samples, records, computer tapes, and waste.
Utilities	• Adequate and properly protected primary power. • Adequate and proper hooks-ups for phones and computer network. • Gas supply area, water purification system.
Environmental	• Air conditioning and ventilation matched to the system heat dissipation of the LC/MS system and to the temperature range specified for proper system operation. • A comfortable temperature for staff.

Table 7.1- (cont.)-Requirements for an LC/MS lab facility

Facilities	Typical Requirements
Safe Work Environment	• A means for dealing with the noise generated by the vacuum pumps. • Adequate and proper filtering/exhausting of all vacuum pump exhausts and LC vapors.
Record-Keeping	• Instrument logbooks: samples and maintenance. • Standard operating procedures (SOP's).
Accessories & Supplies	• LC: hexane, methanol, acetonitrile, distilled deionized water, tools, etc. • MS: vacuum pump oils, tools, etc.

How much space do you need?

Space will be needed for the mass analyzer, vacuum pumps, liquid chromatograph, gas supplies, storage area for general supplies, the data system and printer. For specific dimensions (and weight) on individual instruments, check with the LC/MS manufacturer (see publications about individual instruments for details).

The first step is to draw up a floor plan of your lab (see Figure 7.1), noting present storage areas, counter space, sinks, power outlets, and exhaust hoods.

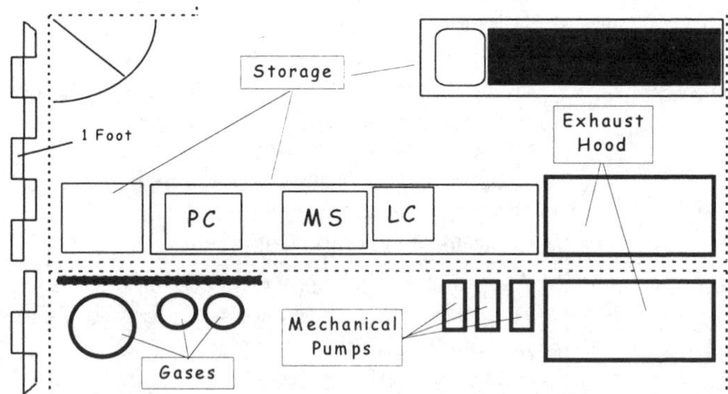

Figure 7.1- Floor Plan for a lab with a "benchtop style" LC/MS system.[2] The chromatograph (LC), mass analyzer (MS) and data system (PC) are in one room; mechanical pumps, exhaust hood and gas supplies are in an adjoining room.

Placement of Your Instrument

LC/MS System: The LC/MS system should be placed in a well-ventilated area, away from direct sunlight and sources of dust, heat, cold, and humidity. The LC/MS system should stand on a flat, stable surface away from places prone to vibration or shock.

The data system should be placed either on the same bench or a table of sufficient strength, or may be placed separately to the extent permitted by the interconnecting cables. Cables supplied with the equipment are of sufficient length to meet the needs of most installations.[3] For service access, a minimum clearance (e.g. 8-10 in, 20-30 cm) should be available behind all parts of the instrument.

Mechanical Pumps: The mechanical pumps[4] (for the mass analyzer and the LC/MS interface) should be placed on the floor. Alternatively, the pumps can be placed inside cabinets underneath the instrument or preferably in an adjoining room.[5] The mechanical pumps should be placed in a convenient location near an exhaust hood (see below).

There is typically one pump associated with the vacuum chamber of the mass analyzer and one or two mechanical pumps required for the LC/MS interface. Typically, they are equipped with vacuum hoses 2-9 meters (6-10 ft) in length- sufficient length to meet the needs of most installations. For service access, the pumps should be placed for easy access to their power switch, oil level site glass, ballast valve, and oil drainage plug.

Exhaust Hood: The mechanical pumps for the vacuum chamber and the LC/MS interfaces exhaust solvent vapors, sample vapors and oil mist. These exhaust vapors must be kept from entering the lab and the environment with appropriate filtering, trapping and/or exhaust system. These vapors can be trapped in any number of absorbing or cold trapping devices and eventually venting the filtered gas through an exhaust hood.[6]

Worksheet 7.0- Lab Facilities Checklist

Evaluate your facilities to ensure adequate space has been set aside for your LC/MS system, storage areas, and supplies. Use the following checklist.

Pre-Installation Facilities Checklist	Yes	No
1. Floor plan drawn up for placement of:		
• Obtained specifications from instrument manufacturer:	❑	❑
• LC/MS system:	❑	❑
• Computer/workstation:	❑	❑
• Mech. pumps: Same room: ❑ Adjoining room: ❑		
• Gas bottles: Same room: ❑ Adjoining room: ❑		
• Gas cylinder stand(s)/wall bracket(s) installed:	❑	❑
• Storage areas:	❑	❑
• MS supplies: ❑ • Vacuum pump oil: ❑		
• LC supplies/fittings: ❑ • LC solvents: ❑		
• Refrigerators: ❑ • Freezers: ❑		
2. Layout facilitates movement of employees and materials:	❑	❑
3. Adequate access to back of instrument:	❑	❑
4. Adequate access to mechanical vacuum pumps:	❑	❑
5. A safe environment for personnel:	❑	❑
• Exhaust hood(s): ❑ • Storage for LC solvents: ❑		
• Air circulation: ❑ • Storage for oil: ❑		
• Air temperature: ❑ • Ambient noise acceptable: ❑		
• Lighting: ❑ • Evacuation plan in place: ❑		
• Fire extinguisher: ❑		
6. Conform to all safety and facility regulations:	❑	❑
• Exhaust hoods rated for LC solvents and oil mist:	❑	❑
• Placement of gas supply lines:	❑	❑
• Gas cylinder storage:	❑	❑
• Liquefied gas storage:	❑	❑
7. Does lab have limited or restrictive access:	❑	❑
8. Provision for future expansion:	❑	❑

Note: Copies of this worksheet can be downloaded from **www.LCMS.com**.

Instrument & Lab Utilities

All analytical instruments require various utilities in order to operate properly. What is typically different about mass analyzers (say from LC equipment) is the requirement of a 220V (24 amp) power line and various gases for operation. This necessitates equipping the lab with new power and gas lines.

Electrical Requirements

While the power requirements differ from instrument to instrument and depending on the accessories installed, the typical LC/MS instrument[7] will be adequately served by the electrical parameter as shown in Table 7.2.

Table 7.2- Electrical requirements for LC/MS

System Component	Typical Requirements
Mass Analyzer	• A single branch outlet rated at 208-250 V, 50/60 Hz, 24 amps or 99-132 VAC, 20 amps outlets.
Liquid Chromatograph	• A 100, 120, 220, or 240 VAC power supply. • A 15 or 20 amp branch circuit is adequate. • Two or three receptacles are required.
Data System	• A 115 VA. or 230 VAC power supply line. • A 15 or 20 amps branch circuit is adequate. Three or four receptacles are required.
Power Lines	• Relatively free from transients and fluctuations. • Properly grounded.
Cables & Plugs[8]	• Cables supplied with the equipment are of sufficient length to meet the needs of most installations. • Power plugs.
Local Power Disconnect	• A local power disconnect near the instrument is also a must.

Gas Requirements

All LC/MS systems require the use of high purity gases for operation of the LC/MS interface, LC system, and the mass spectrometer (see Tables 7.3 and 7.4). Regulated supplies of gas are connected directly to the individual units: LC and MS. Check with your instrument manufacturer about proper hook-ups and attachments.

Table 7.3- Gas requirements for LC/MS

System Component	Typical Requirement
Liquid Chromatograph	• Liquid chromatographic systems may require the use of a high purity grade of helium gas to degas or sparge the LC mobile phases. • A regulated supply of air or nitrogen may be required to operate the LC autoinjector or pumps.
LC/MS Interface	• The operation of most LC/MS interfaces requires a nebulizing, drying gas and/or bath gas.
Mass Analyzer	• Mass analyzers require high purity gases for chemical ionization, collision gas for MS/MS operation, and "cooling" gas for ion traps.
Gas Supply	• Standard cylinder measures 23 cm (9 in) in diameter and 140 cm (55 in) high, and supplies ~250 ft^3 of gas (e.g. "T" size bottle). • Many other sizes are also available (e.g., lecture bottles, cryogenic container). • Labs can also be equipped with a supply of gas from a centrally located area.
Gas Regulators	• Use "two stage" and/or "corrosive resistant" (where specified) regulators.

Table 7.4- Gases typically used for LC/MS

Use	Gas	Purity	Tank Size	Duration	Note
LC					
• Sparge	Helium	99.995	250	1-2 month	
• Other	Nitrogen	99.99	250	1-2 month	
MS					
• CI	Methane	99.97	250	1-2 month	
	Isobutane	99.5	0.02	0.5-1 year	Liquid
	Ammonia	99.99	0.2	0.5-1 year	Liquid anhydrous
• MS/MS	Ar, N_2, He	99.99	250	1 year	
• Ion Trap	Helium	99.995	250	0.5-1 year	
• FTMS	Liquid He	--------	3640	1 month	
	Liquid N_2	--------	3640	1 week	
LC/MS					
• APCI	Nitrogen	99.999	250	0.5-1 week	
	Liquid N_2	--------	3640	1-2 month	Liquid "best buy"
• ES	Nitrogen	99.999	250	1-2 week	
	Liquid N_2	--------	3540	4-6 week	Liquid "best buy"
	SF_6	99.995	32	1-2 month	SF_6/N_2::1/1
• PB	Helium	99.995	250	1-2 week	
• ICP	Argon	99.9999	250	1-2 week	
	Liquid Ar	--------	3640	1-2 month	Liquid "best buy"

Ammonia gas: requires a corrosive-resistant regulator

Duration: based on using the instrument ~8 hr/day, 5 days a week.

Purity: minimum purity recommended

Sulfur hexafluoride: liquid sulfur hexafluoride (100%) is very expensive while a gas mixture of sulfur hexafluoride and nitrogen (99.999%) is relatively inexpensive.

Tank size: is given in ft^3

Worksheet 7.1- Lab Utilities Checklist

To make sure that adequate electrical power and the necessary gases are provided for your LC/MS system, use the following worksheet to evaluate your facilities:

Pre-Installation Utilities Checklist	Yes	No
Have you contacted your manufacturer about electrical and gas utility needs?	❑	❑
<u>Part A:</u> <u>AC Power Requirements</u>:		
1. Mass Analyzer, LC System, Data System, Accessories		
2. Adequate number of Receptacles?	❑	❑
• AC Volts: 120 V: ❑ 220 V: ❑ Other: ❑		
• AC Amps: 15 amp: ❑ 20 amp: ❑ Other: ❑		
3. Lines relatively free from transients and fluctuations:	❑	❑
4. Local power disconnect present:	❑	❑
5. Existing power conforms to local building codes:	❑	❑
6. Properly rated plugs and receptacles:	❑	❑
7. Facilities manager (or qualified electrician) consulted about electrical and possible construction needs?	❑	❑
<u>Part B:</u> <u>Gases:</u>		
1. Adequate space for gases:	❑	❑
LC: ❑ LC/MS Interface: ❑ MS: ❑ MS/MS: ❑		
2. Purchased gases for:	❑	❑
LC: ❑ LC/MS Interface: ❑ MS: ❑ MS/MS: ❑		
3. Two stage regulators for each gas cylinder/supply line:	❑	❑
4. Tubing and fittings:	❑	❑
5. Check the *Manufacturer's Pre-Installation Manual* for information about the specific pressure range, type and purity of gases that would be required to operate your LC/MS system.	❑	❑
6. Facilities manager consulted about storage and possible construction needs?	❑	❑

Note: *Copies of this worksheet can be downloaded from* `www.LCMS.com`.

Environmental Considerations

A comfortable and healthy lab environment is a must. You owe it to yourself and fellow employees to create a lab that is free from potential exposure to harmful or uncomfortable working conditions.[9] Today's LC/MS systems operate quietly and with a minimal impact on the air conditioning demands of most lab facilities.

Ambient Temperature

Temperature in the LC/MS lab should be typically maintained between 50 to 80 °F (10 to 27 °C). Maximum ambient temperature should not exceed 80 °F (27 °C) as measured at several places from the instrument. Operation outside of the specified temperature range could invalidate your instrument warranty. Do not place the system in direct sunlight or, near heating and/or air conditioning outlets which can cause differential heating.

A typical LC/MS system's heat dissipation is ~12,000 BTU/hr (3,500 Watts). Approximately 80% of the heat dissipated by the instrument is attributed to the vacuum pumps (e.g. mechanical and turbomolecular or diffusion pumps) required to maintain a proper vacuum (see Table 7.5). Placing the mechanical pumps in an adjoining room will reduce the need for additional cooling in the room where the instrument resides.

Table 7.5- Vacuum Pump Operating Temperatures

Ambient Temperature	Typical Operating Temperatures
• 50 to 80 °F (10 to 27 °C)	• mechanical pumps: ~100 °C • turbomolecular : ~50 °C • diffusion pumps: ~250 °C

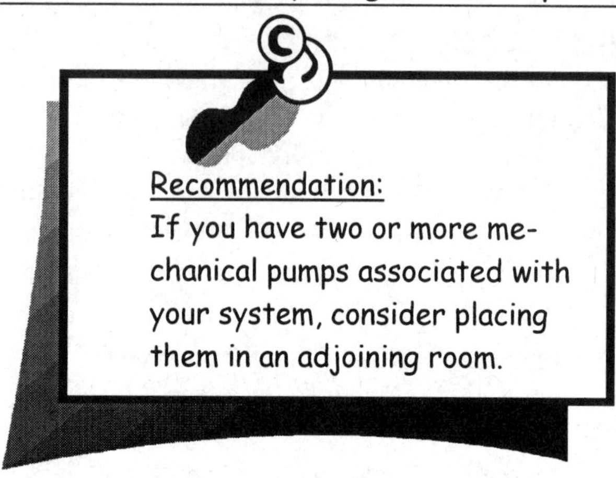

Recommendation:
If you have two or more mechanical pumps associated with your system, consider placing them in an adjoining room.

Air Conditioning

Air conditioning may be required to maintain the temperature within the recommended operating range. Ambient temperature is most important for both the instrument and laboratory staff. Elevated temperatures can cause serious temporary computer malfunctions. The further out-of-specification the temperature rises, the higher the probability of permanent computer system damage, and of reducing the life of vacuum pump(s) and instrument electronics. Running an instrument in a lab at above 95 ^{o}F (23 ^{o}C) is not very comfortable for the operator as well!

Most business facilities have air conditioning today; but the capacity should be checked to ensure that the heat load (BTU/hr or watts) of a given instrument does not exceed the capacity of the current cooling system. In some cases, separate or increased cooling capabilities may be required to maintain temperature requirements. In other cases, simply adding a fan to a stagnant part of the lab may help improve circulation in those areas. Operator comfort may also require an air flow designed to avoid drafts and meet specified indoor air quality requirements. Stable and repeatable instrument performance may also be enhanced by the absence of drafts.

Ambient Noise

The noise associated with all modern LC/MS systems originates primarily from the vacuum pumps (e.g. turbomolecular and mechanical pumps) used to maintain a proper vacuum of the LC interface and mass analyzer. As shown in Figure 7.2 two or more mechanical pumps expose the personnel in the room to a level of noise that may require them to be evaluated for potential auditory impairment.

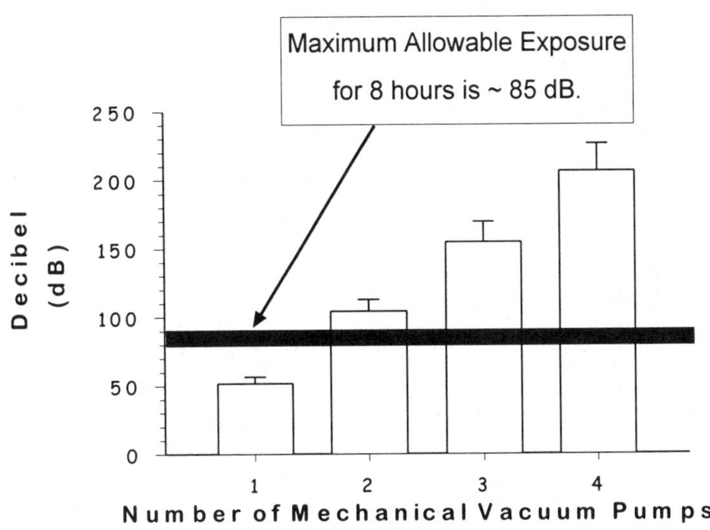

Figure 7.2- The noise level associated with an increasing number of mechanical vane pumps (at ~3 ft). The error bars represent the decibel level when the pumps' ballast are closed (~50 dB, pumps associated with the mass analyzer) and when they are open (~55 dB, pumps associated with LC/MS interface). The maximum allowable exposure for eight hours is ~85 dB as outlined by OSHA.[10]

Diminishing Ambient Noise: The noise associated with the mechanical vane pumps may be diminished by placing the mechanical pumps in a room adjoining your LC/MS lab. In placing the vacuum pumps in an adjoining room you must consider the effect of the length of your vacuum tubing on the pumping efficiency of the vacuum system, both high and low vacuum pumps. The pumping efficiency of the vacuum system will be lessened by increasing the length of the vacuum tubing. As the length of the tubing is increased there is a decrease in the gas flow (conductance) through the tube. In general, (for pressures associated with the mechanical pumps, ≥ 0.1 torr) vacuum tubing of 1-in I.D. (the standard size tubing on most instruments) is good up to 10-15 ft. Further than 15 ft larger I.D. tubing is needed.[11] Figure 7.3 shows the available I.D. vacuum tubing and at what distance the particular tubing should be used. For example, at ~30 ft, 1.5-in I.D. tubing would be appropriate while 2-in tubing would be needed if the distance is greater than 60 ft.

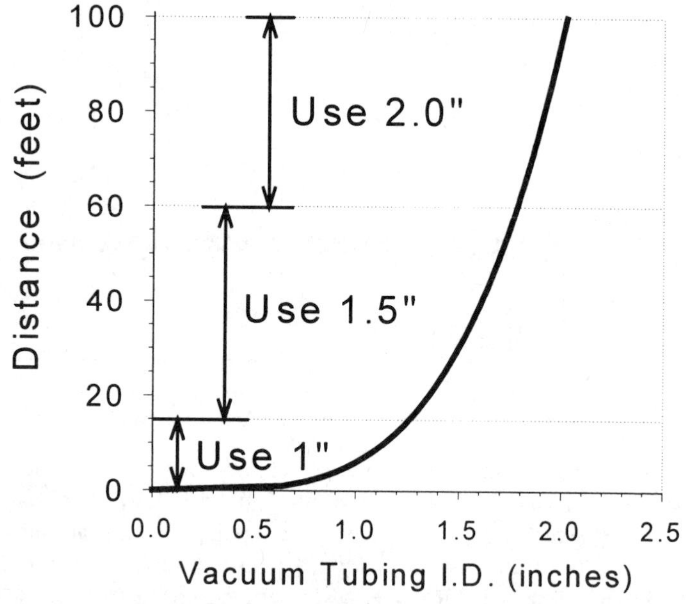

Figure 7.3- Plot of the internal diameter of vacuum tubing needed to maintain the same gas flow (conductance) through the tube at various distances from your LC/MS instrument. The commercially available tubing is marked (e.g., 1-, 1.5- and 2-in I.D. tubing).

Fittings and tubing may be purchased from your LC/MS manufacturer or any of the various after-market suppliers of mass analyzers/vacuum components listed in *Appendix A*.

Mechanical Pump Exhaust

The vacuum system of a typical LC/MS is equipped with vacuum pumps, both high vacuum (such as turbomolecular or diffusion pumps) and low vacuum (such as mechanical vane pumps) pumps backing up the high vacuum pumps. The LC/MS interface is also equipped with vacuum pumps, typically mechanical vane pumps.

The LC system/LC/MS interface can place large liquid and gas loads on the vacuum system (e.g., water and other LC solvents, potentially up to 1 mL/min of liquid; and 500 L/min of air, nitrogen or helium). Most LC/MS vacuum systems are designed to pump this large liquid vapor and gas flow away from the LC/MS system, but ultimately the gases have to go somewhere (hopefully not into your lab!).

The volume of gas (along with any LC solvent) is eventually expelled out of the exhaust ports of the mechanical pumps. This exhaust consists of oil vapors from the pump itself and any gas, LC solvents and potential analytes that have made their way through the mass analyzer and the LC/MS interface into the mechanical pumps. To prevent this exhaust from entering the lab and exposing the lab personnel to potentially harmful situations, the exhaust should be "trapped" and/or "filtered."

<u>Exhaust Mist Trap/Eliminator for Mass Analyzer Vacuum Pumps</u>: Mist eliminators trap and condense the oil mist exiting the exhaust of the mechanical pump, and return the oil to the pump. Newer models of mechanical pumps (e.g., Edwards RV model of pumps[12]) come already equipped with mist filters and drain return lines. The pumps and traps are typically designed to handle water at a rate of ~50-100 g/hr, more than enough for most LC/MS systems. Alternatively, if your system is not equipped with mist traps, they can be purchased from after-market manufacturers (*Appendix A*) and easily added to your pumps.

<u>Exhaust Filters/Purifiers for Mass Analyzer Vacuum Pumps</u>: Once the oil is returned to the pump, any remaining vapors (e.g. water, LC solvents, analytes) must be removed from the exhaust by a purifier before venting the exhaust into an exhaust hood. Typically, these filters are not part of the standard equipment.

Check with the manufacturer of your system or contact after-market manufacturers (see *Appendix A*) for the availability of traps and filters best suited for your instrument.

LC/MS Interface Pumps: Low LC Flow System, 1-100 uL/min: For systems interfaced with a low liquid flow LC/MS system (e.g. electrospray), the mechanical pump, mist eliminator/drain kit and filters described above for the vacuum system would be adequate for the mechanical vacuum pumps associated with the LC/MS interface.

LC/MS Interface Pumps: High LC Flow System, 0.1 -1.5 mL/min: For systems interfaced with a high liquid flow LC/MS system (e.g. API, particle beam, thermospray), we recommend: mechanical pumps with continuous gas ballast, a mist eliminator and an oil return line.[13] This combination continuously recycles oil through the pump's exhaust port, trapping oil in the mist filter and returning any oil to the pump. At the same time all LC solvent vapors are exhausted, allowing for continuous operation at high vapor loads, typically ~200 mL/hr.

Exhaust Hood

The in-house exhaust system should be rated to handle the gas vapor loads from your LC/MS system. Most laboratory hoods can easily handle these gas loads from even the higher-flow LC/MS systems.

When connecting your exhaust into a laboratory hood, keep in mind that the hot solvent vapors will readily condense along the exhaust lines and collect at the lowest point in the line, particularly, in an air-conditioned lab. This is particularly prevalent with particle beam and thermospray LC/MS systems. To avoid this collection in the exhaust lines you should 1) keep your exhaust lines as short as possible, or 2) install a solvent trap near the exhaust port of the pump and transfer only non-condensable vapors to the hood.

Worksheet 7.2-Lab Environment Checklist

To make sure that your lab environment is a safe place to work, use the following worksheet, Evaluation of the Lab Environment, to indicate the items to be discussed with your Facilities Manager.

Pre-Installation Lab Environment Checklist	Yes	No
Part A. Ambient Temperature & Noise:		
1. Floor plan of LC/MS Lab:	❑	❑
• Square feet of lab: ________ # windows: _____		
• # exhaust hoods: ________ # computers: _____		
• Storage Areas: ________ # staff: _____		
• Current air conditioning: ______________		
2. LC/MS System:	❑	❑
# computers: ________ BTU's: ________		
# pumps: ________		
3. Adjoining Room:		
• Current air conditioning: ______________	❑	❑
• Ambient Temperature: ______________	❑	❑
4. Facilities manager or Safety Committee contacted to review the impact on the ambient temperature and noise the LC/MS system will have?	❑	❑
5. Facilities Manager consulted about proposed floor plan, air conditioning needs and any construction needs for lab and adjoining room?	❑	❑
Part B. Mechanical Pump Exhaust & Oil:		
1. Vacuum System & LC/MS Interface Pumps:		
• Equipped with mist traps?	❑	❑
• Equipped with filters?	❑	❑
If not equipped with traps and filters, contact the manufacturer of your system.	❑	❑
2. Is exhaust hood rated for vacuum pumps?	❑	❑
3. Facilities Manager or Safety Committee contacted to arrange for the proper storage and disposal of oil?	❑	❑
4. Facilities Manager consulted about proposed floor plan and any construction needs for lab and adjoining room?	❑	❑

Note: Copies of this worksheet can be downloaded from `www.LCMS.com`.

Safety Considerations

Safety should be a culture, a way of thinking and acting that governs the way people conduct themselves in every lab activity.[14] Establishing and maintaining a safe lab environment is required under federal regulation. The U.S. Occupational Safety and Health Administration "Laboratory Standard" requires that employers train employees to recognize and avoid hazards in the workplace.[15] In addition, the EPA requires generators of hazardous waste to provide annual training to all employees who handle hazardous waste (such as biomedical or chemical waste, used mechanical pump oil). Even though both OSHA and the EPA state that the primary responsibility of safety lies with the employer, employees who are properly and thoroughly trained are the "front line" and are also responsible for safety in the laboratory.

You should set up a meeting with your organization's "Safety Officer" to review the *Pre-Installation Manual* (from your particular instrument manufacturer) and the areas outlined in this chapter to develop a plan that addresses the impact of placing an LC/MS instrument in your lab (e.g. space, construction, exhaust, air ventilation, chemical storage, ambient noise, etc.). The outcome of this meeting should result in a "Plan" to identify any deficiencies or requirements to ensure a safe lab environment. Actions needed to correct any lab deficiencies or meet safety requirements should be itemized and addressed. After the lab facility is set up, the safety plan should meet OSHA, other federal, state, and local regulations as well as be practical for staff.

If your organization does not have a *Safety Officer* you should attempt to establish this position before your instrument arrives. The easiest way to start is to form a "*Safety Committee*." This committee should be composed of both management and laboratory staff. Everyone must feel he or she is involved in the safety of the laboratory. Addressing your safety issues early in the process of implementing your LC/MS lab will create the right attitudes in your lab and prevent potential accidents, injuries and incidents in the future.[16]

Worksheet 7.3-Lab Safety Checklist

The following checklist is intended to prompt an evaluation of your safety issues relating to pre-installation laboratory setup for LC/MS:

Pre-Installation Lab Safety Checklist	Yes	No
1. Floor plan allows safe access to:		
• LC/MS and computer system?	❑	❑
• Mechanical pumps?	❑	❑
• Mechanical pump exhaust?	❑	❑
• Gas bottles/lines?	❑	❑
• Gas cylinder stand(s)/wall bracket(s) installed?	❑	❑
• Exhaust hood(s)?	❑	❑
• Fire alarm?	❑	❑
• Fire extinguishers?	❑	❑
• Storage areas?	❑	❑
• (*other*)	❑	❑
• (*other*)	❑	❑
• (*other*)	❑	❑
• (*other*)	❑	❑
2. Protocols for storage & disposal of:		
• Vacuum pump oil?	❑	❑
• LC solvents?	❑	❑
• LC samples?	❑	❑
• (*other*)	❑	❑
• (*other*)	❑	❑
• (*other*)	❑	❑
• (*other*)	❑	❑
3. Are your exhaust hoods rated for venting oil mist and LC solvents?	❑	❑
4. Is an evacuation plan set up?	❑	❑
5. *Safety Officer* contacted (or one been assigned)?	❑	❑
6. Does your facility have a *Safety Committee*?	❑	❑
7. Are there protocols to limit lab access?	❑	❑

Note: Copies of this worksheet can be downloaded from **www.LCMS.com**.

Record-Keeping Considerations

Maintaining good records is a necessary part of running an efficient lab. The increasing (and changing) number of government regulations makes it virtually impossible to avoid keeping detailed records. But just as important as the need to keep good records for the government and regulators, is the need to keep them for yourself and your organization. The success of your lab depends on them.

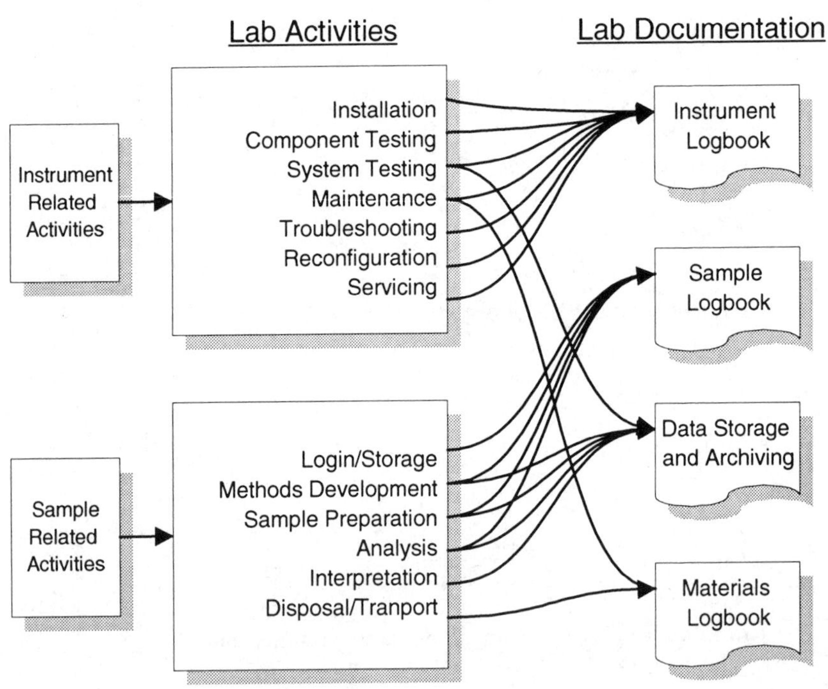

Figure 7.4- Evaluate all activities relating to your LC/MS lab to establish a documentation program.

Worksheet 7.4-Record-Keeping Checklist

To help ensure that your lab transactions are properly recorded and measure the effectiveness of your records system, use the following checklist. A no answer indicates an area that should be examined.

Pre-Installation Record-Keeping Checklist	Yes	No
1. Do you know the reasons for keeping good records?	❑	❑
2. Do you know what records to keep?	❑	❑
3. Do you know what regulations you must follow?	❑	❑
• Good Laboratory Practice (GLP)	❑	❑
• U.S. Occupational Safety and Health Administration	❑	❑
• U.S. Environmental Protection Agency	❑	❑
• (*other*)	❑	❑
• (*other*)	❑	❑
• (*other*)	❑	❑
4. Will you be able to verify the following with your records:		
• Last time maintenance was performed?	❑	❑
• Who performed maintenance?	❑	❑
• What was repaired?	❑	❑
• For how long was the instrument malfunctioning?	❑	❑
• What was the cause of the malfunction?	❑	❑
• Which samples were received?	❑	❑
• Which samples are being analyzed?	❑	❑
• Which samples were disposed of? Where and how?	❑	❑
• Who performed the analysis?	❑	❑
• (*other*)		
• (*other*)		
• (*other*)		
5. Are you familiar with all aspects of your SOP's?	❑	❑
6. Do you have records of *Lab Safety Audits*?	❑	❑
7. Do you have a review process set up for a periodic assessment of records?	❑	❑

Note: Copies of this worksheet can be downloaded from **www.LCMS.com**.

Pre-Installation Training

You should address three types of training before your instrument arrives, particularly if you are new to the field of LC/MS and unfamiliar with the new instrument.

The first type of training is to acquire (if available) a copy of the operators manual for the instrument you are purchasing. Review your *User Documentation* to familiarize yourself with all aspects of the instrument installation, operation, and maintenance. Learn the nomenclature and jargon associated with your instrument and the techniques that you intend to operate. This will assist you in communicating with the instrument company's installation and support staff. It will also allow you to effectively evaluate performance before you are asked to "sign off" on the installation.

The second type of training can occur by obtaining a copy of the software for your new instrument. Even a demonstration version of the software can familiarize you with the features of the software. In general, this exercise may only be applicable for post-acquisition data processing; but it may give you a head start toward becoming proficient at solving problems with your instrument.

The third type of training is to take a course. It is best to attend a course before your instrument arrives (preferably within a month or so of installation). There is typically a lag time (e.g. 3-6 months) between ordering and actual instrument delivery and setup. Attending a course during this period of time will familiarize you with all aspects of the instrument from general maintenance issues, to software, and many of the LC/MS techniques that never make it into the operator's manuals.[17] Read your manual and review the software before you attend a course. When the instrument finally arrives in several months, you will be much better prepared to evaluate and absorb the wealth of the information available at installation. Courses offered by your particular instrument manufacturer[18] are best, but if a course is not offered at the time you want, at minimum you should attend a general course on mass spectrometry interpretation or LC/MS offered by the various professional societies and organizations (*see Appendix A, Sources of LC/MS Information*).

7.2- While Installing Your LC/MS

The day has finally arrived, your LC/MS instrument is going to be set up. All the materials for installation have arrived and are waiting for installation. (It is always a good idea to check the shipping papers to ensure that all materials have arrived and are in good condition.) You have also previously arranged and scheduled an installation date and time. All preinstallation requirements should be addressed by this time so that you do not encounter needless delays and frustration at installation.

Setup

A *Service Engineer* from the instrument manufacturer will arrive at your facility to unpack and setup your instrument. The setup of a benchtop style instrument may take from 1-2 days, while a sector or FTMS-based system may take as long as 7-10 days. Whichever circumstance you are in, you should set aside the entire time period and be present during the entire installation. By not being present, you miss an opportunity to observe an expert (with your system) in action. You will see the assembly of the system, what tools to use, and how to use particular tools and software. You will go through component (subsystem) and system suitability tests, trouble shoot problems, and startup (and shutdown) the workstation and the LC/MS system. In general, you will learn many "tricks of the trade."[19]

Suitability Tests

Suitability Tests are designed to "check-out" a component or your entire LC/MS system to determine if it is operating properly. There are three types of tests: Component Suitability Tests, System Suitability Tests, and Performance Acceptance Tests. If the system passes these tests you are ready to starting running samples. If not, prescribed procedures can remedy the situation.

Some component tests are performed at the factory or by trained Service Engineers, while others can be performed by the operator. You should learn to perform all recommended suitability tests pertaining to your system.

Component Suitability Tests

Component Suitability Tests are designed to test and verify that individual components (e.g. liquid chromatograph, UV/Vis. or PDA detector and mass analyzer) are working as specified. They may require the analyst to check and verify that a particular light is either green or red; or, as is becoming more common today, the analyst initiates a computer program which puts the individual component through a series of steps (e.g. auto-tune or auto-optimization of the mass analyzer, start-up and calibration of the UV/Vis. detector) to verify that the individual component is working properly.

These tests are typically outlined and explained in the *Maintenance* or *Start-Up* sections of the *Users Manual.* These tests should become part of the scheduled maintenance of the instrument. Table 7.6 summarizes some of these tests.

Table 7.6- Component Suitability Tests

Typical Component Suitability Tests	
Liquid Chromatographic	• Flow rates calibration • Pressure suitability • Self diagnostic test (if any) for pump or accessory PDA or UV detector
Mass Analyzer	• Record mass analyzer and LC/MS interface pressures • Self diagnostic test (if any) for vacuum system • Tune ion optics and mass analyzer • Calibrate mass scale • Record and verify voltages/currents for particular lens or elements at a prescribed setting • Auto-tune/auto-calibrate settings for the LC interface and/or ion optics • Auto-gain settings for the detector

System Suitability Tests

System Suitability Tests, unlike the Component Suitability Tests, are performed to verify that the entire instrument is working as specified for a given system configuration. These tests are associated with a particular given setup (e.g. ES, APCI, PB) or a given method or application.

The system generally includes the LC, LC/MS interface, mass analyzer, data system, and data reduction programs. In some cases, System Suitability Tests are performed without a column in-line.[20]

These tests serve as a performance "benchmark" for the system in order to ensure it is working adequately for the intended use. These tests should be performed on a routine basis (e.g. daily or weekly) to verify that the system is in good working order. A typical *System Suitability Test* is summarized in Table 7.7.

Table 7.7- System Suitability Tests

Typical System Suitability Tests	
LC/MS System:	• Performed after the *Component Suitability Tests* are complete. • Apply to any specified system for any specified intended use. • A typical protocol would be: A. Setting up the mass analyzer: • to scan a specified range, at a specified scan rate (50-1000 u, at 1000 u/sec). • in a specified polarity (positive or negative). B. Setting up the LC system: • at a specified flow rate. • specified with or without a column in line. • with a particular column type and mobile phase. C. Configuring the interface (e.g. ES, APCI, PB, etc.): • to specified operating conditions. D. Specifying a mass accuracy over time period (e.g., +/- 0.1 u at m/z 670.3 for compound X/8 hr). E. Injecting a known amount of a component (caffeine) into the system and then recording a response.

Performance Acceptance Tests

Performance Acceptance Tests are specific *System Suitability Tests* that are prescribed by the manufacturer as indicative of their system operating as intended (published specifications). These tests are performed at the manufacturing facility to test a system's performance before shipping to a customers site. These tests are then again performed at the customer's site to verify that the installation is acceptable and the instrument sign-off can occur.

In some cases (not the general rule), the customer may request that a "customer-specified" *Performance Acceptance Test* be performed (in addition to the published specifications) as an acceptance criteria for a given instrument. The terms of this testing is usually negotiated in the *Sales Agreement.* Although this is a form of guarantee that a given instrument will perform the customer's application, it may add variables to the testing which create conflict and uncertainty in the system performance.

For example, many *Service Engineers* are not trained as *Applications Scientists*; rather, they have a strong background in electronics and system troubleshooting. They are quite capable of performing the standard tests, but run into difficulties when asked to perform unfamiliar methods, particularly if they encounter a problem. In some cases an *Applications Scientist* instead of a *Service Engineer* may be required to perform the "customer specified" *Performance Acceptance Test* on-site. This process generally adds time and cost to the installation process. Unless absolutely necessary, we suggest that customer-specified performance issues be sorted out during the instrument evaluation, not at installation!

Performance Acceptance Tests are good indicators of baseline system performance and indicators of performance variation over time. It is always a good idea to repeat these tests periodically to verify that your system is performing at or near the level it was capable of when installed. With proper maintenance, there is no reason to expect the performance of any LC/MS system to degrade in time.

A typical *Performance Acceptance Test* is summarized in Table 7.8.

Table 7.8- Performance Acceptance Test

	Typical Performance Acceptance Tests
LC/MS System:	• Performed after the Component and System Suitability Tests. • Generally specified in the final Sales Agreement. • Usually tests are very specific. • A particular application is specified by: A. Defining the mass analyzer conditions to: • scan a specified range, at a specified scan rate (e.g. 50-1000 u, at 1000 u/sec). • specify a polarity (e.g., positive or negative). B. Defining the LC conditions to: • specify a flow rate. • specify column in- or out-of-line. • specify column type and mobile phase. C. Specify which interface (e.g., Particle Beam, APCI, electrospray, etc.) • specify interface operating conditions. D. Injecting a specified amount of known compound(s) into the LC/MS system and acquiring data under specified acquisition conditions by: • injecting a series of different concentrations. • reducing the data into a report for evaluation. • meeting a detection limit, dynamic range, %RSD and mass accuracy over a specified time, etc.

Itemized List ("Punch List")

A list of items relating to the installation should be compiled during the installation process, including, ordered items that were not received at the customers site, damaged items, back-ordered items, malfunctioning items at installation, and acceptance tests that were not performed because of missing or malfunctioning parts. Make sure you keep excellent documentation of the discrepancies and omissions. Go over the list with the *Service Engineer.*

Before the *Service Engineer* leaves your lab you should establish a plan to reconcile this list of items. Have the list signed by the *Service Engineer* to verify that these open issues exist. Keep a copy of the list for your records. The resolution of these issues may affect the official sign-off date and warranty period of your instrument. Make sure that you and your manufacturer have a clear understanding of the requirements and consequences of each item on the list.

Note, when compiling your punch list, do not expect the *Service Engineer* to sort out every detail on the list. They are usually quite capable of addressing the technical issues; however, issues relating to the *Sales Agreement*, terms, and warranty are usually handled by your *Sales Representative.*

Table 7.9- Make Sure You Have Spare Parts

Typical Spare Parts	
Why include spare parts?	• Items that are hard to come by (you may have to purchase 50 pieces when you only need 1 or 2).
	• Items you will probably use within the first year of operation (LC fittings, tubing, O-rings, vacuum pump oil).
	• Items you can only obtain from the manufacturer.
Spare parts.	• For vacuum components (vacuum pump oil, O-rings)
	• For each LC/MS interface, or source (O-rings, filaments, needles).
	• For liquid chromatograph components

Spare Parts

Spare parts are typically included in the shipped items for new instruments (see Table 7.9). Review the list of spare parts and request that the *Service Engineer* explain the purpose (or function) of each item. In addition to spare parts, you may have purchased "spare items and accessories," such as, additional LC/MS interfaces, filaments, ion gauges, LC UV/Visible lamps, or MS detectors. If you have, make sure they are on the shipping papers.

Tools

A comprehensive listing of the tools you may need during installation (and after installation) for your LC/MS lab is difficult without specifying the exact instrument being installed into your lab. Today more and more instruments do not require tools for assembling or disassembly. We can, however, recommend that you acquire certain tools in addition to your standard set of Allen wrenches and screwdrivers (see Table 7.10).

Table 7.10- Tools for the LC/MS Lab

Tools	Uses
Microscope	• To observe small objects (needles, filaments, apertures). Preferably a dissecting scope with a long focal length and a large depth of field.
Voltmeter	• To test continuity and voltages for the entire LC/MS system. Preferably equipped with an audio output so you can hear and see output.
Flexible Gripper Pick-up Fingers	• To grasp any object that falls into the system. (e.g. small screws, nuts, O-rings, etc.)
Dental Mirror	• To allow you to see out of the way places, particularly inside of the vacuum system.
Gloves	• Cotton for handling vacuum components. Latex or vinyl for chemicals and oil handling.
Tweezers	• To pick up small objects.
Cotton Swabs	• For cleaning components.

Plans for Maintenance

Review the maintenance schedule of the LC/MS system with your *Service Engineer*. With his or her knowledge of the instrument and the schedule specified in the *Instrument Manual*, you should be able to compile a list of diagnostic indicators that would signal when maintenance is needed, and activities and dates when maintenance should be performed.

At this time you can discuss the need for a *Service Contract* or, if a Service Contract[21] has already been included in the final *Sales Agreement*, you can set up a time and date for the first scheduled service visit (sometimes called a PM visit, for *Preventive Maintenance*).

Plans for Record-Keeping

Detailed documentation is crucial to the operation and validation of any analytical instrument; therefore, record all steps involved in the initial installation of your LC/MS system and keep the information in an "instrument logbook." Use any forms or schedules supplied with the instrument for preinstallation, maintenance, or troubleshooting. Incorporate these forms into your record-keeping. Documentation about your LC/MS instrument should also include analytical information (number of analytical sequences run, type and number of samples analyzed, appearance of sample, etc.) and any troubleshooting or future repair work performed on the system.

Taking time to set up and maintain your record-keeping system- will save time and minimize unscheduled downtime. Instead of having to hunt for the information you need you already have it on-hand waiting to be used. Table 7.11 lists some typical logbooks you should have on-hand.[22]

Table 7.11- Logbooks

Typical Logbooks	
• Standard Operating Procedures • Instrument Logbook	• Used Oil & Liquid Waste • Sample Log (login) • Sample Analysis

Instrument Company Contacts

Having the name of someone who knows your instrument is a must. You already have two, your *Sales Representative* and the *Service Engineer* who is installing your instrument. Get their business cards. The instrument manufacturers have entire departments dedicated to supporting customers.[23] Get to know who they are.

In addition, see if there is an LC/MS or Mass Spectrometry discussion group in your area. Here you may find other analysts using your instrument or be introduced to someone who is also trying to solve similar problems (see *Appendix A, Resources*). Do not overlook or discount the experiences of other individuals.

Training at Installation

By being around and asking questions of your *Service Engineer* as your instrument is being set up, you can learn a lot about your instrument. If you have already attended a training course, you can add the *Service Engineer's* knowledge to what you learned previously.

Just by watching how the individual interacts with the instrument (e.g. connecting fittings, pumping the vacuum system down, calibrating the mass analyzer, troubleshooting a problem, etc.) is an invaluable experience. The majority of problems that new users encounter with an instrument are not related to the LC/MS system not working properly; rather they are related to inexperience and lack of experience of the operator![24]

Worksheet 7.5-Installation Checklist

To help ensure that your the installation of your instrument is complete, fill out the following checklist. A "no" answer indicates areas that need to be addressed before the installation is completed and the Service Engineer departs.

Installation Checklist	Yes	No
Part A. Pre-Installation:		
1. Instrument manufacturer contacted about delivery and installation dates?	❑	❑
2. Pre-Installation *Checklists* completed?	❑	❑
3. Have you assigned someone to be present during the instrument installation?	❑	❑
4. Has the *Facility Manager* been contacted about delivery date and any special needs?	❑	❑
Part B. Installation:		
1. Was the instrument set up successful?	❑	❑
2. Did you save the packing containers the system arrived in?	❑	❑
3. Are the vacuum pumps properly exhausted?	❑	❑
4. Are all the required *Performance Acceptance Tests* performed?	❑	❑
• did the system pass all the tests? If not, which one(s) and why?	❑	❑
5. Are there any "customer-specified" *Performance Acceptance Test(s)* that were previously agreed upon in the sales contract?	❑	❑
• if yes, were those tests performed and did they pass? If not, which ones and why?	❑	❑
6. Did you review and compare:	❑	❑
• the original order with the shipping list? ... the spares list? Where there any omissions?	❑	❑
7. Did you go over the maintenance schedule?	❑	❑
8. Did you complete the *Sign-Off Sheet* to indicate a completed and acceptable installation?	❑	❑
9. Did you obtain a business card from the *Service Engineer*?	❑	❑

Note: Copies of this worksheet can be downloaded from **www.LCMS.com**.

7.3-After Installing Your LC/MS

The *Service Engineer* has left, the shipping crates are put away, and your hectic week of installing and learning about a new instrument is over. Now you have time to sit down and organize yourself and your lab, and prepare to use your instrument.

Maintenance

Review the discussions you had with the *Service Engineer* about the recommended maintenance schedule and start a maintenance logbook. Your logbook should include schedules and procedures pertaining to both Scheduled and Routine Maintenance.

Scheduled Maintenance

Scheduled Maintenance procedures address issues on a regular basis: reducing "down time", prolonging system life, and reducing overall operating costs. These procedures are performed at regular intervals. A log book of system performance characteristics and maintenance should be kept so that variance from normal operation can be readily detected and corrective action taken.

Routine Maintenance

Routine Maintenance is typically carried out in conjunction with instrument troubleshooting. Typically a problem arises (e.g. low sensitivity, vacuum chamber pressure too high). After diagnosing the problem, the routine maintenance procedures are used to remedy the problem.

Service Contracts

Service Contracts are available for your instrument from your instrument manufacturer.[25] The price for a Service Contract is typically 5-20% of the price of the instrument.[26] Services offered range from periodic visits by a *Service Engineer* for scheduled maintenance or repair of individual components, to emergency service involving a *Service Engineer* visiting your site within 24 hours to remedy your problem.

Worksheet 7.6-Maintenance Checklist

To help ensure that the maintenance schedule for your instrument is complete, answer the questions in the worksheet, Evaluating Maintenance Schedule. Any "no" answer indicates areas that need to be addressed for a complete Maintenance Schedule.

Pre-Installation Maintenance Checklist	Yes	No
Part A. Scheduled Maintenance:		
1. Changing the oil in the vacuum pumps?	❑	❑
2. Replacing and replenishing calibration fluids?	❑	❑
3. Replacing EI/CI filaments, API discharge, or electrospray needles?	❑	❑
4. Replacing fan filters?	❑	❑
Part B. Routine Maintenance:		
1. Procedures reviewed for:		
• Vacuum System components?	❑	❑
• Electrical components?	❑	❑
• Computer and Network components?	❑	❑
2. Procedures reviewed for Replacing and Installing:	❑	❑
• LC/MS Interfaces?	❑	❑
• Ion Sources?	❑	❑
• Mass Analyzer Detector?	❑	❑
3. Procedures reviewed for Cleaning:	❑	❑
• LC/MS Interfaces?	❑	❑
• Mass Analyzer: Ion Optics?	❑	❑
• Mass Analyzer?	❑	❑
4. Procedures reviewed for Liquid Chromatograph components?	❑	❑
5. Did you obtain a business card from the *Service Engineer*?	❑	❑
6. Did you obtain the phone number of the Service and Application Departments?	❑	❑
• Phone:	❑	❑
• Phone:		

Note: Copies of this worksheet can be downloaded from **www.LCMS.com**.

Safety

Now is the time to schedule a walk-through of your lab with your Safety Officer. You should concern yourself with the completion of your Pre-Installation Safety Check (*Worksheet 7.3- Evaluation of Lab Safety*) and any new problems that you and the Safety Officer discover on the walk-through.

When addressing safety considerations relating specifically to your LC/MS lab, keep in mind that most of the applications in chemistry, biotechnology, and pharmaceutical sciences involve the use of bioactive compounds.[27] Handling these compounds with exposed skin or breathing these compounds from LC/MS aerosols can ultimately affect the health of the lab personnel.

Always check your *Instrument Operators Manual* for *Warning* or potential *Hazards* that you should avoid. Table 7.12 summarizes some general safety considerations.

Table 7.12- Safety Considerations

Item	General Considerations
Hot Surfaces: Vacuum Pumps	Mechanical vacuum pumps operate at ~100 °C.
Hot Surfaces: LC/MS Interfaces	• Newer instruments – all hot surfaces are isolated from the operator. • Older instruments – some hot surfaces are exposed; note warnings. • Atmospheric Pressure Interfaces – APCI: heated nebulizers (100 - 400 °C) – ES: heated spray (200-400°C, TurboSpray) – APCI/ES: heated apertures (25- 100°C) • Particle Beam – Heated nebulizer (25-100°C, thermo-pneumatic) – Heated interface (25 -100°C) – Heated ionization region (50 - 300 °C) • Thermospray – Heated probe (25-100°C) – Heated source (100-300°C)

Table 7.12- Safety Considerations(cont.)

Item	General Considerations
Explosions	• AP LC/MS Interface – spraying organic solvents has been reported to result in explosions[28]
High Voltage	• High voltage is required for operation with: – all mass analyzers – CF-FAB LC/MS Interfaces – API LC/MS Interfaces • Access to high voltage is prevented by interlock systems which when activated (or disabled), turn the voltage off – newer instruments, all voltage sources are isolated from the operator – older instruments, may not have interlocks
Weight	• Mechanical Vacuum Pumps are heavy (~75 lb) • They are not equipped with handles and they should be transported on carts or dollies, or lifted by block and tackle • Mass Analyzers – range ~100 lb to several hundred pounds • LC Equipment – ~100 lb • Computers/Workstations – ~100 lb
Aerosols	• Exposure by inhalation; mist • Are associated with almost every LC/MS interfacing device • Aerosols should be treated with extreme caution since aerosol transport is highly dispersive and an extremely effective means of exposure through the lungs • Spraying LC effluent indiscriminately into your lab is an unsafe practice and should be avoided
Exposure by Touch	• Wear protective clothing, gloves, etc. when handling chemicals
Glass Gauges	• Covered by metal shields or concealed by interlocked doors/covers – newer instruments, all glass gauges are isolated from the operator – older instruments, may not have interlocks or be covered

Worksheet 7.7-Safety Checklist

To make sure that you have set up a safe lab environment, use the following worksheet to evaluate your facilities after the installation of your instrument.

Installation Safety Checklist	Yes	No
1. Pre-installation *Safety Evaluation* completed?	❑	❑
2. Were all deficiencies resolved?	❑	❑
3. Scheduled a walk-through of your lab with your *Safety Officer*?	❑	❑
4. Were any deficiencies found during the inspection?	❑	❑
5. If so, was an action plan composed to address these new areas?	❑	❑
6. Have you put together a *Safety Manual* with required safety procedures?	❑	❑
7. Have you established a *Safety Logbook*?	❑	❑
8. Does anyone on your staff require a review of the *Safety Manual*?	❑	❑
Name:________________________		
Name:________________________		
Name:________________________		
9. If so, schedule a date for *Safety* training:	❑	❑
Date:________________________		
Date:________________________		
10.Schedule future *Safety Inspections* dates:	❑	❑
Date:____________Inspector______________		
Date:____________Inspector______________		
Date:____________Inspector______________		
Date:____________Inspector______________		

Note: Copies of this worksheet can be downloaded from **www.LCMS.com**.

Post-Installation Record-Keeping

Now is the time to start establishing log books. You should also start putting together *Standard Operating Procedures* (SOP's) for the various functions of your lab from how to log in samples, perform LC/MS maintenance and system suitability; storage of waste oils and solvent; to reporting the results and storage of old samples.

Worksheet 7.8-Record Keeping Checklist

To make sure that you have the necessary record keeping and have set up the necessary SOP's, use the following worksheet,

Post-Installation Safety Checklist	Yes	No
1. Pre-installation *Record Keeping Evaluation* completed?	❑	❑
2. Were all deficiencies resolved and systems established?	❑	❑
3. Logbooks were established for:	❑	❑
• maintenance?	❑	❑
• samples?	❑	❑
• safety?	❑	❑
• instrument?	❑	❑
• sample disposal?	❑	❑
4. SOP's were established for:	❑	❑
• system maintenance?	❑	❑
• system suitability?	❑	❑
• sample login?	❑	❑
• methods development?	❑	❑
• system shut down?	❑	❑
• regulations (e.g. GLP, OSHA, EPA)?	❑	❑
• fire drill & lab evacuation?	❑	❑
• oil and solvent storage/disposal?	❑	❑

Note: Copies of this worksheet can be downloaded from **www.LCMS.com**.

Post-Installation Training

If you have not attended the manufacturer's training course before your instrument was installed, now is a good time to plan to attend a course. Call your local *Sales Representative* or *Service Engineer* to inquire about times and dates for the upcoming courses. Also, check out their Home Page- course dates and places are posted (see *Appendix A*).

If you have already attended the manufacturer's course, it is now time to start thinking about the future. It has been suggested that what you know today will be obsolete within 3-5 years.[29] What this suggests is that it is paramount to stay knowledgeable about your specialty. By reading and attending advance courses on LC/MS and your analytical specialty you will stay current on techniques and various applications.

Discussion Questions and Exercises

1. Compile a list of all activities that will occur in your LC/MS laboratory. Then, identify a method for recording that activity.
2. Prepare a to-scale drawing of your laboratory and label the location of all activities occurring in the lab. Does the space allotted for each activity make sense? Is the flow of materials and people through the lab logical? Do people have to needlessly step across power lines? ...communications lines? ...exhaust lines? ...vacuum lines?
3. Itemize all potential health and safety hazards in your LC/MS laboratory. How is each item on this list being addressed?
4. Complete *Worksheets 7.0 to 7.8* to address all aspects of implementing an LC/MS laboratory.

Notes

---

[1] Gause, D.C., Weinberg, G.M., *Exploring Requirements: Quality Before Designs*, Dorset House Publishing: New York (1989). Quote from Preface.

[2] Benchtop style system is defined as an LC/MS system that typically weighs ~100 lb (45 kg) and is designed to sit on a lab bench.

[3] Alternative length power and network cables are usually available for unusual installations.

[4] Mechanical vacuum pumps typically weigh 75-100 lb (34-45 kg). A dolly is required to move pumps from one location to another. Note: mechanical pumps are also referred to as rotary vane or "roughing" pumps.

[5] If pumps are placed in an adjoining room, there will be some need for constructing an access way from the LC/MS lab into the adjoining room for the vacuum tubing and electrical power cables.

[6] Most mechanical pumps sold today for LC/MS systems come already equipped with traps and filters.

[7] Compared to an LC instrument, an LC/MS instrument usually has "one" dedicated power line (@220V, 24 amps) for the mass analyzer. Some instruments require several 120V (20 amps) lines but today this is becoming less common. The power cables for mechanical vacuum pumps may be either plugged into the mass analyzer or to separate outlets. Separate receptacles must be available for the liquid chromatograph (pumps, column heater, accessory detector).

[8] Some of the cables supplied must be specific lengths and type to conform with engineering and safety standards. Replacing those cables may be hazardous, invalidate your warranty and may also degrade the performance of your equipment. Before replacing, check with the manufacturer concerning alternatives. Not all power cords and plugs have the same current rating. Household extension cords do not have overload protection, and are not meant for use with vacuum pumps or computer systems. Contact your facilities manager or a qualified electrician if you are not sure what type of power cord or plug is required for your lab or building. When planning the placement of equipment, remember that each of the individual units (e.g., LC pumps, workstation, printer) may require access by way of a separate power cord to a power outlet.

[9] OSHA Laboratory Standard #1910.1450. It is against the law to expose personnel to harmful situations (Occupational exposure to hazardous chemicals in laboratories). For a copy of the standard, see OSHA's home page: `www.osha.gov`.

[10] OSHA Occupational Noise Exposure (OSHA Standard # 1910.95). For a copy of the standard, see OSHA's home page: `www.osha.gov`.

[11] Even at ~15 feet, larger I.D. tubing is a good idea. Larger tubing (e.g. 1.5-in I.D.) will diminish the chance of oil backstreaming from the mechanical pump to the vacuum chamber and decrease the work the mechanical pump must do to evacuate the vacuum line!

[12] Check out their web site: `www.edwards.boc.com`.

[13] Since the majority of instruments purchased will be equipped with both an electrospray and an APCI interface, we recommend the use of mechanical vacuum pumps that can handle the higher flow rates of 0.1-1.5 mL/min.

[14] Maltz, G.A., "A dedicated chemical hygiene officer is the key to running a safety-compliant lab," Inside Laboratory Management, 1, pages 21-22, 1997.

[15] OSHA Laboratory Standard #1910.1450, *op cit.*

[16] Maltz, G.A., *op cit.*

[17] Tricks of the trade. Many techniques (usually referred to as "tricks of the trade") never made it into the literature, but resided in the "head" of the analyst. But as more analysts use LC/MS instruments and as methods become standardized, these tricks are being written down and disseminated throughout the LC/MS community. But be forewarned, there are many tricks still out there that are not yet fully documented. To find out about these tricks attend LC/MS courses, ask analyst that are currently using LC/MS, and one of the best resources of information is the Application Chemist from the respective instrument companies (they have the most experience with your particular type of instrument).

[18] Three to five day courses offered by the various manufacturers are offered 3-4 times a year and typically cost between $1,500 to $2,500 (room and board not included).

[19] Tricks of the trade, *op cit.*

[20] Testing LC/MS system performance without a column in-line is a common practice using continuous infusion of sample or flow injections. This columnless system configuration allows the fluid delivery and mass detection system to be evaluated without the uncertainty of column-to-column variability.

[21] Service Contract. The cost of a Service Contract for an LC/MS system is typically 5-20% of the initial cost of the equipment. "Many managers have found that with the increasing complexity of instrumentation (increase modularity of instruments, increase in the use of computers and software to control instruments), a service contract is a must. Terms are usually negotiable, but for a contract covering parts and labor for a single major instrument, to take affect at the end of the warran-

tee period, an initial fee of ~10% of the original cost is common." (Milner, O.I., "Chapter 11, Capital Investments," pages 137-147, IN: *Successful Management of the Analytical Laboratory*, Lewis Publishers: Boca Raton (1992)).

[22] Alternatively, today more and more manual documentation is being replaced by computerized forms. If it is available, use it. But there is no good (or perfect) replacement for at least some (or partial) manual documentation.

[23] Today the majority of service staff at most instrument companies are degree-level chemists, acting as problem solvers (Anonymous, "ThermoQuest - on track for $1 billion," *Analytical Instrument Industry Report* 14, page 4 (July 30th, 1997)). But there are situations when the *Service Engineer* installing your instrument has the training for only setting up and running the Component and System Suitability Tests, not solving your experimental problems. You will have to contact the Manufacturer's Application or Service Department for the help you need.

[24] Instrument manufacturers have come to realize that the majority (~85%) of service calls are <u>not to repair a broken instrument</u>; rather what the customer really wants to know is <u>how to do a particular experiment</u> (Anonymous, "ThermoQuest - On track for $1 billion," *Analytical Instrument Industry Report* 14, page 4 (July 30th, 1997)).

[25] For used or older instruments, the original manufacturer and after-market service companies will provide you with a service contract.

[26] Service Contract, *op cit.*

[27] Safety. The tragic death of Karen Wetterhalm from dimethylmercury poisoning (C&EN 75, page 12, June 16, 1997 and Goldberg, C., "Colleagues vow to learn from chemist's death," The New York Times, Page A7, October 3, 1997.)

[28] Always check your *Operator's Manual* and with the instrument manufacturer to inquire about the use of any LC solvent mixture you may have concerns about (particularly with normal phase LC mixtures, see Harbol, K.L., Morgan, D.G., Kitrinos, N.P. Rodriguez, A.A., "Comparison of APCI and electrospray ionization with normal phase LC/MS," Proceedings of the 45th ASMS Conference on Mass Spectrometry and Allied Topics, Palm Springs, California, June 1-5, 1997.) Abstracts can be down-loaded from the ASMS home page: `www.asms.org`.

[29] The occupational half-life- that is, the span of time it takes for one-half of work skills to become obsolete- has declined from seven to fourteen years, to three to five years (Davis, S., Botkin, J., "Chapter 4, Learning Power," pages 84- 108, IN: *Monster Under the Bed*, Simon and Shuster: New York, 1994.)

Part III

You have problems to solve with LC/MS

1. Intelligence

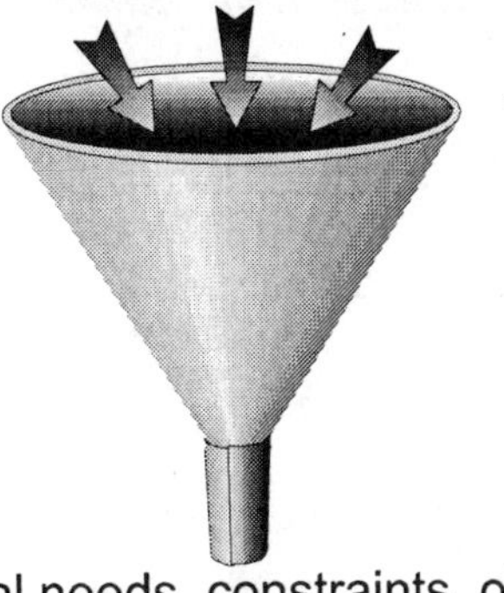

Chapter Eight- The Prelude to Solving Problems (Problem Definition)

Experimental needs, constraints, obstacles

2. Design

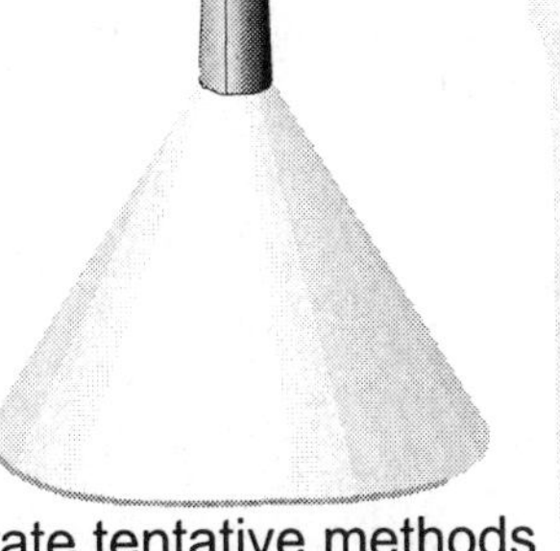

Chapter Nine- Solving Identity Problems

Chapter Ten- Solving Quantification Problems

Generate tentative methods

3. Choice

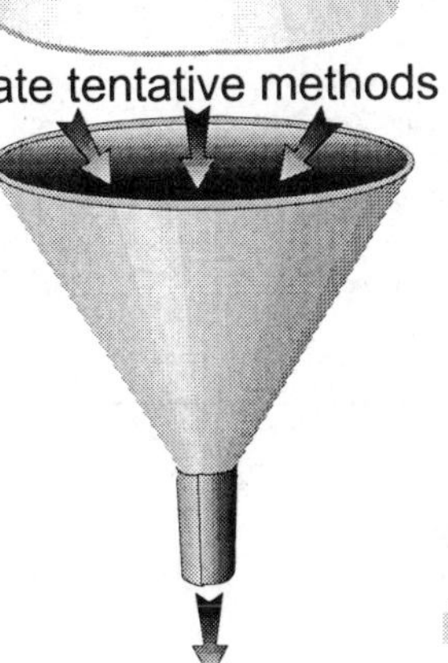

Chapter Eleven- Solving Target Problems

Chapter Twelve- Developing Methods

Implement your LC/MS methods

Chapter Eight- Flow Diagram

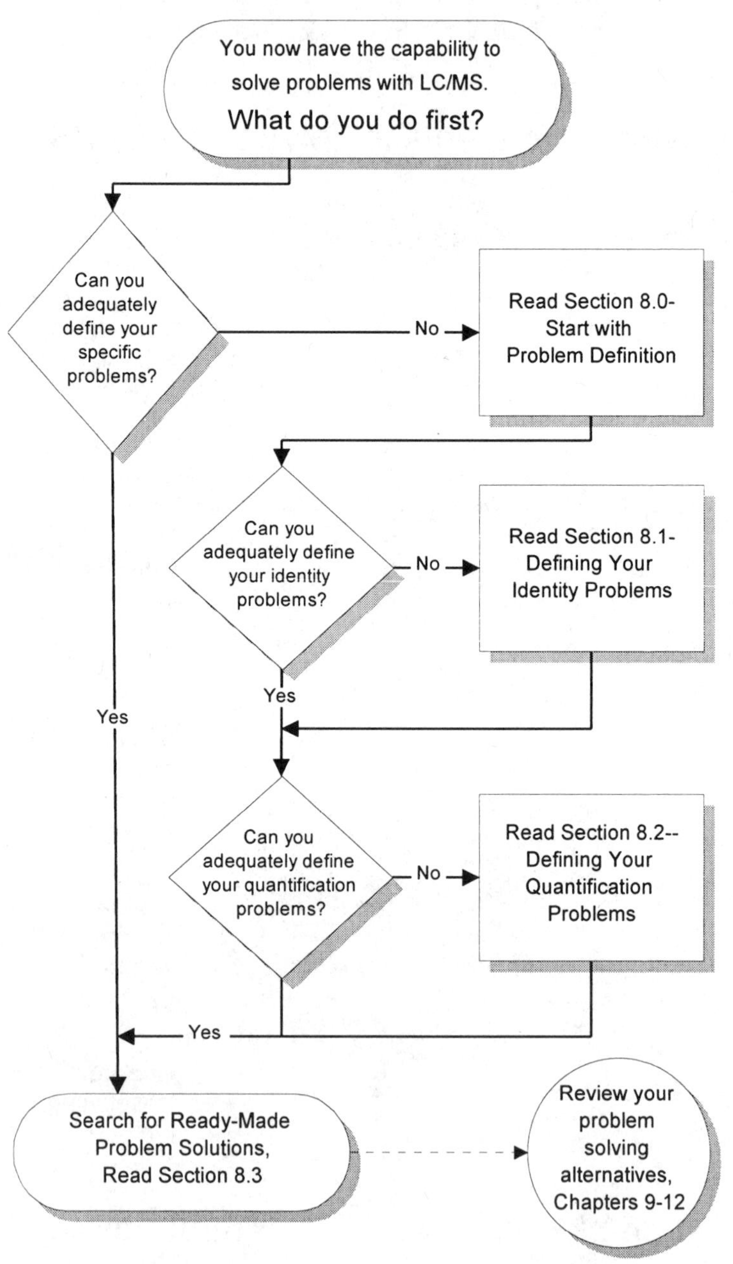

You now have the capability to solve problems with LC/MS.
What do you do first?
Can you adequately define your specific problems?
No
Read Section 8.0- Start with Problem Definition
Can you adequately define your identity problems?
No
Read Section 8.1- Defining Your Identity Problems
Yes
Yes
Can you adequately define your quantification problems?
No
Read Section 8.2-- Defining Your Quantification Problems
Yes
Search for Ready-Made Problem Solutions, Read Section 8.3
Review your problem solving alternatives, Chapters 9-12

Chapter Eight

The Prelude to Solving Problems

(or, The Prelude to Running Samples)

No spreadsheet can create data where there is none. No word processor can help me write better. No on-line database can answer the tough questions.........those which do not yet have answers.[1]

Clifford Stoll, in ***Silicon Snake Oil***

8.0- Start with Problem Definition

You now have the capability to run samples. You may have an instrument in your lab or the ability to send samples outside your lab in order to meet your specific LC/MS problem needs. You have already made the appropriate decisions as to which technologies are generally suited to your specific problems. Armed with these capabilities you are ready to solve problems.

Part III is included in this text to provide you some direction in the process of solving your practical (everyday) problems in LC/MS. How do you deal with real-world problems? How do you deal with real-world samples? Where do you start? The definition process here is not any different than prescribed earlier with respect to selecting technologies and acquiring instrumentation. Always start with a thorough and detailed problem definition. To solve your measurement problems with LC/MS you should first consider requirements of your problem in terms of experimental *needs*, *constraints*, and *obstacles* as outlined in Figure 8.0. An exercise for developing your specific problem definition is provided in Worksheet 8.0. Note, a lack of clear definition may result in significant waste of time and resources, so it is wise to take this process seriously.

Defining Your Specific Problems

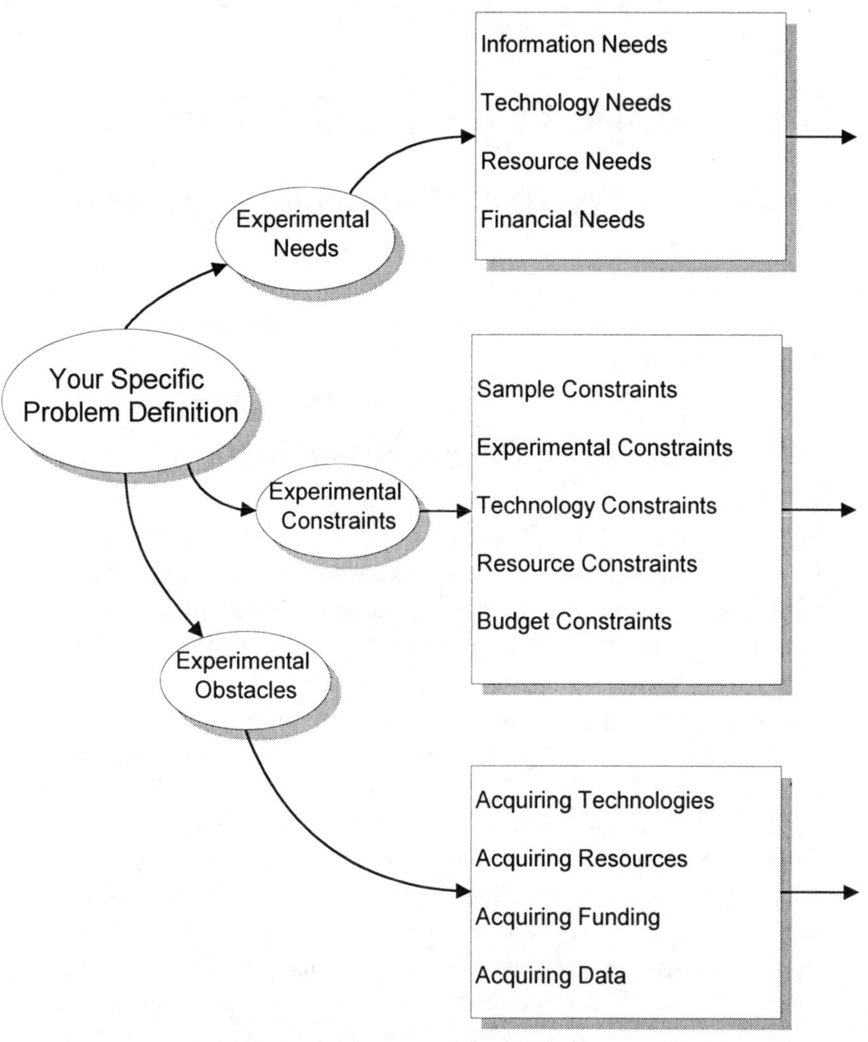

Figure 8.0- Solving your sample specific problems requires that you adequately define your experimental *needs*, *constraints*, and *obstacles* (both pages).

Matching Your Needs to Alternatives

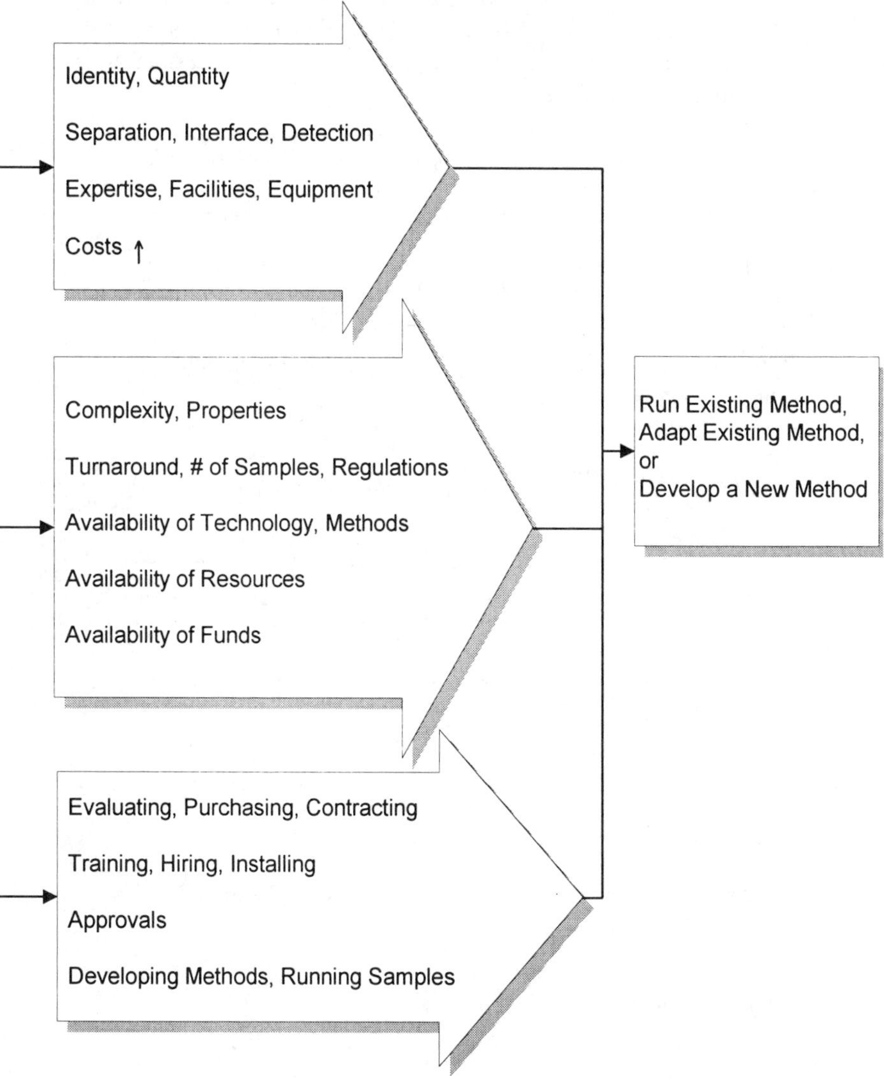

Worksheet 8.0-Defining Your Specific Problems

Part A- What information do you have at the outset of your problem-solving process? List any available information about the nature of your problem, sample, problem stimulus, contacts, sources, customers, etc. This should serve as a starting point.

____

____

____

Part B- Defining your experimental **needs**. Check-off your specific problem requirements and comment on each item.

What are your specific information needs?	Yes	No	Comments
Do you require:			
– identification of analyte(s)?	❑	❑	Which ones?__________
– quantification?	❑	❑	Which ones?__________
– MW information?	❑	❑	Why?__________
– elemental composition?	❑	❑	Why?__________
– structural information?	❑	❑	Why?__________
– compound specificity?	❑	❑	Why?__________
– enhanced response?	❑	❑	How?__________

What are your specific technology needs?	Yes	No	Comments
Do you require:			
– a specific separation?	❑	❑	Why?__________
– a specific inter-face/ionization?	❑	❑	Why?__________
– a specific mass analyzer?	❑	❑	Why?__________

Note: Copies of this worksheet can be downloaded from `www.LCMS.com`.

What are your <u>resource needs</u>?	Yes	No	Comments
Do you require:			
– specialized expertise?	❑	❑	Who?________
– special facilities?	❑	❑	What?________
– special equipment?	❑	❑	What?________

What are your <u>financial needs</u>?	Yes	No	Comments
Do you require:			
– additional funding?	❑	❑	What?________
– special approval?	❑	❑	What?________

Part C- Defining your experimental **constraints**. Check off your specific problem requirements and comment on each item.

What are your <u>sample constraints</u>?	Yes	No	Comments
Do you require:			
– isolation of components?	❑	❑	Why?________
– removal of interferences?	❑	❑	Which ones?________
Do you have:			
– unstable components?	❑	❑	Storage?________
– involatile components?	❑	❑	Volatility?________
– high MW species?	❑	❑	Limits?________
– special handling ?	❑	❑	What?________

What are your <u>experimental constraints</u>?	Yes	No	Comments
Do you have:			
– specific time constraints?	❑	❑	What?________
– equipment constraints?	❑	❑	What?________
– specific sample constraints?	❑	❑	What?________
– regulatory constraints?	❑	❑	What?________
– specific legal constraints?	❑	❑	What?________

Note: Copies of this worksheet can be downloaded from `www.LCMS.com`.

What are your technology constraints?	Yes	No	Comments
Is required LC/MS technology available?	❑	❑	Where? ________
Are existing methods available?	❑	❑	Which? ________
Are existing methods adaptable?	❑	❑	How? ________

What are your resource constraints?	Yes	No	Comments
Is required expertise available?	❑	❑	Who? ________
Are required facilities available?	❑	❑	Which? ________
Is required LC/MS equipment available?	❑	❑	When? ________

What are your budget constraints?	Yes	No	Comments
Do you have a cost/sample constraint?	❑	❑	How much? ________
Do you have a project cost constraint?	❑	❑	What? ________

Part D- Defining your experimental **obstacles**. Check off your specific problem requirements and comment on each item.

What are your experimental obstacles?	Yes	No	Comments
Do you have to develop a method?	❑	❑	When? Who? ________
Can you account for all interferences?	❑	❑	How? ________
Can you get standards?	❑	❑	Where? ________
Can the samples be run?	❑	❑	When? By whom? ________

Part E- Restate your problem in terms of the above **needs**, **constraints**, and **obstacles**. List any additional information required to fully define your problem.

____

____

____

____

____

Note: Copies of this worksheet can be downloaded from `www.LCMS.com`.

The technique diagrammed in Figure 8.0 and evaluated in Worksheet 8.0 for defining your problem is known as *Goal Orientation* which has the primary purpose of clarifying your goals and objective.[2] This process redefines your problem in terms of *needs*, *obstacles*, and *constraints*. It involves writing down a general description of your problem, being sure to include all relevant information; then asking yourself three important questions:

1) What do I want to accomplish? (*determine your needs*)
2) What restrictions must I accept in order to solve the problem? (*determine your constraints*)
3) What might prevent me from accomplishing what I want? (*determine your obstacles*)

The answers to these three questions are then used as a guideline to redefine your original problem statement. This approach is intended to generate a more accurate problem definition. A more accurate problem definition will allow you to more easily focus your resources on the appropriate solution to your problem.

Sample submission forms (Figure 8.1) are often used in laboratories performing instrumental support. These forms are a form of problem definition; however, in many cases they are technique oriented, not problem oriented. Do not be fooled into thinking that filling out a *sample submission form* is a replacement for complete problem definition. Compare Worksheet 8.0 to Fig. 8.1.

The importance of complete, accurate, and up-to-date problem definition cannot be overstated. Your choice of appropriate methods and LC/MS technologies as applied to solving your specific problem is determined primarily by this definition. Also, keep in mind that this process is dynamic, you may receive new information at a later date that forces you to redefine your problem in terms of the new information. You should always be evaluating and reanalyzing your initial problem definition and assumptions throughout the entire problem solving process.

General Problem Types

Problems in LC/MS, as in other fields of endeavor, can be classified into a variety of types. For this general discussion we describe problem types in terms of *well-structured*, *semi-structured*, and *ill-structured* problems. Table 8.0 lists these problem types with respect to their uncertainty, information availability, problem-solving approaches, and keys to success.

FOREST LABORATORIES, INC.
Bioanalytical R+D Department

LC/MS SAMPLE LOG-IN / SAMPLE ANALYSIS FORM

ID#: ____/____/_ **Date Needed:** ☐ **Immediate**
other: ____________
Chrom's attached: ☐ (check)

Sample Sent By:	Telephone#:
Department:	Date:
**Sample ID#: MDD __ __ _	NB#/PP:
Compound Type:	
Number of Samples:	Diluent/Solvent:
Storage Conditions:	
Empirical Formula/Molecular Weight:	
Structure:	Other Details:
HPLC Conditions (if purified by HPLC): *Column Type:* *Size:* *SN#:* *MP:* *Inj. Vol:* *Flow:*	
Sample Solubility:	
Any Special sample handling, hazards, or stability problems:	
MSDS Sheets Provided: Yes/NO	please circle
TO BE COMPLETED BY BIOANALYTICAL R&D STAFF	
Received by:	Date:
Storage:	
Analyst:	Date:
Technique:	
Instrument:	Instrument Book:
Sample Fate:	

___/___ __/ _(Book # / PG # / Spl Run #)(EX: M01016b)

___/_____ / _(MDD/month-day-year/spl #)(ex: MMD073196A)

11/4/96z:\bb_mdd\FORMLCMS.XLS

Figure 8.1- Typical sample submission form for a contract or support lab. Courtesy of Forest Laboratories, Inc.[3]

Well-structured problems are characterized by the availability of all information about the problem and a clear direction toward a solution. Little to no uncertainty is involved. In chemical analysis, we address *well-structured problems* with SOPs, standard methods, and target methods. *Well-structured problems* can effectively be addressed with a "cookbook" approach. These problems are generally routine and repetitive. There are now (and will continue to be) an increasing number of methods in LC/MS to *solve well-structured problems* (See Appendix B). We predict that *well-structured problem* solutions will become the most common application of LC/MS in the not too distant future; particularly since much of the growth in LC/MS is being driven by the pharmaceutical industry where well-structured problem solutions are a prerequisite for product approval. *Semi-structured problems* are characterized by a partial definition of the nature of the problem.

Semi-structured problems contain some level of uncertainty that must be addressed in the problem solving process. The uncertainty about the actual state or desired state; or about how to close the problem gap precludes the use of routine procedures.[4] This type of problem can effectively be solved by protocols or guidelines (Also see Appendix B). Protocols or guidelines give the problem-solver a general framework within which to solve their problem. Usually the problem-solver has the responsibility of filling in missing or new problem information. This could be described as the "fill in the blanks" approach to problem solving. Methods development is inherently a *semi-structured problem* and lends itself well to protocols (See *Chapter Twelve*). Identification of unknowns is generally a *semi-structured problem* (See *Chapter Nine*). Most of the applications for profiling and identifying metabolites of drugs fall into semi-structured problem types.

Ill-structured problems are characterized by little or no information about the problem state or any approach to solving the problem. *Ill-structured problems* are the most difficult and challenging in analytical chemistry. There are many areas of research using LC/MS that could be construed as *ill-structured. Ill-structured problems* in LC/MS are those that involve "finding the needle in a haystack"[5]. In many areas of medical research, we have yet to characterize the cause of a particular disease, consequently, we are left to treat the symptom, rather than cause. LC/MS has become one of many tools researchers are currently using to characterize the chemical basis of disease states. An example might be the recent measurements of protein folding by deuterium exchange and its relationship to Alzheimer's disease.[6]

Note that problem solving is dynamic. Solving a problem adds structure to a given problem. As a consequence of the problem solving process, *ill-structured problems* will generally become *semi-structured*; *semi-structured problems* will generally become *well-structured.*

Defining the structure and uncertainty in your problem will allow you to more easily assess the time, skill (training), personnel (individuals, groups), and equipment requirements for a given problem. LC/MS will play a role in all three problem types.

Table 8.0- Characteristics of Various Problem Types

Problem Type:	Well-Structured	Semi-Structured	Ill-Structured
Uncertainty	Low	Medium	High
Available Prob. Info.	Nearly complete	Partial	Little to none
Problem-Solving Approach	• SOPs • Standard methods • Target meth.	• Protocols • Modifying existing meth. • Methods dev.	• Creative research • Technique dev. • Instrument dev.
Key to Success	• Training	• Training • Education	• Education • Group Work
Solution Charact.	• Routine • Repetitive • Lower Cost	• Range: routine to difficult • Definable	• Non-routine • Ill-defined
Example of Problem Types: Pharmaceutical Industry	• Clinical • Bulk drug • Product • Content uniformity • Stability Phase II, III clinical trials	• Methods dev. • Validations • PK studies • Toxicology • Formulation • Drug metab. • Phase I clinical trials	• Drug discovery • Invention

8.1-Defining Your Identity Problems

Identification is the process of distinguishing one individual component from other components.[7] In chemical systems, the requirements for identity have a great deal to do with the complexity and composition of the components of the system. Retention time is often adequate for identification of well characterized samples. Mass spectra are generally used for identification in more complex samples or in samples where distinguishing characteristics must be provided at the molecular level.

There are three types of identity problems that you generally encounter in LC/MS; namely,

1) **unknowns,**
2) **suspects**, and
3) **targets** (see Figure 8.2.)

Unknowns are sample components to which no chemical structure can be assigned and provides the greatest challenge to the analyst. They generally require the highest level of experimentation, interpretation, and uncertainty. Many problems cannot adequately be defined until the sample components have been identified. In this situation, determining the identity of unknown sample components is an essential part of the problem definition.[8]

Suspects are sample constituents that can be assigned a structure, but the structure has not been confirmed.[9] One example of a *suspect* analyte would be a potential metabolite of a known drug substance observed with LC/MS. Knowledge of the drug and known metabolic pathways may allow the analyst to propose a tentative structure to a compound based on limited information, such as molecular weight measured by mass spectrometry. Arriving at a tentative structure is an important step in the problem-solving process because the analytical process can then shift from interpretation to confirmation. Confirmation is generally accomplished by simply comparing the *suspect* response to a *standard reference material* of the tentative structure.

Targets are sample constituents that have been assigned a structure; but whose presence (or absence) has not been confirmed.[10] With *target* analytes, the question isn't about determining the structure of an analyte; rather, the question is about whether a given structure is present. *Target* analysis involves the selection of experimental parameters that are adequate to distinguish a *target* species from all other sample components. Target analysis requires that you have a *reference standard* of the *target* species.

The information requirements of an experiment are far greater for *unknowns* than *targets*. In some cases, a single ion is adequate information to identify a given *target* from other sample components. It is very difficult to identify an *unknown* without a complete spectral profile; including, molecular ions, isotopic abundances, fragmentation, and accurate masses. When defining your LC/MS related problems, the type of identity problem will determine to a large degree the information requirements of your analysis. A variety of alternatives for solving identity problems are discussed in *Chapter Nine*.

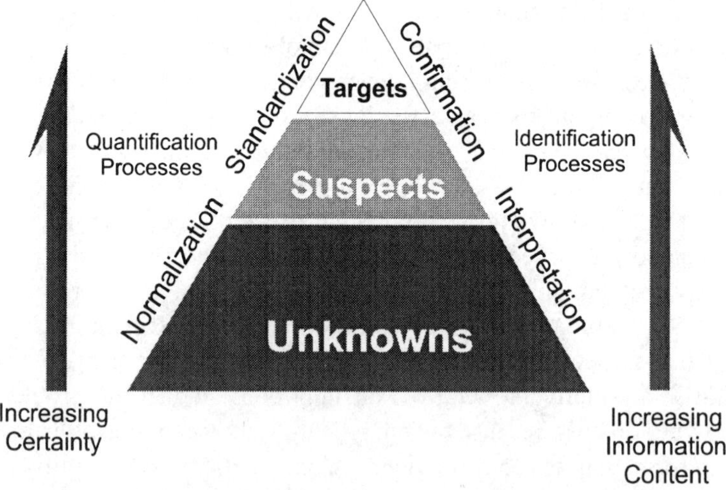

Figure 8.2- Analyte type *-unknown, suspect, or target-* will have a significant effect on your problem-solving alternatives in both quantification and identification. Defining analyte type will assist you in determining which alternatives are appropriate for your specific problem.

8.2- Defining Quantification Problems

Quantification is the process of determining the amount of a given component in a given system. In LC/MS, quantification and identification are inextricably linked. You cannot identify an analyte without a quantitative response. You cannot quantify without distinguishing one response from another (i.e. identity).

Defining your quantification problems in LC/MS is directly related to defining your identity problems. *Unknowns* present completely different obstacles to analysis than do *suspects* or *targets*. By far, the majority of problems requiring quantification in LC/MS are performed using target analysis. This does not presuppose that every sample entering these labs contains only target analytes. To the contrary, these labs may receive samples containing unknowns that have to undergo extensive experimentation to determine identity as a prerequisite to quantification. Subsequent target analysis can then be performed. Because of the importance of target analysis, we have dedicated *Chapter Eleven* to issues relating to various aspects of target analysis.

A systematic treatment of the topic of quantification and the various standardization approaches is presented in *Chapter Ten*. One point that is repeatedly stressed in that chapter is the importance of standards in quantification. Without reference standards the reliability and accuracy of quantitative results is seriously put in question. Conversely, there are many powerful alternatives for very accurate and reliable quantification with LC/MS using standard reference materials. One important class of standards are isotopically labeled standards. The most accurate quantitative results are achieved with labeled standards; particularly when dealing with complex sample treatment and or variable interferences.

When defining your quantification problems, you should keep in mind the type of analytes (Figure 8.2) and the availability of reference standards if analyte identities are known. If quantitative information is required to solve your analytical problem, acquiring *reference standards* as early in the process as possible may contribute to rapid implementation of any given problem solution.

8.3-Search for Ready-Made Solutions

"Don't re-invent the wheel!" After you have adequately defined your problems relating to LC/MS, don't run into the lab and start developing your own method. You should first search for less time and labor intensive alternatives. Has someone already develop a method for your analytes?...Matrices? Has someone already developed a method similar to what you need that can be adapted? The best place to start is in the literature; in-house or public.

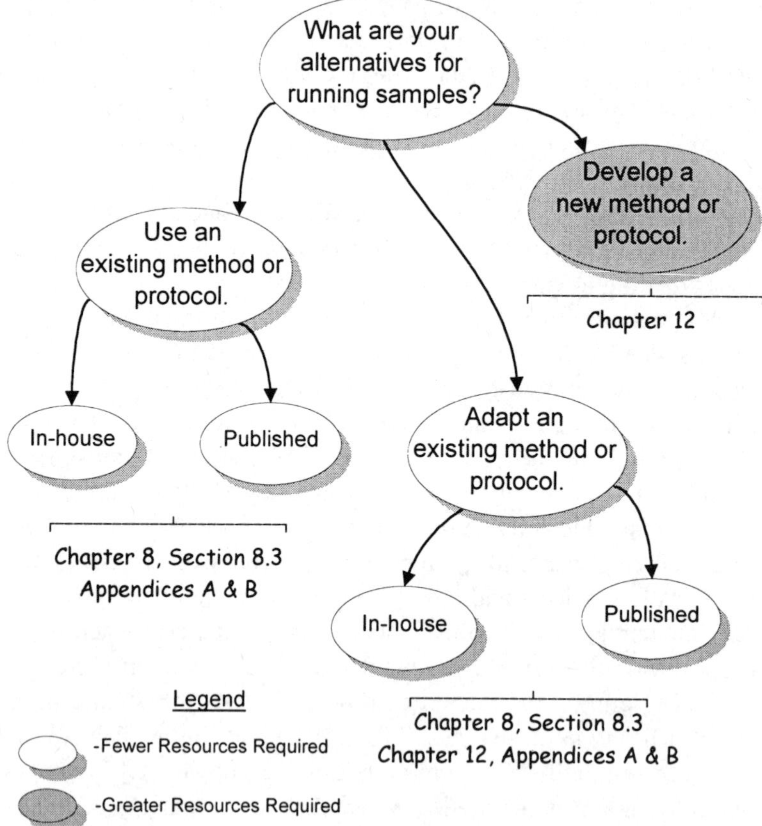

Figure 8.3- There are several alternatives to solving analytical problems with LC/MS. Starting with existing methods or protocols when available will save time and expense.

Your last resort should be starting from scratch and developing a sample specific method. This topic is treated in some detail in *Chapter Twelve*. Developing your own method is time consuming and creates a higher level of uncertainty and risk in your problem-solving process. Treading down uncharted method development paths can be fraught with unanticipated problems that consume time and degrade performance. Following in the footsteps of other analysts will hopefully allow them to clear any landmines along the path before you get there.

Searching for the existence of methods can be accomplished in a variety of ways. You can access in-house methods; published literature in journals; standard methods within your industry; conference abstracts; and in some cases, directly contacting a colleague within your industry may provide you with a method. When starting out, you should seek information from any and all available sources. A summary of common sources of information on LC/MS is provided in Table 8.1. More complete information is provided in Appendices A and B. We also provide continually updated information sources about LC/MS at www.lcms.com on the internet.

Obviously, access to in-house methods will require you to query your internal archives of methods or technical reports. Do not overlook the activities of other labs at remote locations within your organization. Many organizations (particularly, multinational) will develop methods for the same or similar compounds at more than one location because they fail to communicate effectively across large distances. "The left hand doesn't know what the right hand is doing." Good internal communication can save you considerable time and resources.

Searching the literature in LC/MS is in general no different than any other field. A good place to start is in the library or on the internet. STN International (the Scientific and Technical Information Network) offers both a fee based on-line search service that provides accurate, up-to-date, specific information from over 200 scientific, technical, business, and patent databases, and also fee based internet access to selected databases via the new STN Easy service. Some of STN's special features include advanced chemical structure searching, easy cross-searching of complementary databases, chemical reaction information, and computational services.

Today, accessing most information is as easy as logging onto the internet. There is a significant amount of information that is available to you free of charge or at a nominal charge. The fees are generally far less than the cost of gasoline required to drive to your local technical library.

Most of the published applications of LC/MS have occurred in the last five years. Very few applications existed in the literature before 1985. Most of these references are summarized in a variety of comprehensive textual reviews. The most comprehensive text containing information before 1990 are *Liquid Chromatography-Mass Spectrometry: Principals and Applications*, (by Niessen and Van der Greef) and *Liquid Chromatography-Mass Spectrometry: Techniques and Applications* (by Yergey, Edmonds, Lewis, and Vestal).

Table 8.1- Top Sources of LC/MS Information

Information	Source	Location
Abstracts	Chem Abstracts	Library www.cas.org
	Proceedings of the annual meeting of the ASMS.	www.asms.com (downloadable)
Papers	JASMS (Journal of the American Society of	Library www-east.elsevier.com/webjam
	Analytical Chemistry	Library pubs.acs.org/journals/ancham /index.html

It should also be emphasized that searching for ready-made solutions implies that you have identified the components in your samples. Without knowledge of sample composition, you cannot access the wealth of information available to you. A prerequisite to searching the variety of information sources may be screening your samples to identify the components. Identifying unknown sample components can be a prerequisite to complete problem definition.

Discussion Questions and Exercises

1. Complete Worksheet 8.0 for a specific problem that has been encountered by your organization within the previous year.
2. Complete Worksheet 8.0 for an application out of your field of expertise. Extra copies of all worksheets can be downloaded from the internet from `www.lcms.com`.
3. What are the general consequences of inadequate problem definition?
4. What are the consequences of inadequate problem definition in LC/MS?
5. Are most problems in LC/MS well-structured, semi-structured, or ill-structured?
6. Are most problems in your industry well-structured, semi-structured, or ill-structured?
7. Give an example of an ill-structured problem in LC/MS.
8. Give an example of a well-structured problem in LC/MS.
9. What are the primary sources of published analytical methods within your industry? Do you have ready access to this information?
10. What are four sources of information on published methods and protocols in LC/MS?
11. Discuss the relationship between identification and quantification in LC/MS. How are the identification and quantification processes interrelated?

Notes

[1] Stoll, C. *Silicon Snake Oil- Second Thoughts on the Information Highway*, New York: Doubleday (1995). Clifford Stoll, a pioneer of the Internet has written an excellent book cautioning us about the perils and false expectations of the information highway. One could extrapolate his ideas and concerns into the field of chemical analysis where our ability to generate data may be outpacing our abilities to interpret and effectively utilize analytical results to solve substantive problems.

[2] Rickards, T. *Problem-Solving Through Creative Analysis.* Essex, U.K.: Gower Press, 1974.

[3] Reproduced with permission from Forest Laboratories, Inc., 909 Third Avenue, New York, NY 10022. Thanks Seb and Tim.

[4] VanGundy, A.B. *Techniques of Structured Problem Solving.* New York: Van Nostrand Reinhold, 2nd Edition, Page 4 (1988)

[5] Don Hunt's Needle in a Haystack Problem. (a) Hunt. D.F., Henderson, R.A., Shabanowitz, J., Sakaguchi, H., Michel, H., Sevelir, N. Cox, A.D. Appella, E., Engelhard, V.H., "Characterization of peptides bound to the class I MHC molecule HLA-A2.1 by mass spectrometry," Science 255, pages 1261-1263 (1992). (b) Henderson, R.A. Michel, H., Sakaguchi, H., Shabanowitz, J., Appelia, E., Hunt. D.F., Engelhard, V.H., "HLA-A2.1-associated peptides from a mutant cell line: a second pathway of antigen presentation," Science 255, pages 1264-1266 (1992).

[6] Carol Robinson, Presentation at the Montreux Symposium describing the measurement of protein folding with deuterium exchange experiments to measure the stability of variant and normal proteins relating to Alzheimers disease and protein plaque formation. This important and leading edge research illustrates a good example of an ill-structured problem in LC/MS. In these experiments multiple new technologies and methodologies must be incorporated into the experiment to achieve the desired measurements. The creation of new measurement capabilities is by definition an ill-structure problem. Later, when these techniques become routine they will likely fall under semi- or well-structured problems. Robinson, C., "Protein structure analysis by electrospray FTMS," 14th Montreux Symposium on Liquid Chromatography/Mass Spectrometry, Cornell University, Ithaca, New York, USA, July 23-26, 1997. Robinson, C.V., Grob, M., Eyles, S.J., Ewbank, J.J., Mayhew, M., Hartl, F.U., Dobson, C.M., Radford, S.E., "Conformation of GroEL-bound α-lactalbumin probed by mass spectrometry", Nature 372, pages 646-651 (1994). Booth, D.R., Sunde, M., Bellotti, V., Robinson, C.V., Hutchinson, W.L., Fraser, P.E., Hawkins, P.N., Dobson, C.M., Radford, S.E., Blake, C.C.F., Pepys, M.B., "Instability, unfolding and aggregation of human lysozyme variants underlying amyloid fibrillogenesis," Nature 385, pages 787-793 (1997).

[7] Identity is defined in the distinguishing character of an individual. Webster's Ninth New Collegiate Dictionary, Merriam-Webster, Springfield, Mass. 1983.

[8] Many problems cannot begin to be solved without first identifying the chemical species associated with the problem. It should be emphasized that *problem definition* is not simply filling out a sample request form or other questionnaire concerning a problem state. It may involve a longer process which involves screening experiments and identification of unknowns. Only then can a rational problem-solving strategy be implemented.

[9] *Suspects* are sometimes referred to as *tentatively identified compounds* [TIC]; particularly in environmental GC/MS. This term can be confused with *total ion chromatogram* [also shortened to TIC]. We prefer *suspect* because it is short and descriptive. Also note that a library match would create a suspect from an unknown. The identity of an analyte is not *known* until its response is confirmed by comparison with a *reference standard.*

[10] *Targets* may also be referred to as *knowns*. The term *knowns* may actually have a more general connotation; however, with our focus on instrumental analysis and measurement, we prefer using the term *target*.

Chapter Nine- Flow Diagram

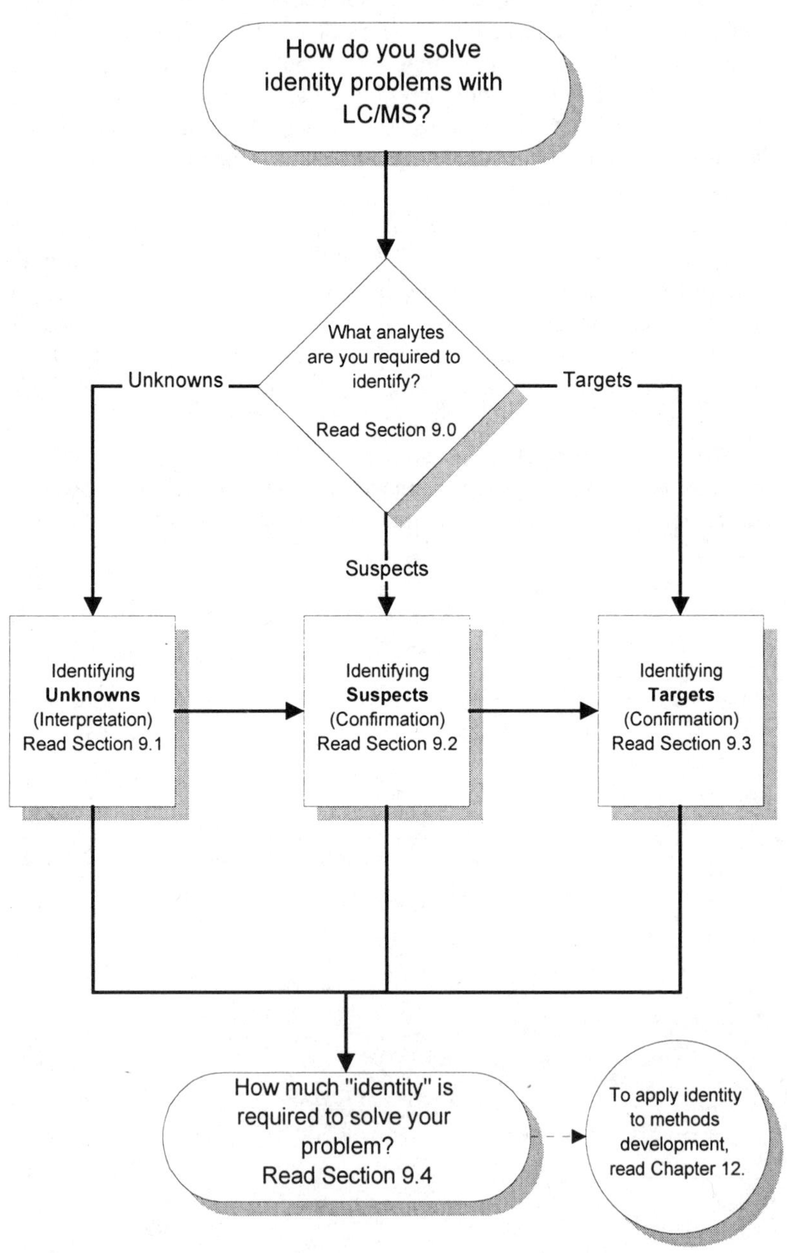

How do you solve identity problems with LC/MS?
What analytes are you required to identify?
Read Section 9.0
Unknowns
Targets
Suspects
Identifying **Unknowns** (Interpretation) Read Section 9.1
Identifying **Suspects** (Confirmation) Read Section 9.2
Identifying **Targets** (Confirmation) Read Section 9.3
How much "identity" is required to solve your problem? Read Section 9.4
To apply identity to methods development, read Chapter 12.

Chapter Nine

Solving Identity Problems

(or, Answering the Question- What is it?)

The molecular ion provides the most valuable information in the mass spectrum; its mass and elemental composition show the molecular boundaries into which the structural fragments indicated in the mass spectrum must be fitted.[1]

Professor F.W. McLafferty,
in ***Interpretation of Mass Spectra***

9.0- Identity Is Ubiquitous

Identification is a prerequisite for solving virtually every problem in chemical analysis. The challenge to the analyst is to define the measurements that are required to provide an unambiguous identification. For simple problems, non-specific measurement techniques are quite acceptable to identify components in a mixture (e.g.retention time). For complex problems, compound identity is usually accomplished with more than one measurement device. Hyphenated techniques such as LC-MS[2] are ideal for these applications because the combination of chromatographic and mass spectrometry provide the essential pieces of the identity puzzle. Compounds that don't pass the identity criteria for both chromatographic and mass spectrometric measurements fail the identity test. This is of particular importance in target quantitation (See *Chapter Eleven*). There is increased reliance on the use of hyphenated techniques for problems in identity because of speed and reliability, as indicated in the recent guidelines of the International Harmonization Conference (IHC).[3] For the first time LC/MS is listed as a potential tool for identity in bulk drugs and formulations.

In *Chapter Eight* we described the three classes of analyte in LC/MS; namely, *unknowns*, *suspects*, and *targets* (knowns). The process of identification will depend, primarily, upon the nature of your specific analytes. This chapter discusses the processes associated with identifying your *unknowns* (Section 9.1), *suspects* (Section 9.2), and *targets* (Section 9.3). At the top level, we describe identity as comprising two very important analytical processes; namely, *interpretation* and *confirmation.* (See Figure 9.0) The first, *interpretation,* is primarily associated with the identity of *unknowns*. The goal of the interpretation process is to acquire and utilize all available information about a given sample in order to elucidate and propose a tentative or suspected structure of the unknown sample components. The second identity process is *confirmation*. Confirmation, as the name implies, is the process of confirming (verifying) the identity of a suspect or target structure. Confirmation is a comparative process and relies on the presence of reference standards for comparison.

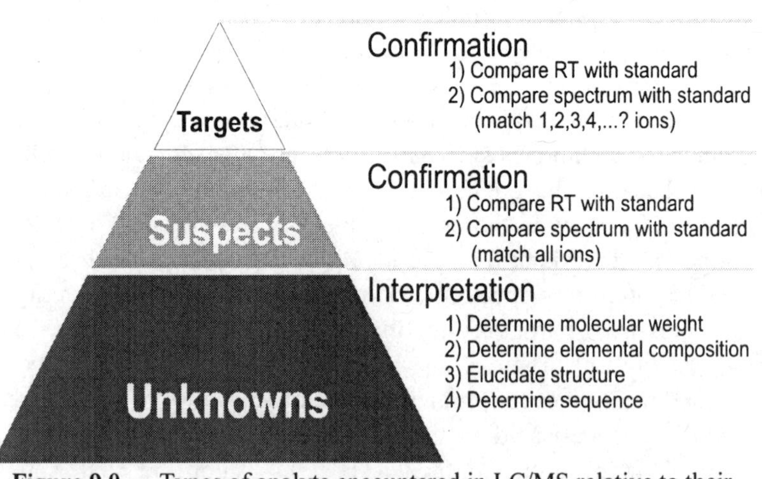
What "identity" techniques should you use?
Targets
Suspects
Unknowns
Confirmation
1) Compare RT with standard
2) Compare spectrum with standard (match 1,2,3,4,...? ions)
Confirmation
1) Compare RT with standard
2) Compare spectrum with standard (match all ions)
Interpretation
1) Determine molecular weight
2) Determine elemental composition
3) Elucidate structure
4) Determine sequence

Figure 9.0- Types of analyte encountered in LC/MS relative to their associated identification processes and respective techniques.

9.1- Identifying Unknowns with LC/MS

The high degree of uncertainty associated with the analysis of *unknown* components can create a considerable challenge in your laboratory. In many career endeavors, this uncertainty is a way of life and is handled in due course. This would be the case for many scientists in drug discovery or organic synthesis. We would be imprudent, however, if we didn't emphasize that there are seldom "complete" unknowns in any laboratory, including yours. You usually have some knowledge about the sample that aids in analysis and interpretation. In general, the techniques that you employ in the interpretation process will usually depend upon your baseline knowledge of your sample and the uniqueness of your unknown.

The interpretation process requires both *inductive* and *deductive* reasoning to be effective. *Deductive* reasoning occurs when your conclusion about a particular result follows from some universal assumption. For example, we cannot assign the elemental composition to a given unknown based on isotopic ratios unless we first assume that the species under investigation has ratios that are consistent with the ratios generally found in nature. Conversely, *inductive* reasoning is required when we draw a general conclusion following from specific or particular observations. We can conclude that a given unknown species has a proposed structure based on the aggregate of many particular observations such as MW, elemental composition, and fragmentation.

The beauty of the interpretation process in LC/MS is that it is absolute. Even one piece of inconsistent data is enough to discount or disqualify a proposed structure from further consideration. Another way of looking at interpretation is that every piece of information has to be consistent with the proposed structure. Whether you interpret a spectrum with two or two hundred ions, each and every ion must match to the substance under investigation. Your task is simply to link each ion to the substance.

We describe identity problems of unknowns with LC/MS in terms of "*the five I's in Identity*"; namely, *Information*, *Instrumentation*, *Ionization*, *Isolation*, and *Interpretation*. We believe that you will have a high level of success with your identity problems by paying strict attention to these five items.

Interpretation: The Five I's In Identity

All problem-solving should start with *information* gathering; unknown identification is no exception. The more you know about a sample, the easier the interpretation. You should make every effort to fully characterize the source of your sample containing *unknowns*. That may mean acquiring actual mass spectral data on some or all of the known components, including any reactants, precursors, or raw materials that have led to the present sample state. Always keep in mind *relative* information is generally easier to interpret than *absolute*. Comparing results between a known and an unknown sample is easier than evaluating a discrete unknown. Mass spectra of starting materials may give clues to the stability of the fragmentation processes associated with your unknown. If given a choice, always choose comparative processes over discrete processes. For example, metabolites are usually compared to parent drugs and degradation products compared to starting materials.

Identification starts with information and ends with interpretation. Somewhere in the middle, samples have to be run. The information content of the results from your sample analysis will depend upon your *instrument*, *ionization,* and *isolation* approaches. First, you must choose an instrument configuration that has the capability of providing you with the needed information (e.g. high resolution/accuracy for accurate mass measurement, MS/MS for fragmentation). At a minimum, tuning values should be checked for acceptable peak shape and calibration checked for accuracy (over the entire mass range of the intended application). In addition, the response of the analyte must be evaluated in order to yield an interpretable result. In general, this means that one or more ionization technique should be capable of producing a spectral component of the molecular mass. This could be a protonated, adducted, or any other characterizable representation of the molecular mass, including, of course, a molecular ion (See MW Definitions). The response should be capable of producing a response from both major and minor isotopic contributions of each spectral component.

A prerequisite to interpretation is the requirement that the results under evaluation be attributable to only one chemical species.

Identifying Unknowns Step-by-Step

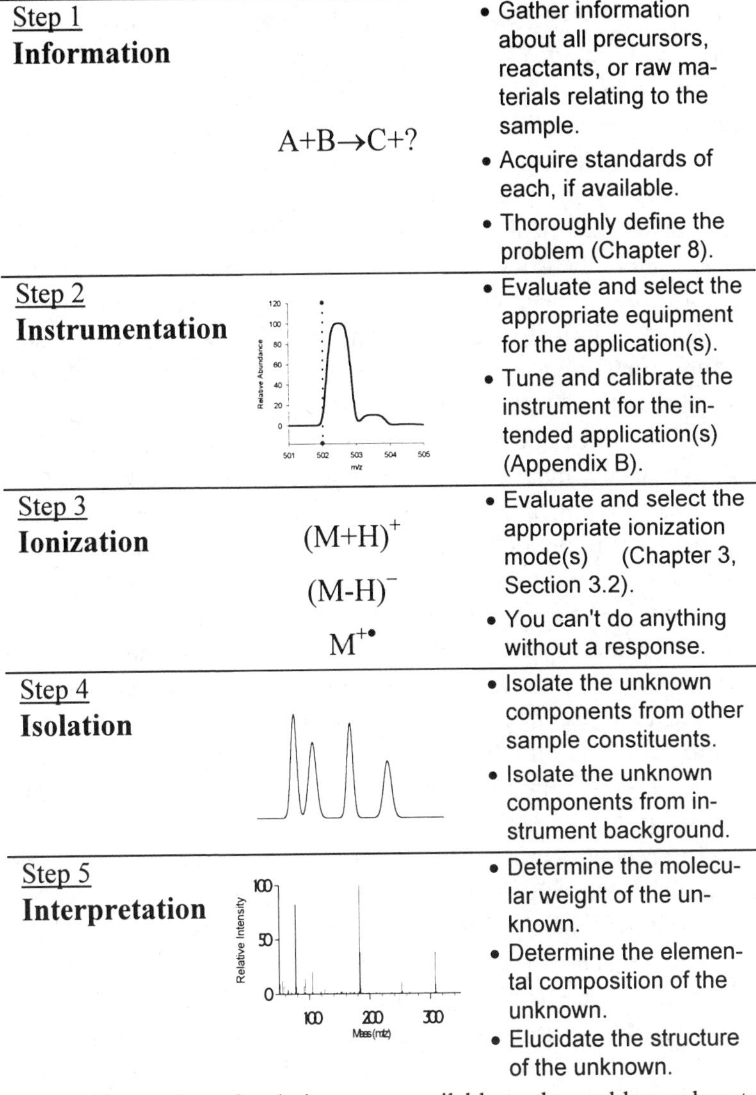

Step 1 **Information**	$A+B \rightarrow C+?$	• Gather information about all precursors, reactants, or raw materials relating to the sample. • Acquire standards of each, if available. • Thoroughly define the problem (Chapter 8).
Step 2 **Instrumentation**		• Evaluate and select the appropriate equipment for the application(s). • Tune and calibrate the instrument for the intended application(s) (Appendix B).
Step 3 **Ionization**	$(M+H)^+$ $(M-H)^-$ $M^{+\bullet}$	• Evaluate and select the appropriate ionization mode(s) (Chapter 3, Section 3.2). • You can't do anything without a response.
Step 4 **Isolation**		• Isolate the unknown components from other sample constituents. • Isolate the unknown components from instrument background.
Step 5 **Interpretation**		• Determine the molecular weight of the unknown. • Determine the elemental composition of the unknown. • Elucidate the structure of the unknown.

A wide number of techniques are available to the problem-solver to *isolate* the response from unknown sample components. These include separations, MS/MS, high resolution, and background subtraction. Each or all of these isolation techniques are widely used where interpretation is required.

Interpretation: Pathways To Structure

The primary goal of the identification of unknowns, using LC/MS, is to match a tentative chemical structure to each unknown in the sample. Here are three approaches that the analyst can use to reach this goal. It is prudent to be aware of all three alternatives on your path to the identification of unknowns.

The first and fastest way to match the response of an *unknown* to a given chemical structure is through *library matching*. The most widely distributed libraries are the Wiley and NIST libraries for electron ionization (see *Appendix A*). These libraries, of course, are limited to electron ionization but will provide instantaneous matching of your acquired spectrum to a huge repository of important industrial, pharmaceutical, and environmental spectra. The search and match process takes seconds and will generally produce a number of ranked matches. In the event that you get an exact match, you can confidently say you have a credible *suspect* and move onto the confirmation process using a putative standard.[4] Libraries are not only limited to EI or public libraries. Most commercial instruments have a library building utility that allows each customer to build their own libraries with no limit to technique or experimental conditions. Many large chemical companies have proprietary libraries with hundreds-of-thousands of compounds that directly relate to their specific products and processes. Identification of unknowns is generally expedited in these laboratory environments.

The second approach to identification of unknowns is the use of problem-specific data as a reference point. For example, simple mass shifts can often be enough to identify a structure of a metabolite. The difference in molecular mass between a drug and an unknown metabolite may be adequate to lead to a suspect structure. Mass shifts for common metabolites are shown in Appendix B. Mass shifts for dozens of observed post translational modifications of proteins are listed on the internet at `www.chait_sgi.rockefeller`. Reference spectra can (and should) be acquired on any material relating to your sample. These sample related reference compounds can directly or indirectly lead to an accurate assignment of the suspect structure of your unknown. You may gain critical problem information from running these compounds than you gain directly from your unknown sample. Don't be afraid to run other reference materials and samples when identifying unknowns.

The last approach is the fundamentals approach through the rigors of interpretation. Generally, interpretation involves determining three important pieces of information about your unknown: 1) molecular weight, 2) elemental composition, and 3) structure. This is an iterative rather than a linear process and may require one or more experiments with LC/MS.

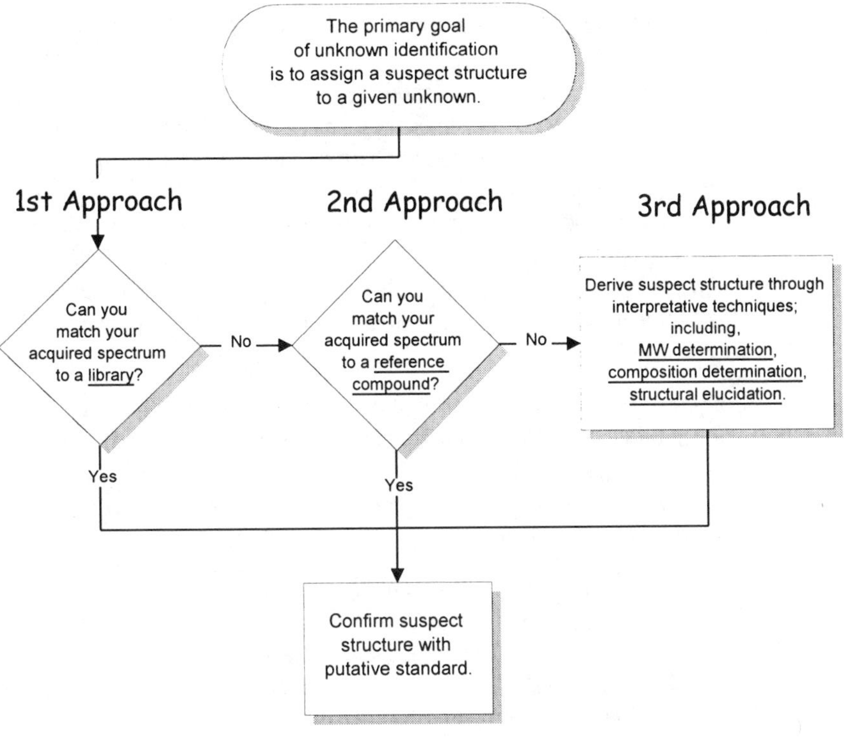

Interpretation: A General Protocol

Here we include a general protocol for interpretation so that you can attack specific interpretation problems in a structured manner.[5]

1. Acquire spectrum of your unknown with the specific intent of generating ions with which you can derive the molecular mass.

   Guideline- Select an ionization mode that is likely to generate intact quasi-molecular or molecular ions. Both positive and negative ionization modes should be considered. This response screening can be performed with flow injection or simple linear gradients. If a separation method exists that has isolated the unknown from other sample components, use that method. (See "MW Determination- *pp. 310-317")*

2. Assign a molecular mass to your unknown species based on one or more ionization modes and their respective acquired spectra.

3. Deduce the elemental composition for the observed (quasi)-molecular ions.

   a. Use isotopic abundance measurements.

      Guideline: Use Mass and Abundance Tables to check your deduction. (See "Elemental Composition Through Isotopic Abundance Measurement-*pp.318-321")*

   b. Use accurate mass measurements compared the theoretical exact mass calculations.

      Guideline: Use Mass and Abundance Tables for exact mass values. (See Elemental Composition Through Accurate Mass Measurement-*pp.322-323")*

5. Compare and reconcile the deduced elemental composition derived from abundance and/or accurate mass measurement with the assigned molecular mass.

   Guideline: The elemental composition must be consistent with the molecular weight determination.

6. Acquire a spectrum under fragmentation conditions.

   Guideline: Acquire fragmentation spectra of unknown under a variety of collision energies. If MS/MS is used, let the entire isotopic cluster (3 to 4 mass units) of the (quasi-)molecular ion through the first analyser. This will allow you to benefit from isotopic abundances of fragment ions. If you select only monoisotopic precursor ions, you will lose fragment isotopic abundance information.

7. Deduce the elemental composition of all fragment ions.
   a. Use isotopic abundance measurements.
      Guideline: Use Mass and Abundance Tables *to check your deduction. (See* Elemental Composition- Through Isotopic Abundance Measurement-*pp.318-321")*
   b. Use accurate mass measurements compared to the theoretical exact mass calculations.
      Guideline: Use Mass and Abundance Tables for exact mass values. (See Elemental Composition- Through Accurate Mass Measurement-*pp. 322-323) Elemental compositions of each fragment should reconcile with both direct measurement of each fragment and the accurate measurement of neutral loss between precursor ion and fragment ion.*

8. Propose structural assignments based on spectral characteristics.
   Guideline: Evaluate neutral loss, important characteristic fragments, and important low mass diagnostic fragments and series. (See Structural Elucidation-*pp. 324-331)*

9. If a structure assignment to your unknown is not obtainable through primary fragmentation processes, acquire spectra under secondary fragmentation conditions.
   Guideline: Acquire fragmentation spectra of key fragment ions under a variety of collision energies. Again, let the entire isotopic cluster of the key fragment ion through the first analyzer. This will allow you to benefit from isotopic abundances of fragment ions. If you select only monoisotopic presuror ions, you will lose fragment isotopic abundance information.

8. Propose sub-structural assignments based on spectral characteristics.
   Guideline: Evaluate neutral loss, important characteristic fragments, and important low mass diagnostic fragments and series. (See Structural Elucidation- pp. 324-331)

10. Compare all assigned mass, abundance, and structural information for consistancy.
    Guideline: <u>All</u> measurements that you attribute to your unknown structure must be consistent with the assigned structure or attributed to an impurity. Any inconsistancies should lead you to consider another structural assignment.

Interpretation: MW Definitions

The use of correct definitions and terminology is essential in LC/MS problem solving. For example, one may use the value of molecular weight from your *Merck Index* for a given compound instead of a calculated "exact mass" from exact mass tables. This can result in serious interpretation errors. Try to understand the subtle differences between each term and their appropriate use and application. Don't assume that everyone you talk to understands these differences.

Molecular Weight

Molecular weight is the sum of the atomic weights of all the composite elements in a given molecule. It is noteworthy to emphasize that this value is averaged over all isotopic species for each element. In contrast, the mass spectrometer has the unique ability to measure the isotopic contributions of each element discretely. The isotope specific measurements in mass spectrometry require that more specific terms be used to apply to molecular weight determination.

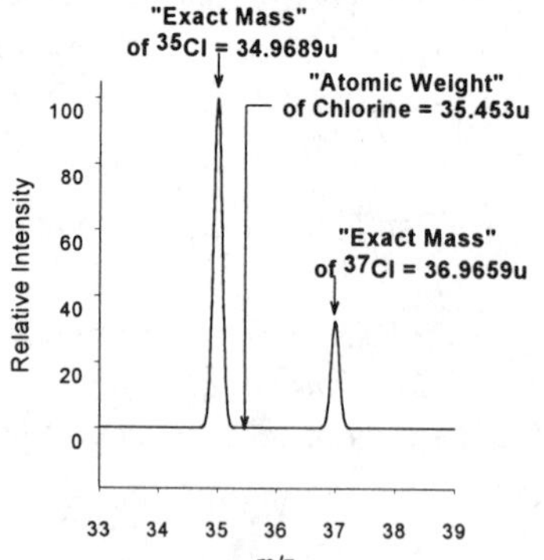

Figure 9.1- Exact masses of chlorine isotopes versus the atomic weight.

Atomic Weight

Atomic weight is the calculated mass determined by averaging the masses over all isotopes of a given element (Figure 9.1). This average is influenced by the relative abundance of each isotope of a given element. Do not confuse atomic weight of an element with exact mass for each isotope of the element. Newcomers to the field of mass spectrometry plug in atomic weights from the periodic table into an empirical formula to calculate an expected mass for a given analyte. This mistake can cause significant error; particularly when estimating the exact mass of high molecular species.

Exact Mass

Exact mass is the calculated mass of individual isotopes that compose a single ion. The masses of each isotope are measured relative to carbon (^{12}C=12.000) where a single unit is 1/12th of the mass of carbon-12 and recorded in a wide variety of mass tables. It is a common mistake to use atomic weight table values instead of exact mass table values.

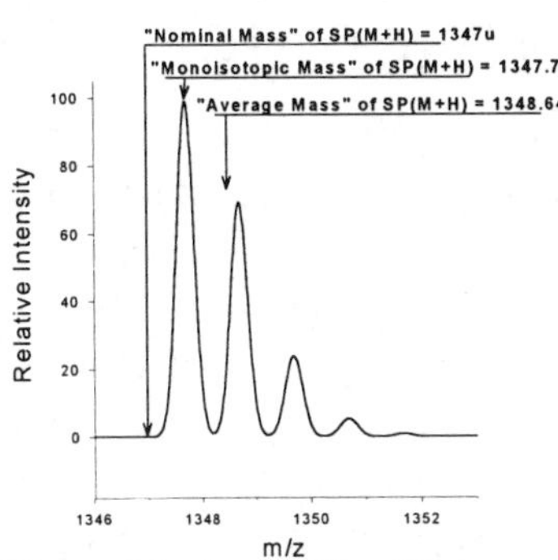

Figure 9.2- Ion profile of Substance P showing *nominal mass* (with mass defect), *monoisotopic mass*, and *average mass*.

Accurate Mass

Accurate mass is the measured mass to high mass accuracy. Accurate mass measurements are compared to exact mass calculations. Mass accuracies in the millimass unit range are generally required for accurate assignment of formulas.

Nominal Mass

Nominal mass is the calculated integer mass of the most abundant naturally occurring stable isotope (Figure 9.2). We do not measure or observe nominal masses with the mass spectrometer. Since the measured masses of ions with the mass spectrometer are not typically integer masses, this serves only as a rough estimate of the expected mass position of a given molecule. Nominal masses are subject to errors due to mass defects. We don't recommend using nomimal mass calculations in LC/MS, especially if you are performing high mass analysis.

Monoisotopic Mass

Monoisotopic mass is the exact mass of the most abundant naturally occurring stable isotope of an element (Figure 9.2). The monoisotopic mass is the sum of the monoisotopic masses of each element in the empirical formula.

Molecular Ion

Molecular ion is the odd-electron ion observed at the *molecular mass* of a given molecule. Molecular ions are formed by electron ionization or electron capture processes.

Protonated Molecule

Protonated molecule is the even-electron ion that results from the addition of a proton to a molecule. Protonation can occur in the liquid or gas-phase.[6]

Quasi-molecular Ion

Quasi-molecular ion is a general term used to denote ions that approximate the position of a molecular ion. A protonated molecule would be a quasi-molecular ion.[7]

Molecular Mass

Molecular mass for this discussion will refer to the mass of a neutral molecule and be denoted by M. By convention, salts would include their associated counterion in the molecular mass. The ion observed for a quaternary amine chloride in positive electrospray would be denoted by $(M\text{-}Cl)^+$.

Interpretation: Electrons, Protons, Neutrons

It is easy to forget; particularly when studying mega-dalton proteins or DNA, that all molecules are made of fundamental building blocks of matter; that is, electrons, protons, and neutrons (See Figure 9.3). When interpreting mass spectra it is useful to keep in mind that mass spectrometers are directly measuring aggregates of *electrons*, *protons* and *neutrons*. The mass spectrometer is capable of tracking the electrons, protons and neutrons comprising compounds introduced for analysis as they undergo chemical reactions or chemical decomposition (fragmentation). Measuring the changes in electronic and atomic configurations allows us to assign unique structures to each piece of a molecule and ultimately "piece" the information together into a complete structure. *The key to effective interpretation is to become a good bookkeeper of the electrons, protons and neutrons measured in each experiment.*

Figures 9-4 through 9-6 are examples of how mass spectrometry can be used to directly measure differences in electrons, protons and neutrons.

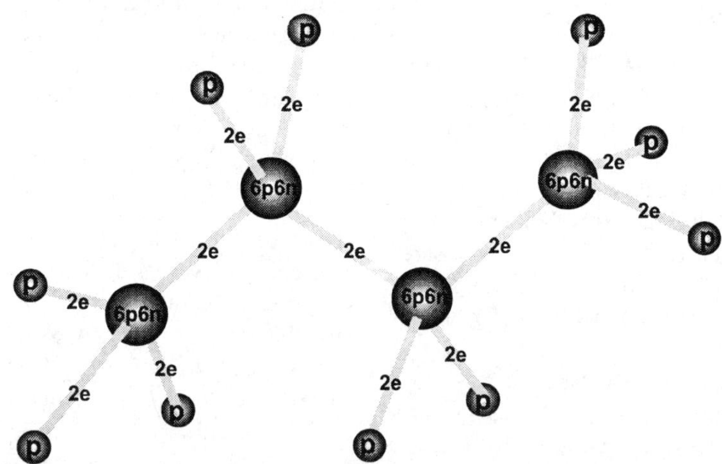

Figure 9.3- Diagram of the relationship between *electrons* (e), *protons* (p), and *neutrons* (n) for a typical organic molecule (butane). Remember, all mass comes from these basic components.

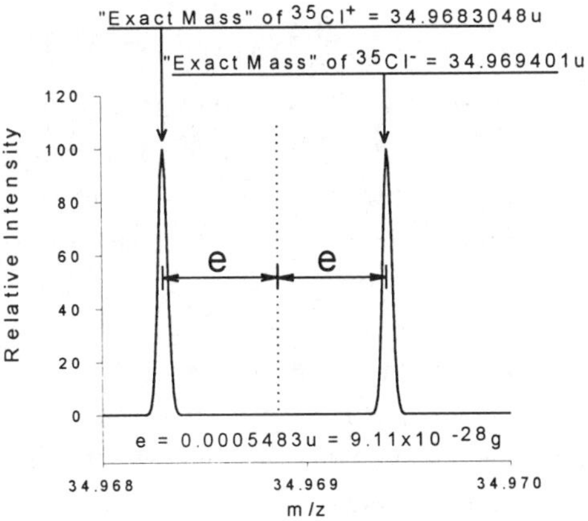

"Exact Mass" of $^{35}Cl^{+}$ = 34.9683048u

"Exact Mass" of $^{35}Cl^{-}$ = 34.969401u

e = 0.0005483u = 9.11x10 $^{-28}$g

Figure 9.4- Direct measurement of *electrons* with mass spectrometry as shown by this representation of FTMS acquisition of ^{35}Cl simultaneously ionized by electron and electron capture ionization. The resulting ions differ by only two electrons- a mass of 1.82x10^{-29}g.

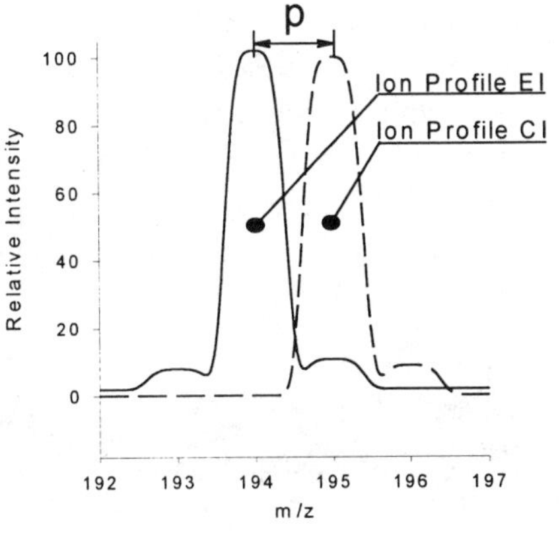

Figure 9.5- Direct measurement of *protons* with mass spectrometry as shown by this overlay of ion profiles of chemical and electron ionization of caffeine. The resulting spectra differ by one proton.

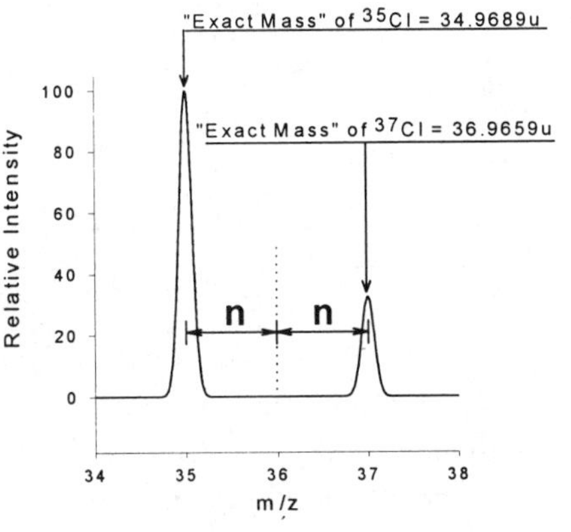

Figure 9.6- Direct measurement of *neutrons* with mass spectrometry as shown by isotope pattern of chlorine showing 35Cl and 37Cl at the naturally occurring ratios. These two ion of chlorine differ by only two neutrons. Their electronic behaviors are identical.

Interpretation: Mass Defects

When does two plus two not equal four? The mass measurements we make in LC/MS can be related to both molecular structures and atomic nuclei. When comparing one molecular structure to another (as will be discussed later in this chapter) mass is conserved. The molecular world obeys the rules of *classical* physics. Conversely, when measuring atomic nuclei, we are observing a *relativistic* world.[8] A relativistic world means that mass is not necessarily conserved; rather the *law of conservation of mass-energy* applies. This simply means that when you put two protons together in a single atomic nucleus, the aggregate mass will not be the sum of two protons, it will be slightly less. This mass difference, referred to as the *mass defect,* is important to the analytical chemist. Every atomic nucleus in this relativistic world has a mass defect associated with it. Why is this useful to the organic chemist? This mass defect allows us to uniquely characterize and identify different combinations of every element in the periodic table by measuring its unique mass. *Mass defect* is of particular importance in accurate mass measurements (also described later in this chapter).

Table 9.0- An example of mass defects at m/z 28.

Element/Molecule	Protons	Neutrons	p+n	Exact Mass
Carbon	6	6		
Oxygen	8	8		
Molecule- CO	14	14	28	27.994914
Nitrogen	7	7		
Nitrogen	7	7		
Molecule- N_2	14	14	28	28.006148
Rest mass (p and n)	1.00782	1.00867		
Sum of rest masses	14.10948	14.12138		28.23086

Table 9.0 shows the difference in mass calculated for CO and N_2. Both molecules have the same number of protons and neutrons, but different nuclei combinations. Each atomic nuclei has a unique mass as illustrated in Figure 9.7. Also note the the profound effect of the large hydrogen numbers in organic compounds in Figure 9.8.

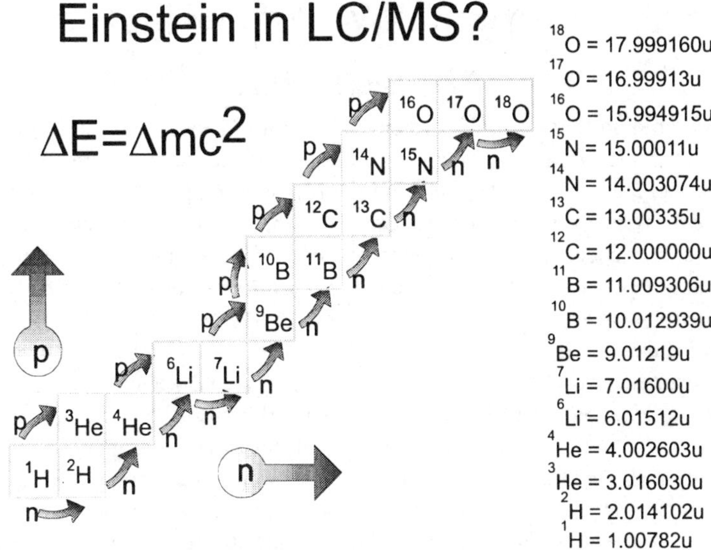

Figure 9.7- An excerpt from the nuclide chart showing the addition of protons and neutrons when forming higher elements. The associated masses are shown at the right. Note the mass defect as you move up the chart. This represents the mass defect in relativistic terms.

Figure 9.8- The effect of *mass defect* of hydrogen as a function of the number of hydrogens in a molecule. In mass spectrometry, mass defect is defined as the difference between the exact mass and the nominal mass of a given isotope. Here, hydrogen at 0.00782u will result in a mass shift at high numbers

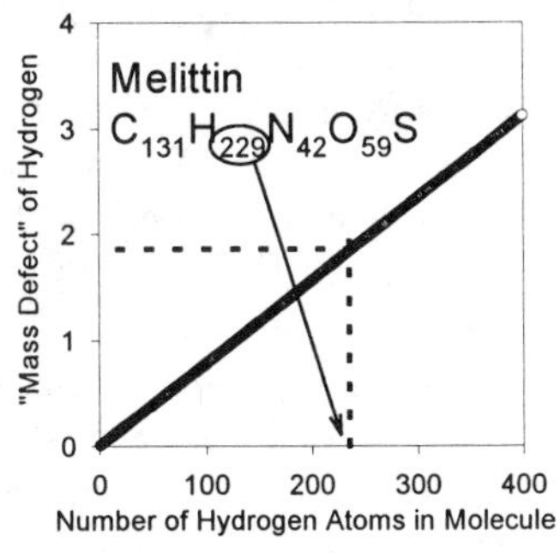

Anyone engaged in organic analysis over 400u should make a consious effort to estimate the effects of hydrogen mass defects. The slightly different definitions of mass defect by mass spectrometrists and nuclear physicists can be confusing but important.

Interpretation: MW Determination

The Big Picture

Determining the molecular weight of a species is an important and widespread procedure in all chemical analysis, particularly in organic chemistry. Traditional techniques, such as titrimetry and electrochemistry, for determining molecular weight involve the measurement of the number of moles of a pure substance for a given measured mass using an analytical balance (e.g. milligrams). These traditional molar measurements produce a value for the molecular weight that is averaged over all isotopic species of a given molecule. The definition of *molecular weight* is therefore, by convention, the average molecular mass over all isotopic species.

In mass spectrometry we can acquire much more accurate mass measurement of an unknown molecule by measuring the mass of the molecule directly with the mass spectrometer rather than measuring molar equivalents. Each isotope can be measured one at a time. Each isotopic species of a given molecule can be resolved to within the mass of a single electron (See Figure 9.4). You don't necessarily have to use pure samples to measure the molecular mass of an unknown with mass spectrometry. More importantly, you don't need to use molar quantities to acquire a reliable result as with many traditional methods. In some cases, mass spectrometry can yield accurate MW information with only a few molecules.

Although we may offend some of the purists, we still refer to this general process in the common (but inaccurate) language "molecular weight determination". Still, we strongly encourage anyone entering the field of LC/MS to be aware of correct terminology and use it appropriately.

The molecular mass (monoisotopic mass) may be measured in a wide variety of ways during the process of identifying an unknown as shown in Figure 9-10. A general discussion of each follows.

Molecular Ion

The simplest to interpret method of determining molecular mass is by measuring the molecular ion. The *molecular ion* is observed at the molecular mass (monoisotopic mass) of the analyte by either electron ionization (removal of one electron in positive ion mode) or electron capture (attachment of one electron in negative ion mode).

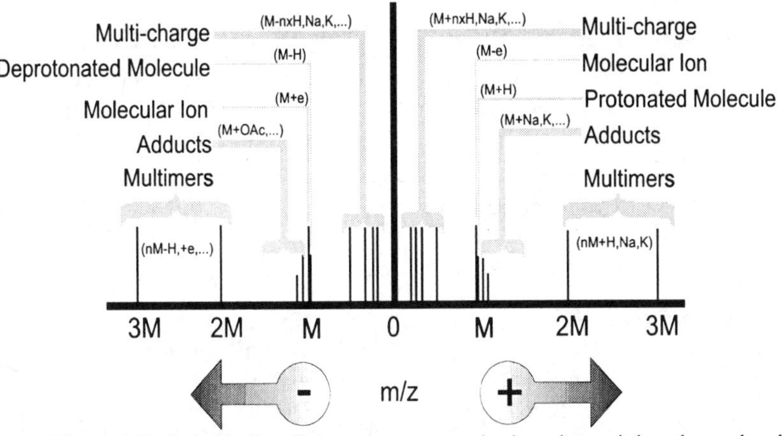

Figure 9.9- Quasi-molecular species measured when determining the molecular mass (M) of a given molecule. You should look everywhere.

(De)Protonated Molecule

Depending on the proton affinity of a given unknown molecule, protons can be added to form the *protonated molecule* or removed to form the *deprotonated molecule*. In the case of protonation ions are observed at one mass unit above the molecular mass in the positive ion mode. Conversely, in the negative ion mode, we observe the deprotonated ion at one mass unit below the molecular mass.

Adduct Molecule

A wide variety of chemical adducts can form with a given molecular species (e.g. Na, K, H2O, MeOH,...) Adducts can come from sample or solvent components and usually involve ionic or hydrogen bonding interactions.[9] These adducts are technique dependent and are listed in Table 9-2.

Multiple-Charge

Electrospray allows us to desorb almost any charge state of a molecule in solution. We can observe multiple charge in both positive and negative modes.

Multimer

Some molecular species will aggregate in solution, on surfaces, or in the gas phase resulting in multimers of the molecular species[10]. These species can be real sample components or artifacts of ionization.

Interpretation: MW Determination

How do you determine the MW of an unknown?

As noted on the previous page, there are a number of places to observe a molecular mass in the mass spectrum. When dealing with unknowns it is best to pick a strategy that suits your instrumentation (interface /ionization) and sample types. Discussed below are three approaches that may be useful for your applications.

The Molecular Ion

The simplest (and most universal) approach to evaluating a molecular mass is to run your sample under electron ionization (EI) conditions. Particle beam is currently the only commercial option in LC/MS to use this approach. Alternatively, do not discount your ability to isolate a fraction from your LC separation and perform EI probe analysis[11] for some samples. All ions in electron ionization are derived from the primary reaction product of electron ionization, $M^{+\bullet}$. The *molecular mass* is directly measured with EI when the molecular ion is present in the spectrum. Simple rules can be used to check the validity of a given $M^{+\bullet}$ assignment.[12]

Rules:

- $M^{+\bullet}$ *must* be the *highest* mass ion in the spectrum.
- $M^{+\bullet}$ *must* be an *odd-electron* ion.
- $M^{+\bullet}$ *must* be capable of yielding important ions in the high mass region of the spectrum through the loss of reasonable *neutral* species.

The observation of $M^{+\bullet}$ in the mass spectrum is limited to compounds that have some volatility to survive the sample evaporation process and that will not fall apart upon ionization. The odd-electron molecular ion will readily undergo both rearrangement and fragmentation immediately after formation. Once a molecular ion is assigned, it is generally confirmed using a softer ionization approach such as CI.

Nitrogen Rule- (As applied to odd-electron ions)

All odd-electron ions (OE^{+}) with an odd number of nitrogens (n=1,3,5,..) will appear at odd nominal masses. All odd-electron ions (OE^{+}) with an even number of nitrogens (n=0,2,4,..) will appear at even nominal masses.

The Acid-Base Approach

A far more general approach to molecular weight determination in LC/MS is to evaluate your unknown in terms of its ability to donate (acids) and accept (bases) protons. The most likely ions you will see with CI, APCI, ES, CFF, and TS will be the result of protonation or deprotonation of your unknown molecular ion. Accepting this as a likely outcome will allow you to run experiments to determine the acid-base properties of your unknown. In complex samples, such as natural products, you may want to perform an acid extract and basic extract, whereby you isolate acidic species by neutralizing them at low pH before extracting into organic solvents, conversely, you isolate basic species by neutralizing them at high pH. Each extract can then be separated and mass analyzed by a variety of LC/MS techniques.

Acidic species will tend to yield deprotonated molecules, $(M\text{-}H)^-$, observed in the negative ion mode, and basic species will tend to yield protonated molecules, $(M\text{+}H)^+$, observed in the positive ion mode. Table 9-11 shows some of the ions that might be observed in the various modes of operation.

Rules:

- $(M\text{+}H)^+$ or $(M\text{-}H)^-$ *should* be the highest mass ion in the spectrum, but *check* for adducts and multimers.
- $(M\text{+}H)^+$ or $(M\text{-}H)^-$ must be an *even-electron* ions.
- $(M\text{+}H)^+$ or $(M\text{-}H)^-$ must be capable of yielding important ions in the high mass region of the spectrum through the loss of reasonable *neutral* species.

Nitrogen Rule- (As applied to even-electron ions)

All even-electron ions (EE^+) with an odd number of nitrogens (n=1,3,5,..) will appear at even nominal masses. All even-electron ions (EE^+) with an even number of nitrogens (n=0,2,4,..) will appear at odd nominal masses.

The Cation-Anion Approach

Not all analytes have exchangeable protons; some chemical species are always ionized in solution and as a consequence are difficult to neutralize, extract, and separate by convention HPLC methods. Examples would be quaternary amines or anionic species such as chloride. For these species *electrospray* is the LC/MS technique of choice for molecular weight determination. They typically respond quite well under electrospray conditions. The only experimental issue that may be questioned is whether the unknown is a cation or anion. This issue can be resolved by running electrospray in both positive and negative ion modes. The convention with cations and anions has M designated as the neutral salt and C is the mass of the counterion.

Rules:

- $(M\text{-}C)^+$ or $(M\text{-}C)^-$ *should* be the *highest* mass ion in the spectrum, but *check* for adducts.
- $(M\text{-}C)^+$ or $(M\text{-}C)^-$ must be an *even-electron* ions.
- $(M\text{-}C)^+$ or $(M\text{-}C)^-$ must be capable of yielding important ions in the high mass region of the spectrum through the loss of reasonable *neutral* species.

Table 9.1- Mass and Abundance of Selected Elements

Element	Symbol	Integer Mass	Exact Mass	Isotopic Abundance
Hydrogen	^{1}H	1	1.00782506	99.99
(Deuterium)	^{2}H or D	2	2.0141	0.01
Carbon	^{12}C	12	12.0000000	98.91
	^{13}C	13	13.0034	1.1
Nitrogen	^{14}N	14	14.00307407	99.6
	^{15}N	15	15.0001	
Oxygen	^{16}O	16	15.99491475	99.76
	^{17}O	17	16.9	0.04
	^{18}O	18	17.9	0.20
Fluorine	^{19}F	19	18.9984046	100
Phosphorus	^{31}P	31	30.9721	100
Sulfur	^{32}S	32	31.9721	95.02
	^{33}S	33	32.9715	0.76
	^{34}S	34	33.9659	4.22
Chlorine	^{35}Cl	35	34.9689	75.77
	^{37}Cl	37	36.9659	24.23
Bromine	^{79}Br	79	78.9183	50.5
	^{81}Br	81	80.9136	49.5

Table 9.2- Molecular Mass Determination

Mode	Process	Some Observed Masses	Designation
EI	Unimolecular	$M^{+\bullet}$	Molecular Ion
CI APCI	Bimolecular Gas-Phase w/ Reagent Gas	$(M+H)^+$	Protonated Molecule
		$(M-H)^-$	Deprotonated Molecule
		$(M+CH_5)^+$,$(M+C_2H_5)^+$,$(M+C_3H_5)$ (for methane CI - +17,29,41,....	Methane Adducts
		$(M+H+nNH_3)^+$ (for ammonia CI- +18,35,52,....)	Ammonia Adducts
	Solvent	$(M+H+nH_2O)^+$ (for water- +19,37,....)	Water Adducts
ES, TS	Desorption of Preformed Ions from Solution	$(M+H)^+$	Protonated Molecule
		$(M-H)^-$	Deprotonated Molecule
		$(M-C)^+$, $(M-C)^-$	Cations, Anions[13]
		$(M+nNa)^+$, $(M+nK)^+$	Alkali Metal Adducts
		$(M+OAc)^-$	Acetate Adducts
		$(M+H+nNH_3)^+$	Ammonia Adducts
		$(M+H+nH_2O)^+$ (for water- +19,37,55....)	Water Adducts
FAB	Desorption of Preformed Ions from Solution	$(M+H)^+$	Protonated Molecule
		$(M-H)^-$	Deprotonated Molecule
		$(M-C)^+$, $(M-C)^-$	Cations, Anions
		$(M+Gly)^+$, $(M+Thio)^+$	Matrix Adducts
		$(M+nNa)^+$, $(M+nK)^+$	Alkali Metal Adducts
		$(M+OAc)^-$	Acetate Adducts
		$(M+H+nNH_3)^+$	Ammonia Adducts
		$(M+H+nH_2O)^+$ (for water- +19,37,55....)	Water Adducts

Interpretation: MW With Multiple Charge

One of the miracles of electrospray has been the observation that this technique is capable of generating intact gas-phase ionic species by associating more than one charge to each individual molecule.[14] As with most single charge mechanisms in LC/MS, deprotonation and protonation are the primary charge generation processes. In order to produce multiple charges in electrospray, a molecule must generally be capable of donating or accepting multiple protons. Many significant biopolymers, such as protein and nucleic acid, meet this criterion and are commonly observed as multiple charge species. The resulting mass spectrum for a given analyte with multiple charge appears as a series of ions representing the distribution of charge states and is illustrated in Figure 9.10. Note that the characteristic *bell-shaped* distribution of ions represents the relative intensity of each charge state.

The **molecular mass**, M, can be calculated from any observed m/z for a multiple-charged ion by using the following equation:

Mass: $$M = n(m_n + m_H) \qquad (9.1)$$

Where M is the molecular mass of the uncharged molecule, m_n is the measured mass from the spectrum, m_H is the mass of a proton (1.00782u), and n is the integer charge on the molecule, (n = ± 1,2,3,4,5,.......) The charge decreases by one charge unit with each successive ion as one moves up the mass range.

There are two ways to determine the unknown charge on a multicharged species. Since we know from experience that adjacent masses in the bell-shaped distribution will always differ by one charge, it becomes a simple problem of two equations and two unknowns. Solving equation 9-1 for two adjacent ions differing by one charge (n=i, n=i+1) yields the general solution:

Charge: $$i = \frac{m_{i+1} - m_H}{m_i - m_{i+1}} \qquad (9.2)$$

The second and most straight forward method for determining charge state is by measuring the mass distance between adjacent isotope peaks for a given ion. For example, the ion at 892.9 could yield ^{13}C isotopes separated by 1.00u. The measure mass difference between ^{12}C and ^{13}C would directly yield charge state (e.g. $n = 1/\Delta m = 1/0.053u = 19$). Unfortunately, a resolving power of 18,000 would be required to make this direct charge determination.

We can use the data from Figure 9-10 to evaluate the charge state of any measured mass by comparing two adjacent masses. For example, we can take measured mass 942.8 in the spectrum as (m_i) and measured mass 892.9 as (m_{i+1}). Using equation 9.2 we can easily solve for the charge stage i of mass 942.8.

$$i = \frac{892.9 - 1.0}{942.8 - 892.9} = 17.9 \sim 18 \text{ @ } 942.8,\ i+1 = 19 \text{ @ } 892.9$$

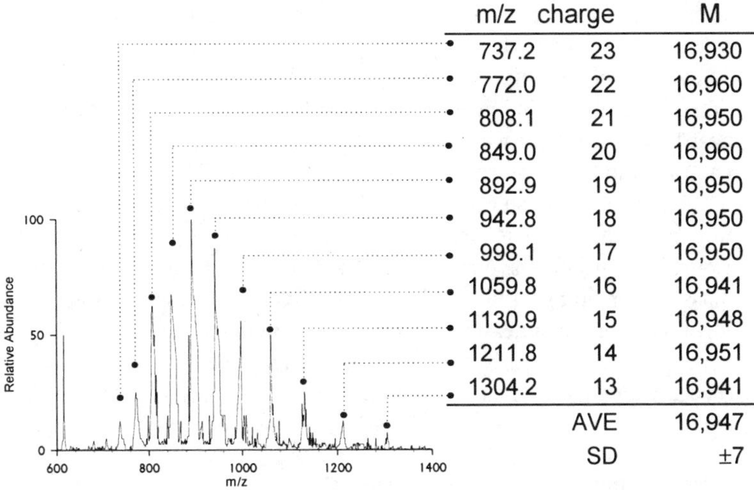

m/z	charge	M
737.2	23	16,930
772.0	22	16,960
808.1	21	16,950
849.0	20	16,960
892.9	19	16,950
942.8	18	16,950
998.1	17	16,950
1059.8	16	16,941
1130.9	15	16,948
1211.8	14	16,951
1304.2	13	16,941
	AVE	16,947
	SD	±7

Figure 9.10- Multi-charge spectrum of Myoglobin (horse heart- MW 16951).

Most practitioners of electrospray have access to one of a variety of computer algorithms to separate and interpret multicharge results, such as those proposed by Mann *et.al.*[15] or entropy algorithms.[16] At least one version is typically available from your instrument manufacturer and may have proprietary components.

Interpretation: Elemental Composition
Through "Isotopic Abundance" Measurements

The rules for mass spectrometry are founded in the physics of the universe. Mass spectrometry essentially measures atomic nuclei. That's where all the matter is, that's where all the mass is. As a consequence, a general understanding of atomic nuclei is essential since its composition will affect your measurement. We have already discussed electrons, neutrons, and protons as the basic building blocks of all matter. We have also discussed the relativistic nuclear world manifested in terms of *mass defect*. Suffice it to say, all nuclei are not alike, some are more stable than others. Some are radioactive, and will disappear in an instant and others will disappear in 10^4 years. The bottom line for the analytical chemist is to know that the abundance of various atomic nuclei are predictable and that many elements can be measured at more than one mass as *isotopes*. *Isotopes* consist of a family of nuclei that have the same number of protons but may vary by one or more neutrons. The abundance of each combination of protons and neutrons (isotope) is a function of its stability and history. The isotopic abundance of most elements hasn't changed significantly since the origin of our solar system. The natural abundances of several elements is listed in Table 9.22.

We can use the isotopic abundance of each element for any ion found in a mass spectrum to predict its elemental composition. Because of isotopic contributions, one elemental formula will be observed at several masses coinciding with the relative abundance of each isotope in the formula. For example, the molecule chloromethane, CH_3Cl, has isotopic contributions from carbon, hydrogen, and chlorine. It is a very straight forward process to relate the abundances of ions from m/z 50-53 to the chemical formula. The pattern is shown on the right.

m/z	Formula Showing Isotopes	Relative Abundance
50	$^{12}C^{1}H_3{}^{35}Cl$	100.0
51	$^{13}C^{1}H_3{}^{35}Cl$	1.1
52	$^{12}C^{1}H_3{}^{37}Cl$	32.5
53	$^{13}C^{1}H_3{}^{37}Cl$	0.3

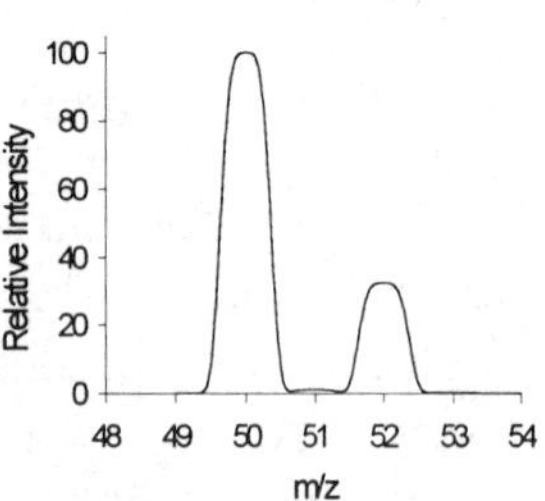

The simple example of chloromethane does not fully indicate the power of isotopic abundance as a tool for determining elemental composition. The relative abundance of each isotope observed for an ion in the mass spectrum is also a function of the number of each element in the molecule. The isotopic cluster around each ion in the mass spectrum is like an elemental "fingerprint". The contribution of each isotope for each element will follow the binomial theorem. Thus, by comparing the measured relative abundances with the theoretical binomial distribution for a given elemental composition the analyst can accurately assign the elemental composition to each pure isotopic cluster in the spectrum.

As shown below, the **Binomial Theorem** is used to calculate the isotopic abundances (P = monoisotopic mass, a = abundance of 1st isotope, b = abundance of 2nd isotope, n = number of isotopic element in the molecule):

$$(a+b)^n = a^n + na^{n-1}b + (n(n-1)/2)\, a^{n-2}b^2 + (n(n-1)(n-2)/6)\, a^{n-3}b^3$$

$$+(n(n-1)(n-2)/6)\, a^{n-3}b^3 + (n(n-1)/2)\, a^{n-2}b^2 + na^{n-1}b$$

n	P	P+1	P+2	P+3	P+4	P+5
1	a	b				
2	a^2	2 ab	b^2			
3	a^3	3 a^2b	3 ab^2	b^3		
4	a^4	4 a^3b	6 a^2b^2	4 ab^3	b^4	
5	a^5	5 a^4b	10 a^3b^2	10 a^2b^3	5 ab^4	b^5
n	a^n	n $a^{n-1}b$	$(n(n-1)/2)\, a^{n-2}b^2$	$(n(n-1)(n-2)/6)\, a^{n-3}b^3$	...n ab^{n-1}	b^n

For Carbon: ^{12}C (a=98.9%), ^{13}C (b=1.1%), Normalized to P=100

n	P	P+1	P+2	P+3	P+4	P+5
1	100	1.1				
2	100	2.2	0.01			
3	100	3.3	0.04	0.0001		
4	100	4.4	0.07	0.0005	0.000001	
5	100	5.5	0.12	0.0014	0.000005	~0
100	100	110*	60.56	22.00	5.93	1.58

*P+1 becomes base peak in isotope cluster at n=91

The expressions for all elements in a complex molecule are slightly more complex, fortunately, the abundances for just about any combination of isotopes that would appear in an organic molecule between 12u and 500u appear in the *Mass and Abundance Tables*.[17] Don't sweat the math, just look up the mass of any ion in the spectrum. Both your observed mass and abundances should match in order to make a valid assignment of elemental composition to an observed ion cluster. We have included abundance tables for carbon, bromine, sulfur, and chlorine in Tables 9.3 to 9.6.

Table 9.3- Carbon Isotopic Abundances

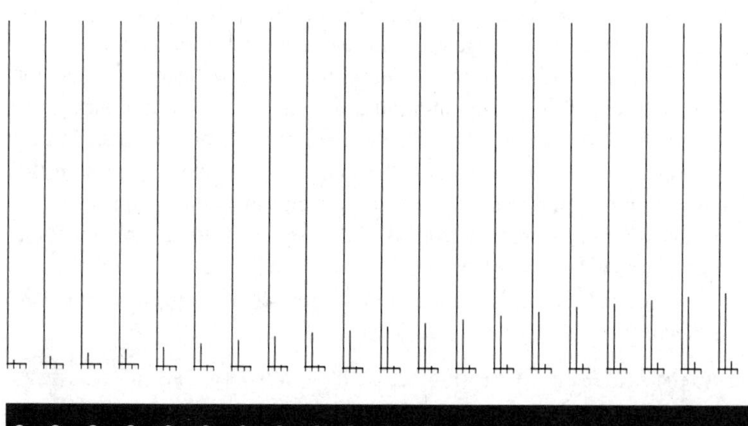

Table 9.4- Bromine Isotopic Abundances

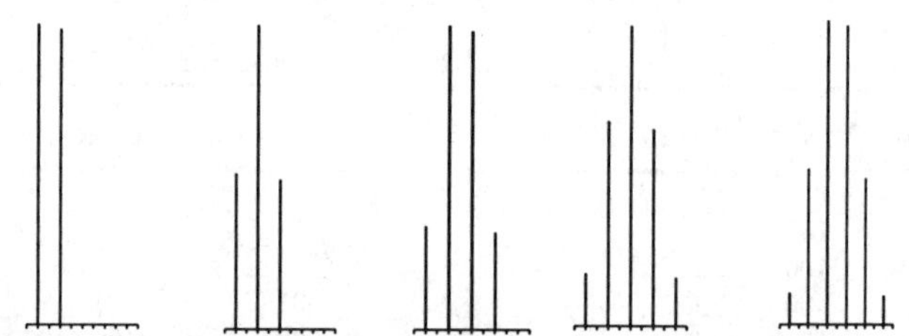

	Br	Br_2	Br_3	Br_4	Br_5
m(P)*	**79**	**158**	**237**	**316**	**395**
P	100.0	51.0	34.0	17.4	10.4
P+2	98.0	100.0	100.0	68.0	51.0
P+4		49.0	98.0	100.0	100.0
P+6			32.0	65.3	98.0
P+8				16.0	48.0
P+10					9.4

***m(P)** is the nominal mass contribution to the mass of the monoisotopic mass of the observed molecule from bromine.

Table 9.5- Sulfur Isotopic Abundances

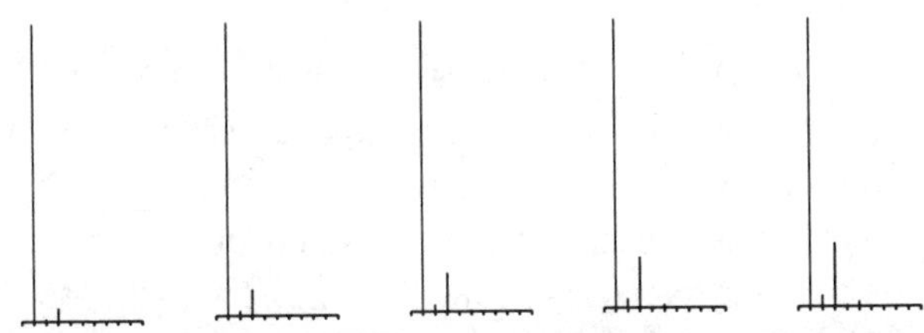

	S	S_2	S_3	S_4	S_5
m(P)*	**32**	**64**	**96**	**128**	**160**
P	100	100	100	100	100
P+1	0.8	1.6	2.3	3.2	4.0
P+2	4.4	8.8	13.2	17.6	22.0
P+3		0.07	0.21	0.42	0.7
P+4		0.19	0.58	1.20	1.9
P+5				0.02	0.05
P+6				0.03	0.09

***m(P)** is the nominal mass contribution to the mass of the monoisotopic mass of the observed molecule from sulfur.

Table 9.6- Chlorine Isotopic Abundances

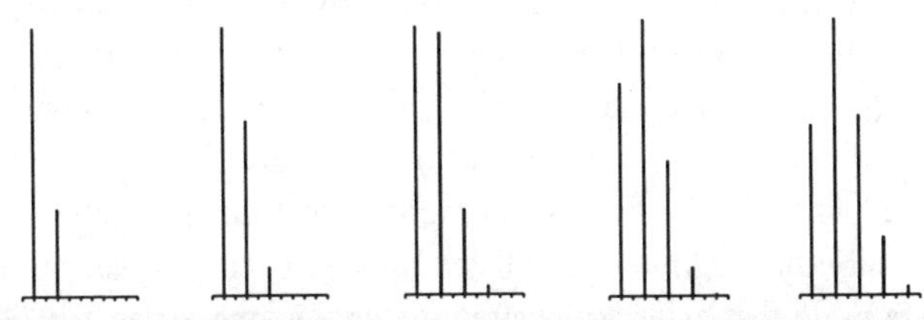

	Cl	Cl_2	Cl_3	Cl_4	Cl_5
m(P)*	**35**	**70**	**105**	**128**	**160**
P	100	100	100	76.9	61.5
P+2	32.5	65.0	97.5	100	100
P+4		10.6	31.7	48.9	65.2
P+6			3.4	10.5	20.9
P+8				0.9	3.4
P+10					0.2

***m(P)** is the nominal mass contribution to the mass of the monoisotopic mass of the observed molecule from chlorine.

Interpretation: Elemental Composition

Through "Accurate Mass" Measurement

Arguably, the most powerful technique available for solving identity problems in LC/MS is accurate mass measurement (Figure 9.11). The accurate measurement of mass position of all ions, quasi-molecular, molecular, or fragments allows the analyst to assign an elemental composition to each ion, and relate the specific composition of each fragment to the entire molecular structure. Accurate mass measurement directly gives you the empirical formula of each ion measured. Accurate measurement of the differences between two ion masses can also provide a direct measurement of the composition of a neutral loss of a specific chemical structure from one ion to the other. So, information from direct measurement or differences between two measurements can be used to provide piecemeal elemental composition of the entire molecule, each fragment, and each neutral loss. These specific and structurally related elemental formulas are useful in creating a complete picture of the entire molecule.

The key to success in accurate mass measurement is calibration. Although each different instrument may be calibrated and tuned slightly differently, all accurate mass experiments require at least two accurately calibrated masses bracketing the measured ions. In some cases, the required mass accuracy is only obtained by scanning a narrow mass range at reduced scan speed. One must realize that the acquisition requirements will vary from instrument to instrument. Some of the common calibration standards are listed in *Appendix B*. One should always keep in mind that calibrations are tune dependent, especially at high mass accuracy. Don't calibrate before you tune your instrument; ALWAYS TUNE BEFORE YOU CALIBRATE. Another common mistake occurs when analysts fail to load their most recent calibration file into their acquisition method. Acquiring data under old calibration conditions will result in miscalibrated masses.

Fortunately, for most analysts, the advent of newer technologies such as FTMS and TOF (as well as the traditional workhorse for accurate mass, the magnetic sector) gives a wide variety of choices for acquiring accurate measurements.

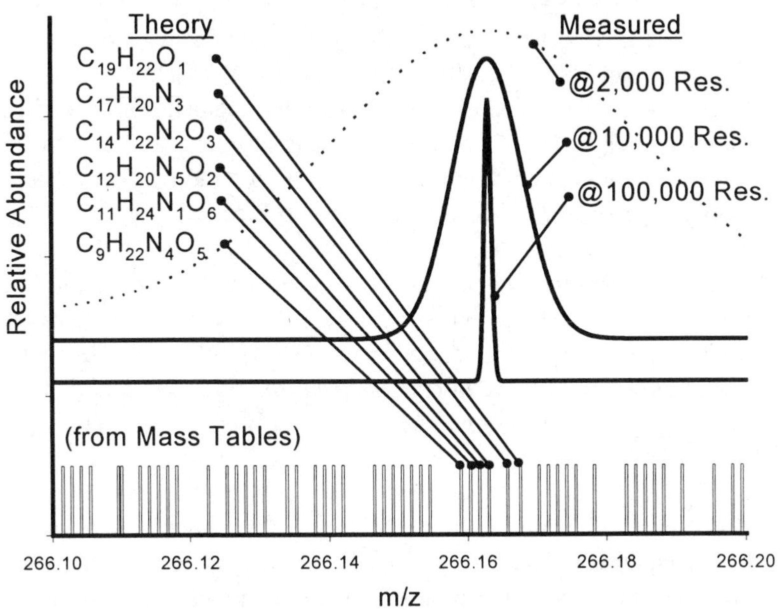

Figure 9.11- Illustration of the comparison between accurate mass data and exact mass values acquired from theoretical values.

Theoretical exact mass values are listed in *Mass and Abundance Tables for use in Mass Spectrometry,* by J. Beynon and A. Williams.[18] All LC/MS labs should own a copy or acquire a software version.[19] Software versions of mass tables are available from all instrument manufacturers of high resolution instruments. Even a lab with a low resolution instrument may be able to narrow their choices of a given assignment if they can acquire their data accurate to 0.1 amu. Combining isotopic abundance and accurate mass measurement is a fullproof method to determine the elemental composition of a given ion cluster. Some of the newer software packages are also integrating this information into a convenient data reduction package[20]

Interpretation: Structural Elucidation Through Fragmentation of "Odd-Electron" Ions

The third and perhaps most challenging component of interpretation, is structural elucidation. We typically elucidate structure by measuring the mass and intensity of various fragment ions observed in our mass spectrum. The first challenge in interpretation is to acquire a fragmentation spectrum. The most common and best documented approach to yield fragment ions is electron ionization (EI). The uniqueness of EI fragmentation processes is found in the nature of it's primary product ion, the *molecular ion.* The formation of molecular ions in EI requires that an electron be removed from the molecule leaving one unpaired electron (radical) with the parent molecule, referred to as an *odd-electron* (OE^+) ion. In addition, EI imparts significant internal energy into the molecule generally resulting in significant fragmentation.

The excited odd-electron molecular ion leads to *unimolecular* decomposition (fragmentation). The pathways of unimolecular decomposition will depend primarily on the amount of internal energy and the structure of the molecule. By accurately controlling the amount of energy in EI, we can probe many fine-points of molecular structure. However, most EI spectra are acquired at *ca.* 70 eV to yield very <u>reproducible</u> fragmentation spectra.

Table 9.7 shows some of the fragmentation pathways associated with decomposition of these odd-electron species. Simple cleavage will result in the formation of a neutral radical (containing the odd-electron) and an even-electron (EE^+) ion. There are many characteristic neutral losses and fragments associated with specific chemical structures and compound types. Fortunately, there are selection rules that govern the fragmentation process. Generally, the unpaired electron will tend to stay with the fragment that has the highest ionization energy (**Stevenson's Rule,**[21] Table 9.8). As a general rule non-bonding electrons are easier to remove than π electrons which are easier to remove than σ electrons[22]; therefore, the unpaired electron will tend to stay with the saturated hydrocarbon moieties while the charge will tend to stay with the structures with extra electrons and resonance capabilities.

Table 9.7- Odd-Electron Decomposition

Ionization	Molecular Ion	1° Fragmentation	2° Fragmentation
ABCD + e →	$ABCD^{+\bullet}$ ⟶	$A^+ + BCD^\bullet$	1 bond cleaved (charge migration)
(EE) (10-100eV)	(OE^+) ⟶	$A^\bullet + BCD^+$	1 bond cleaved (charge retention)
		↳	$BC^+ + D$
	⟶	$D^\bullet + ABC^+$	1 bond cleaved (charge migration)
		↳	$A^+ + BC$
Rearrangement	⟶	$AD^{+\bullet} + BC$	2 bonds cleaved (charge migration)
• Electron interacts with neutral molecule in the 10^{-16} sec timeframe. Energy is partitioned into electronic (not vibrational) transitions.	• Molecular ion is formed by ejection of an excited electron. Excess internal energy is dissipated through the molecule leading to fragmentation. • Rearrangement of the molecular ion can occur.	• Fragmentation will generally occur to form EE^+ fragment and neutral radical. • (Apply Stevenson's Rule) • Rearrangement and multiple bond cleavage can lead to OE^+ fragments. • (Apply Nitrogen Rule)	• Stable EE^+ fragment ions can be diagnostic by having characteristic masses as well as characteristic low mass series.

Some of the most diagnostic fragments are those that involve more than a simple cleavage; rather they can involve multiple bond cleavages and rearrangement of structural entities, particularly protons. The presence of odd-electron fragments is a tell-tale sign of rearrangement. One can identify these OE^+ fragments using the **nitrogen rule**.

Interpreting mass spectra from first principals is an activity that can take a lifetime to master, but only a few days to understand. Anyone with a chemistry background should be able to grasp the basics of interpretation after only a few days of training. Training, practice and experience are the three ingredients for success. We recommend the following steps in order to acquire the basic capabilities to interpret mass spectra.

Before Structural Elucidation

1. Purchase one or more texts on interpretation. We recommend everyone purchase *Interpretation of Mass Spectra* by McLafferty and Turecek. Other texts listed in *Appendix A*.

   Guideline: Every LC/MS lab should have a text with the basic procedures for determining elemental composition, molecular weight and structural elucidation.

2. Take a short-course on mass spectral interpretation. Courses are offered by ACS, ASMS, and other organizations. (*Appendix A*) Some, but not all universities, may also have courses on mass spectrometry available from time to time. You should check at you local university for schedules.

   Guideline: A focused classroom environment is the only way to fully develop good habits and efficient methods. A one or two day course is well worth the investment. If you are considering a university course, make sure that it has a practical focus and includes interpretation. Some academic courses may focus too heavily on theory and not enough on problem-solving.

3. Purchase a text or electronic version of the *Mass and Abundance Tables*. See *Appendix A*.

   Guideline: Make sure the key references are in your lab. Don't be forced to go to the library to look up mass and abundance values.

4. Purchase a text or electronic library for reference. Most of the current data bases are listed in *Appendix A*.

   Guideline: The ability to compare an acquired spectrum with that of a known compound will assist you in comparing and evaluating fragmentation mechanisms, comparing neutral losses, comparing key low mass fragments, and more confidently making structural assignments. If you acquire an electronic version, make sure the data base is capable of being randomly browsed.

Table 9.8-Values for Ionization Energy and Proton Affinity

Selected Compounds	Ionization Energy	Proton Affinity	Type
CH_4	12.5	5.9	RG
C_2H_6	11.5	6.2	
C_3H_8	11.0	6.7	
C_4H_{10}	10.6	7.5	
C_6H_6	9.2	8.2	S
CO	14.0	6.5	
CO_2	13.8	5.7	
H_2O	12.6	7.7	RG,S
O_2	12.1	4.6	
HCOOH	11.3	8.1	B
CH_2O	10.9	7.9	
CH_3OH	10.8	8.2	S
CH_3COOH	10.4	8.5	B
$C_6H_5COCH_3$	9.3	9.1	
CH_3CN	12.2	8.4	S
NH_3	10.2	9.1	RG
$C_2H_5NH_2$	8.9	9.6	
HCl	12.7	6.4	B
C_6H_5Cl	9.1	8.2	
He	24.6	2.1	C
Ar	15.8	4	C
N_2	15.6	5.8	C

These structures are seen throughout LC/MS, from solvents (S), to buffers (B), to ionization reagent gases (RG), to neutral target gases (C), to leaving groups and competitive moieties in decomposition. This table was not intended to be a complete listing, but rather a representative excerpt from the more complete listing in McLafferty and Turecek, *Interpretation of Mass Spectra.*[23]

Interpretation: Structural Elucidation Through Fragmentation of "Even-Electron" Ions

The history, documentation, training and data bases in mass spectrometry have shown the preponderance for evaluating odd-electron ions. Unfortunately, for the vast majority of practitioners in LC/MS, the ions that you will be observing in your spectra will be even-electron. Electron ionization, the basis of odd-electron protocols and mechanisms is not the most frequent and most applicable ionization mechanism in LC/MS. APCI, ES, TS, FAB, and CI(PB) all produce even-electron ions with a few noted exceptions.[24]

We routinely solve structural elucidation problems in LC/MS by collisional dissociation of these even-electron parent (precursor) ions. The translational energy of an ion accelerated toward a neutral target species is partitioned into internal energy. This interaction can result in decomposition of the incident ion. This process is referred to as *collision induced dissociation* (CID) or *collision activated dissociation* (CAD).[25] Typical neutral target species (collision gases) are argon, nitrogen, air, or helium (primarily in sectors). Most people use argon. Nitrogen can be cheaper. Note that all collision gases should be high purity, as prescribed in *Chapter Seven*.

The fragmentation pathways associated with even-electron (EE^+) ion decomposition are somewhat different than those associated with OE^+. Unlike OE^+ processes, there is no radical site in EE^+ to initiate fragmentation. EE^+ ions will generally result from charge migration across a cleavage. The **Even-Electron Rule** states that the neutral fragment in an even-electron dissociation will tend to be the fragment with the lowest proton affinity, as a consequence, the fragment with the highest proton affinity will retain the charge. The energetically favored products of EE^+ from a single bond decompositon are a neutral (EE) species and an EE^+.[26] This can only occur through the charge migration process. Table 9.9 shows some of the primary processes associated with even-electron decomposition.

Table 9.9- Even-Electron Decomposition.

Ionization	Collision	1° Fragmentation	2° Fragmentation
$ABCD^+ \rightarrow$	$ABCD^+ + X \rightarrow$	$ABC^+ + D$	1 bond cleaved (charge migration)
(EE^+)	(EE^+) (5-1000eV)	↳	$AB^+ + C$
	$\longrightarrow$	$ABC^\bullet + D^+$	1 bond cleaved (charge retention)
Rearrangement	$\longrightarrow$	$DA^+ + BC$	2 bonds cleaved (charge retention)
• Most ions generated in LC/MS are even-electron from either protonation or deprotonation to form low internal energy products.	• Ions are accelerated toward neutral target species X. Collisions convert translational energy into internal energy resulting in decomposition. • Rearrangement of the molecular ion can occur.	• Fragmentation will generally occur to form EE fragment and neutral molecule. (Apply Even-Electron Rule) • OE+ products are generally not favored. • Rearrangement and multiple bond cleavage are common in CID spectra.	• Stable neutral molecules are common neutral losses in EE+ fragmentation.

A comparison of OE^+ and EE^+ decomposition of clenbuterol is presented as an excellent example of the dramatic difference between the these two decomposition pathways. The CID fragments observed for clenbuterol show the typical neutral loss pathways that characterized EE^+ decomposition. In contrast, the EI spectrum has as its primary fragmentation pathway α-cleavage to yield the base-peak of m/z 86 and the other high mass fragment m/z 190. The OE^+ process is associated with the molecule splitting in half and revealing a series of structural fragments that can trace to initial radical sites. The EE^+ decomposition is characterized by neutral loss of one attached piece after another with the charged product increasing in resonance stabilization. Since most practitioners today use CID, the purportedly "less diagnostic" EE^+ decomposition is the only alternative.

Interpretation: Structural Elucidation

Even- vs Odd-Electron Fragmentation Pathways

CID Fragmentation of Protonated Clenbuterol[27]	m/z	electron pairing	rings +db[28]	comments
(Cl)₂(H₂N)-phenyl-CH(OH)-CH₂-NH_2^+t-Butyl	277	EE^+	4.5	$(M+H)^+$
↓ $-H_2O$	-18			2rH charge retention
(Cl)₂(H₂N)-phenyl-CH=CH-NH_2^+t-Butyl	259	EE^+	5.5	
↓ -2-methylpropene	-56			rH charge retention
(Cl)₂(H₂N)-phenyl-CH=CH-NH_3^+	203	EE^+	5.5	
↓ $-Cl^•$	-35			loss of Cl radical to form OE^+ is likely followed by rH
HN= ring (Cl, H) -CH=CH-NH_3^+	168	OE^+	6.0	
↓ -HCl	-36			
HN= ring -CH=CH-NH_3^+	132	OE^+	7.0	note the signficant resonance stabilization

EI Fragmentation of Protonated Clenbuterol[29]	m/z	electron pairing	rings +db[27]	comments
[Cl, H_2N, Cl, OH, NHt-Butyl; 86, 190]$^{+\bullet}$	276	OE^+	4.0	$M^{+\bullet}$
↓ - •OH,- •NH_2	-33 -17,-16			inductive cleavage
Cl, H_2N, Cl, $\overset{+}{N}H_2$t-Butyl	259	EE^+	5.5	
$M^{+\bullet}$ ↓ -•CH_2NHt-Butyl	-86			α-cleavage
Cl, H_2N, Cl, $\overset{+}{O}H$	190	EE^+	5.5	
$M^{+\bullet}$ ↓ Cl, H_2N, Cl, OH •	-190			α-cleavage
$H_2C{=}\overset{+}{N}$(H)t-Butyl	86	EE^+	0.5	
$M^{+\bullet}$ ↓ Cl, H_2N, Cl, OH, •NH	-219			inductive cleavage
+t-Butyl	57	EE^+	-0.5	

Interpretation: Structural Elucidation Through "Labeling" Experiments

One of the most powerful tools available to the analyst in LC/MS is the use of isotopic labeling techniques. These techniques are widely used in mass spectrometry to accomplish a variety of objectives, including the tracing of labeled drugs through metabolic pathways, the tracing of labeled reactants through reaction pathways, and the tracing of labeled structural moieties through fragmentation pathways. In general, the goal of all labeling experiments is to produce a chemical equivalent of your primary analyte with some chemical tag or label. There are two common types of labeling experiments each of which has different objectives. The first type of labeling experiment is referred to as **stable label** (See Figure 9-13). A stable label involves the use of one of many available stable isotopes; usually, ^{2}H(deuterium), ^{13}C, ^{15}N, or ^{18}O. All of these examples are used for various applications in organic analysis where tracking one element may have a particular structural, or mechanistic, or temporal implication in the experiment.

The second type of labeling experiment is performed using **radio labeled** analytes. In such an experiment, a radioactive tag is incorporated into the molecule of interest. Tritium (^{3}H) and ^{14}C are the most common radio labels in the world of organic analysis. Radio labeling is sometimes used with mass spectrometry when radioactivity detectors are placed in parallel or series with the mass spectrometer. The high sensitivity and selectivity of the radio-detector response is correlated with the mass spectral results. It should be stressed that introducing radioactive material into a mass spectrometer is a serious endeavor and is only performed by laboratories with the proper equipment and training to deal with the increased danger and liability of radioactive material.

The trend in recent times is to use stable labels for as many tracer applications as possible. This is of particular emphasis when performing Phase I clinical trials where metabolic profile information is of primary importance. A recently developed technology known as CRIMS (Chemical Reaction Interface Mass Spectrometry) has coupled both gas and liguid chromatography to the mass spectrometer.[30]

Isotopic Labeling

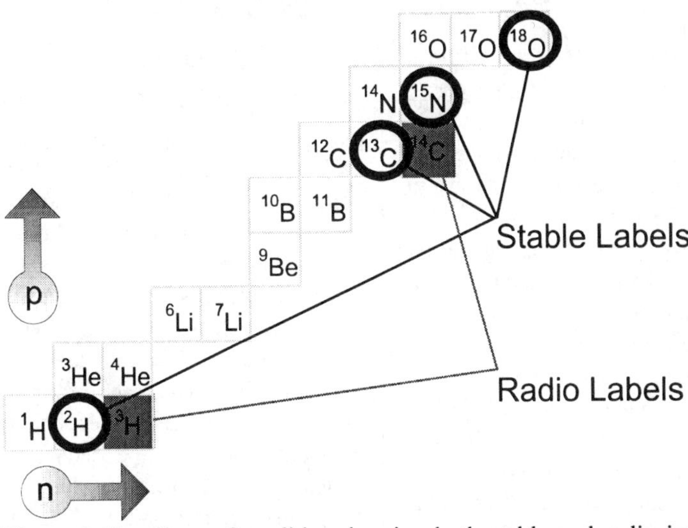

Figure 9-13- Chart of nuclides showing both stable and radio isotopes commonly used in organic analysis (^{2}H, ^{13}C, ^{15}N, and ^{18}O).

This technigue typically breaks down the stable labeled analyte into oxidative combustion products of carbon, nitrogen, and sulfur. The labeled combustion-product retention times are measured with mass spectrometry.

Another significant application of stable labeling has been the use of deuterium exchange with a wide variety of chemical species. Deuterium exchange has been shown to offer insight into the secondary and tertiary structure of proteins. This technique has been used for both static structure determination and dynamic structure determination of proteins and protein multimers.[31]

A comprehensive review of labeling techniques and their application to chemical analysis is found in *Radioactive and Stable Isotope Tracers in Biomedicine: Principals and Practice of Kinetic Analysis* by Robert Wolfe (1992).[32] Many of the techniques and protocols described in this book can be directly applied to solving identity problems in LC/MS.

Interpretation: Structural Elucidation

Proteins and Peptides

More than MW is needed

LC/MS may seem like the perfect tool for protein analysis- just shoot the sample into the mass spec and get a molecular weight with no complex, time consuming, wet chemistry. No doubt, molecular weight is the most basic piece of information needed to identify a protein. But today's biochemist needs more than a molecular weight. Sequence information and changes due to post-translational modifications have emerged as a major focus of biochemical analysis of proteins. This is understandable. The activity of many proteins is related to their degree of phosphorylation or glycosylation (addition of sugar chains). So the demands on MS have gone from molecular weight to sequence and structural features that confer function. Most methods require a combination of techniques (LC, enzymatic cleavage, electrophoresis, CE, mapping protocols etc.). The list below reflects current focus:

- amino acid sequence
- site of phosphorylation
- locaton of disulfide bonds
- extent/location of glycosylation
- amino acid modificatons
- amino acid substitiution
- secondary structural changes
- tertiary structural changes
- epitope mapping

Confirm Structure with Public Data Bases

One of the most important changes in protein chemistry has come about through the construction and use of data bases of protein sequences. Using these data bases it is possible to identify proteins when only having a partial sequence. The database not only includes sequence information but also a second level that has annotations to the sequence which can lend more insight into the nature, identity and function of a protein. A partial list of data bases is below:

Protein Data Bank

National Institutes of Health
`www.mol-bio.info.hin.gov/cgi-bin/pdb`

PROWL
Rockefeller University
`www.chait_sgi.rockefeller.edu`

Biologist Resources Cambridge
`www.bio.cam.ac.uk`

Brookhaven Natl. Laboratories
`www.pdb.bnl.gov`

Human Genome Project
`www.ornl.gov/techresources/human_genome/home.html`

EMBL-European Molecular Biology Labs
`www.mann embl-heidelberg dc/default.html`

The most important choice a protein chemist will have to make before doing and analysis is the method of ionization. In general, this is a choice between MALDI (matrix assisted laser desorbtion) or ESI (electrospray ionization). Each of these techniques has advantages and disadvantages that are a function of your research needs. To make a choice we recommend that you read *Mass Spectrometry for Biotechnology* by Gary Siuzdak.[37]

A Typical Protocol for Protein Sequencing/Mapping

Step	Biochemical Protocol	LC/MS Data Produced
1	Isolate and purify protein	Determine MW, Purity
2	Reduce and alkylate protein	
3	Digest with protease(go to step 4)	Map peptide molecular masses and search databases. Acquire fragmentation data and sequence and search databases.
4	LC/MS separation with CID	Determine mass and sequence of constituent peptides and search databases.
5	Glycosidase and phosphatase digestion (in parallel)	Map peptide molecular masses and search databases. Acquire fragmentation data and sequence and search databases.
6	LC/MS separation with CID	Determine MW and sequence of constituent peptides. Look for mass shifts to determine positions of phosphorylation and glycosylation.

The protocol for the analysis of a protein, even with the advent of LCMS technology has not changed much. In general, the most important information requires peptide maps from protease digestions. By using those maps it is possible to determine which peptides have phosphate or sugar moieties.

Interpretation- Structural Elucidation

Nucleotides

Nucleic Acids by LC/MS

As described previously there is no doubt that the field of protein chemistry is benefiting immensely from the use of mass spectrometers. However, for nucleic acids, the other biological polymer, the use of mass spectrometers is problematic.

DNA consists of only six components, four bases, phosphate and sugar in the familiar double helix structure. From these, the entire living world is constructed. A mass spectrometer can identify deoxyribose (a sugar), phosphate (links the sugars) and the four bases, adenine, guanine, cytosine and thymine. However, the most important information about a DNA molecule is not its composition; but rather, the sequence of the bases. Therefore, the real question is "can mass spectrometers sequence nucleic acids?" The answer is "yes." But, can mass spectrometers do sequencing as quickly and efficiently as the more traditional gel techniques. The answer this time is "no."[33]

This doesn't mean that that MS doesn't have a role in the nucleic acid world, just that it won't be used as extensively (at this time) for the determination of unknown sequences of appreciable length.

The two ionization methods used to sequence oligonucleotides are the same as that used for proteins: MALDI and electrospray. MALDI sequences nucleotides from Sanger dideoxy reaction (those used in conventional nucleotide sequencing) or by the use of exonucleases to generate a sequence ladder. These same methods can be used with electrospray in negative ion mode. Using these techniques, routinely 15-20 mers can be sequenced. Current state of the art is 50-100 mers. Gels, on the other hand, can sequence 1200 mers- an order of magnitude better. There is no doubt that gels require a long prep time and run time[34] but the data is very accurate and requires a minimum of interpretation.

So where does LC/MS fit into the nucleic acid world? There are three places: *identification*, *confirmation* and *interpretation*.

Identification

The gap in performance between gels and MS does not mean there is no use for MS analysis of nucleic acids. Since much of the genome level sequencing is proceeding at a rapid pace the real role for MS may be in the "post-genomic" world. This is the world of sequence databases where much of the sequencing has been accomplished and the real need is to determine the sequence of small pieces of oligonucleotides to compare to a reference genome for identification.[35] LC/MS can be faster, cheaper and just as accurate as gels if you're doing repetitive "target" assays or diagnostics. There is much software being developed today for analyzing data in the context of genomic sequence information and correlating the mass spectra of proteins and nucleic acids to sequences in databases[36] (see databases listed in protein section).

Confirmation

Another use of LC/MS would be in the manufacture of oligonucleotides. During solid phase synthesis of specific nucleotide sequences there are many places where errors can be made. Current technology will produce a 20 mer with only 82% efficiency.[37] This means there will be a need for quick and efficient means to analyze manufactured nucleotides to determine purity. ESI can do this with high accuracy in negative ionization mode. The oligonucleotide manufacturing industry will be the supplier of DNA probes to use in silicon lab-on-a-chip technology.[38] This technology will play a large role in clinical diagnostics.

Interpretation

Even though gel techniques are an order of magnitude better than LC/MS there are problems with gels. The use of LC/MS can avoid compression; the artifact caused by secondary structure changes in a nucleotides that changes its electrophoretic mobility.[39] In addition, LC/MS has the ability to distinguish a false termination in the Sanger reaction because it can distinguish between a deoxy and dideoxy nucleotide. Using LC/MS techniques, modification in RNA can be detected. In fact, the RNA world may be a place that LC/MS can play a large problem-solving role since the sequences are shorter and may contain modifications that confound the more traditional methods.[40]

Worksheet 9.0-Interpretation

This worksheet is intended to prompt you to consider many of the important issues relating to interpretion of unknown sample constituents. This requires you to simultaniously consider both experimental and interpretive aspects of the problem. This is not a linear process; rather, experiment and evaluation should cycle back and forth in a series of affirmations and nullifacations, each proposed structure having to obey the laws of physics, including conservation of matter, mass, and energy.

Part A- Determining Molecular Weight. Check-off your specific capability and result and comment on each item.

	Yes	No	Comments
– Can you acquire EI spectrum of unknown?	❑	❑	How?__________________
– If no, can you acquire spectrum of protonate molecule?	❑	❑	How?__________________
– If no, can you acquire spectrum of deprotonated mol.?	❑	❑	How?__________________
– Do you see any adducts?	❑	❑	Which ones?____________
– Do you see multicharge species?	❑	❑	Which ones?____________
– Do you see multimers?	❑	❑	Which ones?____________
– Can you determine a molecular mass?	❑	❑	What is it? ____________
– Is the observed mass the highest mass in spectrum?	❑	❑	What type? ($M^{+\cdot}$,$(M+H)^+$) __________________
– Is the ion odd- or even-electron? (nitrogen rule)	❑	❑	Which? How do you know? __________________
– Are all observed neutral losses consistant with those from molecular ions?	❑	❑	Which?________________

Part B- Determining Elemental Composition of Molecular Species. Check off and comment on each item.

	Yes	No	Comments
– Can you determine the isotopic abundances of your (quasi-) molecular ion?	❑	❑	What are they? __________
– Can you estimate elemental composition from (quasi-) molecular ion based on abundance?	❑	❑	What is the formula? ______
– Can you deduce abundances for specific isotope patterns?	❑	❑	hat elements? What numbers? __________________

Note: Copies of this worksheet can be downloaded from `www.lcms.com`.

Worksheet 9.0 - Interpretation

	Yes	No	Comments
– Can you measure with high mass accuracy?	❑	❑	What is the accurate mass measured?__________ What is closest exact mass from *Mass Tables*? __________
– Are all your masses and compositions consistant?	❑	❑	Where are they inconsistant? __________
– Can you explain inconsistancies? (impurities, background fragments)	❑	❑	How? __________

Part C- Determining Elemental Composition of <u>Fragment Ions</u>. Check off and comment on each item.

	Yes	No	Comments
– Can you determine the isotopic abundances each fragment ion	❑	❑	What are they? __________
– Can you estimate the elemental composition from each ion based on strictly abundance?	❑	❑	What is their formulas? __________
– Can you deduce abundances for specific isotope patterns (Cl,Br,S,C)?	❑	❑	What elements? What numbers? __________
– Can you measure with high mass accuracy?	❑	❑	What is the accurate mass measured? __________ What is the closest exact mass from *Mass Tables*? __________
– Are all your masses and compositions consistant?	❑	❑	Where are they inconsistant? __________
– Can you explain inconsistancies? (impurities, background fragments)	❑	❑	How?__________

Note: Copies of this worksheet can be downloaded from `www.lcms.com`.

Worksheet 9.0-Interpretation

Part D- Elucidating Structure. Check off and comment on each item.

	Yes	No	Comments
Can you acquire a fragmentation spectra?	☐	☐	How?____________
Can you observe important neutral losses?	☐	☐	Which ones?
Can you confirm the elemental composition of each neutral loss?	☐	☐	What are they?
Can you determine the odd- and even-electron species to each fragment? (mark tentative rearrangement products)	☐	☐	What are they?
Can you evaluate the ionization energy and/or proton affinity to each proposed fragment or neutral loss assignment?	☐	☐	What are they?
Can you apply Stevenson's Rule to any fragmentation pathways?	☐	☐	Which ones?
Can you apply the Even-Electron Rule to any fragmentation pathways?	☐	☐	When? By whom?
Can you calculate rings + double bonds for each fragment?	☐	☐	What are they?
Can you elucidate a complete structure?	☐	☐	hat is it?

Note: Copies of this worksheet can be downloaded from `www.lcms.com`.

9.2-Identifying Suspects

Once you have identified a tentative structure, you must verify that structure through the confirmation process. Unlike interpretation, which may require many experiments and significant experience and reasoning, confirmation is simply a one-on-one comparison between the response of your sample material and a *reference standard material*. Once you have a tentative identification, your first step should be to acquire a standard. The availability of putative standards, of course, will depend on the uniqueness of your standard. You can acquire a standard in a variety of ways; including, purchase, isolate, or synthesize (See Figure 9.14).

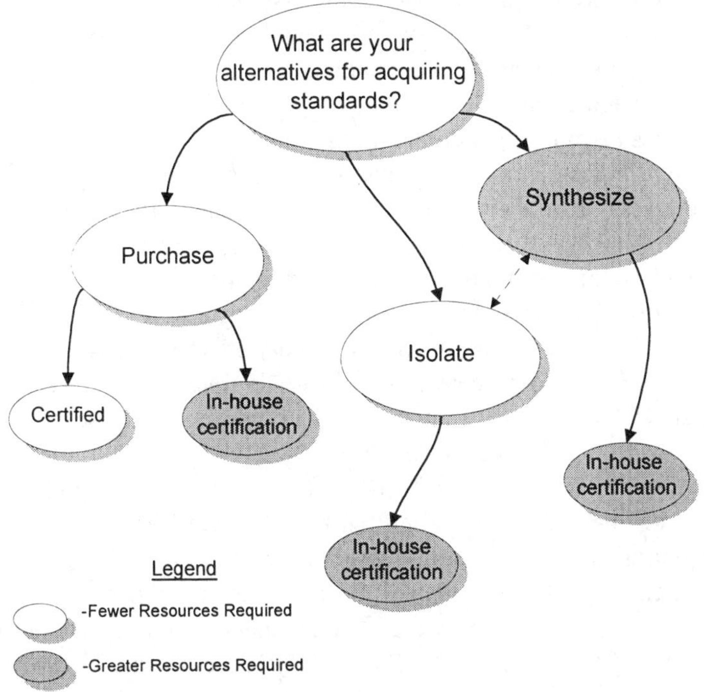

Figure 9.14- Alternatives for acquiring reference standards for suspect and target compounds. Note that your confirmation process will only be as reliable as your standard. Standards should be certified for both purity and composition.

9.3-Identifying Targets

Solving identity problems for targets is much more straight forward than for unknowns; however, it certainly is not trivial. The issue for identity with target species does not involve the rigor of interpretation; rather, it involves determining the signal characteristic(s) of the target species that best select your target species from the population of other sample components (or potential sample components). The rigor in target identification is in isolating the response characteristics of your target from those of your sample. This can be done through prudent selection of separation techniques, ionization techniques, or mass analysis techniques.

Assume that you have available a standard reference material for your target analyte. The confirmation process (Illustrated in Figure 9.15) in the simplest case would be to compare your sample response to that of your standard. The spectral characteristics of the standard material should <u>exactly</u> match the spectral characteristics of your target containing standard.

Attributes of the sample and the solution can, however, affect the response of a given analyte. Sample and solvent related adducts can change the observed characteristics of the target spectrum when compared to the standard material. Other considerations such as pH and ionic strength of the solution can affect the mass spectral response. Every effort should be made in the confirmation process to eliminate all *sample* and *solution* related influences that contribute to differential response between the target analyte and its standard.

The importance of confirmation of targets goes without saying. No method will be valid unless the identical response between standard and sample can be assured. That is the major emphasis of methods development and the major focus of method validation. We have devoted two additional chapters of this book to this important topic; the first, Chapter 11, addresses the broader aspects and requirements of target analysis, and second, Chapter 12, describes the requirements of methods development and validation.

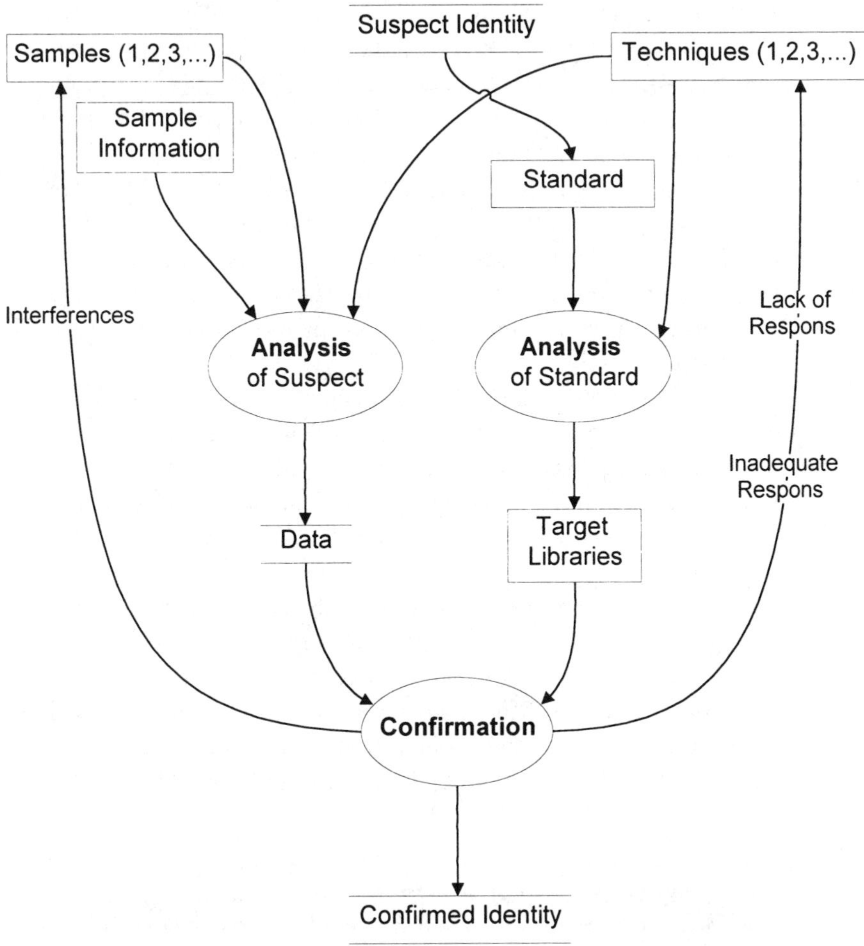

Figure 9.15- Flow diagram of the process of confirming the identity of *suspect* or *target* analytes. To confirm the identity of any sample species, the sample results must be compared directly with the results from a reference standard.

9.4- How Much Identity Is Enough?

An important question, with continuing debate, asks- How much information is required to unambiguously identify an analyte? The answer, of course, depends on the uncertainty associated with your specific problem. The more uncertainty associated with your problem, the more information it takes to confirm the identity of a given sample component. Identity for environmental applications, where the matrix and sample source are highly variable, may require more information to confirm the identity of a target than that rquired for a clinical trial, where matrix and analyte present less uncertainty. A recent workshop entitled "Limits of Confirmation, Quantification, and Detection" (sponsored by the ASMS) was conducted to address some of these issues.[41]

The so-called "three ion rule" is as good place to start any discussion about confirmation requirements because much of our regulatory policy and GC/MS history has endorsed this rationale. It should also be stated up-front that the rationale for the "three ion rule" does not apply directly to most LC/MS practitioners who are doing MS/MS, rather than EI.

The "three ion rule" was proposed by Sphon (1978) in response to the need for GC/MS assays for residue analysis.[42] It was determined that specificity in the confirmatory procedure required at least three ions at restricted abundance ranges for a given analyte. His original study utilized approximately 30,000 compounds in a nominal mass EI data base. Table 9.8 relects an updated version of this study using a modern-day 270,000 spectra data base.

Table 9.8- How many ions does it take for identification?[43]

Ion 1 (% range)	Ion 2 (% range)	Ion 3 (% range)	#Ions	Hits
268 (1-100)	-	-	1	9995
268 (1-100)	239 (1-100)	-	2	5536
268 (90-100)	239 (10-90)	-	2	46
268 (90-100)	239 (50-70)	-	2	9
268 (90-100)	239 (50-90)	145 (5-90)	3	15
268 (90-100)	239 (50-70)	145 (45-65)	3	1

It is clear from Table 9.8 that three ions are needed to unambiguously confirm a target in this data base at a molecular weight near the most populated masses from 200-300 amu. This situation reflects the requirement of separating a target with a single stage of mass analysis at low resolution.

Confirmation with MS/MS or even MS/MS/MS is likely to yield greater selectivity and specificity in a large population of target interferences due to multiple stages of mass analysis. In MS/MS, target specific response is obtained by removing all isobaric interferences. In general, the following criteria should be used for determining your confirmation requirements in MS/MS. Any potential isobaric interference that fragments to your target fragment mass must be removed to assure specificity. This can be accomplished with chromatography, high resolution mass analysis, or sample preparation. If this is accomplished, one unique fragment ion in MS/MS is adequate for confirmation.

Discussion Questions and Exercises

1. Name five types of quasi-molecular ions. Describe how they are formed.
2. List the predominant ions (base peak) found in each of the following ionization techniques in positive ion mode: a) electrospray ionization (ESI), b) APCI, c) thermospray (TSI), d) fast atom bombardment (FAB), and e) electron ionization (EI).
3. Discuss some of the differences between EE^+ and OE^+ decomposition.
4. Predict the identity of the compound that yielded the following exact mass measurement and isotopic ratio. What are the closest five masses and their formulas? Which of the five best matches your compound in both mass and isotopic ratio?

Measured mass	P+1 abundance	P+2 abundance
454.279	30%	5%

5. What is the ratio of P, P+1, P+2, and P+3 for Substance P at molecular weight 1296 and formula ($C_{100}H_{200}O_{30}N_{12}$)?
6. What is Stevenson's rule? What is the even-electron rule? Discuss how they are related. Discuss how they are different.
7. Apply the nitrogen rule to odd-electron ions. Apply the nitrogen rule to even-electron ions.

Notes

[1] This quote appears in Professor McLafferty's book about interpretation and refers primarily to electron ionization mass spectra which contain the molecular ion radical (M^+). This idea is fully applicable to all mass spectra where the molecular mass is represented by protonated or adducted species. defining the molecular boundry is the key analytical distinction between mass spectrometry (MS) and other detection modes. Defining the composition and size of a molecule is essential for identification. In contrast, this type of information is not available in NMR, IR, UV, etc. This key distinction makes MS the primary tool for problem-solving where identity is a critical component of the problem solution. (McLafferty, W. F., Turecek, F., *Interpretation of Mass Spectra*, 4th. ed., University Science Books: Mill Valley, 1993.)

[2] The use of the hyphenation is generally directed toward the instrument configuration while the "slash" (LC/MS) refers the the technique. This distinction is delineated in the excellent ACS Short Course Manual by David Sparkman and Jack Watson. We have chosen to use LC/MS exclusively throughout this book and avoid the subtle difference in terminology. Watson, J.T. and Sparkman, O.D., *Mass Spectrometry: Principals and Practice*, July 1997.

[3] IHC guideline on compound identity testing for bulk drug substances states. This is a clear endorsement by the IHC of LC/MS technologies for problems where identification of impurities and degradation products can have an effect on the quality, safety, and efficacy of drug materials. A complete listing of ICH quidelines is available at **www.ifpma.org**.

[4] A "putative standard" is one that you put into your instrument. In other words, a standard that you can weight, dissolve, and analyze.

[5] There are a number of excellent protocols for spectral interpretation, including, those found in reference 1 and those found in Watson, J.T., *Introduction to Mass Spectrometry* 3rd ed.; Lippincott-Raven: Philadelphia, PA (1997). We have not deviated from the spirit of these well

known protocols, but we have adapted the present protocol to focus more on MS/MS strategy than the traditional EI strategies.

[6] Note that the term protonated molecular ion is considered inappropriate. You do not protonate a molecular ion, rather, you protonate a molecule in the ion generation process.

[7] The term *quasi-* is defined it *Webster's Ninth New Collegiate Dictionary* as: *as if it were, approximately*. In terms of our measurements of ionized molecules of any type, they will tend to be in the proximity of the molecular ion if it were observed in the spectrum. In this context, protonated molecules, ammoniated molecules, sodiated molecules, potassiated molecules, deprotonated molecules would all fall under this general heading. However, multicharge and multimer forms of the ionized molecular species would not fall under this definition of quasi-molecular ion.

[8] Halliday, D., Resnick, R., *Fundamentals of Physics*, Revised Printing, pages 987-1011, Wiley and Sons: New York (1986).

[9] Yergey, A. L., Edmonds, C. G., Lewis, I. A. S., Vestal, M. L., *Liquid Chromatography/Mass Spetrometry*, pages 74-75, Plenum Press: New York (1990).

[10] The term multimer can have several connotations; for example it can refer to a covalently bound species that is observed as a result of two reactant monomers interacting to form a dimer (thymine dimerization in DNA); conversely, multimers can refer to non-covalently bound species such as protein multimers.

[11] "Probe analysis" refers to the sample introduction technique whereby a solid or liquid sample is placed on the end of a sample probe and subsequently introduced through a vacuum interlock into the high vacuum ionization region of the mass spectrometer. This approach can be used for virtually any ionization mode including EI, CI, and FAB.

[12] McLafferty, W. F., Turecek, F., *op cit.*

[13] "C" stands for counterion of the particular anion or cation in solution.

[14] Matthias Mann, Chin Kai Meng, and John B. Fenn, "Interpreting Mass Spectra of Multiply Charged Ions," *Anal. Chem.* 61, 1702-1708 (1989). This article is an excellent overview of the importance and potential of multiple charge ionization with electrospray. Several deconvolution algorithms are proposed for automated data reduction.

[15] ibid.

[16] Reinhold, B.B., Reinhold, V.N. "Electrospray Ionization Mass Spectrometry: Deconvolution by an Entropy-Based Algorithm" *J Am Soc Mass Spectrom* Volume 3, Page 207 (1992)

[17] Beynon, J.H. and Williams, A.E., Mass and Abundance Tables for Use in Mass Spectrometry, Elsevier, New York (1963).
[18] Ibid.
[19] Mass Spec, version 3.0 is a product from Trinity Software that provides mass and abundance information for any compound up to 10,000. The software package generates potential fragment masses and isotopic abundances. Information can be obtained at Village Square, Suite 205, 607 Tenney, Mountain Hwy, Plymouth, N.H., Internet site: `www.trinitysoftware.com`. It's a bargain at less than $150. Other software programs (WWW based, freeware, and commercial) are posted on the Internet site "Mass Spectrometry on the Internet" at: `userwww.service.emory.edu/~kmurray/mslist.html`.
[20] Perseptive Biosystems of Framington Mass. has incorporating accurate mass and isotopic abundance into a single data reduction package.
[21] Stevenson, D.P. Disc. Faraday Soc. 10:35, (1951).
[22] McLafferty, W. F., Turecek, F., page 46-48, *op cit.*
[23] Ibid, page 278.
[24] Several ionization processes in LC/MS are capable of producing odd-electron ions including electron capture associated with any of the CI sources with rich electron plasmas (APCI, TS, CI(PB)). In addition, electrospray is capable of electrochemically producing radicals in solution as described by Gary Van Berkel and coauthors. Van Berkel, G.J., Zhou, Feimeng, and Aronson, J.T. "Changes in bulk solution pH caused by the inherent controlled-current electrolytic process of an electrospray ion source," *Int. Journal Mass Spectrom. Ion Processes*, 162, pages 55-67 (1997).
[25] Price, P. "Standard Definitions of Terms Relating to Mass Spectrometry, A Report from the Committee of Measurements and Standards of the American Society of Mass Spectrometry," *J Am Soc Mass Spectrom.* 2, page 336-348 (1991).
[26] Karni, M., and Mandelbaum, A. Org. Mass Spectrom. 15: 53 (1980).
[27] Fragmentation data for CID of Clenbuterol was obtained from the ACS Short Course manual "Mass Spectrometry: Principals and Practice" J.T. Watson and O. David Sparkman.
[28] Rings plus double bonds (r+db) is used the evaluate the saturation of each ion in the mass spectrum during elucidation. The equation for calculating r+db for a compound of formula $C_xH_yN_zO_n$ is: $r+db = x - \frac{1}{2}y + \frac{1}{2}z + 1$. The general formula from the periodic table relates to groups I,II,III and IV elements of formula $I_yII_nIII_zIV_x$; where examples of group

I elements are H,F,Cl,Br, II elements are O,S, III elements are N,P and IV elements are C,Si)

[29] Fragmentation data for EI of Clenbuterol was obtained from Wiley Mass Spectral Data Base. Spectrum was contributed by J.H. Henion.

[30] Abramson, F.P, "CRIMS: Chemical Reaction Interface Mass Spectrometry," *Mass Spectrometry Reviews*, 13, 341-356 (1994).

[31] Robinson, C.V., Groß, M., Eyles, S.J., Ewbank, J.J., Mayhew, M., Hartl, F.U., Dobson, C.M., and Radford, S.E., "Conformation of GroEL-bound α-lactalbumin probed by mass spectrometry," *Nature*, 372, pages 646-651, (1994).

[32] Wolfe, Robert R., *Radioactive and Stable Isotope Tracers in Biomedicine; Principals and Practice of Kinetic Analysis,* Wiley-Liss: New York (1992).

[33] Henry, C., "Can MS Really Compete in the DNA World," *Analytical Chemistry News and Features,* April 1, 1997.

[34] There is a caveate: preparation time and run time can be compensated for by the parallel nature of the analysis in which many lanes can be run simultaneously, *ibid.*

[35] Yates, J. R., McCormack, A. L., Eng, J., "Mining Genomes with MS," *Analytical Chemistry News and Features*, September 1, 1996.

[36] ibid.

[37] Siuzdak, G., *Mass Spectrometry for Biotechnology*, page 112, Academic Press: New York (1996).

[38] Wade, N., "Meeting of Computers and Biology: The DNA Chip," *The New York Times,* Tuesday, April 8, (997).

[39] Henry, C, *op cit.*

[40] Henry, C., *op cit.*

[41] Baldwin, R., Bethem, R.A., Boyd, R.K., Budde, W.L., Cairns, T., Gibbons, R.D., Henion, J.D., Kaiser, M.A., Lewis, D.L., Matusik, J.E., Sphon, J.A., Stephany, R.W., Trubey, R.K., 1996 ASMS Fall Workshop: "Limits to Confirmation, Quantitation and Detection," *J. Am. Soc. Mass Spectrom.* 8, pages 1180-1190 (1997).

[42] The Three Ion Rule. As proposed by Sphon, J.A., "Use of mass spectrometry for confirmation of animal drug residues," *J. Assoc. Off. Anal. Chem.* 61, pages 1247-1252 (1978).

[43] This table was adapted from reference 41.

Chapter Ten- Flow Diagram

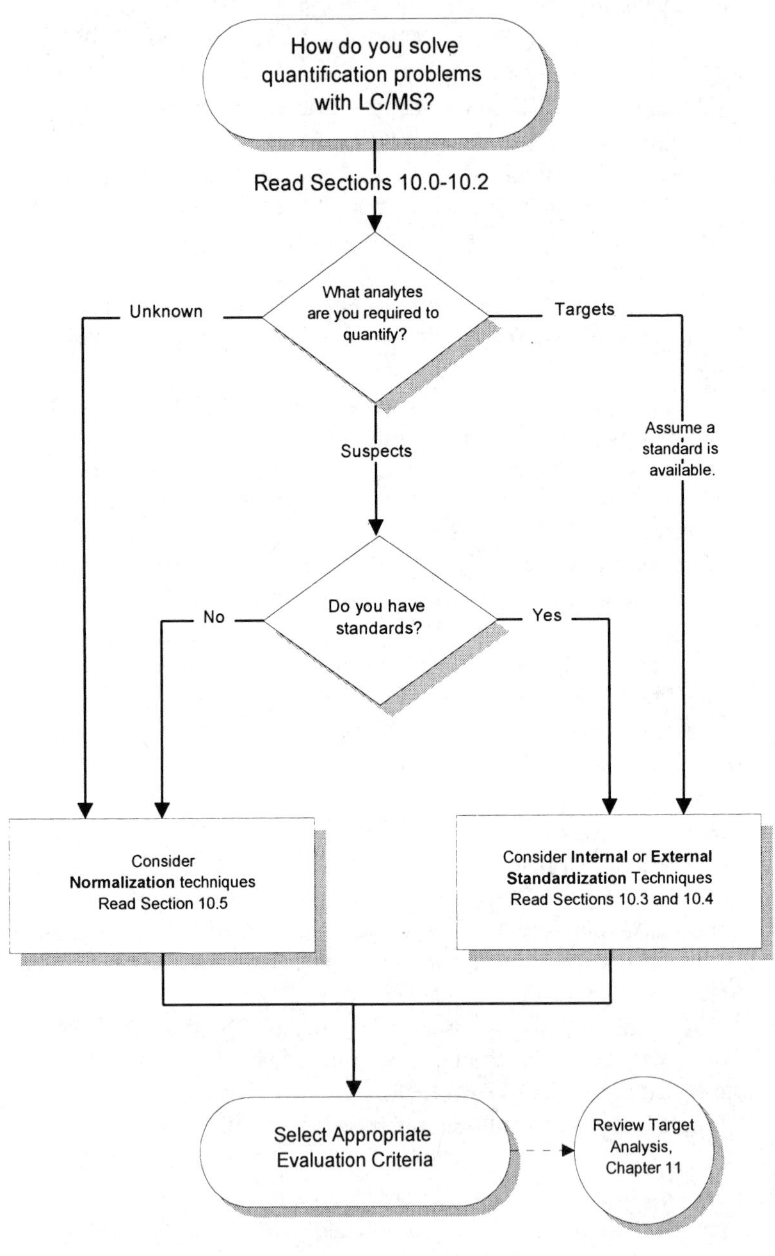
How do you solve quantification problems with LC/MS?
Read Sections 10.0-10.2
What analytes are you required to quantify?
Unknown
Targets
Suspects
Assume a standard is available.
Do you have standards?
No
Yes
Consider Normalization techniques Read Section 10.5
Consider Internal or External Standardization Techniques Read Sections 10.3 and 10.4
Select Appropriate Evaluation Criteria
Review Target Analysis, Chapter 11

Chapter Ten

Solving Quantification Problems

(or, Answering the question- How much?)

Not everything that counts can be counted, and not everything that can be counted counts.[1]

Sign hanging in Albert Einstein's office in Princeton

10.0- What is your response?

The majority of LC/MS systems are currently, and will continue to be, applied to solving quantification problems. Known target compounds originating from complex and variable matrices, such as plasma, urine, and tissue, can be accurately and reliably quantified with LC/MS. The analytical power of LC/MS is demonstrated in these applications because of speed, accuracy, and reliability. This power is derived from the ability of LC/MS in its various configurations to yield a response that is directly traceable to both the concentration of a target in a given sample and a "putative" standard of your target. The key to successful quantification in LC/MS is the correct matching of both sample properties and standards. Not all standards are the same; not all standardization approaches are the same. This chapter will assist you in determining the specific requirements of your sample relative to your standard. It is a testimony to the importance of quantification that the last three chapters of this book are devoted to issues relating to this topic. *Chapter 11* discusses issues relating to selectivity, specificity, and sensitivity in target analysis, while *Chapter 12* discusses the strategies and protocols for developing target methods.

When performing quantitative analysis we are engaged in a comparative process whereby we compare the response of a putative standard to the response of a target species in a sample.[2] Figure 10.0 diagram is an overall scheme for quantitative analysis. Several important criteria can be applied to all quantification in LC/MS; basically, all quantification:

- is comparative, measuring a known response of a standard to the response of an unknown.
- must to be traced back to the weight of a putative standard on an analytical balance (standard purity should be known).
- involves at least two types of analytical runs: one of a standard to determine a response factor, and the second to determine the concentration of an unknown.

The basic tenet of all quantitative analysis is that the response (Area) of a given analyte is proportional to the concentration in the sample. If the response is linear, equation 10.0 will apply (a simple linear expression with a slope of RF (**Response Factor**)).[3]

Response: $$\text{Area} = (\text{RF}) \times \text{Concentration} \qquad (10.0)$$

Since we are considering a comparative process, we must assume that this expression holds for both the standard (subscript std) run and the sample (subscript unk for unknown) run.

Standard Run: $$\text{Area}_{std} = (\text{RF}_{std}) \times \text{Conc}_{std} \qquad (10.1)$$

Sample Run: $$\text{Area}_{unk} = (\text{RF}_{unk}) \times \text{Conc}_{unk} \qquad (10.2)$$

The most important criterion for accurate quantification is the requirement that the response factor of the target standard is identical to the response factor of the target sample. This is the major impetus behind selection of all standardization approaches described throughout this chapter.

Prerequisite: $$\text{RF}_{std} = \text{RF}_{unk} \qquad (10.3)$$

Every practitioner of LC/MS must gain insight into the degree to which instrument and experiment parameters influence these factors. Every link in the analytical chain has an affect on response. Detailed knowledge of each specific parameter along the analytical chain will assist the method developer in optimizing sensitivity (increasing the slope of the analytical curve). In addition, the properties of the sample will have a significant affect on relative response. The good news is, with LC/MS you can account for virtually any discrepancy in relative response between sample and standard by judicious selection of standardization technique.

The most important of the standardization technique in LC/MS involves comparing the response of two components in the same sample relative to one another (**Internal Standardization**). The concentration of one sample component is known (designated by $Conc_{is}$), the other is your unknown target. The ratio of response between these two sample components is established with a standard run containing known concentrations of both internal standard and the target analyte. In this situation the response factor is reported in terms of a **Relative Response Factor** (abbreviated- RRF).

Determining RRF with Standard Run:

$$\frac{Area_{std}}{Area_{is}} = \underbrace{\frac{RF_{std}}{RF_{is}}}_{RRF} \times \frac{Conc_{std}}{Conc_{is}} \tag{10.4}$$

The relative response of two peaks in a given standard run are compared relative to the amount of each standard placed in the standard solution. Here, the standard is a putative reference material of the target analyte- called the target standard peak with $Area_{std}$- and the internal standard is a putative reference material (analog or labeled) with $Area_{is}$.

Sample Runs with RRF:

$$\frac{Area_{unk}}{Area_{is}} = RRF \times \frac{Conc_{unk}}{Conc_{is}} \tag{10.5}$$

$$\frac{Area_{unk}}{Area_{is}} = \frac{RFF}{Conc_{is}} \times Conc_{unk} \tag{10.6}$$

The results from experiments with relative measurements are sometimes plotted as area ratios as a function of concentration ratios as seen in equation 10.5 or as a function of concentration as seen in equation 10.6.

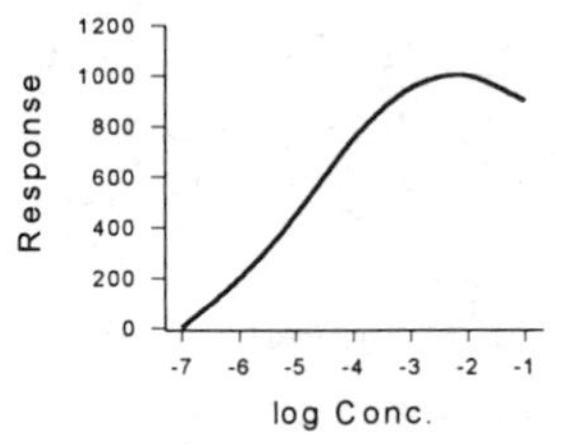

Figure 10.0- An electrospray curve with turnover at the high end of the analytical curve for caffeine.[4]

Nonlinear Response

The preferred performance for quantitative analysis in LC/MS is the generation of linear response as illustrated in Figure 10.1a. Unfortunately, many of the technologies in LC/MS do not produce linear response under all operation ranges and conditions; including electrospray[5] and particle beam.[6] In these situations, higher order treatment (Figure 10.1b) may be required to produce accurate and reliable results. In addition, care must be taken to use multiple levels of standard and multiple replicates in order to get a representative higher order fit. Note that your response factor will also have higher order terms.[7]

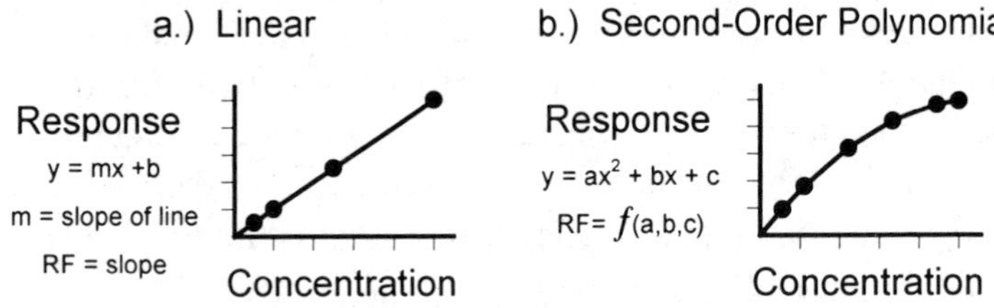

Figure 10.1- Linear versus higher order response.

Another way to address non-linearity issues is to use relative response instead of absolute. Some issues relating to nonlinearity may cancel out with internal standardization. Note that internal standardization is not a universal solution to nonlinear response. Another solution to nonlinear response may also be simply to dilute your sample into a linear range. This may be useful for techniques that lose response at higher concentrations such as electrospray (see Figure 10.0). At higher concentrations the ion desorption process is somehow shut down. The analytical curve can even turn over to yield a negative slope at high concentrations.[8]

10.1- The Primacy Of Standards

Quantitative analysis is based on the ability to compare the response of an analyte to a known reference standard. In most of analytical chemistry the quantity of a substance is rarely (if ever) determined from first principles. Quantity (and identity) are almost always determined from comparisons! The availability and proper use of standards for comparison will determine to a large extent your ability to quantify analytes.

For this discussion, sample components are considered either *targets*, *suspects* or *unknowns* (see *Chapter Nine*). Targets are known compounds for which have an identity and molecular structure established. We assume that a standard exists for target analytes. Suspects have been identified and have a tentative, but unconfirmed, structure. Suspects may or may not have standards. If so, they can be treated similar to targets; if not, they should be treated similar to unknowns. And lastly, for unknowns, you have neither identity nor structure.

Reference standards are substances prepared for use as the standard in an assay, identification, or purity test. The substance may be either the new (compound) or a known impurity. It has a quality appropriate to its use (e.g. UV chromaphore for UV). For a new substance, reference standards are intended for use in assays. Impurities should be adequately identified and /or controlled and purity should be measured by a quantitative procedure.[9] There are three basic types of reference standards:

1) pure analyte,
2) chemical analog, and
3) isotopically labeled analog.

A pure analyte is the compound itself. Isotope analogs are chemically and structurally the same as the analyte but differ in molecular weight. The difference in molecular weight is caused by replacing an atom in the molecule with its heavy stable (non-radioactive) analog. Chemical analogs are compounds chemically similar to the analyte, but structurally different.

Internal vs. External Standardization

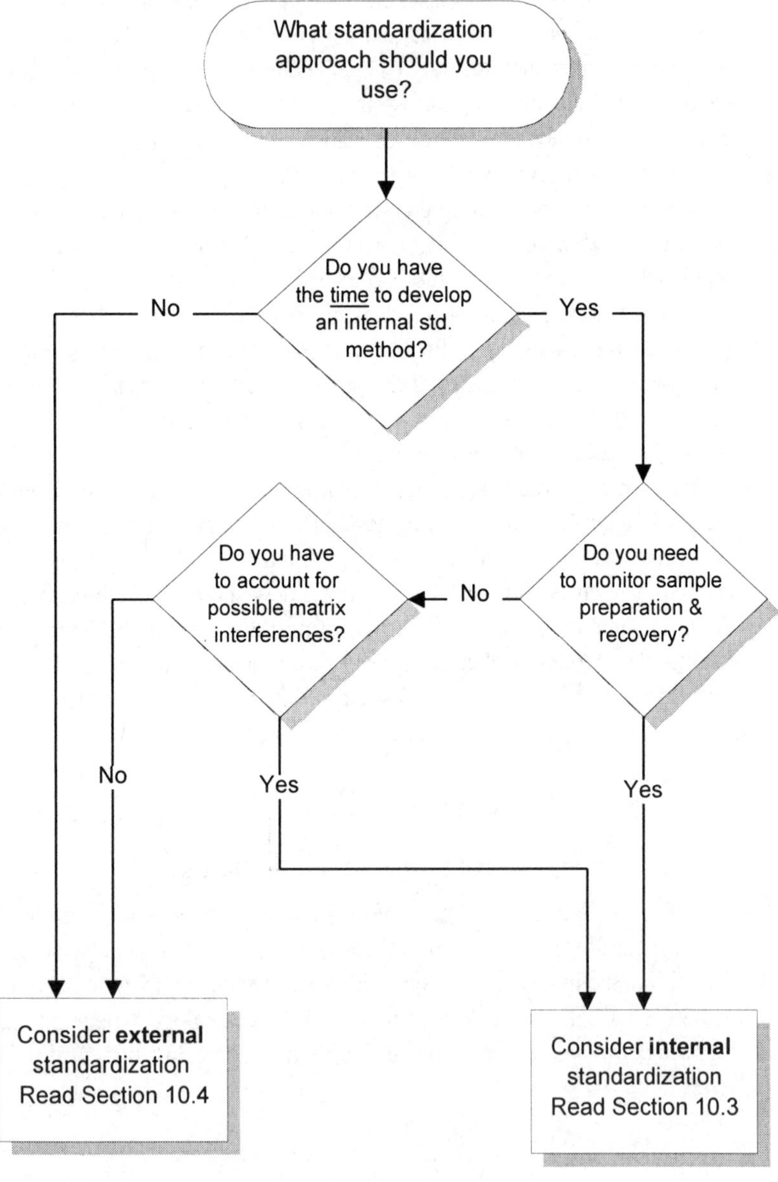

Figure 10.2- Deciding on external versus external.

10.2- What Are Your Alternatives?

In general you have three choices for quantification: internal standardization, external standardization, and normalization. Your choice of any of these will depend on three top-level criteria:

1) Availability of putative standards,
2) Sample characteristics, and
3) Available time.

If you are doing *target* analysis your knowledge of your analyte should be fairly extensive; if not, you likely can gain this knowledge. When you enter the realm of *suspects* the options you have are a little more limited. The information you possess about your sample components is not complete, you only have a tentative structure. *Unknowns* pose a special but not intractable problem. Normalization techniques are used exclusively for the quantification (or semi-quantification) of unknowns.

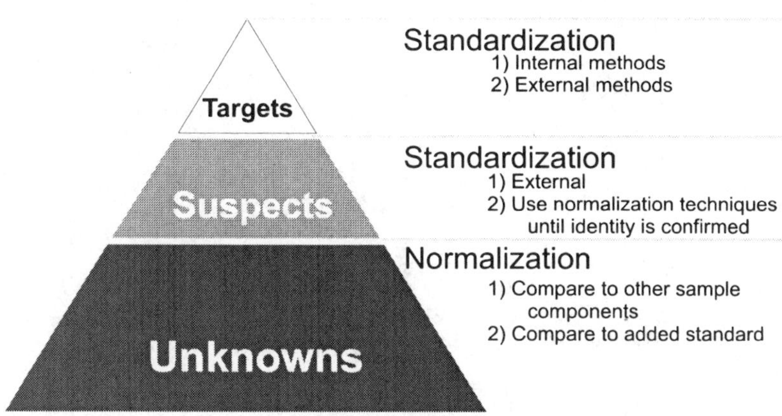
What "quantification" techniques should you use?
Targets
Suspects
Unknowns
Standardization
1) Internal methods
2) External methods
Standardization
1) External
2) Use normalization techniques until identity is confirmed
Normalization
1) Compare to other sample components
2) Compare to added standard

Figure 10.3- Quantification techniques to use with targets, suspects and unknowns.

Deciding Which Internal Standard

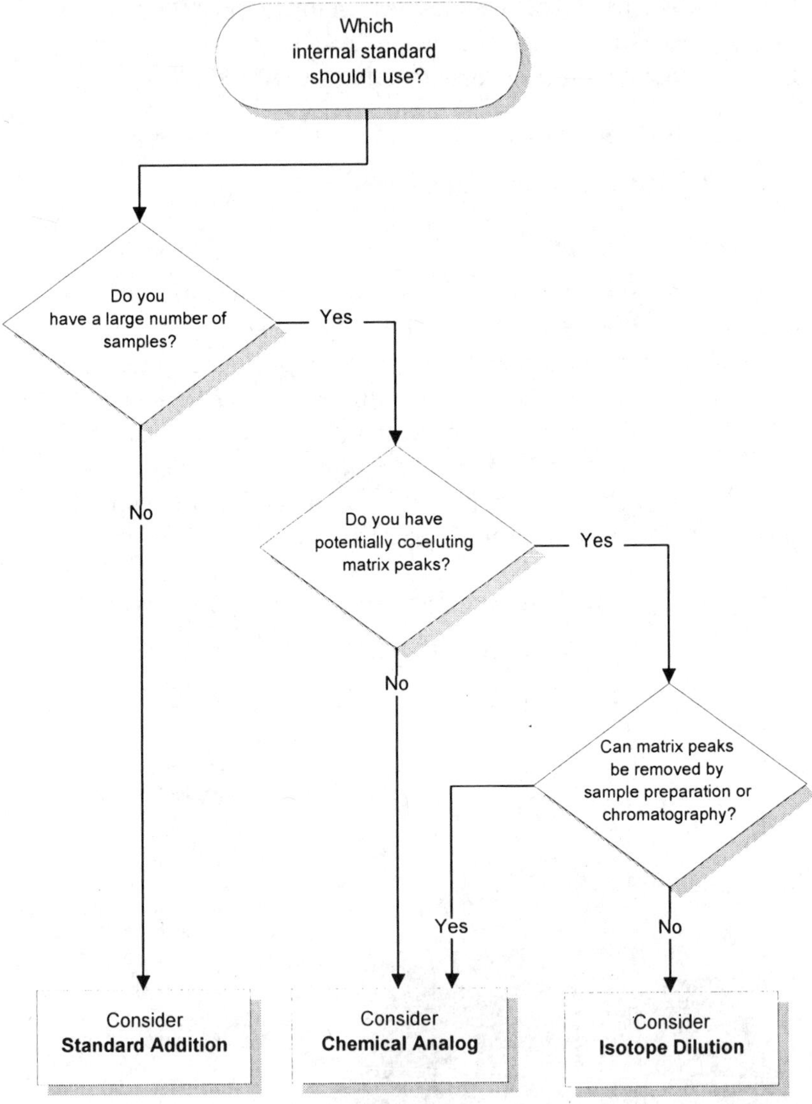

Figure 10.4- Selecting the appropriate internal standardization approach.

10.3- Internal Standardization

The techniques of *internal standardization* utilize a reference compound added to your sample. The response of the other components in your sample are measured against the response of this added standard. All three types of standard are used in internal standardization: *chemical analogs, isotopically labeled analogs,* or *pure analytes*. Some important selection criteria are shown in Figure 10.4.

The implementation and justification for each standardization approach is described in this section. Each topic is presented in the two-page nutshell format that we have used throughout this text. In general, we present a rationale for each standardization approach from an analytical utility perspective. In addition, we present an outline of the two types of acquisition required for quantitative analysis. The first, *standard run,* will generally provide the analyst with the specific response factor for their analyte. The second, *sample run,* will provide unknown target concentration.

The following is a list of terminology, symbols, or abbreviations that we use throughout the next three sections.

Peak Areas	
$Area_{std}$	–peak area of the standard
$Area_{unk}$	–peak area of the unknown
$Area_{is}$	–peak area of internal standard
$Area_{lab}$	–peak area of labeled (enriched) mass
$Area_{unlab}$	–peak area of unlabeled mass

Concentration	
$Conc_{std}$	–concentration of the standard
$Conc_{unk}$	–concentration of the unknown
$Conc_{is}$	–concentration of the internal standard
$Conc_{lab}$	–concentration of the labeled (enriched) standard

Response Factors	
RF	–response factor
RRF	–relative response factor

Standardization: Chemical Analog

Chemical analog standardization is a technique that introduces a chemical analog standard directly into the analytical sample. The response of the target analyte is measured relative to the response of the chemical analog standard. The primary reason for utilizing chemical analog standards is to normalize the response of a given target analyte to the response of the chemical analog component (see Table 10.1). By doing this, variation in injection, sample preparation, and instrumental parameters observed between the standard to the sample run can be accounted for. The prerequisites of a chemical analog standard are outlined below.[10]

- Analogs should have a retention time similar to the analyte.
- Analogs should be stable in the sample matrix.
- Analogs should not be a component in the sample.
- Analog response should be similar to the analyte.
- Analogs should be adequately resolved chromatographic peaks.

The analytical response is determined as follows. A known amount of the *chemical analog* is added to the sample and the standards. The sample contains an unknown amount of the target species, while the standards contain a known amount. The amount added to both sample and standard is generally the same. The standards are generally run at multiple concentration levels of the target *analyte standard* to cover the intended analytical range. The analog standard is generally added in the mid range in concentration of the analytical curve (see Table 10.0).

The standards are run to establish an analytical curve and determine the *relative response factor.* Generally, the sample run follow immediately after the standard runs. Areas are obtained for analyte and analog in both standard and sample runs. The area response from the standard runs is used to calculate the method specific *relative response factor* for the target analyte. The relative response factor is subsequently used to determine the concentration of unknowns based on the area ratios between analyte and analog in the sample runs.

Table 10.0- Internal Chemical Analog Standardization

Standard Runs			Sample Runs		
Step 1	Analysis		Step 3	Analysis	
Run	$Area_{is}$ - $Area_{std}$	Conc*	Run	$Area_{is}$ - $Area_{unk}$	Conc**
1		$Conc_{std1}$			
2		$Conc_{std2}$			
3		$Conc_{std3}$			
4		$Conc_{std4}$	1 to n		$Conc_{unk}$=?
	*all standards contain same $Conc_{is}$			**all standards contain same $Conc_{is}$	
Step 2	Calibration	Product	Step 4	Calculation	Result
	$\frac{Area_{std}}{Area_{is}}$ vs. $\frac{Conc_{std}}{Conc_{is}}$ (points 1, 2, 3, 4)	⇩ RRF		$\frac{Area_{unk}}{Area_{is}} \times \frac{Conc_{is}}{RRF} = Conc_{unk}$	

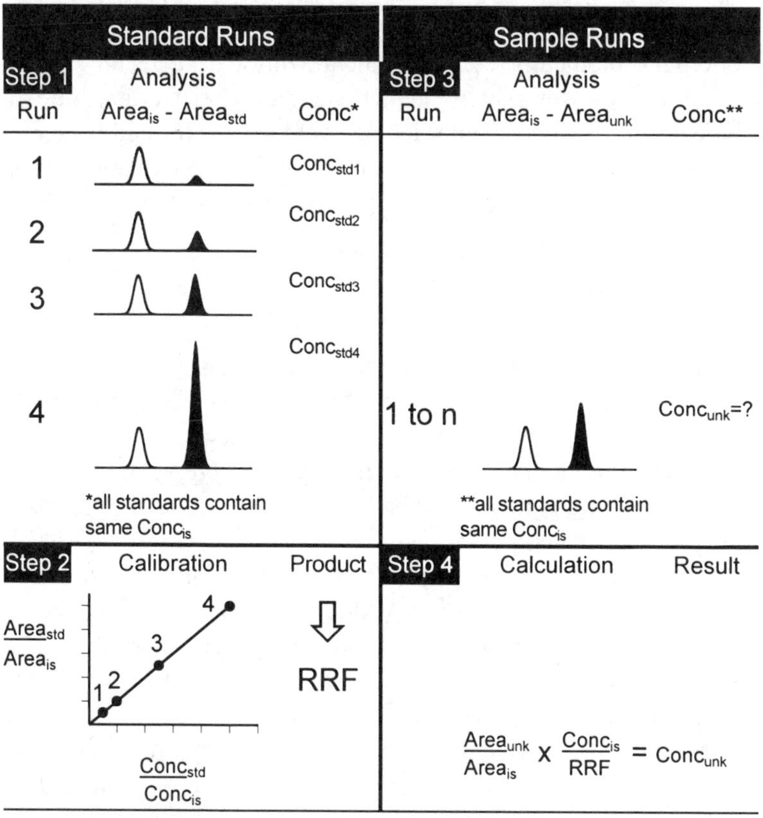

The unshaded peaks are the analog internal standard ($Area_{is}$) while the shaded peaks are the target analyte ($Area_{std}$ or $Area_{unk}$). The area ratios of each injection are used to generate a calibration curve and RRF in Step 2. The RRF is used to calculate unknown concentration in Step 4.

Table 10.1- Utility: Internal Chemical Analog Standardization

Conditions under which Chemical Analog Standards are used.
• Need to monitor sample preparation
• Need for high accuracy
• Cost and time are not important
• A suitable chemical analog standard is available

Standardization: Isotopically Labeled

Internal isotopically labeled standardization is a technique that introduces an isotopically labeled standard directly into the analytical sample. Isotopically labeled analogs are formed by substitution of typically 2-6 of the atoms in a molecule by their "heavy" stable (non-radioactive) analogs (see *Chapter 9*).[11] For example:

- carbon-13 is substituted for carbon-12,
- deuterium (hydrogen-2) is substituted for hydrogen-1, and
- nitrogen-15 is substituted for nitrogen-14.

The primary reason for utilizing a stable isotopically labeled standard is to normalize the response of a given target analyte to the response of an isotopically labeled analog (see Table 10.3). By doing this, variation in injection, sample preparation, and instrumental parameters observed between the standard to the sample runs can be compensated for.

The analytical response is determined as follows. A known amount of the *isotope analog* is added to the sample as well as the standards. The amount added to both sample and standard is generally the same. The standards are generally run at multiple concentration levels of the target *analyte standard* to cover the intended analytical range. The isotope standard is generally added equal to or 1.2X the highest concentration point in the analytical range (see Table 10.2).

The standards are run to establish an analytical curve and determine the *relative response factor* and the *purity factor.* Areas are obtained for unlabeled analyte and labeled analog in both standard and sample runs. The area responses from the standard runs are used to calculate the method specific *relative response factor* for the target analyte. To establish a purity of the isotopically labeled standard, a run is performed of just the isotopically labeled analog (see Run 1, Table 10.2). This is to account for the fact that the isotope is usually not 100% pure; there is some unlabeled amount present in the standard.[12] The relative response factor and purity factor are subsequently used to determine the concentration of unknown based on the area ratios between unlabeled analyte and labeled analog in the sample runs.

Table 10.2- Internal Stable Isotopically Labeled

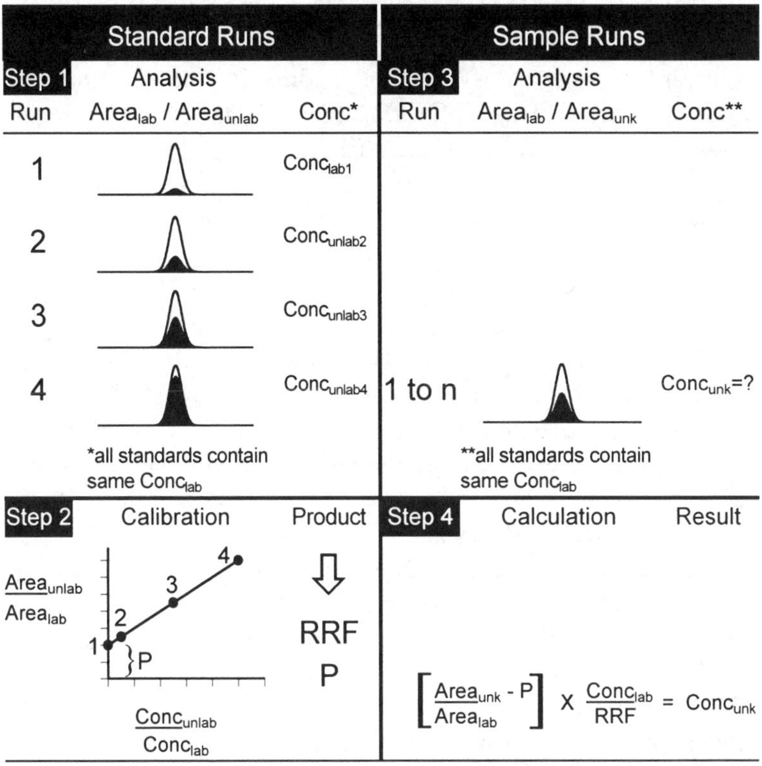
Standard Runs
Sample Runs
Step 1
Analysis
Run
Area_lab / Area_unlab
Conc*
1
Conc_lab1
2
Conc_unlab2
3
Conc_unlab3
4
Conc_unlab4
*all standards contain same Conc_lab
Step 3
Analysis
Run
Area_lab / Area_unk
Conc**
1 to n
Conc_unk=?
**all standards contain same Conc_lab
Step 2
Calibration
Product
Area_unlab / Area_lab
1
2
3
4
P
Conc_unlab / Conc_lab
RRF
P
Step 4
Calculation
Result
[Area_unk / Area_lab - P] x Conc_lab / RRF = Conc_unk

The unshaded peaks are the labeled internal standard ($Area_{lab}$) while the shaded peaks are the target analyte ($Area_{unlab}$ or $Area_{unk}$). The area ratios of each injection are used to generate a calibration curve, RRF and a purity factor (P) in Step 2. The RRF and P are used to calculate unknown concentration in Step 4.

Table 10.3- Utility: Internal Isotopically Labeled

Conditions under which Stable Isotope Standards are used.
• Need to monitor sample preparation • Need for high accuracy • Need for a high speed analysis • Cost is not important[13] • Need to compensate for random variables • Need to correct for nonlinearity • A stable isotope standard is available

Standardization: Standard Addition

Standard addition standardization is a technique that introduces a standard of your target compound directly into the analytical sample. The primary reason for utilizing standard addition standardization is speed (see Table 10.5). A method can be developed for a small number of analytes within a short period of time. The amount present in the sample can be determined in as little as 3 injections (the original sample and two spiked samples, see Table 10.4).

The basis of standard addition method is first estimating the amount of analyte present in the sample. Then by extrapolating the curve from spiked samples, the original amount present in the sample can be determined. This assumes that the response for the particular analyte is linear in the concentration range of the sample. The amount of the standard added to the sample should be reasonably close to the estimated concentration.

The analytical response is determined as follows. The sample is first divided up into several aliquots. One of the aliquots is analyzed to establish a response for the unknown (see Step 1, Table 10.4). The amount present in the sample is either estimated from this response based on previous experience or alternatively the amount present can be determined by making a series of runs of the pure standard at various concentrations to bracket the amount present.[14] Once an estimate of the amount present in the original sample has been established a known amount of the pure standard is then added to the other aliquots. These standards are run at multiple concentration levels of the estimated target concentration.

The standards are run to establish an analytical curve and determine the response factor for the target analyte. Areas are obtained for the analyte in both the original sample and standard runs. The area responses are used to calculate the method specific *response factor* for the target analyte. The response factor is subsequently used to determine the concentration of the unknown based on the peak areas of the unknown and the standards.

Table 10.4- Standard Addition Standardization

Standard Runs			Sample Runs		
Step 1 Analysis			**Step 2** Analysis		
Run	$Area_{unk}$	Conc	Run	$Area_{unk}$ + $Area_{std}$	Conc
1		$Conc_{std}$	1		$Conc_{unk1}$
			2		$Conc_{std2}$
			3		$Conc_{std3}$
Step 3	Calibration	Product	**Step 4**	Calculation	Result
	$Area_{std}$ + $Area_{unk}$ vs. $Conc_{std}$ (points 1, 2, 3)	⇩ RF		$\frac{Area_{unk}}{RF} = Conc_{unk}$	

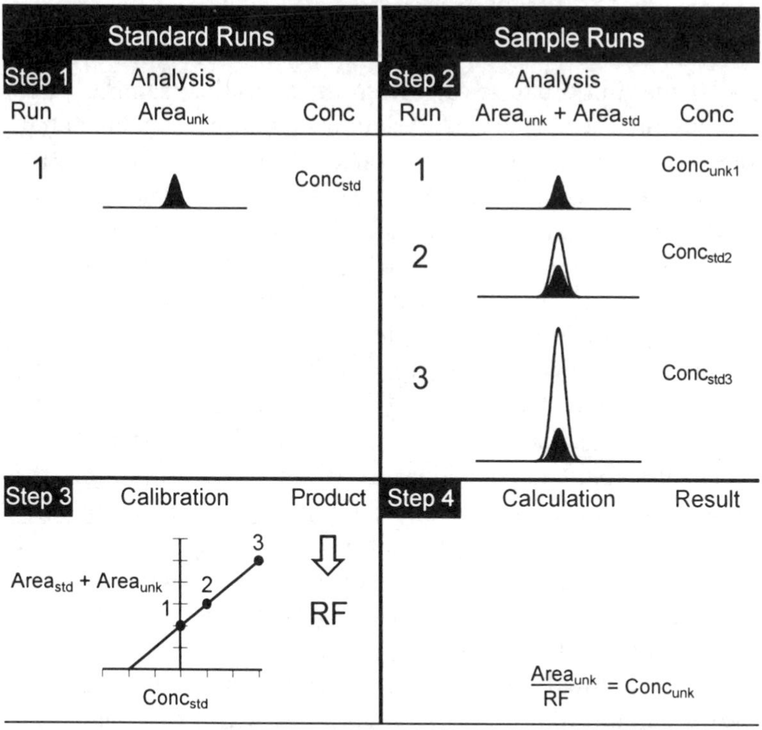

The unshaded peaks are the sum of the unknown and the standard ($Area_{unk}$ + $Area_{std}$) while the shaded peak is the target analyte ($Area_{std}$ or $Area_{unk}$). The area of each injection (standard and unknown) is used to generate a calibration curve and a RF in Step 3. The RF is used to calculate unknown concentration in Step 4.

Table 10.5- Utility: Standard Addition Standardization

Conditions under which Standard Addition is used.
• Complex sample and no matrix blank or matrix standard • Sample preparation and LC conditions can not remove co-eluting matrix peaks that suppress response • Assumes the response is linear (valid close to sample conc.) • Need for a low cost, fast analysis • Small number of samples! • Trace identification and quantification

10.4- External Standardization

External standardization is a technique that analyzes the target standard in separate runs from the analytical sample. The amount of target analyte present in a sample is then determined by comparing the response of the target to the response of a separately analyzed standard. The primary reasons for utilizing external standardization with LC/MS is the need for a one time assay where the method development time is limited (see Table 10.7). By running a set of standards separate from the samples, a calibration curve can be established before the analysis of the samples. Once the calibration curve has been established all the samples can be subsequently analyzed and amounts determined very quickly.

By analyzing the standard separately, no correction for variations due to changes in sample matrix, injection volumes, sample preparation, and instrumental parameters is applied. These variations in instrument parameters can be minimized by preparing and analyzing the standard and sample runs as close in time as possible.

The analytical response is determined as follows. A known amount of the *pure analyte* is added to the standards. The standards are generally run at multiple concentration levels to cover the intended analytical range. The standards are run to establish analytical curve and determine the *response factor* (see Table 10.6). Generally, the sample runs follow immediately after the standard runs.

Areas are obtained for the target in both standard and sample runs. The area response from the standard runs is used to calculate the method specific *response factor* for the target analyte. The response factor is subsequently used to determine the concentration of an unknown present in the sample runs based on the measured area of the target analyte.

Table 10.6- External Calibration Curve

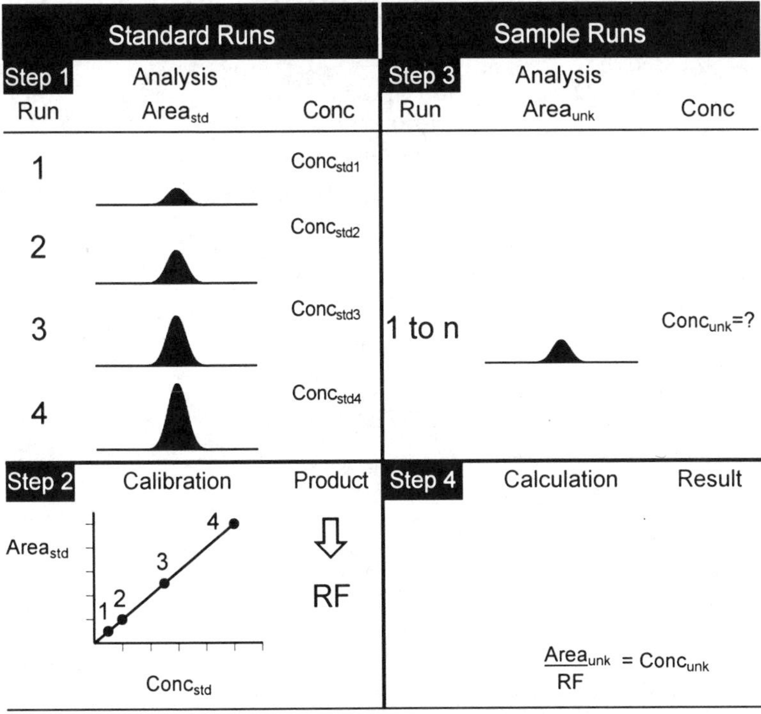

The shaded peaks are the target analyte ($Area_{std}$ or $Area_{unk}$). The area response of each injection is used to generate a calibration curve and RF in Step 2. The RF is used to calculate unknown concentration in Step 4.

Table 10.7- Utility: External Standardization

Conditions under which External Standardization is used.
• Sample with wide dynamic range
• One time assay
• A small number of samples
• No matrix interferences
• When speed is of the essence

No Standard! What can you do?

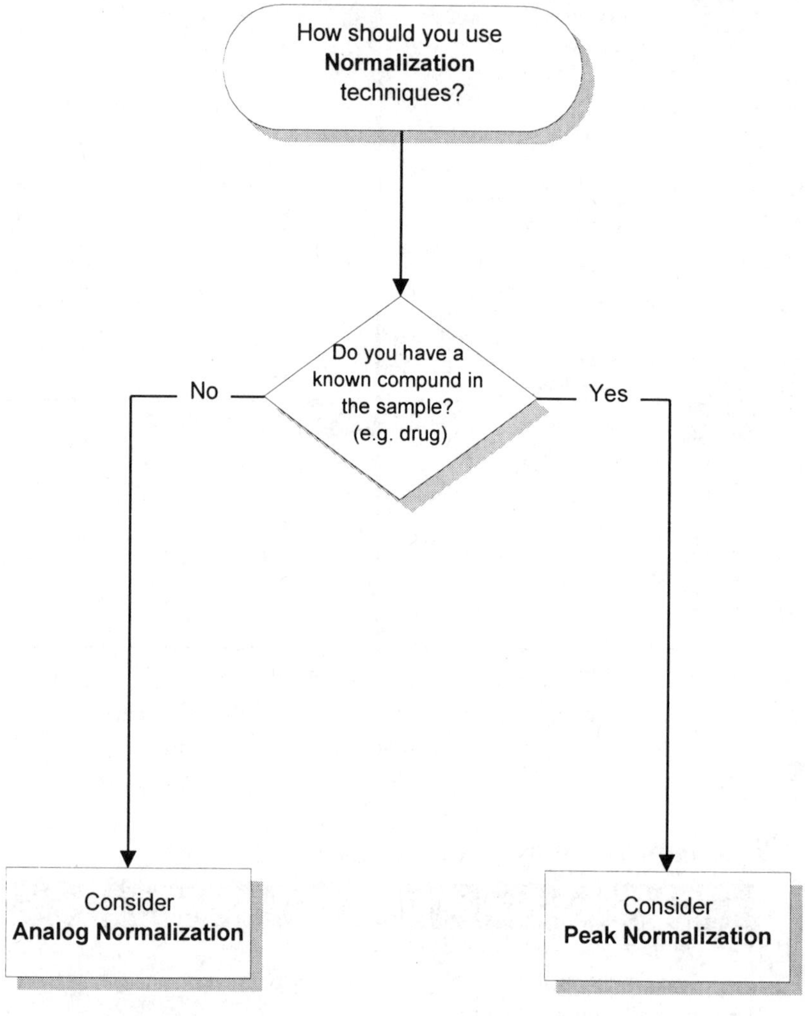

Figure 10.5- Choosing a normalization technique.

10.5- Normalization

What do you do when you don't have a standard? This is not an uncommon occurrence. All metabolites and/or degradation products are unknowns until they are identified. What if you need to estimate an amount present in a sample before you have a standard available or even an identity? The answer is *normalization.* There are three basic types of normalization: *area normalization*, *peak normalization*, and *analog normalization.*[15] All are based on the common theme of comparative analysis.

Area Normalization

Area normalization is used extensively as a QA/QC tool for monitoring batch to batch reproducibility.[16] By determining the area of each peak in the chromatogram relative to the total area of the chromatogram, small discrepancies in the amount present for individual components can be measured. This is used for changes in time or batch to batch in process applications. This is likely not going to be used extensively in LC/MS.

Peak Normalization

Peak normalization is used for those situations where one component in a sample is known and the other components present in the sample are not. The identity and response factors for these unknowns have not been established. Examples of the types of analyses where peak normalization is used are product degradation, shelf-life and metabolic studies. This type of analysis is very common in drug development and has a regulatory basis with the requirement of bulk drug impurities to be specified at 0.1%.[17] How do you determine 0.1%?

Analog Normalization

Analog normalization is used in situations similar to peak normalization. However, when using analog normalization a chemical analog standard is added to the sample before analysis. The responses of unknowns in the sample are then normalized to the analog standard.

Standardization: Normalization

Peak normalization (Case A) is a technique that utilizes the response of a known component in the sample to determine the concentration of an unknown component in the sample. Analog normalization (Case B) is a variation of this approach where a chemical analog standard is quantitatively added to the sample and the response of the unknown components in the sample are normalized to its response. The response of the unknowns in both cases are measured relative to the response of the known species. The primary reason for utilizing peak or analog normalization is to “estimate” the relative concentration of unidentified components in solution (see Table 10.9). By doing this, the true concentration of the unknown in an analytical assay can be determined in the future when the unknown is identified and its response factor is determined. At that time, the identified unknown can be run against the previously used standard to determine a relative response factor. From this information the true concentration of the unknown in the previous sample can be determined. This is the best thing you can do if you don't have a standard. In many ways, it is like internal standardization, only without a target reference standard. In addition, if an analog standard is utilized, variation in injection, sample preparation and instrument parameters observed between the standard and the sample run can be reconciled. Always remember that these measurements are only crude estimates.

Normalization experiments are simple in the sense that data is acquired and the area of all the unknown peaks is normalized to one known component. The concentration of the known sample component can be determined, then the amount of sample unknowns "estimated" in relation to it. In the case of the analog standard, you simply add a known amount of the analog standard to the sample and ratio everything to that standard.

Table 10.8- Peak and Analog Normalization Standardization

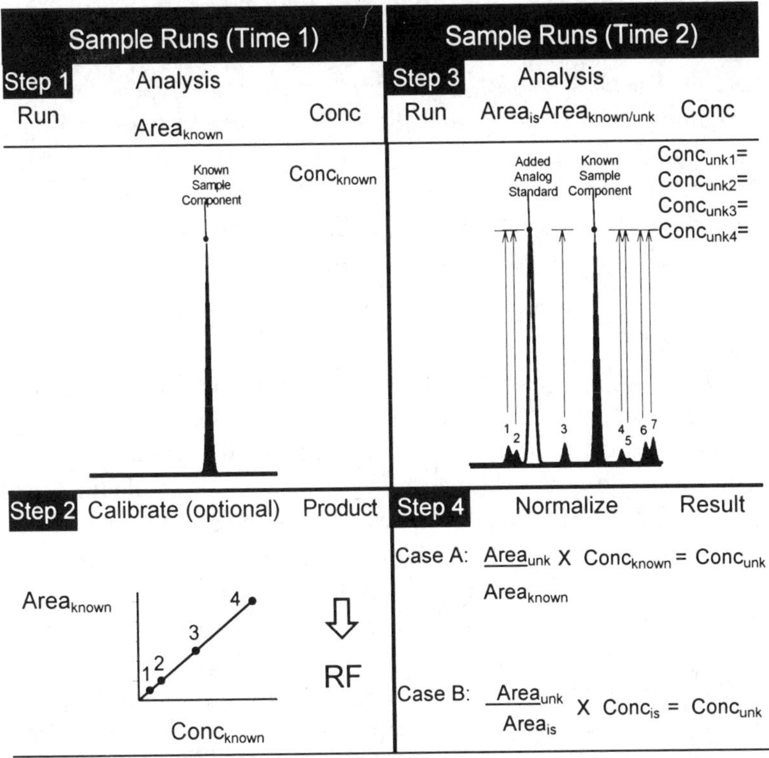

Case A: $\frac{Area_{unk}}{Area_{known}} \times Conc_{known} = Conc_{unk}$

Case B: $\frac{Area_{unk}}{Area_{is}} \times Conc_{is} = Conc_{unk}$

The unshaded peak is the analog standard ($Area_{is}$) while the shaded peaks are the sample components ($Area_{unk/known}$). The area of the unknowns are normalized to the area of either the known sample component in Case A or the added analog standard in Case B.

Table 10.9- Utility: Peak and Analog Normalization

Conditions under which Peak/Analog Normalization is used.
• Unknowns in sample • Presence of unidentified degradation products • No standards available for the unknowns • A response factor has not been determined for the unknown • Precursor analyte disappears in sample (use analog)

10.6- Validation Figures-of-Merit

As was stated at the very beginning of this chapter the ultimate choice of a method will, to a very large extent, depend on the availability of a standard and the type of standard (e.g., analog, isotope). But, on the other hand, if standards are readily available there are other considerations that can affect your ultimate choice. These considerations will be referred to as figures of merit: accuracy, precision, ruggedness and linearity.[18]

Accuracy

Accuracy is the degree of agreement between the best estimated found value and its true value. This can be a confusing term; in practice it is often confused with precision and reproducibility.[19] When you stop to consider accuracy, is it possible to obtain zero inaccuracy? To even hope to approach absolute accuracy we must have some reference point on which to base all our measurements. That reference point, since we are looking at a quantity, best comes from an actual weight of the analyte as determined on a very accurate balance (in our case the mass analyzer).

It is obvious that there is a great deal of error possible at all stages of the method (e.g., sample preparation, dilution, mass measurement). The general rule of thumb is that accuracy should be assessed using a minimum of 9 determinations over a minimum of three concentration levels within a specified range.[20]

Precision

Precision is defined as the extent to which results correspond to each other which are obtained when a definite process (in our case LC/MS) is repeatedly applied to identical samples. In practice it is composed of two closely related features of an analysis: repeatability and reproducibility. Again, these terms may seem like synonyms but for use in quantification they have precise definitions (bad pun intended).

Repeatability refers to the amount of agreement among replicate analyses using the *same* equipment in the *same* laboratory.

Reproducibility refers to the amount of agreement between analyses in *different* laboratories and includes variations in equipment, sources of solvent, auto-injectors, etc. In general the "amount of agreement" is measured as a standard deviation (SD) around an averaged value. In actuality the SD is a measure of imprecision since an increase in the SD reflects a higher imprecision, so like golf, you want the lowest SD. As in determining accuracy, precision should be assessed using a minimum of 9 determinations over a minimum of three concentration levels within a specified range.[21]

Ruggedness (or Robustness)

Even among very lucid authors the definition for ruggedness becomes confused with the previously stated definition of reproducibility. We find it best to think of ruggedness as a measurement of differences caused by the deliberate variation in method parameters.[22] The question to ask is: What are the boundary conditions of a method beyond which accuracy and precision are compromised? The magnitude of the answer to this question, determines the ruggedness of your method.

Ruggedness deals with much broader perturbations in the method and equipment than are considered in reproducibility. These variations can be but are not limited to:

- stability of solutions,
- different equipment,
- different analysts,
- variations in mobile phase composition, pH buffers,
- different columns,
- changes in temperature, and
- changes in flow rate.

Linearity

The linearity of an analytical procedure is its ability to elicit test results, either directly or by a well-defined mathematical transformation, that are proportional to the concentration of the analyte in a sample. Linearity is determined across a specific range of concentrations. The range is the interval between the lower and the upper levels of the analyte that have been determined with precision and accuracy (there are those terms again).

Current guidelines state that linearity should be established by visual evaluation of a plot of responses as a function of concentration (or amount). A minimum of five points is recommended. Once this is established the line can be evaluated for linearity by appropriate statistical analyses including correlation coefficient, y-intercept (best case zero intercept) and the slope.

Peak Height vs. Peak Area

In many LC/UV labs (particularly in the pharmaceutical industry) strip chart recorders are used as back-up in the event of a computer malfunction. This is to insure that the data is not lost. Peak height measurements are generally easier to obtain from a strip chart than peak area. For this reason methods are validated for both peak height and peak area.

For LC/MS strip chart recordings are not a practical alternative method to collect data. The three dimensional nature of LC/MS data makes recording on a two dimensional medium impractical.[23]

Worksheet 10.0- Standardization Approaches

Part A- List all available information about the nature of your sample, standards if any, require methods criteria if any, etc. ________________

____

Part B- Selecting a Standardization Approach. Answer the following questions in each section. If you answer Yes to all of the questions in a particular section, you should consider that particular technique. A No answer to any questions would indicate that you consider another quantitation technique.

Internal Standardization: Chemical Analog	Yes	No
• Do you need to monitor sample preparation?	❑	❑
• Do you have a large number of samples?	❑	❑
• Is a suitable analog of the target analyte available?	❑	❑
• Is a pure standard of the target analyte available?	❑	❑
Internal Standardization: Stable Isotope	**Yes**	**No**
• Do you need to monitor sample preparation?	❑	❑
• Do you need to account for possible suppression of your analyte signal from the matrix?	❑	❑
• Do you have a large number of samples?	❑	❑
• Is a stable labeled standard of the target analyte available?	❑	❑
• Is a pure standard of the target analyte available?	❑	❑
Internal Standardization: Standard Addition	**Yes**	**No**
• Do you only have a few samples ?	❑	❑
• Are you required to develop a method immediately?	❑	❑
• Are you limited in the amount of sample?	❑	❑
• A matrix blank or standard is unavailable?	❑	❑
• Do you have a single analyte?	❑	❑
• Is a pure standard of your target analyte available?	❑	❑
External Standardization:	**Yes**	**No**
• Do you have a small number of samples?	❑	❑
• Are you required to develop and validate a method immediately?	❑	❑
• Will this assay be used only once or twice?	❑	❑
• Do you require an assay that has a wide dynamic range?	❑	❑
• Is a pure standard of the target analyte available?	❑	❑
Normalization	**Yes**	**No**
• Are you required to quantify unknown sample components?	❑	❑
• Is a target standard unavailable?	❑	❑

Note: A copy of this worksheet can be downloaded from **www.LCMS.com**.

Discussion Questions and Exercises

1. Why are standards one of your most important considerations when solving quantification problems?
2. Why should you choose internal standardization even if other methods are appropriate?
3. Why is stable isotope dilution so powerful?
4. List two situations in which isotope dilution standardization should be used.
5. Definitions for reproducibility and ruggedness are sometimes confused. What is the difference?

Notes

[1] From the website (Copyright: Kevin Harris 1995):
`rescomp.stanford.edu/~cheshire/EinsteinQuotes.html`

[2] Milliard, B., *Quantitative Mass Spectrometry*, Heyen: London (1978). A handy book for anyone performing quantitative LC/MS analysis.

[3] This simple expression assumes a zero intercept.

[4] This example of the electrospray analytical curve was acquired with caffeine over six orders of magnitude in concentration at the laboratory of the authors while developing electrospray source configurations.

[5] Raffaelli, A., Kostiainen, R., Bruins, A.P., "Interference from sodium salts and sample adsorption in electrospray experiments," page 310, *Proceedings of the 39th ASMS Conference on Mass Spectrometry and Allied Topics*, Nashville, TN, May 19-24, 1991. Raffaelli, A., Bruins, A.P., "Factors affecting the ionization efficiency in electrospray/ion spray mass spectrometry," page 126, *Proceedings of the 38th ASMS Conference on Mass Spectrometry and Allied Topics*, Tucson, Arizona, June 3-8, 1990.

[6] Behymer, T.D., Budde, W.L., Ho, J.S., Bellar, T.A., "Mass transport and calibration in liquid chromatography/particle beam/mass spectrometry," page 1456, *Proceedings of the 40th ASMS Conference on Mass Spectrometry and Allied Topics*, Washington, DC, May 31-June 5, 1992.

[7] Cuthbert, D. and Wood, F.S. *Fitting Equations to Data; Computer Analysis of Multifactor Data*, John Wiley, New York (1980).

[8] Analytical curve for electrospray.

[9] International Conference on Harmonization. Specifications: Test Procedures and Acceptance Criteria for New Drug Substances and New Drug

Products: Chemical Substance, July 1997. Document available from the ICH internet site: `www.ifpma.org/ich5q.html`.

[10] Boyd R.K., "Quantitative trace analysis by combined chromatography and mass spectroscopy using external and internal standards", *Rapid Commun. Mass Spectrom.* 7, pages 257-271 (1993).

[11] The amount of enrichment of a stable isotope should be greater than three, preferably four to six. This insures that there is very little (or no) contribution or "cross talk" between the Parent+3 isotope peak of the unlabeled molecule and the Parent peak of the isotopically enriched compound.

[12] If the isotope analog was available at a purity of 100% the calculation to determine the amount of unknown present would reduce to the equation used in chemical analog standardization!

[13] The major drawbacks of isotope dilution is the availability of an isotope standard for your analyte and the expense of synthesizing and purifying an isotopic analog. Suppliers of stable isotopes are listed in *Appendix A*.

[14] An estimation of the amount of analyte present can be determined by performing a separate series of analysis of your pure standard (0.25x, 0.5x, 2x, 5x, .. 10x the amount you suspect is present) to bracket or triangulate the amount present.

[15] In general, there is not much written about normalization techniques. Its primary use, in the past, was with refractive index (RI) detectors or with GC-FID. Today in LC/MS, it is used extensively during the drug discovery phase of drug development. Normalization is used to give a quick answer to the question, how much of a metabolite or new component is present? You will hear about normalization techniques more and more.

[16] Robards, K., Haddad, P.R., Jackson, P.E., *Principles and Practice of Modern Chromatography Methods*, San Diego: Academic Press (1994).

[17] International Conference on Harmonizaton, *op cit.*

[18] International Conference on Harmonization, *op cit.*

[19] Szepesi, G., *How to use Reverse-Phase HPLC*, VCH Publishers: New York (1992).

[20] Krull, I., Swartz, M., "Introduction: National and International Guidelines," LC-GC 15, pages 534-540 (1997). "Validation of Analytical Procedures: Methodology," from the ICH internet site, `www.ifpma.org/ich5q.html`.

[21] Krull, I., *et al.*, page 536, *op cit.*

[22] Krull, I., *et al.*, page 539, *op cit.*

[23] With the significant integration of computers in the control of the instrument, a malfunctioning computer will result in more serious problems than data acquisition. Your instrument will not be operational!

Chapter Eleven- Flow Diagram

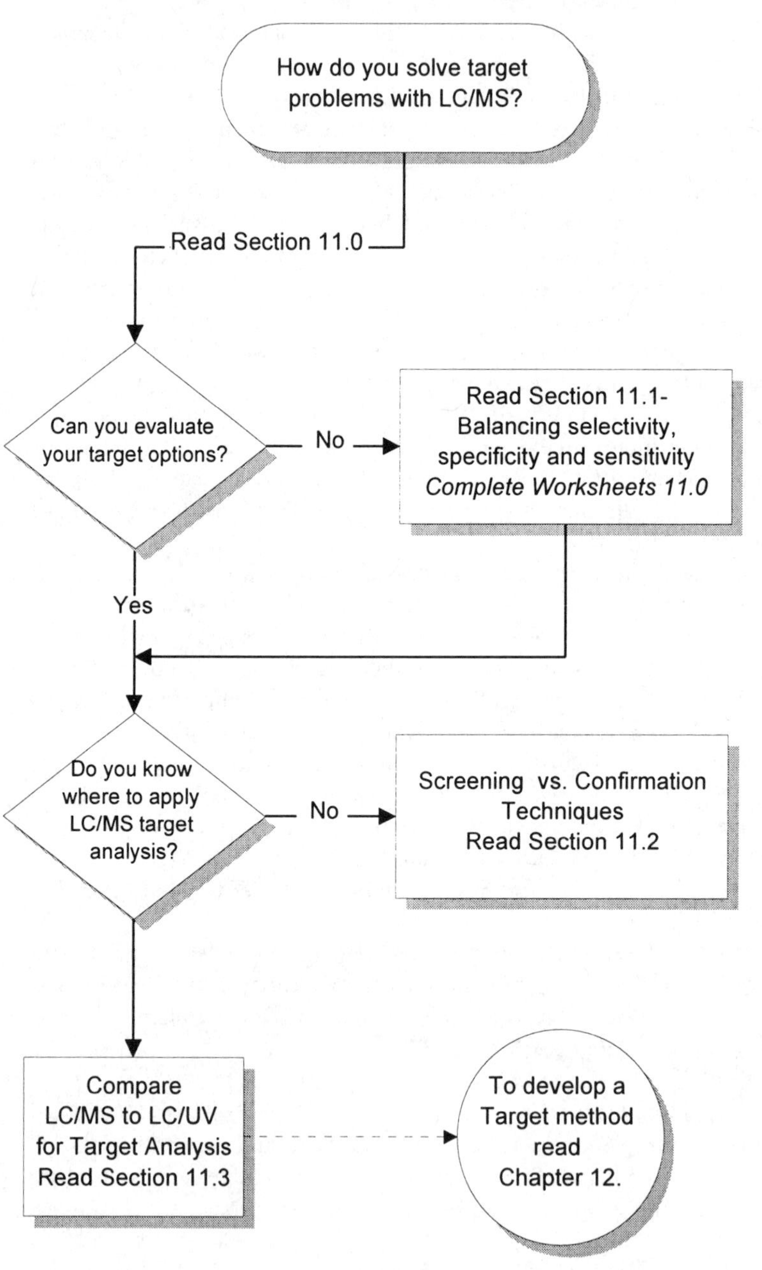
How do you solve target problems with LC/MS?
Read Section 11.0
Can you evaluate your target options?
No
Read Section 11.1- Balancing selectivity, specificity and sensitivity
Complete Worksheets 11.0
Yes
Do you know where to apply LC/MS target analysis?
No
Screening vs. Confirmation Techniques
Read Section 11.2
Compare LC/MS to LC/UV for Target Analysis
Read Section 11.3
To develop a Target method read Chapter 12.

Chapter Eleven

Solving Target Problems

(or, Finding the needle in the haystack (fast!))

> *...it takes about two to four months to develop and validate an HPLC method. If this is combined with an MS detector the same method requires only three to four weeks. ... using LC/MS/MS reduces the method development time to two weeks or less.*[1]
>
> *Robert Stevenson*
> ***American Laboratory News Edition***

11.0- LC/MS:The Ultimate Target Technique

Mass spectrometry is the most powerful tool for quantitative analysis if evaluated on the basis of speed, specificity, and reliability. This is supported by the dominance of GC/MS in target analysis for almost two decades; primarily fueled by the realization of society at large that environmental problems were significant. Environmental initiatives in GC/MS required "broad-based"[2] analytical methods to screen for, or confirm, a large number of priority pollutants originating from very diverse matrices. These initiatives were largely successful. LC/MS provides similar analytical capabilities to GC/MS; however, the differences between these two hyphenated technologies will likely result in the typical end-users being quite different. LC/MS has emerged primarily in applications areas in health sciences and biotechnology where the problem needs are somewhat different.

LC/MS has become the ultimate target analysis technique for quantitative determination of target species because of the broad analytical range (as described throughout *Chapter Two*), its sensitivity (as described in *Chapter Three*), and its speed in both methods development and analysis.[3]

This chapter discusses the important criteria and trade-offs required for target analysis with various LC/MS technologies.

11.1- Evaluating Your Target Options

The power of target analysis is found in the ability of LC/MS to isolate the response of a target component and quantify that response relative to some standard (as described in *Chapter Ten*). Even when a suitable standardization approach has been evaluated and selected for your specific target species, you are still faced with evaluating the many optional instrumental approaches in LC/MS. Deciding on which target technique to use is a balancing act between selectivity, specificity, and sensitivity (see Figure 11.0 and Table 11.0). Referring to a technique as selective is to describe its ability to *isolate* the response of an individual component from that of other sample components. Specificity refers to the characteristics of the response of an individual component that make it *unique*. One response characteristic may not be unique (e.g. MW), but two or three may provide adequate information to allow you to characterize a chemical component. Sensitivity pertains to the amount of analyte detected in the instrument.[4]

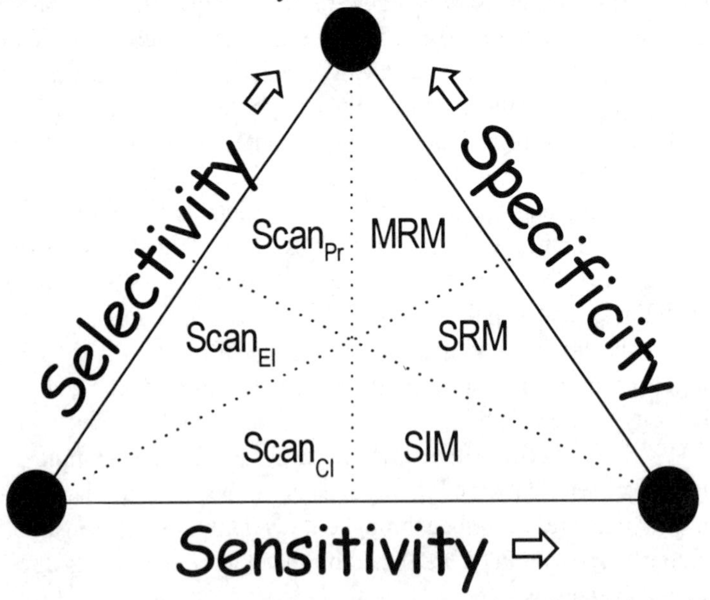

Figure 11.0- The relationship of sensitivity, selectivity, and specificity; with mass spectrometric modes of analysis.[5]

Table 11.0- Balancing LC/MS Criteria.

Criteria	Techniques	Opportunity or Benefit
Selectivity	Ionization	Selective ionization process (e.g., EI, ES, etc. and polarity (+/-))
(pertains to isolation)	Fragmentation	Increased number of fragment to chose from
	Mass Resolution	Select specific mass from isobaric masses
	MS/MS	Removes spectral interferences
	LC Separation	Retention (capacity), separation (α), and efficiency (plate number) allow chromatographic isolation of target from other components
Specificity	Ionization	Generally yield quasi-molecular ion
	Fragmentation	Compound-specific fragments
(pertains to individual	Mass Accuracy	Yields elemental composition and formula
character-istics)	MS/MS	Moiety-specific fragment information
	LC Separation	Compound-specific retention time
Sensitivity	Ionization Efficiency	How well analytes (neutral or ionic) are converted into gas-phase ions
(pertains to collective efficiencies)	Interface Efficiency	How well analytes (neutral or ionic) are transported through the LC/MS interface to the mass analyzer
	Mass Transport Efficiency	How well ions are transported through the mass analyzer
	Detection Efficiency	How well ions, once through the mass analyzer, are detected
	Dwell Time	The length of time the mass analyzer records the intensity of an individual mass
	Duty Cycle	Percent of time spent on a given mass (SIM = 100%)
	Chromatographic Efficiency	Narrow LC peaks (e.g., separating adjacent peaks, lower extra-column effect)
	S/N	Noise reduction, averaging, filtering

The Balancing Act

Beside balancing selectivity, specificity and sensitivity for a particular method, one must also consider other criteria, such as, time, deciding what is good enough, and money There is no "right" mixture of these items, it depends on your circumstances.

In certain situations, time is important. For example you may have the need to analyze 2,000 clinical samples within a two week period. Developing an LC/MS/MS method coupling the techniques of fast-LC, APCI, and selected reaction monitoring (SRM) tandem mass spectrometry can allow you to perform one analysis every 2-5 minutes- completing the work within the allotted time period. In this example, you have given up LC selectivity and specificity, while gaining MS selectivity, specificity, sensitivity and time! The speed of analysis (with reliability) that is gained is generally only accomplished when using the ruggedness and reliability of isotopic labeled internal standardization (*Chapter Ten*).

In other situations, broad-based methods are required. For example, you are required to identify and quantify 30 environmental target components plus possible contaminants- that are not ionic in solution and have molecular weights below 1,000 u. Developing a method using the techniques of liquid chromatography, particle beam LC/MS, with the universal detection capabilities of EI (while scanning), will allow you to develop a method to chromatographically separate all the components within 15-20 minutes, confirm the identity of each target, suggest possible identities of any contaminants present by comparison to a commercial mass spectral data base, and then quantifying (or semi-quantify) all components [targets and (contaminants)] detected. In this example you have given up MS selectivity and sensitivity, while gaining LC selectivity; and LC and MS specificity.

In some situations cost is a concern. Figure 11.1 illustrates the relationship between cost of analysis, and reliability (% error, selectivity). Noteworthy with this evaluation is the difference between one (e.g. LC/UV) and two step (LC/MS) analyses. Clearly, to decrease error and increase reliability, we are required to utilize more stages of analysis, including LC/MS/MS. What is not intuitive is that the cost for the final result is lower with multiple steps compared to a single step. If you need reliable results, LC/MS or LC/MS/MS can be the most cost effective way to achieve those results.

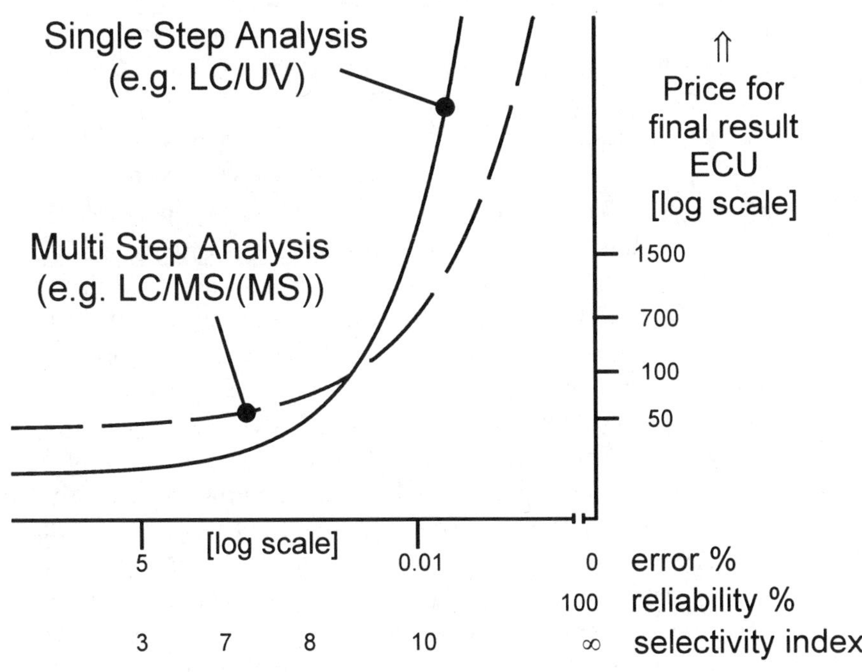

Figure 11.2- Comparison of cost per analysis with reliability, expected error, and selectivity for single and multi-component analytical schemes (adapted from reference 6 with permission).

More Nutshells- Target Criterion

Two-page nutshells are presented on selectivity, specificity and sensitivity. After determining which criteria are important for your "balancing act," use Worksheet 11.0 to select a target technique that best fits your needs. Use additional sheets if required.

Target Criterion: Selectivity

Selectivity can be achieved with LC/MS using both the liquid chromatographic and the mass analyzer stages. The LC column isolates the sample components in time, while the MS isolates sample components from each other according to mass.

Target Selectivity from LC

LC can play a very important role in isolating target components in LC/MS. Whether a crude or a fine separation is implemented, column separations are an essential part of target analysis in that they physically remove chemical interferences, as well as, spectral interferences. Note: Mass spectro-metry, even in its highest performance modes does not remove chemical interferences that plague ionization and transport of analyte in the mass spectrometer. In addition, the isolation of your target response from other components can improve the ability and accuracy of integration by ensuring baseline to baseline integrations.

Target Selectivity from MS

There are many options with mass spectrometry to improve selectivity by utilizing the wide variety of instrumental attributes. We can use resolution, selective MS/MS masses, selective collision energy, and multiple stages of MS/MS to isolate target species from other sample components. In general, the analyst wants a mass spectral response that is isolated from the response of other sample components (or whose response is not effected by other sample components). MS is ideal for the first, (not the latter).

Target Selectivity from LC/MS

Through judicious selection of either interface or ionization conditions, it may be possible to enhance the selectivity of an anlayte. For example, by adjusting the pH of the LC mobile phase for optimal electrospray ionization of your target you may suppress the response of an interference by neutralizing that component while keeping your target ionized.

Table 11.1 summarizes and ranks the aspects of LC/MS that are utilized to increase selectivity in target analysis. MS/MS is the most commonly used means of increasing selectivity, followed by mass resolution. Table 11.2 summarizes the different types of LC/MS acquisitions [or modes] and the corresponding instruments that are utilized for target analysis that have high and low selectivity.

Table 11.1- Selectivity Criteria for LC/MS Target Analysis

Selectivity Enhancement Criteria	MS SIM	MS Scan	MS/MS SRM/MRM	MS/MS Scan
Ionization	★	★	★/★	★
Fragmentation	★	★★	★★/★★★	★★★
Mass Resolution	★	★★	★★/★★★	★★★
MS/MS	n.a.	n.a.	★/★★	★★★
LC Separation	★	★	★/★★	★

★★★ .... significant advantage
★★ .... some advantage
★ .... no advantage
n.a. .... not applicable

Table 11.2- Preferred Instrumentation with Selectivity Emphasis

Selectivity (Emphasis)	Target Mode	Instrumentation	Comments
High	MRM (MS^n)	QQQ, Sectors, Hybrids, (IT), (FTMS)	MS^n gives the highest selectivity and has shown applicability even with structural isomers
	Scan-Prod	QQQ, IT, Q-TOF, FTMS, Sectors, Hybrids	Library matching MS/MS spectra can improve selectivity, similar to EI libraries
	SRM	QQQ, Sectors, Hybrids, IT[7]	Any mode of MS/MS will generally have higher selectivity than single analyzer modes
Low	Scan-EI	Q, Sectors	Reverse search algorithms allow selection of target analytes in the presence of background matrix
	SIM	Q, Sectors	Least selective, requires separation for selectivity
	Scan-CI	Q, Sectors, TOF	Least selective, requires separation for selectivity

See notes for abbreviations.[8]

Target Criterion: Specificity

Target specificity, like selectivity, is achieved with both the liquid chromatographic and the mass analyzer stages.

Target Specificity from LC

With LC, selectivity and specificity are sometimes used interchangeably; however, the component of specificity with LC that is key to target analysis in LC/MS is the compound-specific retention time. The identification of a target species in a specified retention time window is one of the most important elements of "identity" in target analysis.

Target Specificity from MS

Mass spectrometry offers a large number of alternatives for specificity. Generally, the quest for a unique response for a target species is one of the major criteria for target method development. The developer can use every available tool to acquire a unique response; such as, specific fragmentation ion(s), specific multiple fragmentation ion(s) (MS^n) with tandem MS, and mass/abundance ratios such as those argued by the "three ion rule" (see page 344). Although not generally exploited for specificity reasons in quantification, high mass accuracy measurements (with high resolution) can provide a compound-specific response- in some cases it may be a useful tool. Generally, high resolution is used to enhance selectivity in the case of isobaric interferences.

Target Specificity from LC/MS

Most interfaces in LC/MS produce simply a quasi-molecular ion that has *insufficient* specificity for most (but not all) target applications, particularly for target compounds in the range from 100 to 500u. These techniques all require MS/MS fragmentation to achieve the required specificity. One noted exception is particle beam that provides EI fragmentation patterns that can be highly specific to an individual target and matched against target libraries.

Table 11.3 summarizes and ranks the aspects of LC/MS that are utilized to increase specificity in target analysis. Table 11.4 summarizes the different types of LC/MS acquisitions [or modes] and the corresponding instruments that are utilized for target analysis that have high and low specificity.

Table 11.3- Specificity Criteria for LC/MS Target Analysis.

Specificity Enhancement Criteria	MS SIM	MS Scan	MS/MS SRM/MRM	MS/MS Scan
Ionization	★	★(★★)	★/★	★(★★)
Fragmentation	★	★(★★)	★/★★	★★★
Mass Accuracy	★	★★	★/★★	★★★
MS/MS	n.a.	n.a.	★/★★	★★★
LC Separation (Usually a prerequisite)	★	★	★/★	★

★★★ .... significant advantage- parenthesis indicate EI.
★★ .... some advantage
★ .... no advantage
n.a. .... not applicable

Table 11.4- Preferred Instrumentation with Specificity Emphasis

Specificity (Emphasis)	Target Mode	Instrumentation	Comments
High	Scan-Prod	QQQ, IT, Q-TOF, FTMS, Sectors, Hybrids	Compound specific fragmentation observed in MS/MS or MS^n gives highest specificity
	Scan-EI	Q, Sectors	Universal response and compound specific response
	MRM (MS^n)	QQQ, IT, FTMS, Sectors, Hybrids	Multiple compound specific ions increase method specificity
Low	SRM	QQQ, Sectors, Hybrids	Programmed SRM of 1 or 2 ions at each retention time is commonplace
	SIM	Q, Sectors	Programmed SIM of 1 or 2 ions at each retention time is commonplace
	Scan-CI	Q, Sectors, TOF	Limited to MW determination

Target Criterion: Sensitivity

For sensitive target analysis in LC/MS you must pay strict attention to everything; the LC, interfaces, ionization, mass analysis, even data processing.

Target Sensitivity from LC

In general, narrow chromatographic peaks mean higher sensitivity. For trace target analysis care must be taken at every point of the separation system; from injection to the interface. Any component of the separation system that contributes to the broadening of the chromatographic peak will potentially compromise your sensitivity. Care must be taken to eliminate extra-column and or on-column dispersion of any type. Columns should well packed with small particle size. All tubing and fittings should be matched and precut. All tubing diameters and lengths should be minimized. Gradient separations can compress the peaks on-column to enhance sensitivity. In addition, on-column enrichment can give significant enhancement in method sensitivity when sample volumes are large and time for loading the sample on the LC column is available.

Target Sensitivity from MS

Sensitivity with mass spectrometry is simply a game of trying to get all of your ions detected, all of the time. This is difficult in practice; however, for target analysis, you really don't care about all of the ions, only your target ions; consequently, SIM and SRM are ideal alternatives for sensitivity. Simply measure one ion, 100% of the time. This will generally enhance both transmission and ion statistics for improvements of orders of magnitude over scanning techniques. Resolution is also a factor with sensitivity. With the exception of TOF, increased resolution will decrease sensitivity; therefore, to obtain enhanced sensitivity one can open the resolution (e.g. some analysts open their Q1 in triple quadrupole to several mass units wide giving up selectivity to gain sensitivity.) (See Fundamentals: Quadrupoles, Appendix D).

Target Sensitivity from LC/MS

Table 11.5 summarizes and ranks the aspects of LC/MS that are utilized to increase sensitivity in target analysis. Table 11.6 summarizes the different types of LC/MS operating modes relative to sensitivity.

Table 11.5- Sensitivity Criteria for LC/MS Target Modes.

Sensitivity Enhancement Criteria	MS SIM	MS Scan	MS/MS SRM/MRM	MS/MS Scan
Ionization Efficiency	☆	☆	☆/☆	☆
Interface Efficiency	☆	☆	☆/☆	☆
Transmission Efficiency	☆☆	☆	☆☆/☆☆	☆
Detection Efficiency	☆☆	☆	☆☆/☆☆	☆
Duty Cycle	☆☆☆	☆	☆☆☆/☆☆	☆
LC Efficiency	☆	☆	☆/☆	☆
S:N	☆☆	☆	☆☆☆/☆☆	☆☆

☆☆☆ .... significant advantage
☆☆ .... some advantage
☆ .... no advantage
n.a. .... not applicable

Table 11.6- Preferred Instrumentation with Sensitivity Emphasis

Sensitivity (Emphasis)	Target Mode	Instrumentation	Comments
High	SRM	QQQ, Sectors, IT	Triples and sectors combine high duty cycle with good S:N
	MRM	QQQ, Sectors, IT	As you add target ions you have tradeoff with duty cycle
	SIM	Q, Sectors	High duty cycle, with higher noise contribution
Low	Scan-Prod	QQQ, IT, Q-TOF, FTMS, Sectors, Hybrids	Scanning instruments lose the battle for duty cycle, while ion storage devices are used for increasing duty cycle
	Scan-CI	Q, TOF, Sectors, Hybrids	Single scanning instruments tend to have lower voltage sources and higher relative noise
	Scan-EI	Q, Sectors, Hybrids	Same as CI, limited to PB

See notes for abbreviations.[9]

Worksheet 11.0- Selecting Target Techniques

Part A- The first step in selecting a target technique is to consider the impact of LC and MS selectivity, specificity and sensitivity. List any available information about any restrictions, customer requests, regulatory requirements, etc. to consider. This should serve as a starting point.

____

____

____

Liquid Chromatography Techniques you are considering to enhance:

Selectivity & Specificity		Sensitivity
Reverse Phase LC	Sample Prep.	Small I.D. Columns
Normal Phase LC	• Liquid-Liquid	Fast-LC
Ion Exchange LC	• SPE	On-line Sample Prep.
Column Switching	• Affinity	Trace Enrichment

Mass Spectrometric Techniques you are considering to enhance:

Selectivity	Specificity	Sensitivity
Ionization	Ionization	use SIM
Fragmentation	Fragmentation	use SRM or MRM
Mass Resolution	Mass Accuracy	
MS/MS	MS/MS	

Part B- Selecting a MS or an MS/MS Technique.

Circle LC/MS target techniques you are considering:

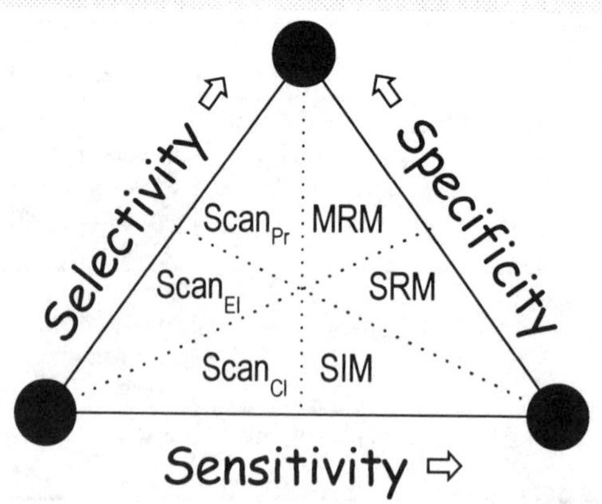

Note: A copy of this worksheet can be downloaded from **www.LCMS.com**.

11.2- Screening vs. Confirmation

There are a large number of applications for "target methods' that seem to fall into two categories; namely, screening methods and confirmation methods.

In general, a *screening method* would be one that affords the analysts the ability to quickly and reliably determine the presence (or absence) of a target species in a given sample. In general, screening methods answer yes/no or positive/negative questions. The need for a screening methods may not be absolute accuracy (threshold ranges may suffice), but reliability. Reliability in screening methods means never having false positives or false negatives. For example, LC/MS technologies have been developed into a cost effective way to screen for neonatal metabolic diseases.[10]

Confirmation methods on the other hand are target methods that have to have optimal components of both "identity" and "quantitative accuracy". These methods are generally required for any measurement where the information derived from that measurement is essential to an important decision or problem. These include methods that have to withstand the scrutiny of the regulatory and judicial environments.

There is seldom any debate over which methods apply to a given problem although one can state that virtually all confirmation methods are adequate for screening; although they may not have the desired attributes of speed and cost per sample that a screening method would possess. Conversely, it should also be said that many screening methods would not hold up to legal or regulatory scrutiny and are often followed by a confirmatory method.

How Good Is Good Enough?

Target methods are generally developed for some intended purpose, whether regulatory or otherwise.[11] The uncertainty of most target methods does not usually come from the target species, rather it comes from the variability and nature of the sample. The information derived from a target assay is usually quite simple, -Is it there? [yes/no], and -How much is there? [concentration]. LC/MS is generally capable of answering these questions very well. The challenge to the analyst in LC/MS is to select or develop target methods that are appropriate for their intended use. This generally requires the method to be validated for it's intended purpose (Section 12.3).

To answer the question- How good is good enough?, one must rephrase the question to- Does the target method meet my validation criteria? Most industries; particularly those with a regulatory base or requiring regulatory approval have guidelines for validation or methods that are already validated with established method criteria.

Validation requirements are always under continuing discussion and debate. Various groups, such as professional societies (e.g., ASMS, AOAC), governing bodies (e.g., EPA, FDA, USP, EC, ICH), and industries (e.g., pharmaceutical, environmental) are all active in establishing criteria for confirmation, quantitation and detection. If you are interested in more information we suggest you contact your industry-specific agencies.

One resource in the pharmaceutical industry is the International Conference of Harmonization (ICH). The purpose of the ICH is to make recommendations on ways to achieve greater harmonization in the interpretation and application of technical guidelines and requirements for product registration of new medicines. Their home page contains documents outlining methods and validation guidelines (**www.ifpma.org/ich1.html**).

A second resource are journals. For example, the journal "Analytical Chemistry" in late 1996 published an article discussing the changes in the Federal Rules of Evidence affecting the admissibility of scientific evidence in court.[12] The authors discussed the four factors that must be considered in determining whether the analysis on which testimony is based falls within the range of accepted standards, such as, "whether the technique or method has been tested (thus revealing its fallibility or validity)", "whether the known error rate of the technique is acceptable" (uncertainty), "whether the technique or method has been subjected to peer review" (been published) ,and "whether the technique is generally accepted by the scientific community."

A third resource are workshops. For example, in the fall of 1996 the *American Society of Mass Spectrometry*[13] conducted a workshop entitled: "Limits to Confirmation, Quantitation and Detection." One topic of discussion was "how much selectivity can be sacrificed without seriously comprising the confidence levels for analyte identification." Three points of view were presented outlining what established confirmation- the number of ions required (see Figure 11.3)[14],[15]; prescriptive guidelines[16]; and the use of a grading system to assign selectivity indices to the entire "analytical chain" from sample preparation to analysis.[17]

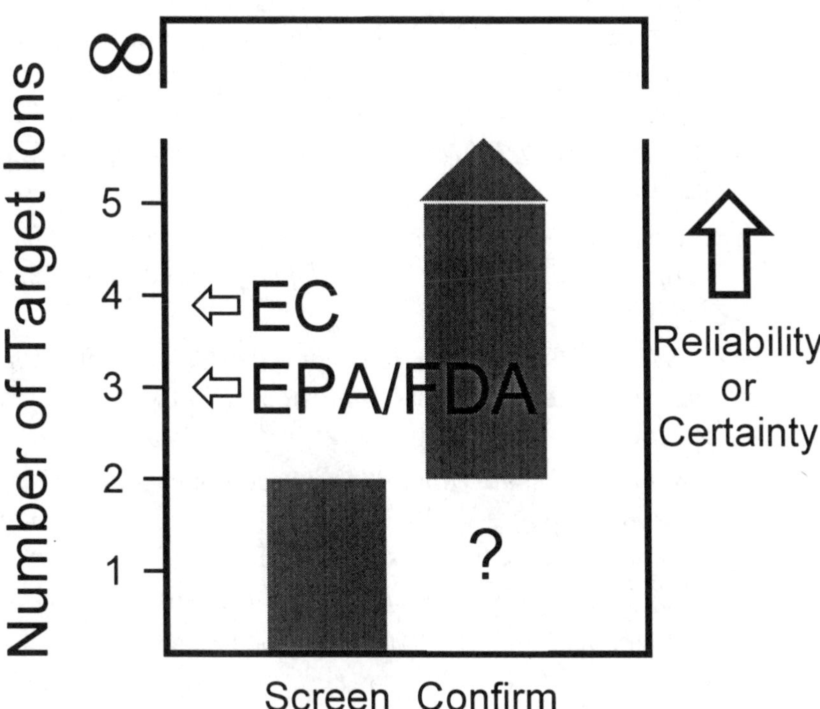

Figure 11.3- Number of ions required for screening and confirmatory MS analysis. EC[18]: European Community; EPA/FDA[19]: US Environmental Protection Agency/Food and Drug Administration.

Where to Find Target Methods

Target methods can be found in a variety of sources from books, articles, data-bases, in-house reports, etc. Search them all. Today with more and more resource information on-line computer based searches for currently published methods is very easy.

Appendix A lists books, journals, data bases, etc. where you can find examples of published methods. We recommend at a minimum every lab should have access to following:

- *Applications of LC/MS in Environmental Chemistry*, edited by D. Barcelo;
- *Biochemical and Biotechnological Applications of Electrospray Ionization Mass Spectrometry*, edited by A. Peter Snyder;
- *Electrospray Ionization Mass Spectrometry: Fundamentals, Instrumentation, and Applications*, edited by Richard B. Cole;
- *Liquid Chromatography-Mass Spectrometry: Applications in Agricultural, Pharmaceutical, and Environmental Chemistry*, edited by Mark A. Brown;
- *Liquid Chromatography-Mass Spectrometry: Principles and Applications*, by W.M.A. Niessen and J. Van der Greef;
- *Liquid Chromatography-Mass Spectrometry: Techniques and Applications*, by Alfred L. Yergey, Charles G. Edmonds, Ivor A.S. Lewis, and Mervin L. Vestal; and
- the journal, LC/MS Update and Soft Ionization, by HD Science.

Examples of LC/MS Target Methods

The following sections describe various LC/MS target methods. We picked these applications not because they are the only examples [they are not] of target methods but because they exemplify a particular aspect of the use of the selectivity and specificity that LC/MS affords you.

Target Applications: Electrospray

Electrospray in Accelerated Drug Development

Mike Lee and his colleagues[20] at Bristol-Myers Squibb Pharmaceuticals Research Institute describes approaches to make LC/MS techniques the cornerstone of accelerated drug development, and the identification of impurities, degradants and biomolecules.

Electrospray in Clinical Diagnosis

David S. Millington, Donald Chase and their associates[21] at Duke University Medical Center describe semi-automated electrospray LC/MS/MS methods, based on isotope dilution, for the analysis of amino acids and acylcarnitines in human whole blood, plasma and urine. Complete metabolic profiles of target compounds were generated in less than one minute per sample. They predict that utilizing LC/MS/MS will successfully detect hypermethioninemias with very low rates for false positives and false negatives.

Electrospray in Integrity of Recombinant Proteins

Ragulan Ramanathan, Walter Zielinski and coworkers[22] from Washington University and the U.S. FDA describe a protocol to test the integrity of recombinant proteins and the consistency from batch to batch of their preparations. They used the techniques of electrospray to assay the rate of H/D exchange of recombinant insulins. The assay could be conducted at one time point for a sample size of less than 2 micrograms.

Electrospray with Chemical Derivatization

Martin Quirke and co-workers[23] from Oak Ridge National Laboratory describe a protocol for activating non-ionic compounds to ionic or solution-ionizable derivatives to form "electrospray active" forms of the compounds. By means of derivatizing they were able to enhance the specific and selective mass spectral [and if necessary the optical] detection of simple alkyl halides, alcohols and amines.

Target Applications: APCI

APCI with Drug Analysis

Gary Bowers and colleagues[24] at Glaxo Wellcome, Inc. describe a totally automated approach for combining solid phase extraction with flow injection analysis and atmospheric pressure chemical ionization LC/MS/MS to analyze several thousand samples in support of Phase I clinical studies.

APCI with Pharmacokinetic Studies

Tom Covey and co-workers from New York State College of Veterinary Medicine[25] describe a method combining APCI and tandem mass spectrometry offering a combination of selectivity, sensitivity, and speed. By utilizing a fast-LC method, they were able to analyze one sample every minute.

Timothy Olah and colleagues[26] at Merck Research Laboratories describe the use of LC with tandem MS for determining mixtures of drug candidates as part of a high-throughput bioanalysis involving pharmacokinetics and metabolic stability studies. This approach relied on simple isolation methods, isocratic chromatography, and LC/MS/MS conditions of atmospheric pressure chemical ionization and Mixed Reaction Monitoring that can be adapted and applied to numerous agents. In a single analysis, they were able to determine the plasma concentration of 12 drug candidates. Using this method the authors were able to screen more than 400 compounds in a 6 month period.

APCI with Screening Small-Molecule Libraries

Jack Henion and associates[27] at New York State College of Veterinary Medicine describe a method combining the selectivity and specificity of immunoaffinity and atmospheric pressure ionization LC/MS for screening a combinatorial libraries of 20-30 closely related benzodiazepines.

Target Applications: Particle Beam

Drug Residue Analysis

David Heller and Frank Schenck from the FDA[28] describe a particle beam LC/MS method for the determination of ivermectin in bovine milk and liver. They noted that in the past, the presence of ivermectin residues was confirmed by a lengthy method based on partial hydrolysis, derivatization, HPLC and fluorescence detection. As an alternative to this method they developed an LC/MS method using negative chemical ionization. The specificity required for a regulatory confirmation procedure was achieved by monitoring the molecular ion and four fragment ions.

Pesticide Analysis

Tom Behymer and associates from the U.S. EPA[29] describe the utility of particle beam LC/MS for analyzing semivolatile contaminants- benzidines and nitrogen-containing pesticides, which may be present in drinking water or drinking water sources.

Achille Cappiello and his associates at the Istituto di Scienze Chimiche, Universita di Urbino[30] describe a particle beam LC/MS method for the determination of 32 base/neutral and 13 acidic pesticides in water. They utilized a particle beam interface they developed which allows the introduction of lower mobile-phase flow rates, ~ 1 microliters/min., into the mass spectrometer ion source.

Response Factors and Detection Limits

Jack Northington and co-workers at West Coast Analytical Service, Inc.[31] describe the particle beam LC/MS retention times and detection limits of approximately 50 compounds. They analyzed amines, organometallics, and organic acids. Although the detection limits for the analytes were variable, many had detection limits of 5-50 nanograms in a full scan (electron ionization) acquisition. These detection limits were equivalent to 0.25 - 2 micrograms/mL (in a 20 microliter injection volume), which is similar to the sensitivities commonly required in the U.S. EPA Method 8270 using GC/MS.

Target Applications: Thermospray

Pesticide Analysis

Tom Bellar and Bill Budde from the U.S. EPA[32] describe a thermospray LC/MS method for 52 pesticides and other compounds of environmental interest.

Dietrich Volmer and Karsten Levsen from the Fraunhofer Institute of Toxicology[33] describe a method that employs off-line and on-line solid-phase extraction and thermospray LC/MS analysis with time-scheduled selected ion monitoring for environmental monitoring for a series of 51 nitrogen- and phosphorus-containing pesticides. On column detection limits range from 40 - 600 picograms.

Target Methods

Al Yergey and associates[34] have compiled a bibliography of LC/MS applications up to 1988. The majority of the applications utilize thermospray LC/MS for the analysis of biomolecules, industrial residue, and pharmaceuticals. Since that time API interfaces (e.g., APCI and electrospray) have replaced thermospray due to their unrivaled combination of enhanced selectivity, specificity, and ease of use.

11.3- The Target Experiment

Experimentally, target analysis involves a number of steps to confirm the presence (or absence) of a target analyte with LC/MS. It is the culmination of all the analytical capabilities described throughout this book. It simultaneously solves both identity and quantification problems. After balancing all the experimental parameters relating to speed, selectivity, sensitivity, and specificity, an experiment has to be conducted. This section describes the mechanics of a typical LC/MS target experiment. This process cannot come about without the existence of a valid method (Chapter 12). We include this description for newcomers to the field who may not have an appreciation for what actually occurs during a target experiment.

Standard Runs: Acquiring Libraries, RT, and RF

The first requirement of a target method is the calibration process. During calibration, standards are run as described in *Chapter Ten* in accordance with the specified method standardization approach (e.g. add chemical analog and target standards for internal analog method). Standards are usually treated in the same manner as the samples. This means they are taken through the same sample preparation and separation process as the target samples. (For some target applications, a certified matrix standard[35] is utilized in the calibration process.) Generally, multiple levels of standard are analyzed under the experimental method experimental conditions. Typically, only "target" masses are acquired with target methods.[36] The compound-specific masses of the target and standards are measured. From these measurements one is able to obtain three important types of information;

1) identity information is supplied from standard runs by providing the retention times of both target and standard,
2) mass/intensity ratios for both target and internal standard can be established (used for library matching),
3) the response factor of both target and standard can be determined in the analytical range of the method.

These spectral and response information are stored by the instrument computer for subsequent use during and after sample analysis. If you have more than one target, spectral results may be manually or automatically stored in a target library.

Samples Runs: Acquiring Sample Data

The next stage of the target experiment is sample analysis. Samples are treated in the same manner as the standards. Internal standards are added to the samples before sample treatment. The samples are analyzed as per the method. The logic behind target data treatment is outlined in Figure 11.4. The mass spectral data from the sample run is compared with that of the standard run. For a target to be confirmed it must meet the identity criteria of the method. First, the response of a potential target in the sample must occur at the same retention time as the standard run. If not, no target exists in the sample. If the retention time conditions are met, then the mass spectrum of the peak must be consistent with the standard. If these conditions are met, then the target is present in the sample. At this time, the area from the target species in the sample can be used along with the response factor acquired from the calibration process to calculate the concentration of the target analyte in the sample.

LC/MS target methods differ from a typical target method in LC/UV in that the confirmatory step in LC/MS involves matching the response of the target to a mass spectral library. This library matching result in the high specificity of target analysis with LC/MS compared to LC/UV. Even in the simplest case where the mass spectral library contains of two or three ions, this far exceeds the specificity of UV.

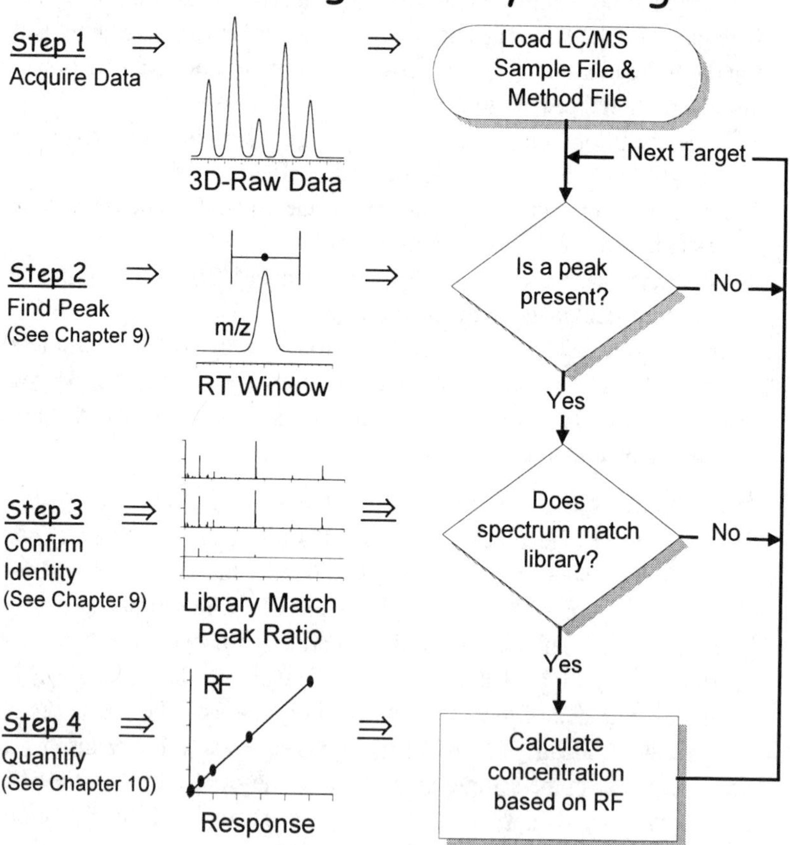

Figure 11.4- Typical LC/MS Target Method.[37]

For LC/UV target methods, the UV spectral of a particular component is not typically used for identification, but for detection.[38] If a chromatographic peak is detected in a prescribed retention window, the intensity is recorded and the amount quantified and reported. If a peak is not detected, the target is reported as not found. No decision process takes place to qualify the identity of the target. If a peak is detected, an amount is calculated.

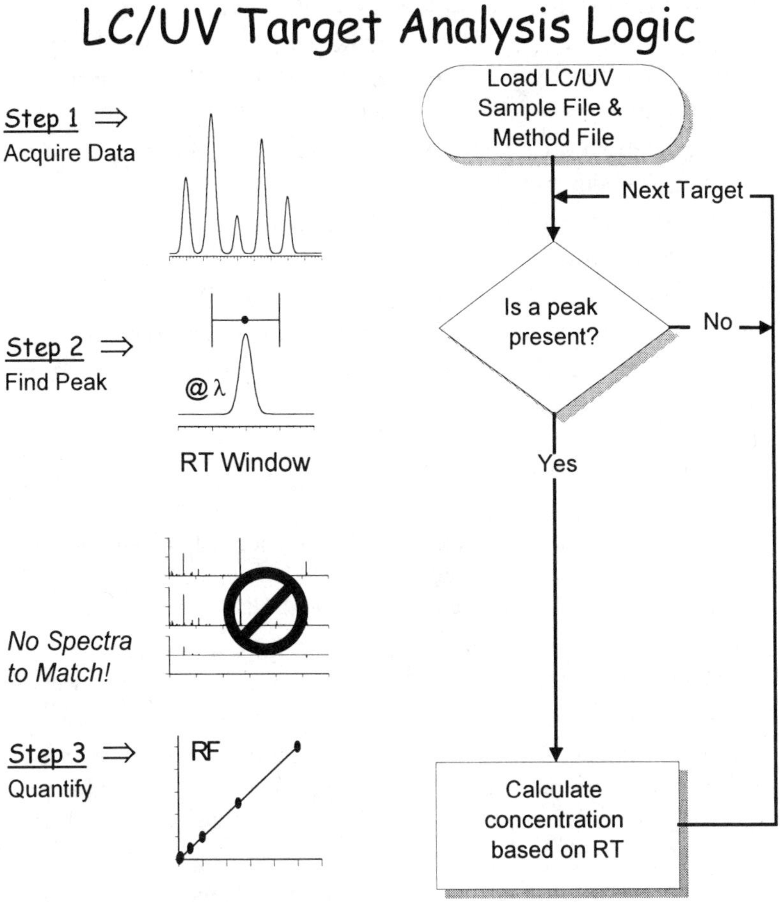

Figure 11.5- Typical LC/UV Target Method.[39]

Discussion Questions and Exercises

1. What are the *instrumental* parameters in LC/MS that have the greatest affect on your target analysis? How do these parameters affect selectivity, specificity, and sensitivity?
2. What are the *sample* parameters that have the greatest affect on your LC/MS target analysis? How do these parameters affect selectivity, specificity, and sensitivity?
3. Complete Worksheet 11.0 for a target analysis problem that has entered your organization in the last year.
4. What are the major differences between <u>LC/UV</u> and <u>LC/MS</u> based target analysis? ... in terms of reliability? ... in terms of ruggedness? ... in terms of speed?
5. What are two differences between a screening and a confirmatory LC/MS method?
6. Are most target analysis methods in your industry screening or confirmatory methods?
7. Give an example of a screening method in LC/MS.
8. Give an example of a confirmatory method in LC/MS.
9. Why does LC/MS method development take less time than LC/UV method development?

Notes

[1] Stevenson, R., "The World of Separation Science: HPLC '96 puts the spotlight on new detectors and columns; MS continues to evolve," American Laboratory, News Edition 29, pages 4-8 (1997).

[2] Behymer, T.D., Bellar, T.A., Budde, W.L., "Liquid chromatography/ particle beam/mass spectrometry of polar compounds of environmental interest," Anal. Chem. 62, pages 1686-1690 (1990).

[3] Phoenix International is performing over a thousand samples per week per instrument. See their website at `www.pils.com` for sample information. The rate of 200 samples per day is becoming a standard for throughput in LC/MS high throughput applications.

[4] Sensitivity. The use of recorded current per mass (coulomb/microgram or ampere/Pascal) is recommended by the ASMS when referring to sensitivity (Price, P, "Standard definitions of terms relating to mass spec-

trometry: A report from the committee on measurements and standards of the American Society for mass spectrometry," J. Am. Soc. Mass Spectrom. 2, pages 336-348 (1991).). These two terms are rarely used in GC/MS and LC/MS analysis. Instead the terms, signal-to-noise and signal-to-background are commonly used. Signal-to-noise is the measure of the signal in the presence of noise- a measure of precision. While signal-to-background is a measure of detectability- referring to the measure of the signal strength for a sample compared to a measure of the signal strength when no sample is present. It is often referred to incorrectly as signal-to-noise. (Watson, J.T., Sparkman, O.D., ACS Short Course, "Mass Spectrometry: Principles and Practice," July 1997.)

[5] Abbreviations

FTMS-	fourier transform mass spectrometer
Hybrids-	BE-TOF (include in early 1998)
IT-	ion trap
MRM-	multiple [or mixed] reaction monitoring, MS/MS
Q-	single quadrupole
Q-TOF-	quadrupole-time-of-flight hybrid
QQQ-	triple quadrupole
$Scan_{CI}$-	see Scan-CI
$Scan_{EI}$-	see Scan-EI
$Scan_{Prod}$-	see Scan-Prod
Scan-CI-	chemical ionization (APCI, +/- ES, +/- thermospray, +/- CI, +/- continuous flow FAB) and scanning a mass range
Scan-EI-	electron ionization (EI) and scanning a mass range
Scan-Prod-	product ion scan, tandem mass spectrometry (MS/MS) [also referred to as daughter scan]
Sectors-	magnetic, electrostatic and sector hybrids (e.g., EBQQ)
SIM-	selected [or single] ion monitoring, MS
SRM-	selected [or single] reaction monitoring, MS/MS

[6] Stephany, R.W., Presentation: "European Union Regulatory Residue Analysis for Veterinary Drugs and Banned Anabolic Agents: A Strategic System, Its Tools and Reliability Targets," in the report from the 1996 ASMS Fall Workshop: Baldwin, R., Bethem, R.A., Boyd, R.K., Budde, W.L., Cairns, T., Gibbons, R.D., Henion, J.D., Kaiser, M.A., Lewis, D.L., Matusik, J.E., Sphon, J.A., Stephany, R.W., Trubey, R.K., "Limits to Confirmation, Quantitation and Detection," J. Am. Soc. Mass Spectrom. 8, pages 1180-1190 (1997). ECU: European Community Units.

[7] Ion trap. Some instruments can be operated in a SRM or MRM model, but it is felt that a product ion scan is more appropriate for an ion trap instrument

[8] Abbreviations, *op cit.*

[9] *ibid.*

[10] Millington, D.S., Chace, D.H., Stevens, R.D., "Rapid, automated analysis of diagnostically important metabolites in physiological matrices by LC/ESI-MS/MS," Proceedings of the 12th LC/MS Montreux Symposium, Hilton Head Island, SC, November 1-2, 1995.

[11] Baldwin, R., "Regulatory Credibility in the Arena of Consumer Protection," in the report from the 1996 ASMS Fall Workshop: Baldwin, R., Bethem, R.A., Boyd, R.K., Budde, W.L., Cairns, T., Gibbons, R.D., Henion, J.D., Kaiser, M.A., Lewis, D.L., Matusik, J.E., Sphon, J.A., Stephany, R.W., Trubey, R.K., "Limits to Confirmation, Quantitation and Detection," J. Am. Soc. Mass Spectrom. 8, pages 1180-1190 (1997).

[12] Kuffner, C.A., Jr., Marchi, E., Morgado, J.M., Rubio, C.R., "Capillary electrophoresis and Duabert: Time for admission," Anal. Chem. 68, 241A-246A.

[13] 1996 ASMS Fall Workshop. Baldwin, R., Bethem, R.A., Boyd, R.K., Budde, W.L., Cairns, T., Gibbons, R.D., Henion, J.D., Kaiser, M.A., Lewis, D.L., Matusik, J.E., Sphon, J.A., Stephany, R.W., Trubey, R.K., Report from the 1996 ASMS Fall Workshop: "Limits to Confirmation, Quantitation and Detection," J. Am. Soc. Mass Spectrom. 8, pages 1180-1190 (1997). Copies of the Workshop materials are available from the authors of the report:

Robert A. Bethem: (`bbethem@altalab.com`),

Robert K. Boyd: (`robert.boyd@nrc.ca`), and

K. Trubey: (`trubey@esvax.email.duPOnt.com`)

[14] *ibid.* Presentation: Matusik, J., Sphon, J.A., "FDA Regulatory Confirmation Criteria."

[15] 1996 ASMS Fall Workshop, *op cit.*, Presentation: Cairns, T., "Development of Guidelines or Detection, Confirmation, and Quantitation by GC/MS, LC/MS, and MS/MS."

[16] *ibid.*

[17] Stephany, R.W., op cit.

[18] de Ruig, W.G., Stephany, R.W., Dijkstra, G.J., "Criteria for the detection of analytes in test samples," J. Assoc. Off. Anal. Chem. 72, pages 487-490 (1989).

[19] The Three Ion Rule. As proposed by Sphon, J.A., "Use of mass spectrometry for confirmation of animal drug residues," J. Assoc. Off. Anal. Chem. 61, pages 1247-1252 (1978).

[20] Lee, M.S., Kerns, E.H., Hail, M.E., Liu, J., Volk, K.J., "Recent applications of LC-MS techniques for the structure identification of drug metabolites and related compounds," LC-GC 15, pages 542-558 (1997).

[21] Millington, D.S., Chace, D.H., Stevens, R.D., "Rapid, automated analysis of diagnostically important metabolites in physiological matrices by LC/ESI-MS/MS," Proceedings of the 12th LC/MS Montreux Symposium, Hilton Head Island, SC, November 1-2, 1995.

[22] Ramanathan, R., Gross, M.L., Ziellinski, W.L., Layloff, T.P., "Monitoring recombinant protein drugs: A study of insulin by H/D exchange and electrospray ionization mass spectrometry," Anal. Chem. 69, pages 5142-5145 (1997).

[23] Quirke, J.M.E., Adams, C.L., Van Berkel, G.J., "Chemical derivtization for electrospray ionization mass spectrometry. 1. Alkyl Halides, alcohols, phenols, thiols, and amines," Anal. Chem. 66, pages 1302-1315 (1994).

[24] Bowers, G.D., Clegg, C.P., Hughes, S.C., Harker, A.J., Lambert, S., "Automated SPE and tandem MS without HPLC columns for quantifying drugs at the picogram level," LC-GC 15, pages 48-53 (1997).

[25] Covey, T.R., Lee, E.D., Henion, J.D., “High speed liquid chromatography/tandem mass spectrometry for the determination of drugs in biological samples.” Anal. Chem. 58, pages 2453-2460 (1986). This article is the first example of the use of fast-LC, APCI and tandem mass spectrometry for pharmacokinetic studies.

[26] Olah, T.V., McLoughlin, D.A., Gilbert, J.D., "The simultaneous determination of mixtures of drug candidates by liquid chromatography atmospheric pressure chemical ionization mass spectrometry as an in-vivo drug screening procedure," Rapid Commun. Mass Spectrom. 11, pages 17-23 (1997).

[27] Wieboldt, R., Zweigenbaum, J. Henion, J., "Immunoaffinity ultrafiltration with Ion Spray HPLC/MS for screening small-molecule libraries," Anal. Chem. 69, pages 1683-1691 (1997).

[28] Heller, D.N., Schenck, F.J., "Particle beam liquid chromatography/mass spectrometry with negative ion chemical ionization for the confirmation of ivermectin residue in bovine milk and liver," Biol. Mass Spectrom. 22, pages 184-193 (1993).

[29] Behymer, T.D., Bellar, T.A., Ho, J.S., Budde, W.L., "Method 553: Determination of Benzidines and Nitrogen-Containing Pesticides in Water by Liquid-Liquid Extraction or Liquid-Solid Extraction and Reverse Phase High Performance Liquid Chromatography/Particle Beam/Mass Spectrometry," pages 173-212, IN: Methods for the Determination of Organic Compounds in Drinking Water, Supplement II; US-EPA Report, EPA/600/R-92/129; Order No. PB92-207703, 270 pages, 92(23), Abstract No. 266,774 (1992). Also see Note 2.

[30] Cappiello, A., Famiglini, G., Bruner, F., "Determination of acidic and basic/neutral pesticides in water with a new microliter flow rate LC/MS particle beam interface," Anal. Chem. 66, pages 1416-1423 (1994).

[31] Northington, D.J., Hovanec, B.M., Shelton, M., "Particle beam LC-MS in environmental analysis," Am. Env. Lab. 2, pages 34-41 (1990).

[32] Bellar, T.A., Budde, W.L. "Determination of nonvolatile organic compounds in aqueous environmental samples using liquid chromatography/mass spectrometry," Anal. Chem. 60, pages 2076-2083 (1988).

[33] Volmer, D., Levsen, K., "Mass spectrometric analysis of nitrogen-containing pesticides by liquid chromatography-mass spectrometry," J. Am. Soc. Mass Spectrom. 5, pages 655-675 (1994).

[34] Yergey, A.L., Edmonds, C.G., Lewis, I.A.S., Vestal, M.L., *Liquid Chromatography-Mass Spectrometry: Techniques and Applications*, Plenum Press: New York (1990).

[35] Many bioassays in the pharmaceutical industry use spiked matrix standard for all calibration runs. This is to ensure that the sample and standard responses are closely matched.

[36] Broad-based methods such as U.S. EPA methods are usually scanning all masses. See reference 29.

[37] Abbreviations

2D-	two dimensional information [data]: time and intensity
3D-	three dimensional information [data]: time, intensity, and mass spectrum
@m/z-	trace of a particular m/z mass value versus time, a chromatographic peak
@λ-	trace of a particular wavelength versus time, a chromatographic peak
RF-	response factor
RT-	retention time

[38] PDA [photodiode array] has been used to perform spectral matching to identify components. But the spectral specificity of a UV spectrum is low. For example analogs, such as, metabolites (see *Chapter One, Figure 1.1*) or cortiosteroids (Dorschel, C., "The role of particle-beam LC-MS in separation development," LC-GC 15, pages 950-959 (1997).) are difficult to distinguish from each other based on their UV spectrum.

[39] *ibid.*

Chapter Twelve- Flow Diagram

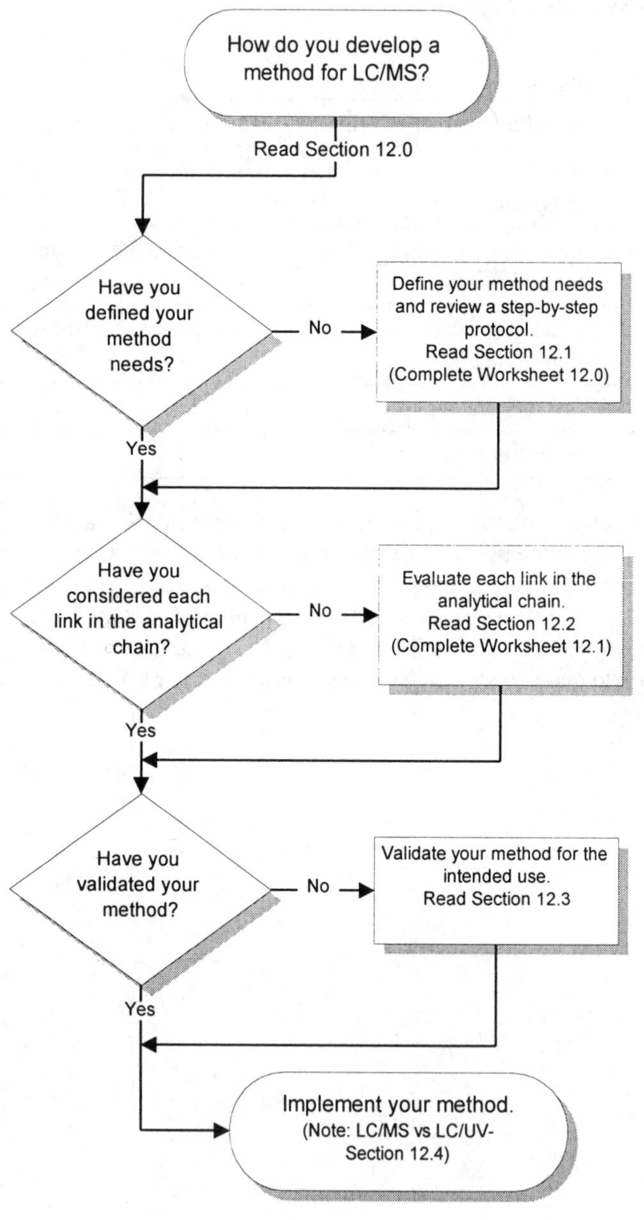

Chapter Twelve

Developing Methods with LC/MS

(or, Running samples with impunity.)

Enforcement of regulations requires sound, reliable and validated analytical procedures as the basis of those regulations governing adulterations, contaminations, additives, as well as the composition and efficacies of FDA-regulated products. Rulings will not withstand legal scrutiny without validated analytical procedures as their basis.

Harold C. Thompson, Jr., FDA[1]

12.0-Eliminating Uncertainty

Here we are at the last chapter of this book and to this point have not addressed one of the most important topics relating to LC/MS; that of course, is methods development. No assay will ever be valid unless all of the uncertainty has first been hammered out in the method development process. Throughout this entire text we have equated LC/MS with problem-solving. The *potential* problem solving capabilities of LC/MS go without question. We can find examples of the utility of LC/MS for almost any application you may conceive. But it is the method development process that assures that each and every result that is acquired is reliable. Since uncertainty and reliability are inversely related, it is the systematic and complete elimination of uncertainty at every point in an analysis that gives us confidence in a reliable result.

There is some credence to the old adage "that more parts means more can go wrong." LC/MS is not immune from this adage. If we move from link to link in the analytical chain we can clearly see that the chain belonging to LC/MS has many links; any one potentially breaking (see Figure 12.0). The method development process should address each link in a systematic progression with the primary goal of meeting the intended information needs of the analytical problem. Sample, separations, interfaces, ionization, mass analysis, and data reduction all must be throughly evaluated in the process. The uncertainty associated with each component in this chain cannot be overlooked in order for a method to hold up to the rigors of validation; and, THERE IS NO METHOD WITHOUT VALIDATION!

In *Chapter Eight* we consumed considerable ink discussing the importance of problem definition as the basis for successful problem-solving (Review *Worksheet 8.0*). Problem definition is of particular importance in methods development. You should spend significant effort and energy in order to define the needs, constraints, and obstacles that you face in your development. Remember, launching down the pathway to methods development without a clear definition of the exact objectives and intended use of a method can waste time, manpower, and resources. We end this chapter with a discussion on some of the differences and similarities between methods development in LC/MS versus LC/UV.

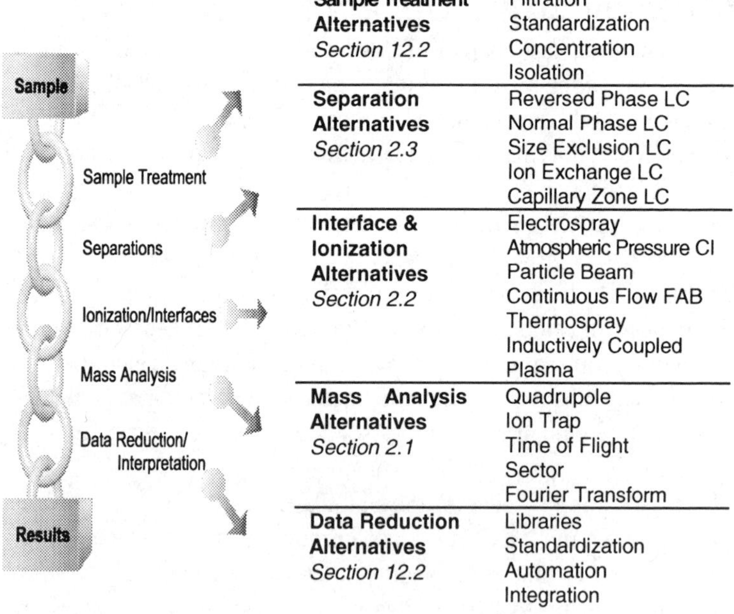

Sample
Sample Treatment
Separations
Ionization/Interfaces
Mass Analysis
Data Reduction/
Interpretation
Results
Sample Treatment
Alternatives
Section 12.2
Filtration
Standardization
Concentration
Isolation
Separation
Alternatives
Section 2.3
Reversed Phase LC
Normal Phase LC
Size Exclusion LC
Ion Exchange LC
Capillary Zone LC
Interface &
Ionization
Alternatives
Section 2.2
Electrospray
Atmospheric Pressure CI
Particle Beam
Continuous Flow FAB
Thermospray
Inductively Coupled
Plasma
Mass Analysis
Alternatives
Section 2.1
Quadrupole
Ion Trap
Time of Flight
Sector
Fourier Transform
Data Reduction
Alternatives
Section 12.2
Libraries
Standardization
Automation
Integration

Figure 12.0- Analytical chain and methods development.

12.1-A Step-by-Step Protocol

Methods development is not necessarily linear with the analytical chain; its a cyclical problem-solving process just like all the others described throughout this text (see Figure 12.1). The process begins with definition (see Worksheet 12.0) and ends with implementation.

We will outline in the rest of this chapter using a step-by-step approach covering the aspects of evaluating targets, matrix and separations conditions (see the following page and Table 12.0). How one integrates all these components will be largely hardware specific. How one validates with be largely industry or regulation specific. But do not be discouraged, there are lots of resources out there that can help you with integration, automation, and validation (e.g., instrument manufacturers, ICH, EPA protocols, professional organizations; see *Appendix A*.).

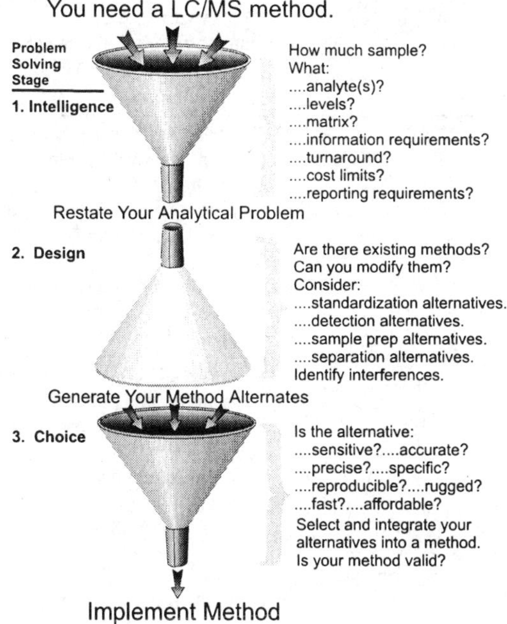
You need a LC/MS method.
Problem Solving Stage
1. Intelligence
How much sample?
What:
....analyte(s)?
....levels?
....matrix?
....information requirements?
....turnaround?
....cost limits?
....reporting requirements?
Restate Your Analytical Problem
2. Design
Are there existing methods?
Can you modify them?
Consider:
....standardization alternatives.
....detection alternatives.
....sample prep alternatives.
....separation alternatives.
Identify interferences.
Generate Your Method Alternates
3. Choice
Is the alternative:
....sensitive?....accurate?
....precise?....specific?
....reproducible?....rugged?
....fast?....affordable?
Select and integrate your alternatives into a method.
Is your method valid?
Implement Method

Figure 12.1- Methods development in a structured problem-solving scheme.

LC/MS Method Development-Step-by-Step

Step 1
Definition

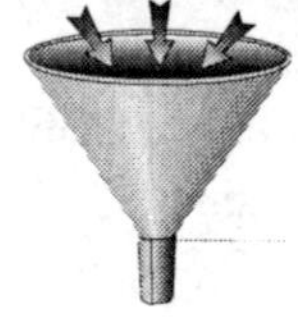

a) Gather information about all targets and matrices (Worksheet 12.0).
b) Acquire standards of all targets and matrix if available.
c) Thoroughly define the problem (Worksheet 8.0).
d) Establish method validation criteria (Worksheet 12.0).

Step 2
Evaluate Targets

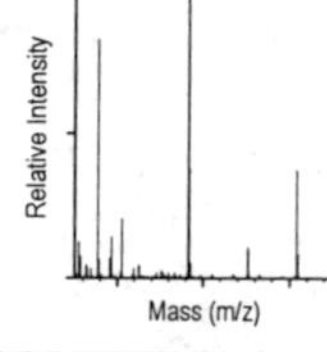

a) Evaluate the response of each target under varying ionization conditions.
b) Evaluate the fragmentation processes associated with each target.
c) Determine figures-of-merit (FOM).

Step 3
Evaluate Matrix

a) Evaluate all potential interferences.
b) Develop a scheme to eliminate possible interferences (e.g. extraction, filtration, etc.).
c) Select standardization approach (Worksheet 10.0).
d) Concentrate target analyte as required.
e) Compare Step 2 FOM to expected sample response.

Step 4
Evaluate Separations

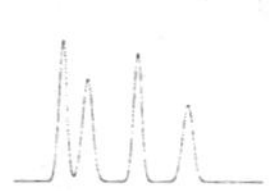

a) Establish separation requirements.
b) Evaluate internal standard if required (Worksheet 10.0).
c) Reconcile compatibility between interface and separation flows, solvents, and buffers.
d) Develop a scheme to separate target analyte from interferences.
e) Concentrate target as required.

Step 5
Method Integration

a) Integrate all components of the analytical scheme into coherent method.
b) Develop integration routines and automation.
c) Establish quantification methods and libraries.
d) Write a draft method.

Step 6
Method Validation

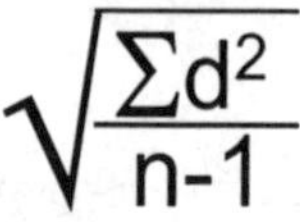

a) Conduct intra-laboratory validation.
b) Conduct inter-laboratory validation (if required).
c) Revise and issue method.
d) If validation criteria are not met, go to **Step 1**.

Table 12.0- Target and Matrix Information is Useful at Every Step

Step 1 Definition	Step 2 Response	Step 3 Treatment	Step 4 Separation	Step 5 Integration	Step 6 Validation
Target Information					
1. Name	☑	☐	☐	☐	☐
2. Formula	☑	☐	☐	☐	☐
3. MW	☑	☑	☑	☐	☐
4. Monoisotopic Mass (exact)	☑	☐	☐	☐	☐
5. Nominal Mass	☑	☐	☐	☐	☐
6. CAS #	☑	☐	☐	☐	☐
7. Structure	☑	☑	☑	☑	☐
8. Solubility	☑	☑	☑	☐	☐
9. pKa	☑	☑	☑	☐	☐
10. Stability (light, solvent, heat, air, etc.)	☑	☑	☑	☑	☑
11. Storage (refrigerate, desiccate, freeze)	☑	☑	☑	☑	☑
12. Availability of Standard	☑	☑	☑	☑	☑
13. Availability of Methods	☑	☑	☑	☑	☑
Matrix Information					
1. Type(s)	☑	☑	☑	☑	☐
2. Variability	☑	☑	☑	☑	☑
3. Solubility	☑	☑	☑	☑	☐
4. Heterogeneity	☑	☑	☑	☑	☑
5. State of Matter (solid, liquid)	☑	☑	☑	☑	☑
6. Availability of Standards	☑	☑	☑	☑	☑

Checked boxes indicate the stage in method development where this information may be useful. All available information about targets and matrices should be acquired from the earliest stage in method development.

Worksheet 12.0-Defining Your Method Needs

Part A- Defining your method **needs**. Check-off your specific method requirements and comment on each item.

Target Information

1. Name ________________ ❑
2. Formula ________________ ❑
3. MW ________________ ❑
4. Monoisotopic Mass (exact) ________________ ❑
5. Nominal Mass ________________ ❑
6. CAS # ________________ ❑
7. Structure (note functional groups) ________________ ❑
8. Solubility ________________ ❑
9. pKa ________________ ❑
10. Stability ________________ ❑
11. Storage ________________ ❑
12. Availability of Standard ________________ ❑
13. Availability of Methods ________________ ❑

Matrix Information

1. Type(s) ________________ ❑
2. Variability ________________ ❑
3. Solubility ________________ ❑
4. Heterogeneity ________________ ❑
5. State of Matter (solid, liquid) ________________ ❑
6. Availability of Standards ________________ ❑

What are your specific <u>method needs</u>?	Comments
What are the target method concentration levels?	What?________________
What are the target method concentration ranges	What?________________
What is the required number of target ions?	Why?________________

What are your <u>technology needs</u>?	Comments
Do you require a specific separation?	Why?________________
Do you require a specific interface/ionization?	Why?________________
Do you require a specific mass analyzer?	Why?________________

What are your <u>resource needs</u>?	Comments
Who will develop the method?	Who?________________
Do you require special facilities?	What?________________
Do you require special equipment?	What?________________

Note: Copies of this worksheet can be downloaded from `www.LCMS.com`.

Worksheet 12.0-Defining Your Method Needs

Part B- Defining your method **constraints**. Check off your specific method requirements and comment on each item.

What are your sample constraints?	Yes	No	Comments
Do you require isolation of components?	❑	❑	Why?__________
Do you require removal of interferences?	❑	❑	Which ones?__________
Do you have unstable sample components?	❑	❑	Storage?__________
Do you have involatile components?	❑	❑	Volatility?__________
Do you have high MW species?	❑	❑	Limits?__________
Do you have special handling requirements?	❑	❑	What?__________

What are your experimental constraints?	Yes	No	Comments
Do you have specific time constraints?	❑	❑	What?__________
Do you have equipment constraints?	❑	❑	What?__________
Do you have specific sample constraints?	❑	❑	What?__________
Do you have specific regulatory constraints?	❑	❑	What?__________
Do you have specific legal constraints?	❑	❑	What?__________

What are your technology constraints?	Yes	No	Comments
Is required LC/MS technology available?	❑	❑	Where?__________
Are existing methods available?	❑	❑	Which?__________
Are existing methods adaptable?	❑	❑	How?__________

What are your standards constraints?	Yes	No	Comments
Do you have a suitable analog standard?	❑	❑	Which?__________
Do you have a labeled standard?	❑	❑	Which?__________
Can you acquire a labeled standard?	❑	❑	Where?__________
How much standard is available?	❑	❑	How much?__________
Is the available standard pure?	❑	❑	How pure?__________

What are your resource constraints?	Yes	No	Comments
Is required expertise available?	❑	❑	Who?__________
Are required facilities available?	❑	❑	Which?__________
Is required LC/MS equipment available?	❑	❑	When?__________

What are your validation requirements?	Yes	No	Comments
What is your required accuracy?	❑	❑	What?__________
What is your required precision?	❑	❑	What?__________
What is your required LOD?	❑	❑	What?__________
What is your required LOQ?	❑	❑	What?__________
What is your required linearity?	❑	❑	What?__________
Are other criteria required?	❑	❑	What?__________

Note: Copies of this worksheet can be downloaded from `www.LCMS.com`.

12.2- Step-by-Step Considerations

Once you have thoroughly defined your method requirements, you can begin the laboratory part of method development. (Don't forget to scour the literature for a ready-made solution, before going into the lab!) Whether you are adapting a published method or starting from scratch you should start your lab work by first evaluating your target responses.

There are literally thousands of details that a method developer must address. Some are target specific, some sample specific, and some are instrument specific. Any one issue can have a positive or negative effect on your results. The key to success in the method development process with LC/MS is to address the important considerations first, all the details and refinements can follow later. We will try to point out some of the important questions you should address in each step along the way. Learning which parameters might have the most beneficial effect on your response will expedite your development process and generally result in sensitive and rugged methods.

Another important piece of advice- Don't over-develop your methods! If your early experiments give adequate results, don't spend months trying to find the optimum method. Validate what you have and move on to other problems.

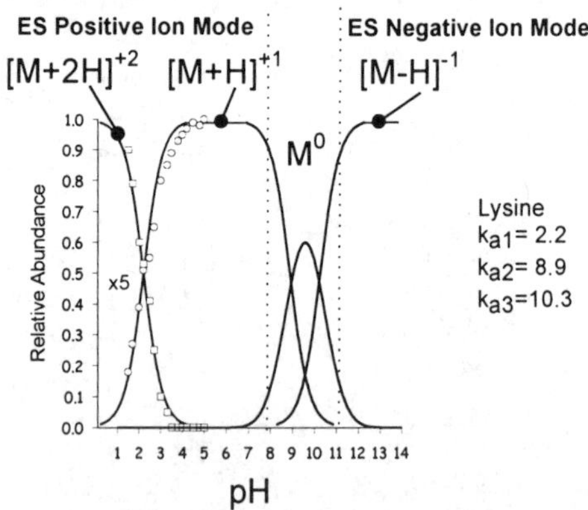

Figure 12.1- Relative abundance versus pH of the various charge states of the triprotic amino acid, lysine. The circular and square symbols represent measured abundance of lysine as a function of pH from 1-5 pH units. The squares represent the +2 charge state and the circles represent the +1 charge state. Analyte charge state in solution will effect the response in virtually all LC/MS techniques; particularly electrospray.[2]

Method Development: Target Response

Each target in your method should be thoroughly evaluated under ionization conditions available with your mass analyzer. Since the majority of method developers in LC/MS today are operating either atmospheric pressure chemical ionization (APCI) or electrospray (ES) we have outlined some rules-of-thumb for evaluating target response with these particular LC/MS interfaces. But these same processes may be also be applied to particle beam (PB), thermospray (TS) and continuous flow-FAB (CFF). The following are suggested for optimization of target analyte response:

1. Determine Response Characteristics-

Infuse or flow inject a solution of your target compound into the ion source of your mass spectrometer to optimize the response.

> *Guidelines- Important experimental parameters that should be evaluated are pH, ionic strength, solvents, ion polarity, temperatures, dimensions, and flow rates. The goal of this effort should be to obtain a response that is representative of the molecular mass (e.g. protonated molecule). It is desirable to have sensitive response with minimal fragmentation (deal with sensitivity first, fragmentation later).*

pH Considerations

For most analytical problems in LC/MS, your targets will likely be either acids, bases, or polyprotic species with the ability to accept or donate one or more protons. Knowing the pK_a for each proton on your target analyte will allow you to predict the state of ionization of your target as a function of pH (see Figure 12.1 for a triprotic amino acid). You should be aware that the response in all LC/MS techniques is highly dependent upon pH. Techniques that require the sample to be vaporized such as PB and APCI will favor vaporization of neutral species in the liquid phase. This is due to the higher volatility of neutral versus charged species. With these techniques, there is strong evidence that adjusting the pH to a value where neutral species exist will aid in their volatilization. Conversely, for desorption techniques such as ES, the desorption process requires that the analyte be charged in order to be desorbed. With these techniques, adjusting the pH to promote the charged state of the analyte over its neutral specie will enhance the response in the instrument. Figure 12.3 illustrates the regions that favor desorption in contrast to vaporization.

You can optimize the response of your analyte by adjusting the pH of your sample solution or mobile phase. Simply adjust the pH into a region where you get the best response. This is generally 2 pH units away from the pKa. In the case of acids, 2 units lower in pH will yield best response for APCI, PB, and TS; and 2 units higher in pH will yield the best response for electrospray. For bases, the opposite is true, 2 units higher in pH will yield favorable conditions for APCI, PB, and TS; while 2 units lower in pH will yield enhanced conditions for ES (see Table 12.3).

Ultimately, you may have to use a buffered mobile phase in order to maintain the appropriate ionization state of your analyte. But with some situations, the ideal pH for interface response may not be ideal for chromatographic separation. In these situations, post column addition or removal[3] of a buffer may accommodate the ideal pH for both separation and gas-phase ion production. A number of buffers are suggested in Table 12.1 for various pH ranges. In LC/MS, generally volatile buffers are preferred.

Table 12.1- Some Buffers used in LC/MS

Buffer	Proton	pKa	Range
Citrate	1	3.1	2.1 - 4.1
	2	4.7	3.7 - 5.7
	3	5.4	4.4 - 6.4
Formate*	1	3.8	2.8 - 4.8
Acetate*	1	4.8	3.8 - 5.8
Tris[4]	1	8.3	7.3 - 9.3
Ammonia*	1	9.2	8.2 - 10.2
Borate	1	9.2	8.2 - 10.2
Diethylamine	1	10.5	9.5 - 11.5

* Represents volatile buffers.

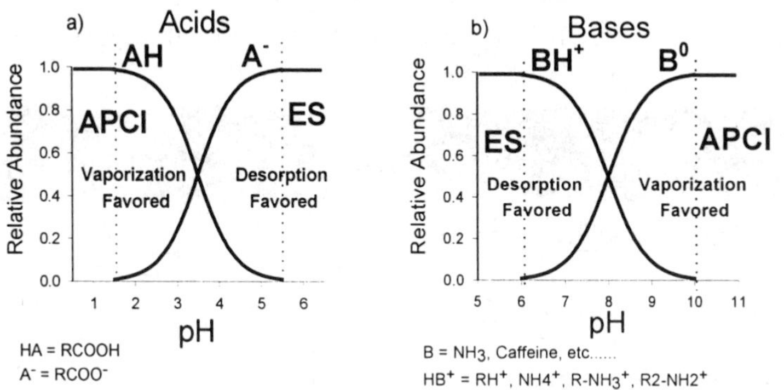

Figure 12.3- Favored pH regions for ES and APCI: desorption and vaporization. Relative abundance of an acid (a) and its conjugate base, and of a base (b) and its conjugate acid at various pH.

For electrospray, even if you know the pKa's of your targets and adjust the pH of the solution for optimum performance, there may be certain situations where the pH at the electrospray needle region may not reflect the pH of the bulk LC mobile phase. It has been noted that due to the redox reactions that take place in the metal capillary of an electrospray ion spray, the solution pH may change significantly- by as much as 4 pH units![5] This could have significant effect on the equilibrium ion distribution of many basic and acidic analytes. These effects can be moderated by buffering the mobile phase. Some of the reduction potentials vs. SHE are listed in Table 12.2 for potential redox reaction in various solvent systems.

This effect was noted to be most prominent in situations with low flow rate interfaces (e.g., nano-spray and sheathless capillary electrophoresis ES/MS) and non-buffered solutions. Therefore for optimum response the mobile phase should be buffered (e.g., ammonium acetate or formate).

Table 12.2- Electrolytic reactions that may alter solution pH in ES (adapted from ref. 5 with permission).

Reaction	Solution	E^0_{red}	pH Chang
Oxidation- Positive Ion Mode			
$4OH^- \leftrightharpoons 2H_2O + O_2 + 4e^-$	water	0.401	↓
$2H_2O \leftrightharpoons O_2 + 4H^+ + 4e^-$	water	1.229	↓
$2NH_4^+ \leftrightharpoons N_2H_5^+ + 3H^+ + 2e^-$	aqueous ammonium	1.275	↓
$NH_4^+ + H_2O \leftrightharpoons NH_3OH + 2H^+ + 2e^-$	aqueous ammonium	1.35	↓
Reduction- Negative Ion Mode			
$CH_3OH + 2H^+ + 2e^- \leftrightharpoons CH_4 + H_2O$	MeOH/H_2O	0.58	↑
$2H_2O + O_2 + 4e^- \leftrightharpoons 4OH^-$	aqueous	0.401	↑
$2H^+ + 2e^- \leftrightharpoons H_2$	aqueous	0.0	↑
$CH_3COOH + 2H^+ + 2e^- \leftrightharpoons CH_3CHO + H_2O$	acetic acid	-0.13	↑
$CH_3OH + H_2O + 2e^- \leftrightharpoons CH_4 + 2OH^-$	MeOH/H_2O	-0.25	↑
$2H^+ ([H^+] = 10^{-7}\ M) + 2e^- \leftrightharpoons H_2$	aqueous	-0.414	↑
$2NH_4 + 2e^- \leftrightharpoons 2NH_3 + H_2$	aqueous ammonium	-0.55	↑
$2H_2O + 2e^- \leftrightharpoons H_2 + 2OH^-$	aqueous	-0.828	↑

Table 12.2- Optimizing Your Target Response

ES	1. Adjust pH for acids 2 pH units above pK_a, for bases 2 pH units below pK_a. 2. If you don't know pKa, try at least three buffer regions and both positive and negative ion modes to establish ionizability. 3. Removes salts that may suppress signal. 4. Removes salts that may form adducts. (Na, K) 5. Use solvents that are hydrophobic so target ions will readily move to surfaces.[6] 6. Lower pH to protonate anything that may ion pair with your target. 7. Optimize the flow rate. 8. Increase the rate of evaporation. (e.g. Turbospray)[7] 9. Adjust position of spray relative to sampling aperture. 10. Adjust temperatures and voltages to minimized fragmentation resulting from thermal degradation or interface (source) CID. 11. Use spray chamber gas that will suppress discharge in negative ion modes (e.g. oxygen, SF_6).
APCI	1. Adjust pH for acids 2 pH units below pK_a, for bases 2 pH units above pK_a. 2. Adjust nebulization temperature to maximized signal and minimize fragmentation. 3. Evaluate solvents for optimal response. More volatile solvents may augment the production of gas-phase neutrals. 4. Adjust your discharge voltage for optimal and stable signal. 5. Adjust the position of your discharge needle relative to the ion sampling aperture. 6. Clean your needle if fowled from the plasma.

2. Determine Fragmentation Characteristics-

Once the response from the target molecule has been optimized for the molecular species, you must address selectivity and specificity requirements of the method by evaluating the fragmentation of your target molecule with MS/MS or interface (or source) collision induced dissociation (CID, for API sources) experiments. (See *Chapter 11* for discussion on selectivity and specificity.)

> *Guidelines- Under the conditions of optimal response, acquire spectra under varying fragmentation conditions. This can be done by infusing your analyte or flow injecting. For MS/MS the fragmentation can be effected by adjusting the collision energy, the collision gas composition, and the collision cell pressure. Typically, you will standardize on a specific collision gas and pressure, and get most of your selectivity (for mass) by varying the collision energy. In triple quadrupole instruments, vary the collision energy from 5 to 50 volts incrementing the voltage every 5-10 volts. For ion traps, vary the excitation voltages in accordance with the manufacturers protocol (waveform, tickle voltage, etc.) With sectors, higher energy collisions and different fragmentation pathways can be examined. For interface CID the fragmentation can be effected by adjusting the potential on the interface skimmers.*

3. Acquire Analytical Figures-of-Merit-

Inject your target analyte under optimized instrument conditions in order to establish a baseline of analytical figures-of-merit.

> *Guidelines- Acquire results from replicate (at least 3) injections over the intended analytical range. At least one point per decade of dynamic range should be acquired to establish linearity (or determine linearity range) and detection limits. The signal should be acquired in the flow injection mode. No LC column is necessary at this point. Note that this information is only giving the response under ideal conditions, not the conditions that might occur under normal sample treatment and separation. This baseline information is valuable for you to estimate the requirements that you will have to know in developing both sample treatment (concentration) and separation steps for your method.*

Confirming the identity of your target may require that you measure the response from three or more ions (Discussed in *Section 9.4*). This requirement of identity can be a challenge to the method developer who has to evaluate, primarily through MS/MS experiments, the fragmentation pattern of a given molecule. The criteria for selection of a given fragment ion will be a combination of specificity, sensitivity, and selectivity. As discussed in *Chapter 11*, the ion (or ions) you choose have to be compound-specific enough to use for the identity step of target analysis, they have to be selective enough to isolate your target response from that of other sample species, and they have to be abundant enough to meet the sensitivity requirements of the method. For example, most quantitative assays in the pharmaceutical industry are single ion methods run under the selected reaction monitoring (SRM) operating conditions to achieve a high degree of both sensitivity and selectivity.

Unfortunately, it is sometimes difficult to find three ions of sufficient intensity without degrading the sensitivity of the method (Requiring a method to use a minor ion will limit both sensitivity and dynamic range of the method). There are several techniques that can be applied to this problem in order to maintain sensitivity. One approach is to use multiple collision energies in the same acquisition to yield higher intensity fragments throughout the energy range. Another approach to use MS/MS/MS (MS^n) to yield a larger number of high intensity fragments. Suffice it to say, your method is only as good as the weakest response in the method.

Another trick to deal with low abundance fragments (in selected ion monitoring (SIM) mode or SRM) is to dwell on the lower intensity ions for a significantly longer period of time than the more intense ones. For example, you may dwell for a second on a weak signal and a 100 ms for a stronger signal. In this manner your relative response equalizes while you fulfill the "multiple target ion" requirements of your method.

Method Development: Sample Treatment

The highest level of uncertainty associated with an analytical method will likely come from the sample matrix. Sample matrices can contain every known type of interference (spectral and ionization suppression). These might include spectral interferences with isobaric species, signal suppression with salts and other species, and proton scavengers that "steal" protons away from your analyte in the gas phase. Other possible interferences from solvent and background contamination can cause the same effects, but sample related interferences are less predictable and cannot be removed by simply changing solvent grades or cleaning your ion source. Sample interferences must be identified, removed, or at a minimum, accounted for.

1. Determine Strategy for Sample Treatment

Address the important topics first; such as:

- acquiring a representative matrix standard (certified or otherwise),
- removing solids, and
- obtaining a homogeneous sample solution.

Without a representative matrix standard you will have a difficult time developing a method. Under these conditions, you may be better off using *standard addition* approaches (see *Section 10.3*) if your sample number is small. The treatment of various heterogeneous and solid samples is outlined in Figure 12.4. Solid materials or macromolecules can be removed in a variety of ways; including, precipitation, filtration, centrifugation, and dissolution. A homogeneous solution is required before you can add your target standard to the matrix (called a matrix spike). In the best case, this solution could be analyzed. In the case of plasma samples, many labs simply precipitate plasma proteins with acetonitrile and inject the supernatant directly onto an analytical column. This is the quick and dirty approach. Unfortunately, this approach may not remove unwanted interferences; the consequence being long chromatographic run times to separate the matrix components from the targets, unreliable results, or short column life.

When dealing with aqueous samples, many analysts will use extractions to isolate their target analyte and internal standard from the sample components. This is an excellent way to remove unwanted salts and ionic species that may interfere with your response; especially with electrospray. Liquid-liquid or solid phase extraction (SPE) are effective extraction strategies, with SPE being more amenable to automation. Figure 12.5 illustrates some of the strategies for treating liquid samples.

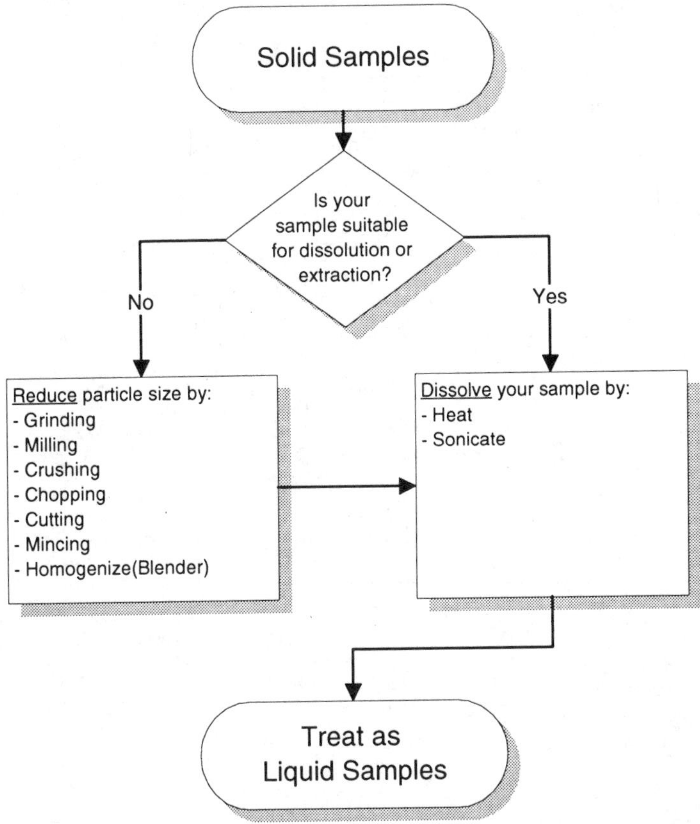

Figure 12.4- Decision tree for pretreatment of solid samples.[8]

There are a number of strategies for isolating your target and internal standard form matrix interferences. One should always keep in mind that we are still dealing with acids and bases. Figure 12.6 reminds you of the pH dependence upon extraction efficiency.

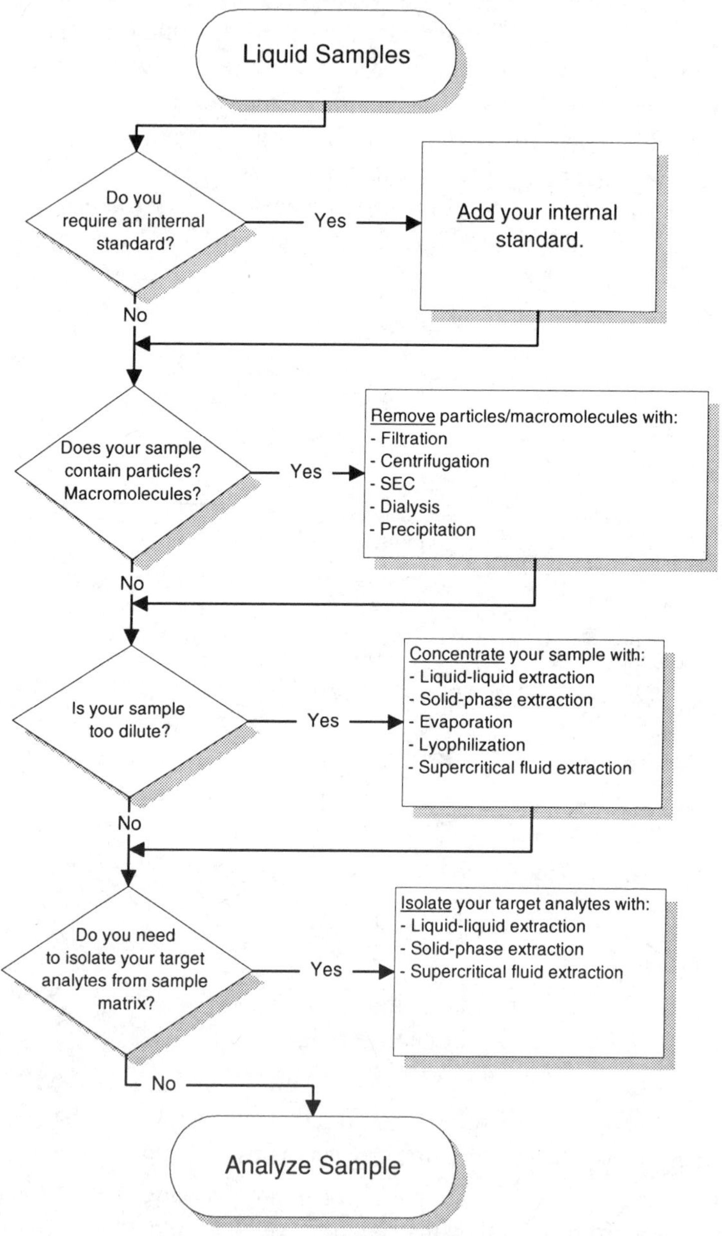

Figure 12.5- Decision tree for pretreatment of liquid samples.[8]

If you are extracting acids from aqueous samples by means of an organic solvent or SPE, you need to neutralize the molecules by lowering the pH approximately 2 pH units below the pK_a. Conversely, if you are extracting bases, you need to neutralize the molecules by raising the pH 2 units above the pK_a.

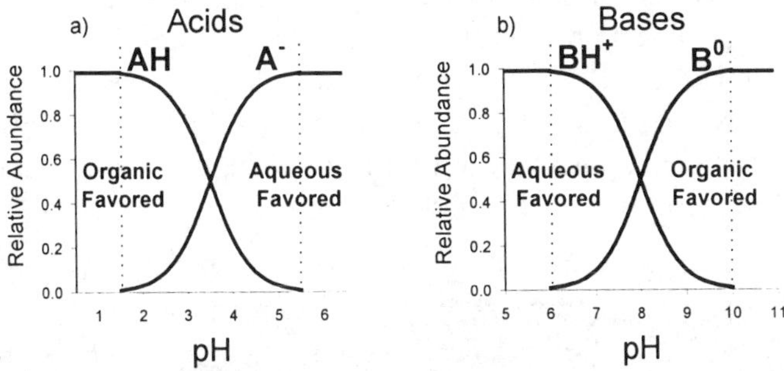

Figure 12.6- a) Relative abundance of acid and conjugate base at various pH values. b) Relative abundance of base and conjugate acid at various pH values. Favored regions for aqueous and organic extraction are labeled.

2. Determine Standardization Approach.

Chapter Ten detailed the criteria for selecting standardization approaches. The decision as to which approach is suitable for your problem is made during method development. If time permits, internal standardization is preferred for most methods in LC/MS. If labeled standards are available they are the best to use because they will behave exactly like your target. Chemical analogs are the next best approach if they behave in a similar manner to your target. For example, internal labeled (stable isotope) methods can even be used when no separation is implemented. As long as no spectral interference interferes with quantification this approach is perfectly valid. Internal standardization with chemical analogs is always used with a chromatographic separation to ensure the relative response between standard and target is constant. Complete *Worksheet 10.0* on choosing standardization approaches.

3. Determine Concentration Requirements.

Based on your initial target response experiments and the resulting figures-of-merit, you should be able to estimate the ability of your instrument to measure the prospective sample target concentrations. A factor of 60 has been stated as a good rule-of-thumb for sensitivity loss through sample treatment and separation compared to flow injection of neat target standard.[9] The basis for the factor-of-60 is a 10-fold loss for on-column injection, a 2-fold loss for recovery, and a 3-fold loss for system dispersion. With this rule of thumb, you can estimate the enrichment or concentration requirements of your method (see Table 12.4).

Table 12.4- Enrichment Requirements for LC/MS Method[9]

Sample (Lowest Level)	Neat Standard LOD (Step 2)	Factor-of-60	Method Enrichment Requirement
1 pg/mL	500 pg/mL	60	30,000
1 pg/mL	1 pg/mL	60	60
200 pg/mL	1 pg/mL	60	none needed

Enrichment is accomplished by a number of techniques. With liquid-liquid extraction one can extract a large volume into a small volume, dry down, and reconstitute into a smaller volume. This is generally labor intensive.

Alternatively, one can enrich significantly on-column; with either SPE or utilizing the analytical LC column. On-column enrichment is a balance of time and volume. Large volumes of sample can be concentrated on-column to gain significant enrichment, but you may have to wait for the sample to load onto the column. This can become a restriction when operating in the capillary or microbore flow regime. It may be better to enrich a large volume of sample onto an SPE cartridge at high flows (and short time-intervals) then subsequently introduce the enriched band of target analyte and standard onto a lower flow analytical column.

Method Development: Separations

There can be significant overlap between the sample preparation and analytical separations part of a method. In general, we would say that particulates and salts should be removed in the sample preparation stage of the analysis.

Separations are performed in LC/MS to increase the specificity of the method for identity based on a unique retention time, to increase the selectivity of the method by isolating the target from other sample components, and to enrich the analyte and thus enhance sensitivity.

The separation requirements for LC/MS will vary from zero to hour long separations. There are a number of well established methods for optimizing separations that are fully applicable to LC/MS. There are several excellent texts, including, *Practical HPLC Method Development* (2nd. edition) by L.R. Synder, J.J. Kirkland, J.L. Glajch (1997).[10,11] All of the principals of separation and optimization that apply to the general LC methods development, also apply in LC/MS.

There are significant pressures and incentives for method developers in LC/MS to shortcut or skip some of the requirements for LC separation because of the higher selectivity and specificity of LC/MS.[12] It is noteworthy to say that people with the least development time seem to have the most targets or suspects (drug metabolism, PK, n-in-1 dosing). Unfortunately, in the real world, time-consuming development protocols do not match the throughput demand for many applications.

The ability to use labeled internal standards gives the method developer the most rugged standardization approach known to chemistry. This standardization capability significantly reduces separation requirements to the point where no separation is justified in certain situations.[13]

The following is one strategy for developing a separation for LC/MS analysis. The depth and time that the method developer should devote to this topic will be highly dependent up the number of targets species and the available time.

1. Determine Separation Needs

a) Do you have to remove known interferences? Particularly isobaric or known suppressants. (Salts)
b) Do you need "improved selectivity" as insurance against undetermined interferences? (Ruggedness)
c) Do you need improved detection limits? (to present the detector with better peak shape, less noise, or narrower band.)
d) Do you need to concentrate large volume of analyte into a narrow band? (Factor-of-60, enrichment)
e) Do you need to change solvents after extraction? (Dry down organic, reconstitute in aqueous buffer for separation.)
f) Do you need high specificity in retention time? (e.g. multi-component mixture, containing structural isomers, etc.)
g) Do you need a two minute separation?

Select a separation strategy that will match your separation needs to your instrumentation. Most LC/MS developers will likely utilize an isocratic method with a reversed-phase column at flow regimes for 2 mm or 4.6 (3.9) mm I.D. columns, and electrospray (pneumatic assisted, ion spray) or APCI. Table 12.5 outlines some recommended practices for LC/MS separations.

2. Scouting Run of Your Targets & Standards

Run a gradient elution separation of your target and internal standards to scout their relative retention on a reversed-phase column. Typically this is accomplished with methanol and water gradients from 10% to 100% methanol. If your target analytes elute within a narrow band an isocratic method is most likely preferred with your method. To convert to isocratic conditions, simply note the mobile phase composition during the elution of your target band and subtract 2.5 times the void volume of your column to determine your approximate isocratic strength.[14,15] Adjust the solvent strength, composition, column selection, and buffer conditions to achieve required separation. Figure 12.7 illustrates the effects of pH on the elution and retention of acids and bases.

To this point in the discussion concerning method development, the method developer was required to consider pH three times; namely, for ionization/evaporation for optimization of the mass spectrometer response, for extraction, and now for separation. It is very important that the method developer have a clear understanding of the charge state of their targets and standards during each stage of the analysis.

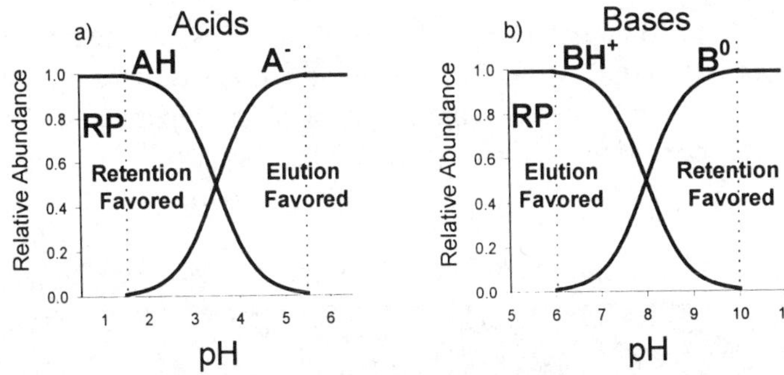

Figure 12.7- a) Relative abundance of acid and conjugate base at various pH values. b) Relative abundance of base and conjugate acid at various pH values. Favored regions for retention and elution are labeled.

3. Survey Gradient Run of Your Spiked Matrix

Once you have performed and adapted a separation to meet the requirements of your target and standards the matrix must be evaluated for interference effects on either target or standard. The same separation conditions that resulted from your target separation should be applied to the separation of spiked matrix standards. Under suitable separation conditions for your targets you must determine if your target or standard response is degraded by either:

1) isobaric interferences at your target or analog quantification masses, or
2) chemical interferences that suppress (or enhance) the response of your target or analog standard quantification.

One of the major incentives for methods development in LC/MS is *shortened runtime* derived from higher selectivity and specificity. Many of the current methods are operating at a k' of 2 to 5. Under these conditions the peaks are quite narrow and most of the sample salts have passed through the column. Care must be taken in this accelerated world to ensure that the separation is rugged. This generally means running a significant number of different matrix standards in the methods development process. Some laboratories reported running up to 15 different matrix standards in order to fully characterize interferences.[16]

If the scouting separation of spiked matrices has adequately removed interferences, convert the method to isocratic conditions. If not, adjust the solvent strength, composition, column selection, and buffer conditions to achieve required separation.

Table 12.5- Recommended Practices for LC/MS Separations

LC Parameter	Recommendation
Buffers	Use volatile buffers or membrane suppression to remove involatile buffers.
Solvents	Use clean and filtered solvents. Low residue preferred.
Plumbing	Minimize extra-column dead volume. Use minimum tubing lengths, minimum I.D. (0.003 - 0.005-in), and precut tubing. Minimize the number of unions.
Columns	Use guard columns. Low bleed columns are now available for LC/MS. Low flow is always better for vacuum applications. Microbore columns are a practical dimension.
Detectors	High pressure cells are required for any back pressure technique such as thermospray, particle beam, and APCI. Ancillary detector is very useful in methods development when screening for interferences.

12.3-Validation- It ain't over till....

At the beginning of this chapter we mentioned that we were going to leave method validation up to you because it is so dependent on your individual circumstances. But a few words should be said about validation. A method is not fully developed until it has been validated. The validation process is a critical and culminating part of the method development process. Your final published method to a large extent will be bounded by the results of your validation, so in a real sense the method is the validation. All the effort in method development may be forgotten while the results and utility of the final implementation are cemented in time by the validation process.

Validation is also complicated by the lack of formal guidelines.[17] Fortunately, Ira Krull and Michael Swartz have recently completed an excellent text entitled, *Analytical Method Development and Validation* which should be a mandatory reference for every methods development lab. One important point that the authors emphasize is that validation is not confined to only the LC or LC/MS method. Software, hardware, method, and system suitability are all requirements for validation. This is of particular importance to the method developer when they have to deal with sample complexities, multiple components, integration, timing, and data reduction issues relating to LC/MS.

Although, there exists a general lack of validation guidelines, several industry-specific organizations or agencies have taken the initiative to present guidelines that the developers may use. The International Conference Harmonization (ICH) has put forth a number of documents in various stages of review and acceptance. Among these are ICH-Q2A: *Test on Validation of Analytical Procedures*[18] and ICH-Q2B: *Validation of Analytical Procedures: Methodology.*[19] In addition, the Environmental Protection Agency (EPA) has provided guidelines for method development and validation for the *Resource Conservation and Recovery Act.*[20] A summary of these guidelines is provided in Table 12.6.

Table 12.6- Guidelines for Method Validation.

Parameter	ICH	EPA[21]
Specificity	• suitable identification test • assay and impurity test	• prescribed by the method
Linearity	• a minimum of 5 concentrations is recommended.	• 30%
Range	• 80-120 % of the test concentration • for content uniformity, 70-130% of the test concentration	• as prescribed by the method
Accuracy	• should be established across the specified range of the analytical procedure	• 70-130% and the RSD should be <30%
Precision	• a minimum of 9 determinations covering the specified range for the procedure • a minimum of 6 determinations at 100% of the test concentration	• RSD should be <30%
Detection Limit	Based on any of the following: • visual evaluation • signal-to-noise; between 3 to 2:1 • standard deviation of the response and the slope, =3.3 sigma/slope • standard deviation of the blank • the calibration curve	• Method Detection Limit (MDL)[22] MDL = tS t = student t value, S = SD of the replicates
Quantitation Limit	Based on any of the following: • visual evaluation • signal-to-noise; 10:1 • standard deviation of the response and the slope; = 10 sigma/Slope • standard deviation of the blank • the calibration curve	Based on any of the following: • prescribed by the method • mass spectral match
Robustness	• influence of variations of pH in a mobile phase • variations in mobile phase • different column lots • temperature • liquid flow rate	Based on any of the following: • prescribed by the method • control charts

12.4- LC/MS vs. LC/UV

Methods development in LC has been the subject of extensive analysis in several excellent reviews[23] and books.[24,25] In general, the approach to development in LC has focused on developing adequate method selectivity and specificity through adjustment and optimization of separation parameters (e.g. column, solvent strength, pH, temperature, etc.). This approach is partly due to the lack of compound specificity associated with most conventional LC detectors.

In contrast, the high compound specificity and selectivity associated with mass spectrometry leads methods developers in LC/MS to begin their development process with a much higher emphasis on the mass analysis capabilities and less emphasis on the separation capabilities of their system. Table 12.7 illustrates some of the important differences between LC and LC/MS when considering methods development. These differences can lessen the requirement for developing highly selective and efficient separations with LC/MS. They can also lead to shorter sample run times without compromising reliability with LC/MS. In addition, these differences can also allow shorter method development times with LC/MS. It should be noted, however, that most of the method development protocols for LC are completely applicable to LC/MS.

In the extreme, some screening methods in LC/MS require no chromatographic separation at all.[26,27] Adequate specificity and selectivity is achieved with the use of direct introduction of sample into an LC/MS interface and subsequent mass analysis. Critics can correctly argue that this is not truly LC/MS; however, we prefer to think of problem-solving in terms of the quality of the results, not in terms of what components are stuck together to get the desired result. We consider running an LC/MS interface without a column in-line as simply one limit or boundary of your many alternatives in the field of LC/MS. Note that columnless LC/MS is the exception, not the rule. Most analysts require some separation to achieve the desired selectivity and ruggedness in their methods. You will too.

Table 12.7- Method Development: LC/UV vs. LC/MS.

Criteria	LC	LC/MS
Method Specificity	Most of the specificity of an LC method derives from the separation.	Most of the specificity of LC/MS derives from the mass analyzer.
Method Selectivity	Most of the selectivity of an LC method derives from the separation.	Selectivity derives from both the mass analyzer and the separation in LC/MS.
Quantification	Baseline separations are required for reliable quantification.	No separation is required; some separation is recommended for improved reliability of quantification.
Standardization	LC method usually use external standardization.	LC/MS methods usually use internal standardization.
Calibration Curves	Single point calibration is commonplace in LC/UV.	Single point calibration is seldom used in LC/MS.
Buffers	Buffers are generally added to the sample to suppress ionization and enhance solubility.	Buffers can have a variety of functions; including, • ionization suppression for vaporization, • ionization enhancement to promote desorption in ES.
Effects from Chemical Interferences	Baseline separations are required to isolate analytes from interferences.	• Labeled internal standards account for effects from interferences. • Baseline separations are an alternative to removing chemical interferences.

Discussion Questions and Exercises

1. Discuss the difference in method development approach between a single target method (such as a clinical assay) and a broad-based method (such as an environmental method)?
2. Name five types of interference that can be associated with a sample matrix?
3. How would you remove the interferences from your method?
4. How would you prove they were gone?
5. Why is method development in LC/MS reported to be as much as ten times faster than conventional LC/UV method development.
6. In your own words, what is the purpose of the validation process?
7. What would be the APCI operating conditions you would choose for a target compound with a pK_a of 4.0?
8. What would be the ES operation condition you would choose for a target compound with pK_a of 4.0?
9. How would you extract this target compound from a tissue sample?
10. Propose a extraction, separation, and detection scheme for this compound.

Notes

[1] This is part of the *Introduction* to the FY goals of the National Center for Toxicological Research (NCTR) for 1997 authored by Director of Chemistry Harold C. Thompson dated February 17,1997. This statement is an excellent summary of the scope and incentive behind the methods development and validation processes within the regulatory jurisdiction of the FDA (see Internet site: `www.fda.gov/nctr/resplan/chem.html`)

[2] Willoughby, R. and Sheehan, E., "Fundamentals of Electrospray," Presented at the 14th Montreux Symposium on LC/MS, Cornell University, Ithaca, NY, USA, July 23-25, 1997.

[3] Stevenson, R., "PBA '97, The era of triple-stage rockets," American Biotechnology Laboratory 15 (December 1997). It was reported that Prof. Norbit Lamers of Eindhoven Univeristy (Eindhoven, The Netherlands) and Dr. Alexander J. Debets of N.V. Organnon (Oss, The Netherlands) used a modified electrosuppression device (from Dionex) to

continuously remove bad ions (e.g., phosphate) from the sample stream after separation. This improve the detection of basic drugs with MS and UV detection

[4] Tris(hydroxymethyl)aminomethane

[5] van Berkel, G.J., Zhou, F., Aronson, J.T., "Changes in bulk pH caused by the inherent controlled-current electrolytic process of an electrospray ion source," Int. J. Mass Spectrom. Ion Processes 162, pages 55-67 (1997).

[6] Mann, M., Fenn, J.B., "Chapter 1, Electrospray Mass Spectrometry, Principles and Methods," pages 1-35, IN: *Mass spectrometry, Clinical and Biomedical Applications, Volume 1*, Desiderio, D.M. (Ed.), Plenum Press: New York (1992).

[7] Turbo IonSpray is a tradename of PE/Sciex for a high temperature version of their proprietary Ion Spray technology. By heating their spray to 600°C the ion generation process is enhanced by the rapid evaporation and an order of magnitude signal enhancement has been observed.

[8] Majors, R., "Guide to Sample Preparation," a special project supplement published by LC-GC magazine (1997). This flow diagram was adapted for LC/MS from the general flow diagrams developed for the column: Sample Prep Perspectives, column editor Ronald E. Majors. This reference is an excellent overview of sample preparation and should be required reading for all analysts.

[9] Lowes, S., "LC/MS/MS Training," Presented at the 14th Montreux Symposium on LC/MS, Cornell University, Ithaca, NY, USA, July 23-25, 1997. Dr. Lowes presented an excellent overview of the method development protocol used in one of the top LC/MS labs in the world, Advanced BioAnalytical Services. Their broad experience is the foundation of the factor-of-60 rule.

[10] McMaster, M.C., *HPLC- A Practical Users Guide*, UCH Publishers, New York, page 20 (1994).

[11] Snyder, L.R., Kirkland, J.J., Glajch, J.L., *Practical HPLC Method Development*, John Wiley & Sons: New York (1997). The most recent comprehensive text on LC methods development.

[12] Stevenson, R., "The World of Separation Science: HPLC '96 puts the spotlight on new detectors and columns; MS continues to evolve," American Laboratory, News Edition 29, pages 4-8 (1997). "...it takes about two to four months to develop and validate an HPLC method. If this is combined with an MS detector the same method requires only three to four weeks. ... using LC/MS/MS reduces the method development time to two weeks or less."

[13] Bowers, G.D., Clegg, C.P., Hughes, S.C., Harker, A.J., Lambert, S., "Automated SPE and tandem MS without HPLC columns for quantifying drugs at the picogram level," LC-GC 15, pages 48-53 (1997).

[14] Snyder, L.R., Kirkland, J.J., "Chapter Sixteen, Gradient Elution and Related Procedures," pages 662-719, IN: *Introduction to Modern Liquid Chromatography* (2 ed.), John Wiley and Sons: New York (1979).

[15] Kirkland, J.J., "Practical method development strategy for reversed-phase HPLC of ionizable compounds," pages 486-500, LC-GC 14, (1996). Develops a strategy to handle the separation of acids and bases.

[16] Reported during general discussions at the 14th Montreux Symposium on LC/MS, Cornell University, Ithaca, NY, USA, July 23-25, 1997.

[17] Krull, I., Swartz, M., "Validation Viewpoint: Introduction: National and international guidelines," LC-GC 15, pages 534-540 (1997).

[18] ICH-Q2A: *Test on Validation of Analytical Procedures.* (see the Internet site: `www.ifpma.org/ich5q.html`)

[19] ICH-Q2B: *Validation of Analytical Procedures: Methodology.* (see the Internet site: `www.ifpma.org/ich5q.html`)

[20] Environmental Protection Agency (EPA) has provided guidelines for method development and validation for the *Resource Conservation and Recovery Act* (see the Internet site: `www.epa.gov`).

[21] Methods for the Determination of Organic Compounds in Drinking Water (Supplement II) "Method 553. Determination of benzidine and nitrogen-containing pesticides in water by liquid-liquid extraction or liquid-solid extraction and reverse phase high pressure liquid chromatography/particle beam/mass spectrometry".

[22] MDL: is defined as the statistically calculated minimum amount that can be measured with 99% confidence that the reported value is greater then zero.

[23] Kirkland, J.J., "Practical method development strategy for reversed-phase HPLC of ionizable compounds," pages 486-500, LC-GC 14, (1996). Develops a strategy to handle the separation of acids and bases.

[24] Snyder, L.R., *et al., op cit.*

[25] McMaster, M.C., *op cit.*

[26] Millington, D.S., Chace, D.H., Stevens, R.D., "Rapid, automated analysis of diagnostically important metabolites in physiological matrices by LC/ESI-MS/MS," Proceedings of the 12th LC/MS Montreux Symposium, Hilton Head Island, SC, November 1-2, 1995.

[27] Bowers, G.D., *et al., op cit.*

Appendix A- Flow Diagram

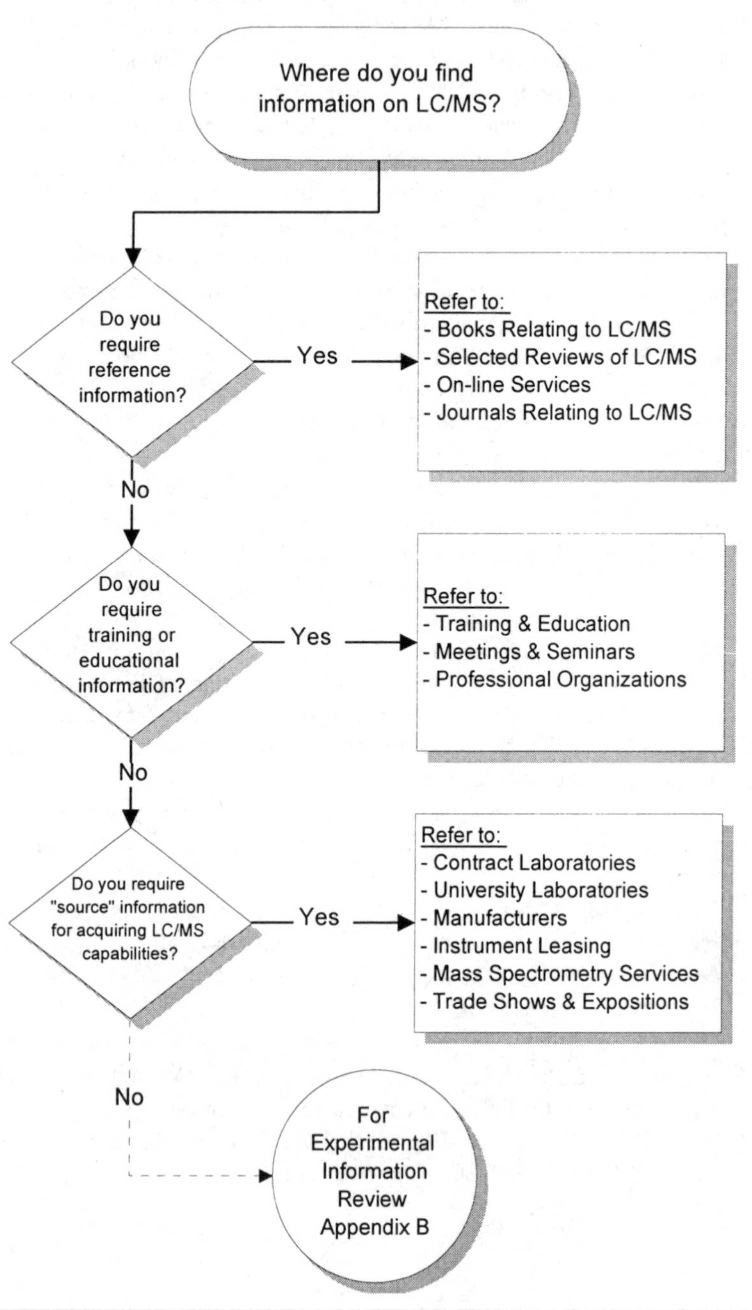

Appendix A

Sources for LC/MS Information

(or, Resources for LC/MS)

There are many ways to profit from knowledge. Some are very lofty & some rather mundane. Some come from experience & some necessitate study. Some require an understanding of how to deal with people & some of how to deal with things.

Stan Davis & Jim Botkin
in ***The Monster Under the Bed***[1]

Resources: Where do you find them?

Successful problem-solving in LC/MS starts with accessing and gathering information related to your problem and your needs. Appendix A is intended to provide you with a comprehensive listing of sources and resources for any information relating to LC/MS. We provide lists of books, reviews, on-line services, journals, meetings, manufacturers, contract labs, and much more. We also concede that the printed medium is ineffective at keeping up with changes in this type of information. To address this dilemma, we have created an Internet site that will continually be updating and expanding this information.

We have, in some cases, placed editorial comments about specific sources of information to allow the reader to better evaluate it's utility. We apologize ahead of time for the small font, since we have made an effort to keep this text to a manageable size.

Resources: Books Relating to LC/MS

Concerning LC/MS

Applications of LC/MS in Environmental Chemistry
edited by D. Barcelo. Journal of Chromatography Library Volume 59.
Elsevier Science: Amsterdam (1996).

Biochemical and Biotechnological Applications of Electrospray Ionization Mass Spectrometry
edited by A. Peter Snyder provides an application driven approach in areas of immunology, biochemistry & natural toxins.
ACS: Washington, D.C. (1996).

Electrospray Ionization Mass Spectrometry: Fundamentals, Instrumentation, and Applications
edited by Richard B. Cole presents a wide view of the current knowledge of the basic aspects of electrospray ionization mass spectrometry.
John Wiley: New York (1997).

Global View of LC/MS: How to Solve Your Most Challenging Analytical Problems
by Ross Willoughby, Edward Sheehan & Sam Mitrovich provides a problem solving guide to all aspects of LC/MS.
Global Publishing: Pittsburgh (1998).

Liquid Chromatography-Mass Spectrometry: Applications in Agricultural, Pharmaceutical, and Environmental Chemistry
edited by Mark A. Brown is a collection of seventeen applications dealing with agricultural chemicals; pharmaceuticals & metabolism; & environmental analysis presented at the 197th National Meeting of the American Chemical Society, 1989.
ACS: Washington, D.C. (1990).

Liquid Chromatography-Mass Spectrometry: Principles and Applications
by W.M.A. Niessen & J. Van der Greef focuses on principles & strategies of interfacing a liquid chromatographic system to a mass spectrometer.
Marcel Dekker: New York (1992).

Liquid Chromatography-Mass Spectrometry: Techniques and Applications
by Alferd L. Yergey, Charles G. Edmonds, Ivor A.S. Lewis, & Mervin L. Vestal describes the techniques of primarily thermospray with some mention of particle beam, APCI, MALDI, & electrospray. Approximately 30% of the book is dedicated to an exhaustive bibliographic listing of LC/MS interfaces, apparatus, applications, patents, & reviews up to 1988.
Plenum Press: New York (1990).

Tandem Techniques
by Raymond P.W. Scott describes the techniques of LC/MS & CE/MS in: Chapter 9, Liquid Chromatography-Mass Spectrometry (LC/MS) Tandem Systems (pages 329-404) & Chapter 13, Electrophoresis-Spectrosc. Tandem Syst. (pages 483-516).
John Wiley & Sons: New York (1997).

Related to LC/MS

Analysis of Neuropeptides by Liquid Chromatography and Mass Spectrometry
by Dominic M. Desiderio is the 6th volume in the series Techniques & Instrumentation in Analytic Chemistry.
Elsevier Science Pub.: Amsterdam (1984).

Analytical Method Development and Validation
by Michael E. Swartz & Ira S. Krull present method validation protocol (USP & the International Conference on Harmonization), system suitability, method development, & optimization, & validation approaches.
Marcel Dekker: New York (1997).

Biological Mass Spectrometry
edited by A.L. Burlingame & James A. McCloskey is a collection of topics presented at the Second international Symposium on Mass Spectrometry in the Health & Life Sciences, San Francisco, CA, August 27-31, 1989.
Elsevier Science Pub.: Amsterdam (1990).

Biological Mass Spectrometry Present and Future
edited by Takekiyo Matsua, Richard M. Caprioli, Michael L. Gross, & Yousuke Seyama is a collection of papers from the

Kyoto '92 International Conference on Biological Mass Spectrometry.
John Wiley & Sons: Chester (1994).

Forensic Applications of Mass Spectrom.
by Jehuda Yinon covers the applications of mass spec. techniques to forensic analyses.
CRC Press: Boca Raton (1995).

Guide to Mass Spectrometry
by Kenneth L. Busch & Thomas A. Lehman is an encyclopedic handbook that provides comprehensive listings for all concepts, techniques & terminology used in mass spectrometry.
VCH Publishers: New York (1997).

Mass Spectrometry
edited by A.M. Lawson describes the techniques of mass spectrometer used in clinical biochemistry.
Walter de Gruyter: Berlin (1989).

Mass Spectrometry of Biological Materials
edited by Charles N. McEwen & Barbara S. Larsen contains papers on the subject of mass spectrometry of proteins & other biological matrices.
Marcel Dekker: New York (1993).

Mass Spectrometry in the Biological Sciences: A Tutorial
edited by Michael L. Gross contains papers describing instruments, methods & applications presented at the meeting NATO Advanced Study Institute on Mass Spectrometry in the Molecular Sciences, Cetraro, Italy, June, 1990.
Kluwer Academic: Dordrecht (1992).

Mass Spectrometry for Biotechnology
by Gary Siuzdak describes mass spectrometers, ionization, detection & selected applic.
Academic Press: San Diego (1996).

Mass Spectrometry for the Characterization of Microorganism
edited by Catherine Fenselau.
ACS: Washington, D.C. (1994).

Mass Spectrometry, Clinical Biomedical Applications Volume 1 and 2
edited by Dominica M. Desiderio describes the techniques of mass spectrometry, LC/MS interfaces & their applications.
Volume 1 (1992) & Volume 2 (1994).
Plenum Press: New York.

Mass Spectrometry: Principles and Applications
by Edmond de Hoffman, Jean Charette, & Vincent Stroobant provides an introduction to modern mass spectrometry for students & scientists. Wiley: Chichester (1996).

Methods in Enzymology, Volume 193: Mass Spectrometry
edited by James A. McCloskey is a review of LC/MS up to 1990 describing the techniques of mass spectrometry: chemical ionization, TS, flow-FAB; & their applns.
Academic Press: San Diego (1990).

Methods in Enzymology, Volume 270: High Resolution Separation and Analysis of Biological Macromolecules, Part A Fundamentals
edited by Barry L. Karger & William Hancock is an up to date review of the technologies of liquid chromatography & electrophoresis combined with mass spectrom. Academic Press: San Diego (1996).

Methods in Enzymology, Volume 271: High Resolution Separation and Analysis of Biological Macromolecules, Part B Applications
edited by Barry L. Karger & William Hancock is an up to date review of applications using the techniques of liquid chromatography & electrophoresis combined with mass spectrom.
Academic Press: San Diego (1996).

Time of Flight Mass Spectrometry
edited by Robert J. Cotter is a collection of twelve papers dealing with the history, several applications, description of basic principles, electrospray & MALDI time of flight mass spectrometry.
ACS: Washington, D.C. (1990).

Time-of-Flight Mass Spectrometry: Instrumentation and Applications in Biological Research
by Robert J. Cotter presents the basic principls of time-of-flight mass spectrometry with a strong emphasis on applications in biological reasearch, peptides/proteins, oligonucleotides, and other biological macromolecules. ACS Professional Reference Books: Washington, D.C. (1997).

Chromatography

Chromatographic Separations Based on Molecular Recognition
edited by Kiyokatsu Jinno provides an in-depth understanding of the molecular recognition mechanism as implemented for chromatographic separation.
VCH Publishers: New York (1996).

Handbook of Capillary Electrophoresis
by James P. Landers pulls together diverse area & applications of CE for analyst involved in biotechnology, clinical chemistry, pharmaceutical, biochemistry, & chemical sciences. CRC Press: Boca Raton (1996).

High Performance Liquid Chromatography: Principles and Methods in Biotechnology
edited by Elena Katz describes various aspects of HPLC such as: principles of separation, column selection, sample preparation in biochromatography.
John Wiley & Sons: New York (1996).

HPLC Methods on Drug Analysis
by M.K. Ghosh describes more than 1,000 HPLC methods for 650 common drugs.
Springer-Verlag: New York (1992).

Introduction to Modern Liquid Chromatography
by L.R. Snyder & J.J. Kirkland covers the state of LC up to 1979.
John Wiley & Sons: New York (1979).

Practical HPLC Method Development
(2nd. edition) by L.R. Synder, J.J. Kirkland, J.L. Glajch present up-to-date insights to choosing seaparation conditions.
John Wiley & Sons: New York (1997).

Mass Spectrometry

A Guide to Materials Characterization and Chemical Analysis
edited by John P. Sibilia describes the essentials of analytical techniques for the student, researcher, & business manager that are needed to evaluate the advantages & disadvantages of various techniques.
VCH Publishers: New York (1996).

Interpretation of Mass Spectra
by Fred W. McLafferty & Frantisek Turecek is in its forth edition. This book is "the handbook" on the interpretation of mass spectra. A must for any mass spec. lab.
University Science Books: Mill Valley (1993).

Interpretation of Mass Spectra of Organic Compounds
by Mynard C. Hamming & Norman G. Foster. Academic Press (1972).

Mass Spectrometry, Applications in Science and Engineering
by F. A. White and G. M. Wood.
John Wiley & Sons: New York (1986).

Practical Aspects of Ion Trap Mass Spectrometry: Volume I: Fundamentals of Ion Trap Mass Spectrometry; Volume II: Ion Trap Instrumentation; Volume III: Chemical, Environmental and Biochemical Applications.
Edited by R. E. March & J. F.J. Todd.
Volume 1, Volume 2 & Volume 3.
CRC Press: Boca Raton (1995).

Practical Organic Mass Spectrometry: A Guide for Chemical and Biochemical Analysis
(2nd edition) by J.R. Chapman covers the topics of mass spectrometry up to 1995.
John Wiley & Sons: New York (1995).

Quadrupole Mass Spectrometry and Its Applications
edited by Peter H. Dawson.
American Institute of Physics Press: Woodbury (1995).

Quadrupole Storage Mass Spectrometry
by Raymond E. March & Richard J. Hughes covers the historical development of the ion trap, theory & a description of first commercially available ion traps.
John Wiley & Sons: New York (1989).

Quantitative Mass Spectrometry
by Brian Milliard covers the details of quantatitive analsysis in mass spectrometry.
Heyen: London (1978).

Tandem Mass Spectrometry
edited by Fred W. McLafferty is a collection of topics dealing with tandem mass spectrometry (MS/MS).
John Wiley & Sons: New York (1983).

Techniques of Combining Gas Chromatography-Mass Spectrometry: Applications in Organic Analysis

by William McFadden is a comprehension book on the state of the art of gas chromatography-mass spectrometry in the early 1970's that are still pertinent for students of analytical chemistry.
John Wiley & Sons: New York (1973).

Trace Analysis by Mass Spectrometry
edited by Authur J. Ahearn is a collection of topics covering the state of the art of trace analysis up to 1972.
Academic Press: New York (1972).

Zen and the Art of Motorcycle Maintenance
by Robert M. Pirsig is a must for trouble-shooting and maintenance of instruments (and motorcycles).
William Morrow: New York (1974).

Collections of Spectra

The Wiley Static SIMS Library
By J. C. Vickerman and D. Briggs.
Wiley: New York (1996).

Identification of Essential Oil Compounds by Gas Chromatography/Mass Spectrometry by R.P. Adams.
Allured: Carol Stream (1995).

Instrumental Data for Drug Analysis
by T. Mills, III. Volumes 1-7.
(originally published by Elsevier)
CRC: Boca Raton (1993).

A Mass Spectral and GC Data of Drugs, Pollutants, Pesticides and Metabolites
by K. Pfleger, W.W. Weber.
VCH: New York (1992).

CRC Handbook of Mass Spectra of Environmental Contaminants by R.A. Hites.
CRC: Boca Raton (1992).

Important Peak Index of the Registry of Mass Spectral Data
by F.W. McLafferty and D.B. Stauffer.
Wiley: New York (1991).

The Eight Peak Index of Mass Spectra
The Royal Society of Chemistry: Cambridge (1991).

The Wiley/NBS Registry of MS Data
by F.W. McLafferty and D.B. Stauffer.
Wiley: New York (1989).

Electron Capture Negative Ion Mass Spectra of Environmental Contaminants and Related compounds
by E.A. Stemmler and R.A, Hites.
VCH: New York (1988).

EPA/NIH Mass Spectral Data Base, Supplement 2
by S.R. Heller, G.W.A. Milne and L.H. Gevantman
National Standard Reference Data System, NBS, Dept. of Commerce, USA (1983)

EPA/NIH Mass Spectral Data Base, 4 Volumes (1978); Supplement 1 (1980)
by S.R. Heller, and G.W.A. Milne
National Standard Reference Data System, NBS, Dept. of Commerce, USA.

Compilation of Mass Spectral Data
by A. Cornu, Heyden: Philadelphia (1975).

Laboratory Management

Developing a Chemical Hygiene Plan
by Jay A. Young, Warren K. Kingsley & George H. Wahl, Jr. outlines what you need to know to comply with the
ACS Publications: Washington, DC (1990).

Good Laboratory Practice Standards: Applications for Field and Laboratory Studies
edited by Willa Y. Garner, Maureen S. Barge & James P. Ussary give concrete ideas for establishing a compliance program to meet the EPA's Good Laboratory Practice Standards regulations.
ACS Publications: Washington, DC (1992).

The Internet: A Guide for Chemists
edited by Steven M. Bachrach presents the basics of the Internet along with instruction on becoming an information provider.
ACS Publications: Washington, DC (1996).

Quality in the Analytical Laboratory
by F. Elizabeth Prichard introduces the reader to the concept of quality assurance.
John Wiley & Sons: New York (1995).

Successful Management of the Analytical Laboratory
by Oscar I. Milner reviews the operation of analytical laboratory from the standpoint of the manager's responsibilities.
Lewis Publishers: Boca Raton (1992).

Writing and Designing Manuals
by Gretchen H. Schoff & Patricia A. Robinson is a practical guide to technical publication writing & design.
CRC Press: Boca Raton (1991).

Writing the Laboratory Notebook
by Howard M. Kanare covers the technique of how to keep a proper & permanent laboratory notebook.
ACS Publications: Washington, DC (1985).

Laboratory Safety

American Chemical Society
offers several videos. Contact ACS Media Office at 1-800-227-5558 or 1-202-872-6311, fax inquires to 1-202-872-6336 or check out their home page at www.chemcenter.org. For address see professional organizations below.

Chemical Health & Safety
a bimonthly magazine published jointly by the American Chemical Society & Division of Chemical Health & Safety. For a free issue contact American Chemical Society (see professional organizations below) or send an e-mail to p_commodore@acs.org.

CRC Handbook of Laboratory Safety
(3rd edition) by A. Keith Furr is a comprehensive reference to guide the laboratory personnel in working with safety & health professional to implement effective health & safety programs in their facilities.
CRC Press: Boca Raton (1990).

Handbook of Lab. Health and Safety
(2nd edition) by R. Scott Stricoff & Douglas B. Walters presents a feasible, easy to implement approach to providing a safe workplace.
John Wiley & Sons: New York (1995).

Improving Safety in the Chemical Laboratory: A Practical Guide
(2nd edition) edited by Jay A. Young.
John Wiley & Sons: New York (1991).

The Laboratory Environment
edited by Rupert Purchase deals with activities & communication needed to provide a safe environment for personnel working in a laboratory.
The Royal Society of Chemistry: Cambridge (1994).

Laboratory Health and Safety Handbook
by R. Scott Stricoff & Douglas B. Walters.
John Wiley & Sons: New York (1990).

Learning by Accident
published by the Laboratory Safety Workshop. E-mail: labsafe@aol.com.
Laboratory Safety Workshop: Bnatick (1997).

Prudent Practices in the Laboratory: Handling and Disposal of Chemicals
published by the National Research Council, For information telephone: 1-800-624-6242. National Research Council (1995).

Resources: Selected Reviews of LC/MS

There are numerous reviews of the various technologies in (or related to) LC/MS. The following is a list of excellent reviews that cover most of the important areas of LC/MS; including, fundamentals & technology overviews, applications overviews, & instrument availability.

LC/MS Applications

LC/MS Update and Soft Ionization
(a bimonthly international journal) by HD Science, 242 N. James Street, Tower Office Park, Suite 202, Newport, Wilmington, DE 19804-3168, USA, Phone: 302-994-8026, Fax: 302-994-8837. Home page: www.hdscience.com/hdupdate.htm#update A good up-to-date comprehensive review of applications listing interfaces, liquid chromatography, & mass spectrometer used by the authors.

ES: Technology

Capillary Liquid Chromatography Mass Spectrometry
by Tomer, K.B., Moseley, M.A., Deterding, L.J., Parker, C.E., Mass Spectrom. Rev. 13, pages 431-457 (1994).

Development of the Electrospray Ionization Technique
by Hamdon, M., Curcuruto, O., Int. J. Mass Spectrom. Ion Processes 108, pages 93-113 (1991).

Electrospray and Ionspray; Chapter 10, In: Liquid Chromatography-Mass Spectrometry: Principles and Applications
by Niessen, W.M.A., van der Greef, J., pages 229-245, Marcel Dekker: New York (1992).

Electrospray Ionization-Principles and Practice
by Fenn, J.B., Mann, M., Meng, Ch.K., Wong, Sh.F., Whitehouse, C.M., Mass Spectrom. Rev. 9, pages 37-70 (1990).

Interpreting Mass Spectra of Multiplied Charged Ions
by Mann, M., Meng, C.K., Fenn, J.B., Anal. Chem. 61, pages 1702-1708 (1989).

Liquid Chromatography-Mass Spectrometry and Related Techniques via Atmospheric Pressure Ionization
by Wachs, T., Conboy, J.C., Garicia, F., Henion, J.D., J. Chromatogr. Sci. 29, pages 357-366 (1991).

Liquid Chromatography-Mass Spectrometry. General Principles and Instrumentation
by Niessen, W.M.A., Tinke, A.P., J. Chromatogr. A 703, pages 37-57 (1995).

Mass Spectrometry with Ion Sources Operating at Atmospheric Pressure
by Bruins, A.P., Mass Spectrom. Rev. 10, pages 53-77 (1991).

Multiple Charging in Electrospray Ionization of Poly(ethylene glycols)
by Wong, S.F., Meng, C.K., Fenn, J.B., J. Phys. Chem. 92, pages 546-550 (1988).

ES: Applications

Electrospray: Its Potential and Limitations as an Ionization Method for Biomolecules
by Mann, M., Org. Mass Spectrom. 25, pages 575-587 (1990).

Electrospray and Ionspray; Chapter 10, In: Liquid Chromatography-Mass Spectrometry: Principles and Applications
by Niessen, W.M.A., van der Greef, J., pages 229-245, Marcel Dekker: New York (1992).

Electrospray Ionization for Mass Spectrometry of Large Biomolecules
by Fenn, J.B., Mann, M., Meng, Ch.K., Wong, Sh.F., Whitehouse, C.M., Science 246, pages 64-71 (1989).

New Developments in Biochemical Mass Spectrometry: Electrospray ionization
by Smith, R.D., Loo, J.A., Edmonds, Ch. G., Barinaga, Ch. J., Udseth, H.R., Anal. Chem. 62, pages 882-899 (1990).

Principles and Practice of Electrospray Ionization-Mass Spectrometry for Large Polypeptides and Proteins
by Smith, R.D., Loo, J.A., Ogorzalek Loo, R.R., Busman, M., Udseth, H.R., Mass Spectrom. Rev. 10, pages 359-451 (1991).

ES: Instrumentation

Electrospray Continues to Evolve
by Henry, C., Anal. Chem. 69, pages 427A-432A (1997).

Electrospray MS: Charging Full Stream Ahead
by Voress, L., Anal. Chem. 66, pages 481A-486A (1995).

APCI: Technology

Atmospheric Pressure Ionization Mass Spectrometry
by Carroll, D.I., Dzidic, I., Horning, E.C., Stillwell, R.N., Appl. Spectrosc. Rev. 17, pages 337-406 (1981).

Atmospheric Pressure Ionization Mass Spectrometry: Studies of Negative Ion Formation for Detection and Quantification Purposes
by Horning, E.C., Carroll, D.I., Dzidic, I., Lin, S.N., Stillwell, R.N., Thenot, J.P., J. Chromatogr. 142, pages 481-495 (1977).

Atmospheric Pressure Ionization Mass Spectrometry: Corona Discharge Ion Source for Use in Liquid Chromatograph-Mass Spectrometer-Computer Analytical System
by Carroll, D.I., Dzidic, I., Stillwell, R.N., Haegele, K.D., Horning, E.C., Anal. Chem. 47, pages 2369-2373 (1975).

Atmospheric Pressure Ionization (API) Mass Spectrometry. Solvent Mediated Ionization of Samples Introduced in Solution and in a Liquid Chromatograph Effluent System
by Horning, E.C., Carroll, D.I., Dzidic, I., Haegele, K.D., Horning, M.G., Stillwell, R.N., J. Chromatogr. Sci. 12, pages 725-729 (1974).

Atmospheric Pressure Ionization Mass Spectrometry II: Applications in Pharmacy, Biochemistry and General Chemistry
by Bruins, A.P., Trends Anal. Chem. 13, pages 81-90 (1994).

Chemical Ionization in LC-MS; Chapter 14, In: Liquid Chromatography-Mass Spectrometry: Principles and Applications
by Niessen, W.M.A., van der Greef, J., pages 311-350, Marcel Dekker: New York (1992).

Determination of Sulfa Drugs in Biological Fluids by Liquid Chromatography/Mass Spectrometry/Mass Spectrometry
by Henion, J.D., Thomson, B.A., Dawson, P.H., Anal. Chem. 54, pages 451-456 (1982).

LC-MS Interfacing: A General Overview; Chapter 4, In: Liquid Chromatography-Mass Spectrometry: Principles and Applications
by Niessen, W.M.A., van der Greef, J., pages 81-115, Marcel Dekker: New York (1992).

Liquid Chromatography-Mass Spectrometer-Computer Analytical Systems: A Continuous Flow System Based on Atmospheric Pressure Ionization Mass Spectrometry
by Horning, E.C., Carroll, D.I., Dzidic, I., Haegele, K.D., Horning, M.G., Stillwell, R.N., J. Chromatogr. 99, pages 13-21 (1974).

Liquid Chromatography-Mass Spectrometry, General Principles and Instrumentation
by Niessen, W.M.A., Tinke, A.P., J. Chromatogr. A 703, pages 37-57 (1995).

Liquid Chromatography-Mass Spectrometry and Related Techniques via Atmospheric Pressure Ionization
by Wachs, T., Conboy, J.C., Garicia, F., Henion, J.D., J. Chromatogr. Sci. 29, pages 357-366 (1991).

Mass Spectrometry with Ion Sources Operating at Atmospheric Pressure
by Bruins, A.P., Mass Spectrom. Rev. 10, pages 53-77 (1991).

APCI: Applications

Atmospheric Pressure Ionization Mass Spectrometry II: Applications in Pharmacy, Biochemistry and General Chemistry by Bruins, A.P., Trends Anal. Chem. 13, pages 81-90 (1994).

Atmospheric Pressure Ionization LC/MS. New Solutions for Environmental Analysis
by Voyksner, R.D., Environ. Sci. Technol. 28, pages 118A-127A (1994).

LC-MS Interfacing: A General Overview, Chapter 4, In: Liquid Chromatography-Mass Spectrometry: Principles and Applications
by Niessen, W.M.A., van der Greef, J., pages 81-115, Marcel Dekker: New York (1992).

APCI:Instrumentation

Atmospheric Pressure Ionization Mass Spectrometry II: Applications in Pharmacy, Biochemistry and General Chemistry by Bruins, A.P., Trends Anal. Chem. 13, pages 81-90 (1994).

PB:Technology

Aerosols as Microsample Introduction Media for Mass Spectrometry
by Browner, R.F., Winkler, P.C., Perkins, D.D., Abbey, L.E., Microchem. J. 34, pages 15-24 (1986).

Development and Applications of the MAGIC LC/MS Interface
by Winkler, P.C., Ph.D. Thesis, Georgia Institute of Technology, Atlanta, GA (1986).

Investigation of Enhanced Ion Abundances from a Carrier Process in High Performance Liquid Chromatography Particle Beam Mass Spectrometry
by Bellar, T.A., Behymer, T.D., Budde, W.L., J. Am. Soc. Mass Spectrom. 1, pages 92-98 (1990).

Liquid Chromatography-Mass Spectrometry, General Principles and Instrumentation
by Niessen, W.M.A., Tinke, A.P., J. Chromatogr. A 703, pages 37-57 (1995).

Liquid Chromatography/Particle Beam/Mass Spectrometry of Polar Compounds of Environmental Interest
by Behymer, T.D., Bellar, T.A., Budde, W.L., Anal. Chem. 62, pages 1686-1690 (1990).

Mass Transport and Calibration in Liquid Chromatography Particle Beam Mass Spectrometry
by Ho, J.S., Behymer, T.D., Budde, W.L., Bellar, T.A., J. Am. Soc. Mass Spectrom. 3, pages 662-671 (1992).

Micro Flow Rate Particle Beam Interface for Capillary Liquid Chromatography/Mass Spectrometry
by Cappiello, A., Bruner, F., Anal. Chem. 65, pages 1281-1287 (1993).

Monodisperse Aerosol Generation Interface for Coupling Liquid Chromatography with Mass Spectroscopy
by Willoughby, R.C., Browner, R.F., Anal. Chem. 56, pages 2626-2631(1984).

The Particle-Beam Interface; Chapter 9, In: Liquid Chromatography-Mass Spectrometry: Principles and Applications
by Niessen, W.M.A., van der Greef, J., pages 219-228, Marcel Dekker: New York (1992).

Performance of an Improved Monodispersed Aerosol Generation Interface for Liquid Chromatography/Mass Spectrometry
by Winkler, P.C., Perkins, D.D., Williams, W.K., Browner, R.F., Anal. Chem. 60, pages 489-493 (1988).

Studies with an Aerosol Generation Interface for Liquid Chromatography-Mass Spectrometry
by Willoughby, R.C., Ph.D. Thesis, Georgia Institute of Technology, Atlanta, GA (1983).

PB:Applications

The Particle-Beam Interface; Chapter 9, In: Liquid Chromatography-Mass Spectrometry: Principles and Applications
by Niessen, W.M.A., van der Greef, J., pages 219-228, Marcel Dekker: New York (1992).

Particle Beam Liquid Chromatography-Mass Spectrometry: Instrumentation and Applications. A Review
by Creaser, C.S., Stygall, J.W., Analyst 118, pages 1467-1480 (1993).

PB:Instrumentation

Particle Beam Liquid Chromatography-Mass Spectrometry: Instrumentation and Applications. A Review
by Creaser, C.S., Stygall, J.W., Analyst 118, pages 1467-1480 (1993).

TS:Technology

Combined Liquid Chromatography Mass Spectrometry. Part II. Techniques and Mechanisms of Thermospray
by Arpino, P., Mass Spectrom. Rev. 9, pages 631-669 (1990).

Ionization Techniques for Nonvolatile Molecules
by Vestal, M.L., Mass Spectrom. Rev. 2, pages 447-480 (1983).

Liquid Chromatography/Mass Spectrometry: Techniques and Applications

by Yergey, A.L., Edmonds, C.G., Lewis, I.A.S., Vestal, M.L., Plenum Press: New York (1990).

The Thermospray Interface; Chapter 7, In: Liquid Chromatography-Mass Spectrometry: Principles and Applications
by Niessen, W.M.A., van der Greef, J., pages 157-201, Marcel Dekker: New York (1992).

TS:Applications

Combined Liquid Chromatography Mass Spectrometry. Part III. Applications of Thermospray
by Arpino, P., Mass Spectrom. Rev. 11, pages 3-40 (1992).

Liquid Chromatography/Mass Spectrometry: Techniques and Applications
by Yergey, A.L., Edmonds, C.G., Lewis, I.A.S., Vestal, M.L., Plenum Press: New York (1990).

The Thermospray Interface; Chapter 7, In: Liquid Chromatography-Mass Spectrometry: Principles and Applications
by Niessen, W.M.A., van der Greef, J., pages 157-201, Marcel Dekker: New York (1992).

TS:Instrumentation

Combined Liquid Chromatography Mass Spectrometry. Part II. Techniques and Mechanisms of Thermospray
by Arpino, P., Mass Spectrom. Rev. 9, pages 631-669 (1990).

Thermospray; Chapter 4, In: Liquid Chromatography Mass Spectrometry. Techniques and Applications
by Yergey, A.L., Edmonds, C.G., Lewis, I.A.S., Vestal, M.L., pages 31-85, Plenum Press: New York (1990).

CFFAB:Technology

Continuous-Flow Fast Atom Bombardment; Chapter 8, In: Liquid Chromatography-Mass Spectrometry: Principles and Applications
by Niessen, W.M.A., van der Greef, J., pages 203-217, Marcel Dekker: New York (1995).

Continuous-Flow Fast Atom Bombardment Mass Spectrometry
by Caprioli, R.M., Trends Anal. Chem. 7, pages 328-333 (1988).

Continuous-Flow Fast Atom Bombardment Mass Spectrometry
by Caprioli, R.M., Anal. Chem. 62, pages 477A-485A (1990).

Continuous-Flow Fast Atom Bombardment Mass Spectrometry
by Caprioli, R.M. (Ed.), Wiley: Chichester (1990).

Liquid Chromatography/Fast Atom Bombardment Mass Spectrometry
by Stroh, J.G., Rinehart, K.L., Top. Mass Spectrom. 1, pages 287-311 (1994).

CFFAB: Applications

Continuous-Flow Fast Atom Bombardment; Chapter 8, In: Liquid Chromatography-Mass Spectrometry: Principles and Applications
by Niessen, W.M.A., van der Greef, J., pages 203-217, Marcel Dekker: New York (1995).

CFFAB:Instrumentation

FABMS: Still Fabulous?
by Henry, C., Anal. Chem. 69, pages 625A-627A (1997).

MS: Quadrupoles

MS/MS Flexes Its Muscles
by D. Noble, D., Anal. Chem. 67, pages 265A-269A (1995).

Mass Spectrometry/Mass Spectrometry. Techniques and Applications of Tandem Mass Spectrometry
by Busch, K.L., Glish, G.L., McCluskey, S.A., Wiley: New York (1988).

The Quadrupole Mass Filter: Basic Operating Concepts
by Miller, P.E., Denton, M.B., J. Chem. Educ. 7, pages 617-622 (1986).

Tandem Mass Spectrometry
by McLafferty, F.W. (Ed.), Wiley: New York (1983).

MS: Ion Trap

An Introduction to Quadrupole Ion Trap Mass Spectrometry
by March, R.E., J. Mass Spectrom. 32, pages 351-369 (1997).

Practical Aspects of Ion Trap Mass Spectrometry Volume 1: Fundamentals of Ion Trap Mass Spectrometry (448 pages); Volume 2: Ion

Trap Instrumentation (352 pages); Volume 3: Chemical, Environmental & Biochemical Applications (554 pages).
by March, R.E., Todd, J.F.J. (Eds.), CRC Press: Boca Raton (1995, 1995 & 1996).

MS/MS Flexes Its Muscles
by Noble, D., Anal. Chem. 67, pages 265A-269A (1995).

Mass Spectrometry/Mass Spectrometry. Techniques and Applications of Tandem Mass Spectrometry
by Busch, K.L., Glish, G.L., McCluskey, S.A., Wiley: New York (1988).

Tandem Mass Spectrometry
by McLafferty, F.W. (Ed.), Wiley: New York (1983).

MS: Time-of-Flight

Perfect Timing: Time-of-Flight Mass Spectrometry
by Guilhaus, M., Mlynski, V., Selby, D., Rapis Commun. Mass Spectrom. 11, pages 951-962 (1997).

Time of Flight Mass Spectrometry: Instrumentation and Applications in Biological Research
by Cotter, R.J. (Ed.), ACS: Washington D.C. (1997).

Principles and Instrumentation in Time-of-Flight Mass Spectrometry: Physical and Instrumental Concepts
by Guilhaus, M., J. Mass Spectrom. 30, pages 1519-1532 (1995).

Time of Flight Mass Spectrometry and Its Applications
by Schlag, E.W. (Ed.), Elsevier: New York; 420 pages (1994); also previously published in the journal, Int. Jr. Mass Spectom. Ion Processes, 131 (1994).

Time-of-Flight Mass Analyzers
by Wollnick, H., Mass Spectrom. Rev. 12, pages 89-114 (1993).

Time of Flight Mass Spectrometry for the Structural Analysis of Biological Molecules
by Cotter, R.J., Anal. Chem. 64, pages 1027A-1039A (1992).

The Renaissance of Time-of-Flight Mass Spectrometry
by Price, D., Miles, G.J., Int. J. Mass Spectrom. Ion Processes 99, pages 1-39 (1990).

MS: FTMS

Fourier Transform Ion Cyclotron Resonance Mass Spectrometry
by Marshall, A.G. (Ed.), Int. J. Mass Spectom. Ion Process 157/158, pages 1-410 (1996).

Fourier Transform Mass Spectrometry of High-Mass Biomolecules
by Buchanan, M.V., Hettich, R.L., Anal. Chem. 65, pages 245A-246A, 248A-250A, & 252A-259A (1993).

FT-ICR-MS: Analytical Applications of Fourier Transform Ion Cyclotron Resonance Mass Spectrometry
by Asamoto, B. (Ed.), VCH Publishers: New York (1991).

Fourier Transform Mass Spectrometry: Evolution, Innovation and Applications
by Buchanan, W.V. (Ed.), ACS: Washington D.C. (1987).

MS: Sectors

Introduction to Mass Spectrometry
by Watson, J.T., Lippincott-Raven: New York (1997).

Mass Spectrometry for Chemist and Biochemist
by Johnstone, R.A., Rose, M.E., Cambridge University Press: Cambridge (1996).

Practical Organic Mass Spectrometry. A Guide to Chemist and Biochemical Analysis
by Chapman, J.R., Wiley: New York (1995).

MS/MS Flexes Its Muscles
by Noble, D., Anal. Chem. 67, pages 265A-269A (1995).

Mass Spectrometry/Mass Spectrometry. Techniques and Applications of Tandem Mass Spectrometry
by Busch, K.L., Glish, G.L., McCluskey, S.A., Wiley: New York (1988).

Mass Spectrometry
by Davies, R., Frearson, M., Wiley: New York (1987).

Tandem Mass Spectrometry
by McLafferty, F.W. (Ed.), Wiley: New York (1983).

Resources: On-Line Services

Everyday more & more companies are going on-line & placing a Home Page of their business on the World Wide Web advertising their services & products. Some are listed here.

Manufacturers On-Line

All manufacturers are listed on
`www.LCMS.com`

MS Abstracts

Analytical Abstracts on CD-ROM
The Royal Society of Chemistry: Cambridge, UK.

CASurveyor: Mass Spectrometry and Applic.
info.cas.org/ONLINE/CD/SURVEYOR/surveyor.html

Currents Contents on CD ROM, Physical, Chemical & Earth Sciences
Institute for Scientific Info., Philadelphia, PA.

Identification of Essential Oil Compounds by Gas Chromatography / Mass Spectrometry
www.clickit.com/touch/abp/member/allured.htm

Instrumental Data for Drug Analysis
www.crcpress.com

LC/MS Update
www.hdscience.com/hdupdate.htm#update

A Mass Spectral & GC Data of Drugs, Pollutants, Pesticides & Metabolites
www.vchgroup.de/cc/index.html

Mass Spectrometry Bulletin
The Royal Soc. of Chemistry: Cambridge, UK.

The Wiley Static SIMS Library
www.wiley.com

Government

US Government
For access to one of the world's largest catalogues, refer to the home page: www.gsa.gov

US EPA
Home page: www.usepa.gov

US FDA
Home page: www.fda.gov

US-NIDA
Home page: www.nida.nih.gov

US-OSHA
Home page: www.osha.gov

Laboratory

American Society for Testing and Materials
www.astm.org.

GLP/cGMP Documents and Guidelines
see EPA & FDA sites. Also see home page: www.anachem.umu.se/jumpstation.htm.

Laboratory Certification
Public Health Laboratories, see home page: epsilon.doh.wa.gov/phl/cert_cla.html

Laboratory Equipment Directory
All products & manufacturers listed in their yearly directory can be accessed through their home page: www.labequipmag.com

Laboratory Internet Directory
Complete details on products advertised in "American Laboratory", refer to their home page: www.iscpubs.com

Material Safety Data Sheets
Vermont SIRI MSDS Collection & links to other internet MSDS & hazardous chemical archives, refer to their home page: siri.org/msds/index.html

ProcureNet
Internet mall offering maintenance, repair & operating materials; scientifc, safety & office supplies, refer to their home page: www.procurenet.com

Quality Assurance & Control
Michael E. Mispagel at the University of Georgia's Quality Assurance Unit (Office of the Vice President for Research) has complied Good Laboratory Practice (**GLP**) standards for the EPA & FDA; & Quality Assurance (**QA**) or Quality Control (**QC**) can be found at the home page: www.ovpr.uga.edu/qau/startqau.html.

Sci.Trak
The company is committed to providing the scientific community with the most comprehensive Buyers Guide of information on scientific instrumentation, products & services, refer to their home page: www.scitrak.com.

News Groups

sci.bio.* sci.med.*
sci.env.* sci.chem.analytical
sci.chem.labware scitechniques.mass-spec
comp.soft-sys.stat.systat

Searching Newsgroups

Deja News
www.dejanews.com

SPSS Science
www.spss.com/tech/listserves.html

Stanford University
sift.stanford.edu

Scientific Databases

Protein Data Bank
National Institute of Health.
Home page: molbio.info.hin.gov/cgi-bin/pdb

European Molecular Biology Laboratory
Protein & Peptide Group, home page: macmann6.embl-heidelberg.de

STN Easy
by STN International finds & displays information about chemistry, life sciences, patents, pharmaceuticals, physics, mathematics, computer science, engineering, & general science. Home page, North America: stneasy.cas.org.
Home page, Europe: stneasy.fiz-karlsruhe.de.
Home page, Japan: stneasy-japan.cas.org

Scientific Computing & Automation
contains various directories of Web Sites for chemistry, biology, and the life sciences, home page: www.scamag.com

Search MassRef
by Rockefeller University is a collection of abstracts that pertain to mass spectrometry from Currents Contents, MEDLINE, their own database (PROWL). Home page: chait_sgi.rockefeller.edu/cgi-bin/massref

WindowChem Software
by Cambridge Corporation has selected chemical databases for sale. Home page: www.windowchem.com

Trinity Software
www.trinitysoftware.com

Science News Services

AAAS
www.eurekalert.com

Science
on-line news service of *Science* magazine. Home page: www.sciencenow.org

Chemistry: on Internet

American Chemical Society
Provides a collection of chemistry-related information: journals, databases, information about ACS, professional services, educational materials, discussion & news groups, refer to their home pages: www.ChemCenter.org; www.ace.org, pubs.acs.org, acsinfo.acs.org

CambridgeSoft
provides an index of small molecules on the Web, refer to their home page: www.camsoft.com

The Chemical Educator
journals.springer-ny.com/chedr/samparticle.html

Chromatographic Web Pages
chrom.tutms.tut.ac.jp/JINNO/ENGLISH/RESEARCH/LC_HOME/lc_home.html

Imperial College of Science, Technology and Medicine
www.ch.ic.ac.uk

International Conference on Harmonisation
www.ifpma.org/ich1.html

LC-GC
www.lcgcmag.com

Michigan Technological University
www.chem.mtu.edu/~qahabib/chemsites.html

National Institute of Standards & Technology
www.nist.gov

Periodic Table of Elements
periodic table on the World-Wide Web. University of Sheffield, England. Home page: www.shef.ac.uk/~chem/web-elements

Queens Univeristy, Belfast
a wide range of links. Home page: boris.qub.ac.uk.edward

Rensselaer Polytechnic Institute
index of chemistry resources on the internet. Home page: www.rpi.edu/dept/chem/cheminfo/chemres.html

Stanford University
Stanford Yahoo Chemistry Index. Information on conferences, indices, ans other resources. Home page: www-sul.stanford.edu/catdb/sci.html

Trends in Analytical Chemistry
www.wlsevier.nl/locate/trac

UCLA
WWW Virtual Library Chemistry. Index of direct links to chemistry servers around the world. Home page: www.chem.ucla.edu/chempointers.html

University of Southern California
heat.usc.edu/links/chemlinks.html#classes

Umea University
www.anachem.umu.se/jumpstation.htm

Life Science:Internet

Biologists Resources Cambridge
www.bio.cam.ac.uk

BiomedicalProducts
www.bioprodamg.com

Brookhaven National Laboratories
provides a protein database. Home page: www.pdb.bnl.gov

Human Genome Project
www.ornl.gov/TechResources/ Human_Genome/home.html

Queens Univeristy, Belfast
a wide range of links. Home page: boris.qub.ac.uk.edward

Technical Tips On-Line
by Elsevier Trends Journal. A database of techniques & information for the life scientist. Home page: www.elsevier.com/locate/tto
In Europe: www.elseveier.nl.locate/tto

Mass Spec: Internet

The Analytical Chemistry Springboard
www.anachem.umu.se/jumpstation.htm

Murray's Mass Spectrometry Page
a collection of links to mass spectrometry internet sites by Kermit Murray at Emory University. Home page: userwww.service.emory.edu/~kmurray-/mslist.html

NIST/EPA/NIH Mass Spectral Data Base
Standard Reference Data Program
National Institute of Standards and Technology, refer to their home page: www.nist.gov/srd.

PROWL
by protein mass spectrometry groups at New York University & Rockefeller University. A resource for protein chemistry & mass spectrometry, refer to their home page: chait_sgi.rockefeller.edu

Queens Univeristy, Belfast
boris.qub.ac.uk.edward

Science Hypermedia, Inc.
A Web & CD-ROM-based continuing education in analytical chemistry & instrumentation, including chromatography, spectroscopy, mass spectrometry, electronics, & optics; refer to their home page: www.scimedia.com/index.htm

Wiley Mass Spectral Data Base
www.palisade.com/mass-spec.

Resources: Journals Relating to LC/MS

Primary LC/MS Journals

Analytical Chemistry
published semi-monthly by the American Chemical Society. Contains papers which consider. Home page: pubs.acs.org

International Journal of Mass Spectrometry and Ion Processes
published bi-monthly by Elsevier Science, Inc. contains papers which consider fundamental aspects of mass spectrometry & ion processes, & the application of mass spectrometric techniques to specific problems in chemistry, physics, biology. Home page: www.elsevier.com

Journal of the American Society for Mass Spectrometry
published monthly by Elsevier Science, Inc. Members of the American Society for Mass Spectrometry receive the journal as a benefit of membership. Home page: www.elsevier.com/locate/jasmsonline

Journal of Chromatographic Sciences
published monthly by Preston Publications.

Journal of Chromatography A
published monthly by Elsevier Science, Inc. Home page: www.elsevier.com

Journal of Chromatography B: Biomedical Applications
a companion journal to Journal of Chromatography A. Published monthly. Home page: www.elsevier.com

Mass Spectrometry Bulletin
published monthly by The Royal Society of Chemistry.

Mass Spectrometry Reviews
published bimonthly by John Wiley.
Home page: www.wiley.com
Home page (Europe): www.wiley.co.uk

Rapid Communications in Mass Spectrometry
published monthly by John Wiley & Sons Limited.
Home page (USA): www.wiley.com
Home page (Europe): www.wiley.co.uk

Special Topics Journals

Critical Reviews in Analytical Chemistry
published bi-monthly by CRC Press.

Journal of Analytical Toxicology
published by Preston Publications.
Published bi-monthly.

Journal of AOAC International
published bi-monthly by AOAC International-GR, Contains collaborative study data on which validation of AOAC Official Methods are based.
Home page: www.aoac.org

Journal of Pharmaceutical Sciences
co-published monthly by American Chemical Society & American Pharmaceutical Association.

Elsevier "Trends Journals"
published monthly by Elsevier Science, Inc.
Home page: www.elsevier.com

Magazines, Newsletters

Analytical Consumer
published monthly by J.R.Jordan.
Home page: www.labx.com/acorder.htm

High Tech Separations News
A BCC, Inc. Publication, 25 Van Zant St.
Norwalk, CT 06855-1781 USA
Phone: 1-203-853-4266
Fax: 1-203-853-0348

LC-GC
published monthly by Advanstar Communications is one of the best resources for the chromatographer. If you pick one journal to read cover to cover, pick LC-GC. Subscription is free to all who are involved in chromatography.
P.O. Box 6168
Duluth, MN 55806-6168 USA
Phone: 1-218-723-9477
Home page: www.lcgcmag.com

Laboratory Equipment
c/o Gordon Publications
301 Gibraltor Drive, P.P. Box 650
Morris Plains, NJ 07950-0650 USA

The Mass Spec Source
c/o Scientific Instrument Services, Inc.
1027 Old York Road
Ringoes, NJ 08551-1039 USA

Resources, For Chromatography
c/o LC Resources, Inc.
2930 Camino Diablo, Suite 110
Walnet Creek, CA 94596 USA
Phone: 510-930-9043

Resources: Training & Education

American Chemical Society
sponsors various courses from "Interpretation of Mass Spectra" to "Liquid Chromatography/Mass Spectrometry: Fundamentals and Applications" which are held 10-12 times a year throughout the United States. Contact: American Chemical Society, 1155 16th Street, NW Washington DC, 20036, USA. Telephone: 1-202-452-2113.; Fax: 1-800-227-5558 (Fax on demand). E-mail: education@acs.org. Home page: www.chemcenter.org and www.acs.org

American Society of Mass Spectrometry
sponsors a series of short courses (provides continuing education in the field of mass spectrometry & introductory courses for those entering the field) that precede their annual meeting; A Fall Workshop, that facilitates exchange & provides instruction on emerging techniques (such as LC/MS); & The Sanibel Conference & the Asilomar Conference, that bring together researchers in an informal setting to discuss specialized topics with the intent of stimulating new ideas. Contact: ASMS, 1201 Don Diego Ave., Santa Fe, NM 87505, USA. Telephone: 1-505-989-4517. Fax: 1-505-989-1073. Home page: www.asms.org

Chem-Space Associates
The authors of this book have developed a one day, on-site short course detailing the topics covered throughout this book. A primary emphasis is placed on practical problem-solving approaches. The course can be customized to meet the needs of your organization. Books and handouts are included with the course fee for each attendee. This is a cost effective way to familiarize your organization with the practical side of LC/MS.
Contact: Ross Willoughby, Chem-Space Associates, P.O. Box 111384, Pittsburgh, PA 15238. 1-412-963-6881, Fax 1-412-963-6882, E-mail: ross@LCMS.com
Home Page: www.LCMS.com

Hyphen
provides in-house training is offered by Dr. Wilfried Niessen, co-author with Jan v.d. Greef of Liguid Chromatography-Mass Spectrometry
Contact: Wilfried Niessen, De Welstraat 8, 2332 XT Leiden,
The Netherlands
+31 71-5768628 Phone & fax
Home Page:ourworld/compuserve.com/homepages/Wilfried_Niessen

LC/MS Symposium
a new conference organized in 1997 around LC/MS technology & applications. Contact: Janet Oxford, Telephone: +01-76-324-31-69.
E-mail:101445.1311@compuserve.com

Montreux Symposium on Liquid Chromatography/Mass Spectrometry
sponsors a two-day short course that proceeds the annual meeting. The presentations are of selected topics with a focus on practical applications in the industrial, biomedical, environmental, & pharmaceutical/clinical areas. For 1997, see home page: www.baka.com/webpages/LCMS. For 1998, see listing in *Section G. Meetings & Seminars.*

TNO Pharma
sponsors a 5 day 'hands-on' course. Contact: Dr. Rob J. Vreeken (LC-MS specialist & course organizer) TNO Pharma. Telephone: +31-30-694-42-76. Fax: +31-30-695-67-42. E-mail: vreeken@voeding.tno.nl. Home Page: www.tno.nl

Resources: Meetings & Seminars

Concerning LC/MS

The ASMS Conference on Mass Spectrometry and Allied Topics

sponsor by the American Society of Mass Spectrometry, is held annually to serve as an open forum for the presentation of scientific research & the development of new applications in mass spectrometry. Contact: ASMS, 1201 Don Diego Avenue, Santa Fe, NM 87505, USA. Telephone: 1-505-989-4517. Fax: 1-505-989-1073.
E-mail: asms@asms.org
Home page: www.asms.org

International Symposium on Mass Spectrometry in the Health and Life Sciences

a 5 day conference. The meeting for 1998 will be held in San Francisco. Phone 1-415-476-4893, fax: 1-415-502-1655.
E-mail: sf98ms@itsa.ucsf.edu
Home page: rafael.ucsf.edu/sumposium98

LC/MS Symposium

a new conference organized around LC/MS technology & applications. Contact: Janet Oxford, Telephone: +01-76-324-31-69.
E-mail:101445.1311@compuserve.com.

Montreux Symposium on Liquid Chromatography/Mass Spectrometry

is short three day meeting held annually, alternating between the United States (odd years) & Europe (even years). If there is one meeting you have a chance to attend, this should be the meeting. US home page: www.baka.com/webpages/LCMS. For 1999, search the Internet for "Montreux" or "LC/MS meeting". For Europe 1998, contact: Workshop-Office, International Association of Environmental Analytical Chemistry, Postfach 46, CH-4123, Allschwil 2 , Switzerland. Telephone: +41-61-481-27-89. Fax: +41-61-481-08-05.

Related to LC/MS

AOAC International

holds an annual meeting & exposition. Contact: AOAC International, 481 North Frederick Ave., Suite 500, Gaithersburg, MD 20877-2504, USA. Telephone: 1-800-379-2622 or 1-301-924-7077. Fax: 1-301-924-7089.
E-mail: aoac@aoac.org.
Home page: www.aoac.org.

FACCS

Contact: FACCS National Office, 1201 Don Diego Ave., Santa Fe, NM 87505, USA. Telephone: 1-505-820-1648. Fax: 1-505-989-1073.

ComTech

is a conference & exposition designed to discuss the latest developments in combinatorial mass spectrometry, lead optimization & libraries. Contact: Scitec, Conference Coordination Office, Av. de Provence 20, CH-1000 Lausanne 20, Switzerland. Telephone: +41-21-626-46-30. Fax: +41-21-624-15-49.E-mail: symposia@worldcom.ch
Home page: www.scitec-robotics.com

HPLC

(International Symposium on High Performance Liquid Phase Separations & Related Techniques) is an international conference & exposition held annually. For 1997, Contact: Scretariat, Universal Conference Consultants, China Court Business Centre, China Court, Ladywell Walk, Birmingham B% 4RX, UK. Telephone: +44-121-622-3644. Fax: +44-121-622-2333.
E-mail: 101464.764@compuserve.com
Home page: sponsored by Waters for 1997, see www.waters.com

NanoTech

is a new conference designed to address the emerging field of micro & nanoscale technologies as they are related to drug screening & combinatorial chemistry, pharmaceutical sciences & biotechnology. Contact: Scitec, Conference Coordination Office, Av. de Provence 20, CH-1000 Lausanne 20, Switzerland. Telephone: +41-21-626-46-30 or +41-21-626-23-53. Fax: +41-21-624-15-49.
E-mail: symposia@worldcom.ch
Home page: www.scitec-robotics.com

NMHCC Bio/Technology

sponsors short two day meetings centered around pharmaceutical sciences & biotechnologies (including the use of LC/MS & MALDI/MS for characterization & analysis). Contact: NMHCC, Inc., 71 Second Ave., 3rd Floor, Waltham, MA 02154, USA. Telephone: 1-888-446-6422 or 1-617-663-6000. Fax: 1-617-663-6411.
E-mail: biotech@nmhcc.com
Home page: biotech.nmhcc.org

Resources: Professional Organizations

Write for an application & inquire about benefits & dues. Many societies publish a journal &/or newsletter.

American Chemical Society
1155 Sixteenth Street, N.W.
Washington, D.C. 20036 USA
Telephone: 1-800-227-5558 or 1-202-872-4600
Fax: 1-202-872-6337
Home page: www.acs.org

American Council of Independent Labs.
1629 K Street, NW
Washington, DC 20006 USA
Telephone: 202-887-5872
Fax: 202-887-0021
Home page: www.acil.org

American Society of Mass Spectrometry
1201 Don Diego Avenue
Santa Fe, NM 87505 USA
Telephone: 1-505-989-4517
Fax: 1-505-989-1073
Home page: www.asms.org

Analytical Laboratory Managers Assoc.
P.O. Box 258
Montchanin, DE 19710

Federation of Analytical Chemistry and Spectroscopy Societies-FACSS
1201 Don Diego Avenue
Santa Fe, NM 87505 USA
Telephone: 1-505-820-1548
Fax: 1-505-989-1073
Home page: www.facss.org/info.html

Association for Laboratory Automation
PO Box 572, Health Sciences Center
Multistory Building, Room 6171
University of Virginia
Charlottesville, VA 22908 USA
Telephone: 804-924-9430
Fax: 804-924-5718
Home page: labautomation.org

International Association of Environmental Testing Laboratories
505 Wythe Street
Alexandria, VA 22314 USA
Telephone: 1-703-739-2188
Fax: 1-703-739-2556
Home page: www.iaetl.org

The Protein Society
9650 Rockville Pike
Bethesda, MD 20852-3998 USA
Phone: 301-530-7120
Fax: 301-530-7049
Refer to home page: www.faseb.org.

Society of Forensic Toxicologist
P.O. Box 5543
Mesa, AZ 85211-5543 USA
Phone/Fax: 602-839-9106

Local User Groups
Mass Spectrometry
For information about local or regional Mass Spectrometry user groups in the United States contact the American Society of Mass Spectrometry.

Resources: Contract Laboratories

Here is an alphabetical list of the contract laboratories specializing in LC/MS capabilities in the United States (& Europe). For a broader list of companies (in the US & worldwide) that supply mass spectrometry/laboratory services, see the American Society for Testing and Materials's home page: www.astm.org and *1997 Directory of Mass Spectrometry Manufacturers and Suppliers*, by Stephen A. Lammert in the journal, Rapid Communication in Mass Spectrometry, Vol 11, pages 821-845 (1997).

ABC Laboratories
Missouri Offices
7200 East ABC Lab.
Columbia, MO 65202 U.S.A.
Telephone: 1-573-474-8579
Fax: 1-573-443-9033
European Offices
38 Castleroe Road
Coleraine, BT51 3RL North Ireland
Telephone: + 44-01265-320639
Fax: + 44-01265-320653
Home page: www.abclabs.com

Advanced Bioanalytical Services, Inc.
15 Catherwood Road
Ithaca, NY 14850 U.S.A.
Telephone: 1-607-266-0665
Fax: 1-607-266-2749
E-mail: kurz@abs-lcms.com
Home page: www.abs-lcms.com

Alta Analytical Laboratory
California Offices
5070 Robert J. Mathews Parkway
El Dorado Hills, CA 95762 U.S.A.
Telephone: 1-916-933-1040
Fax: 1-916-933-0940
New Jersey Offices
48 Blue Ridge Avenue
Green Brook, NJ 08812 U.S.A.
Telephone: 1-908-752-2691
E-mail: bbethem@altalab.com
Home page: www.altalab.com

ANAPHARM
2050 Rene-Levesque
Sainte-Foy, Quebec, G1V 2K8, Canada
Telephone: 1-418-527-4000
Fax: 1-418-527-3456
E-mail: mlebel@anapharm.com
Home page: www.anapharm.com

Apollin Co.
PO Box 266
Broomall, PA 19008-0266
Telephone: 1-610-325-7617
Fax: 1-610-325-7617
E-mail: banchan@erols.com

Axelson & Kwok Research Assoc., Inc.
Box 1125, Station A
Delta, B.C., V4M 3T2 Canada
Telephone: 1-604-943-3715
Fax: 1-604-943-7391
E-mail: iroberts@wimsey.com
Home page: www.asi.bc.ca/bcba

Cedra Corp.
8609 Cross Park Drive
Austin, TX 78754 U.S.A.
Telephone: 1-512-834-7766
Fax: 1-512-834-7767
E-mail: rchacon@cedracorp.com
Home page: www.cedracorp.com

Harris Laboratories
624 Pach Street
P.O. Box 80837
Lincoln, NE 68502 U.S.A.
Telephone: 1-402-476-2811; 1-800-776-1716
Fax: 1-402-476-7598
Home page: www.dataedge.com

Kansas City Analytical Services, Inc.
12700 Johnson Drive
Shawnee, KS 66216
Telephone: 1-913-268-5238
Fax: 1-913-268-3240

Keystone Analytical Laboratories, Inc.
113 Dickerson Road
North Wales, PA 19454
Telephone: 1-212-699-8899
Fax: 1-215-699-8848

LAB Pharmacological Res. International
(formerly: I'Institut de Recherche en Pharmacie Industrielle, IRPI)
1000 St. Charles
Vaudreuil (Montreal) Quebec, J7V 8P5
Canada, Telephone: 1-514-455-4522
Fax: 1-514-455-4276

E-mail: champagn@lab-cdn.mds.Compuserve.com

Magellan Laboratories, Inc.
P.O. Box 13341
Research Triangle Park, NC 27709 U.S.A.
Telephone: 1-919-481-4855
Fax: 1-919-481-4908
E-mail: info@magelabs.com
Home page: www.magellanlabs.com

Maxxam Analytics, Inc.
(a Chemex Labs Alberta/NOVAMANN International Partnership)
5540 McAdam Road
Mississauga, Ontario L4Z 1P1, Canada
Telephone: 1-905-890-2555, 1-800-563-6266, Fax: 1-905-890-0370
E-mail: info@on.novamann.ca
Home page: www.maxxam.ca

Microbiological Associates, Inc.
Life Sciences Center
9900 Blackwell Road
Rockville, Maryland 20850 U.S.A.
Telephone: 1-301-738-1000
Fax: 1-301-738-1036

MRI
(Midwest Research Institute)
425 Volker Boulevard
Kansas City, MI 64110-2299 U.S.A.
Telephone: 1-816-753-7600
Fax: 1-816-753-8420
Home page: www.mriresearch.org

M-Scan, Inc.
US Offices
137 Brandywine Parkway,
West Chester, PA 19380 U.S.A.
Telephone: 1-610-696-8210 or 1-610-696-8240
Fax: 1-610-696-8370
Home page: www.m-scan.com
European Offices
M-Scan Ltd.
Silwood Park, Sunninghill,
Ascot Berks, SL5 7PZ England
Telephone: 44 (0)1344 27612;
44 (0)1344 27072
Fax: 44 (0)1344 872709

Northwest Bioanalytical
1141 East 3900 South
Salt Lake City, UT 84124-9906 USA
Telephone: 1-801-268-2431
Fax: 1-513-626-5600
E-mial: rodger@nwtinc.com
Home page:
www.xmission.com/~nwt/bio.htm

OMS Laboratories, Inc.
911 Western Avenue
Suite 412
Seattle, WA 98104-1031 USA
Telephone: 1-206-622-8353
Fax: 1-206-622-4623
E-mail: labOMS@aol.com

Oneida Research Services, Inc.
One Halsey Road
Whitesboro, NY 13492 U.S.A.
Telephone: 1-315-736-3050
Fax: 1-315-736-2460

Panlabs, Inc.
11804 North Creek Parkway South
Bothell, WA 98011-8805 U.S.A.
Telephone: 1-206-487-8200
Fax: 1-206-487-3787
E-mail: panlabs@panlabs.com
Home page: www.panlabs.com

PPD Pharmaco, Inc.
2244 Dabney Road
Richmond, VA 23230 U.S.A.
Telephone: 1-804-359-1900
Fax: 1-804-353-1860
or
8500 Research Way
P.O. Box 620650
Middleton, WI 53562-0650 U.S.A.
Telephone: 1-608-827-9400
Fax: 1-608-827-8807 E-mail:
jennifer.branin@richmond.ppdi.com
Home page: www.ppdpharmaco.com

PharmaKinetics Laboratories, Inc.
302 W. Fayette Street
Baltimore, Maryland 21201 U.S.A.
Telephone: 1-410-385-4500
Fax: 1-410-385-1957
Home page: www.pharmakinetics.com

Phoenix International Life Science, Inc.
Canadian Offices
2350 Cohen Street
Montreal, Q.C., H4R 2N6 Canada
Telephone: 1-514-333-0033
Fax: 1-514-333-8861
USA Offices
Telephone: 1-908-781-9666
FAX: 1-908-781-9667
Home page: www.pils.com

Quanterra Environmental Services
Advanced Technology Group
4995 Yarrow Street
Arvada, CO 80002 U.S.A.
Telephone: 1-303-421-6611
Fax: 1-303-940-1525

Quintiles, Inc.
USA Offices
6707 Democracy Boulevard
Suite 106
Bethesda, MD 20817 U.S.A
Telephone: 1-301-530-9222
Fax: 1-301-571-1726
European Offices
Quintiles Scotland Ltd.
Research Avenue South
Heriot-Watt University
Research Park, Riccarton,
Edinburgh EH14 4AP, Scotland
Telephone: + 44-131-45-5511
Fax: + 44-131-451-2062
E-mail: sales@qred.quintiles.com
Home page: www.quintiles.com

Ricerca, Inc.
7528 Auburn Road
Painesville, OH 44077-1000 U.S.A.
Telephone: 1-440-357-3300
Fax: 1-440-354-4415
E-mail: duane_c@ricerca.com
Home page: www.ricerca.com

RTI
(Research Triangle Institute)
Dreyfus Lab
Room 114
PO Box 12194
3040 Cornwallis Road
Research Triangle Park, NC 27709 U.S.A.
Telephone: 1-919-541-6697
Fax: 1-919-541-7208
E-mail: rdv39@rti.org
Home page: www.rti.org

R&W Bio Research
Mech Elaar Straat 3
4903 Re Oosterhout, The Netherlands
Telephone: + 31-162-436910
Fax: + 31-162-460827
E-mail: info@rwbio.nl
Home page: www.rwbio.nl

SRI
(Stanford Research Institute)
333 Ravenswood Avenue
Menlo Park, CA 94025-3493 U.S.A.
Telephone: 1-415-859-4771
Fax: 1-415-859-4325
E-mail: inquiry_line@sri.com
Home page: www.sri.com

Taylor Technology, Inc.
107 College Road East
Princeton, NJ 08540 U.S.A.
Telephone: 1-609-951-0005
Fax: 1-609-951-0080
E-mail: sag@taytech.com

TexMS Corporation
15701 West Hardy Road
Houston, TX 77060 U.S.A.
Telephone: 1-281-447-2472
Fax: 1-281-447-2473
E-mail: texms@ghgcorp.com
Home page: www.ghg.net/texms

Triangle Laboratories, Inc.
801 Capitola Drive
Durham, NC 27713-4410 U.S.A.
Telephone: 1-919-544-5729
Fax: 1-919-544-5491
E-mail: harvan@compuserve.com

TNO Pharma
Utrechtseweg 48
PO Box 360
Zeist, 3700 AJ The Netherlands
Telephone: +31-30-694-48-44
Fax: +31-30-694-48-45
E-mail: pharma-office@pharma.tno.nl
Home page: www.tno.nl/instit/pharma

Xenobiotic Laboratories, Inc.
Agriculture/Pharmaceutical Contract Research Services
107 Morgan Lane
Plainsboro, NJ 08536 U.S.A.
Telephone: 1-609-799-2295
Fax: 1-609-799-7497
E-mail: xbl@pluto.njcc.com
Home page: www.xbl.com

Resources: University Laboratories

Here is a list of universities, colleges, & institutes specializing in LC/MS contract analysis in the United States, Canada, Europe, Asia & Australia. With the advent of budgetary constraints & a desire to keep down operating cost, colleges & university have gone into the business of performing chemical analyses, on a limited basis. The home page of the individual labs are provided. More & more labs are advertising on the Internet. For an updated compilation of home pages see the home page www.LCMS.com.

USA

Arizona

University of Arizona, Tucson
The Wysocki Research Group
www.chem.arizona.edu/faculty/wyso/index.html

California

California Institute of Technology
Environmental Analysis Center
www.caltech.edu/~eac/home.html
University of California
UC-Berkeley: Mass Spectrometry Facility,
www.cchem.berkeley.edu/College/Facilities/masspec/masspec.html
UC-Berkeley: Evan Williams' Group
www.cchem.berkeley.edu/~erwgrp
UC-Davis: Carlito Lebrilla's Group
www.chem.ucdavis.edu/groups/lebrilla
UC-Los Angeles: Center for Molecular & Medical Sciences Mass Spectrometry
www.chem.ucla.edu/dept/Mass_Spectr.html
UC-Riverside: Mass Spectrometry Facility
www.chem.ucr.edu/facilities/ms/ms.html.
UC-San Francisco: Mass Spectrometry Facility
rafael.ucsf.edu
The Scripps Research Institute, La Jolla
Mass Spectrometry Laboratory
masspec.scripps.edu

District of Columbia

Naval Research Laboratory, Washington DC
Analytical Chemistry & Mass Spectrometry Section
chem1.nrl.navy.mil/analytical

Florida

University of Florida, Gainesville
Mass Spectrometry Services
www.chem.ufl.edu/groups/yost/powell.html

Florida State University, Tallahasse
National High Magnetic Field Laboratory
gamma.magnet.fsu.edu/cimar/ICR

Georgia

University of Georgia, Athens
Chemical & Biological Sciences Mass Spectrometry Facility
www.chem.uga.edu/research.html#MS
Center for FT-ICR Research
rcarlson@ccrc.uga.edu
Complex Carbohydrate Research Center
amstersgi.chem.uga.edu
Emory University, Atlanta
Mass Spectrometry
userwww.service.emory.edu/~fstrobe/eumsc.html
Jeanett Adams' Group
www.chem.emory.edu/~faculty/Adams/adams.html
Georgia Institute of Technology, Atlanta
Ken Busch's Group
www.chemistry.gatech.edu/faculty/busch/busch.html

Illinois

University of Illinois, Urbana-Champaign
School of Chemical Sciences
www.scs.uiuc.edu/~msweb

Indiana

University of Notre Dame, South Bend
Mass Spectrometry Laboratories
www.nd.edu/~massspec

Iowa

Iowa State University, Ames
Mass Spectrometry Facility
www.public.iastate.edu/~kamel

Louisiana

Louisiana State University, Baton Rouge
Mass Spectrometry Facility
chrs1.chem.lsu.edu/htdocs/people/palimbach/pal.html

Maryland

Johns Hopkins, Baltimore
Bob Cotter's Lab
infonet.welch.jhu.edu/admin/spons_project/handbook/appendi2--20.html
University of Maryland, Baltimore
Center for Structural Biochemistry
research.umbc.edu/~smith/chem/fenselau/sbc.html
National Institutes of Health, Bethesda
Structural Mass Spectrometry Group
sx102a.niddk.nih.gov

Massachusetts

Northeastern University, Boston
Barnett Institute
www.barnett.neu.edu/ms/homepage.htm
Harvard University, Cambridge
Microchemistry Facility
golgi.harvard.edu/microchem
Massachusetts Institute of Technology, Cambridge
Division of Toxicology
web.mit.edu/toxms/www/index.htm

Missouri

Washington University, St. Louis
Mass Spectrometry Resource
wunmr.wustl.edu/~msf

Nebraska

University of Nebraska, Lincoln
Nebraska Center for Mass Spectrometry,
wwitch.unl.edu/ncms/instru.html

New Jersey

Rutgers University, Piscataway
Center For Advanced Food Technology
Mass Spectrometry Facility
foodsci.rutgers.edu/caft/massspec.htm

New York

State University of New York, Buffalo
ESI/FT-ICRMS Group
www.chem.buffalo.edu/Fac_Res_Int/Wood/index.html
Cornell University, Ithaca
Analytical Toxicology
jdh4@cornell.edu
Baker Laboratory
www.chem.cornell.edu
Rockefeller University, New York
The Mass Spectrometry Group
pdtc.rockefeller.edu/pdtcmain/ms/ms1.html

North Carolina

North Carolina State University, Raleigh
Mass Spectrometry Facility
ch9000.chem.ncsu.edu/MS_Home.html
National Institute of Environmental Health Sciences, Research Triangle Park
Mass Spectrometry Group
www.niehs.nih.gov/dirlsb/tomer.htm

Pennsylvania

University of Pennsylvania, Philadelphia
Center for Experimental Therapeutics
sunspark.med.upenn.edu/~cetweb/ms.html
Carnegie Mellon University, Pittsburgh
Center for Molecular Analysis
www.chem.cmu.edu/cma

South Carolina

Medical University of South Carolina, Charleston
Biomolecular Mass Spectrometry Facility
www2.musc.edu/Pharm/MS_folder/ms.html
University of South Carolina, Columbia
Mass Spectrometry Laboratory
www.cosm.sc.edu/chem/research.html

Texas

University of Texas, Houston
Analytical Chemistry Center
www.bmb.med.uth.tmc.edu/UTHACC/accjjs.html

Virginia

Virginia Commonwealth University, Richmond
Muddiman Research Group
www.owt.com/users/davekim

Washington

Pacific Northwest National Labs, Richland
Advanced Organic Analytical Methods Group
www.www.pnl.gov/aoam/aoam.htm
Separations & Mass Spectrometry Group
www.emsl.pnl.gov:2080/docs/msd/fticr/advmasspec.html
University of Washington, Seattle
Biochemistry Mass Spectrometry Laboratory
128.95.12.16/WalshLab.html
Biological Mass Spectrometry
thompson.mbt.washington.edu

Canada

Alberta

University of Alberta, Edmonton
Liang Li's Mass Spectrometry Group
www.chem.ualberta.ca/~liweb

Ontario

McMaster University, Hamilton
The McMaster Regional Centre for Mass Spectrometry
www.chemistry.mcmaster.ca/facilities/mrcms.html

Ottawa-Carleton University, Ottawa
Mass Spectrometry Centre
www.chem.uottawa.ca/azloto/uoms.html

Europe

Belgium

University of Antwerp
The Nucleoside Research & Mass Spectrometry Unit
wcc.ruca.ua.ac.be/NucleoMS

University of Gent
Laboratory of Protein Biochemistry
allserv.rug.ac.be/~jvbeeume/spect.html

University of Leuven
Rega Institute for Medical Research
www.farm.kuleuven.ac.be/farm/medchem

France

University of Liège
Mass Spectrometry Laboratory
www.ulg.ac.be/mslab/index.html

Germany

European Molecular Biology Laboratory
Protein & Peptide Group
mac-mann6.embl-heidelberg.de

Fraunhofer Institute for Food Technology and Packaging
Mass Spectrometry Laboratory
www.ilv.fhg.de/ms.html

University of Konstanz
Przybylski's group
www.ag-przybylski.chemie.uni-konstanz.de

Max-Planck-Institutes in Mulheim
Mass Spectrometry Laboratory
www.mpimuelheim.mpg.de/stoecki/mpikofo_mass.html

The Netherlands

Utrecht University
Department of Mass Spectrometry
www.chem.ruu.nl/amsmass/www/amsmass.html

Switzerland

Swiss Federal Institute for Environmental Science & Technology
Mass Spectrometry Facility
www.eawag.ch/dept/inter/ms/index.htm

United Kingdom

Cambridge University
Centre for Molecular Recognition
www-methods.ch.cam.ac.uk/meth/ms/msindex.html

Ludwig Institute for Cancer Research and University College London
Mass Spectrometry Facility
falcon.ludwig.ucl.ac.uk

University of Manchester
Michael Barber Centre for Mass Spectrom.
uchsg11.ch.umist.ac.uk/cms/index.htm

Nottingham Trent University
Creaser's Group
euler.ntu.ac.uk/anenvchem.html

Asia

Korea

Korea Basic Science Institute
Mass Spectrometry Research
comp.kbsc.re.kr/mass/mass.html

Australia

Association of Biomolecular Resource Facilities
www.medstv.unimelb.edu.au/ABRF.html

Melbourne

Royal Melbourne Institute of Technology
Mass Spectrometry Facility
minyos.its.rmit.edu.au/~rcmfa/index.html

Sydney

University of New South Wales
Biomedical Mass Spectrometry Unit
www.unsw.edu.au/clients/bmsu/home.html

University of Sydney
Mass Spectrometry Analytical Facility
www. pharm.su.oz.au/msf

Resources: Manufacturers

Here is an alphabetical list of the current manufacturers of LC/MS systems in the United States. For a broader list of companies (in the US & worldwide) that supply mass spectrometers, mass spectrometry components, liquid chromatography interfaces, instrument upgrades & services, see *1997 Directory of Mass Spectrometry Manufacturers and Suppliers*, by Stephen A. Lammert in the journal, Rapid Communication in Mass Spectrometry, Vol 11, pages 821-845 (1997).

ABB Extrel
(a Division of ABB Process Analytics, Inc.; supplier of APCI, particle beam, thermospray LC/MS interfaces for existing mass analyzers)
575 Epsilon Drive
Pittsburgh, PA 15238, USA
Telephone: 1-412-963-7530
Fax: 1-412-963-6578
E-mail: qms@extrel.com
Home page: abb.com/extrel

Altanta Research Lab. Supplies, Inc.
(supplier of electrospray LC/MS interface for existing mass analyzers)
3480 Tanbark Ct. N.E.
Atlanta, GA 30319, USA
Telephone: 1-404-255-4817
Fax: 1-404-847-9892

AMD Intectra USA
(supplier of electrospray & continuous flow FAB LC/MS interfaces for existing mass analyzers)
4940 Capitol Blvd.
Suite I-R
Raleigh, NC 27604, USA
Telephone: 1-919-876-1099
Fax: 1-919-876-1099

Analytica of Branford
(also a supplier of APCI & electrospray LC/MS interfaces for existing mass analyzers)
29 Business Park Drive
Branford, CT 06405, USA
Telephone: 1-203-488-8899
Fax: 1-203-481-0433
E-mail: aob.com

Ash Instruments
(supplier of continuous flow FAB LC/MS interface for existing mass analyzers)
5 Heather Close
Lyme Green Business Park
Maeclesfield, Cheshire SK11 0LR, UK
Telephone: 0-625-616431
Fax: 0-625-612494

BCP Instruments
(supplier of continuous flow FAB LC/MS interface for existing mass analyzers)
Burd-Club Lyon
47 Rue Maurice Flandin
Lyon 69003, France
Telephone: 72-31-66-27
Fax: 78-05-23-38

Buker, Daltonics
(Bruker Analytical Systems, Inc.)
19 Fortune Drive
Manning Park
Billerica, MA 01821-3991, USA
Telephone: 1-508-667-9580
Fax: 1-508-667-5993
E-mail: info@bruker.com
Home page: www.bruker.com

Finnigan Corporation
(A ThermQuest Company)
355 River Oaks Parkway
San Jose, CA 95134, USA
Telephone: 1-408-433-4800
Fax: 1-408-433-4823
E-mail: info@finnigan.com
Home pages: www.finnigan.com
and www.finnigan.co.uk

Hewlett-Packard Co.
1601 California Avenue
Palo Alto, CA 94304-1111, USA
Telephone: 1-415-857-6100, 1-800-227-9770
Fax: 1-415-852-8011
Home page: www.hp.com/go/chem

Hiden Analytical, Inc.
(supplier of continuous flow FAB LC/MS interface for existing mass analyzers)
75 Hancock Road, Suite D
Peterborough, NH 03458-1100, USA
Telephone: 1-603-924-5008
Fax: 1-603-924-5009

Hitachi Instruments, Inc.
3100 N. First Street
San Jose, CA 95134, USA
Telephone: 1-408-432-0520; 1-800-548-9001
Fax: 1-408-432-0704
E-mail: info@hii.hitachi.com
Home Page: www.hii.hitachi.com

IonSpec Corporation
16 Technology Drive, Suite 122
Irvine, CA 92618, USA
Telephone: 1-800-438-3867
Fax: 1-714-753-9121
E-mail: info@ionspec.com
Home page: www.ionspec.com

JEOL U.S.A., Inc.
11 Dearborn Road
Peabody, MA 01961-6043, USA
Telephone: 1-508-535-5900
Fax: 1-508-536-2205
E-mail: ms@jeol.com
Home page: www.jeol.com/ms

Kratos Analytical, Inc.
(supplier of continuous flow FAB LC/MS interface for existing mass analyzers)
535 E. Crescent Ave.
Ramsey, NJ 07446, USA
Telephone: 1-201-825-7500
Fax: 1-201-825-8659
E-mail: dsurman@kratos.com

Mass Evolution
(supplier of continuous flow FAB LC/MS interface for existing mass analyzers)
330 Meadowfern
Suite 113
Houston, TX 77067, USA
Telephone: 1-713-875-6219
Fax: 1-713-875-9658

Mass Spectrometry International Ltd.
(supplier of electrospray, particle beam & thermospray LC interfaces for existing mass analyzers)
12 Westpoint, Clarence Avenue, Trafford Park
Manchester, M17 1QS, UK
Telephone: 0-161-877-2402
Fax: 0-161-877-2403
E-mail: msi@dial.pipex.com

Micromass, Inc.
(purchased by Waters, Corp. in late 1997)
100 Cummings Center
Suite 407N
Beverly, MA 01915-6101, USA
Telephone: 1-508-524-8200; 1-800-390-4660
Fax: 1-508-524-8210
E-mail: 72551.1113@compuserve.com
Home page: www.micromass.co.uk

M-Scan, Inc.
(supplier of electrospray & continuous flow FAB LC/MS interfaces for existing mass analyzers)
137 Brandywine Parkway
West Chester, PA 19380, USA
Telephone: 1-215-696-8210
Fax: 1-215-696-8370

New Objective, Inc.
(supplier of nanospray sources, PicoTip, as an add-on for existing mass analyzers)
763 D Concord Avenue
Cambridge, MA 02138-1044, USA
Phone: 1-617-576-2255
Fax: 1-617-576-2266
E-mail: newobj@ix.netcom.com

Perkin-Elmer Corporation
761 Main Avenue
Norwalk, CT 06859-0001, USA
Telephone: 1-203-762-1000; 1-800-762-4000
Fax: 1-203-762-6000
E-mail: info@perkin-elmer.com
Home page: www.perkin-elmer.com/sc

PerSeptive Biosystems, Inc.
(Vestec Biospectrometry, also a supplier of thermospray LC/MS interfaces for existing mass analyzers. Merged with Perkin-Elmer in late 1997.)
500 Old Connecticut Path
Framingham, MA 01701, USA
Telephone: 1-508-383-7700; 1-800-899-5858
Fax: 1-508-383-7885
E-mail: info@pbio.com
Home page: www.pbio.com

Phrasor Scientific
(supplier of continuous flow FAB LC/MS interface for existing mass analyzers)
1536 Highland Ave.
Duarte, CA 91010, USA
Telephone: 1-818-357-3201
Fax: 1-818-357-3203

Premier American Technologies Corp.
(supplier of electrospray, particle beam, thermospray, & continuous flow FAB

LC/MS interfaces for existing mass analyzers)
1155 Zion Road Suite 101
Bellefonte, PA 16823, USA
Telephone: 1-814-353-0602
Fax: 1-814-353-0605

Protana
(The Protein Analysis Company, supplier of nanospray sources, Nano-Electrospray, as an add-on for existing mass analyzers)
Staermosegaardvej 16
DK-5230 Odense M, Denmark
Telephone: +45 63-157-215
Fax: +45-63-157-217
E-mail: protana@protana.com
Home page: www.protana.com

Sensar
(a division of Larson Davis, Inc.)
1652 West 820 North
Provo, UT 84601 USA
Telephone: 801-357-0177
Fax: 1-801-343-3617
E-mail: wrwest@lardav.com
Home Page: www.lardav.com

Scientific Instrument Services, Inc.
(supplier of continuous flow FAB LC/MS interface for existing mass analyzers)
1027 Old York Road
Ringoes, NU 08551-1039, USA
Telephone: 1-908-788-5550
Fax: 1-908-806-6631
E-mail: sismspec@aol.com
Home page: www.sisweb.com

Shimadzu
Shimadzu Scientific Instruments, Inc.
7102 Riverwood Drive
Columbia, MD 21046 USA
Phone: 1-410-381-1227 & 1-800-477-1227
Fax: 1-410-381-1222
E-Mail: webmaster@shimadzu.com
Home page: www.sii.shimadzu.com

Waters Corporation
34 Maple Street
Milford, MA 01757 USA
Telephone: 1-800-252-4752
Fax: 1-508-872-8890
E-mail: info@waters.com
Home page: www.waters.com

World Precision Instruments
(supplier of precision-made micropipettes for nanospray, as an add-on for existing mass analyzers)
175 Sarasota Center Blvd.
Sarasota, FL 34240-9258 U.S.A.
Phone: 1-941-371-1003
Fax: 1-941-377-5428
E-mail: wpi@wpiinc.com

Resources: Reconditioned Instruments

Daley-Hodkin Corporation
Autioneers
135 Pinelawn Road
Melville, NY 11747-3144
Telephone: 1-516-293-0200
Fax: 1-516-293-0328

Ford City Equipment Co.
Mac Industrial Park
P.O. Box 189
Ford City, PA 16226-0189
Telephone: 1-412-763-8321
Fax: 1-415-763-2065

International Equipment Trading Ltd.
960 Woodlands Parkway
Vernon Hills, IL 60061 USA
Telephone: 1-847-913-0777
Fax: 1-847-913-0785
E-mail: info@ietltd.com
Home page: www.ietltd.com
See also *LC-GC Buyers Guide 1997-1998* (Vol. 15, 1997) under Mass Spectrometric Detectors, LC.
See Lammert, S.A., "1997 Directory of Mass Spectrometry Manufacturers and Supplies," Rapid Commun. Mass Spectrom. 11, pages 821-845 (1997).

Resources: Instrument Leasing

LC/MS Manufacturers
Leasing plans are available from most manufacturers.

Mass Spectrometers
Listing of brokers, rental & lease companies. See Lammert, S.A., "1997 Directory of Mass Spectrometry Manufacturers and Supplies," Rapid Commun. Mass Spectrom. 11, pages 821-845 (1997).

Resources: LC/MS Services & Supplies

Custom Manufacturing/Modifications, Components, Small Parts Supplier, Instrument Upgrades, Training
See Lammert, S.A., "1997 Directory of Mass Spectrometry Manufacturers and Supplies," Rapid Commun. Mass Spectrom. 11, pages 821-845 (1997).

Chromatography & Mass Spectrometry
see *Laboratory Equipment's* yearly directory published every January. All products & manufacturers listed in their yearly directory can be accessed through their home page. Home page: www.labequipmag.com

Stable Isotopes

Cambridge Isotope Laboratories
50 Frontage Road
Andover, MA 01810 USA
Telephone: 1-800-322-1174; 1-508-749-8000
Fax: 1-508-749-2768
E-mail: cilsales@isotope.com
Home page: www.isotope.com

Dr. Ehrenstorfer GmbH
Bgm.-Schlosser-Str. 6 A
D-86199 Augsburg
Fed. Rep. of Germany
Telephone: +49-821-906080
Fax: +49-821-9060888
Home page: www.analytical-standards.com

Europa Scientific Inc.
259 Industrial Drive
Franklin, Ohio 45005-4429 USA
Telephone: 1-513-743 7211
Fax: 1-513-743 7217
Email: sales@europa-us.com
Home page: www.europa-uk.com

Martek Biosciences Corporation
6480 Dobbin Road
Columbia, MD 21045
Phone: (410) 740-0081
Fax: (410) 740-2985
Home page:
www.bio.com/home/martek/martek.html

Omicron Biomedical, Inc.
1347 N. Ironwood Drive
South Bend, IN 46615-3566 USA
Telephone: 1-219-287-6910
Fax: 1-219-287-7165
E-mail: omicron@skyenet.net
Home page: www.omicronbio.com or www.skyenet.net/omicron

Iostec, Inc.
3858 Benner Road
Miamisburg, OH 45342 USA
Telephone: 1-800-448-9760; 1-937-859-1808
Fax: 1-937-859-4878
E-mail: isosales@isotec.com
Home page: www.isotec.com

Toronto Research Chemicals, Inc.
2 Brisbane Road
North York, ON M3J 2J8 Canada
Telephone: 1-416-665-9696
Fax: 1-416-665-4439
E-mail: info@trc-canada.com
Home page: www.trc-canada.com

ULTRA Scientific
250 Smith Street
North Kingstown, RI 02852 USA
Telephone: 1-800-338-1754
Fax: 1-401-295-2330
E-mail: ultra@mail.ultrasci.com
Home page: www.ultrasci.com/ultrasci

Resources: Expositions & Trade Shows

See above. Most of the meetings sponsored by the various professional organizations, set aside an area for the various manufacturers & suppliers of mass spectrometric equipment to display their equipment.

Analitica - Latin America'98
International Exhibition and Conference for laboratory Technology and Biochemical and Instrumental Analysis. Their 1988 meeting will be held in Sao Paulo, Brazil (Jume 30 - July 2). Contact in the USA, T&G Services, Inc., Michelle Wolfson, 4220 Commercial Way, Glenview, IL 60025, USA. Telephone: 1-847-635-9960. Fax: 1-847-635-6801. E-mail: tgingred@aol.com. Contact in Brazil, Miller Freeman do Brazil, Luis Augusto Malandrino, R Traipu 657, CEP: 01235-000, Sao Paulo-SP-Brazil. Telephone: 00-55-11-3662-2021. Fax: 00-55-11-826-4458.

Pitt Con
(The Pittsburgh Conference on Analytical Chemistry and Applied Spectroscopy) is the largest gathering of instrument manufacturers in the United States. Their annual meeting is held in March. Contact: Pitt Con Office, 300 Penn Center Blvd., Suite 332, Pittsburgh, PA 15235-5503, USA. Telephone: 1-412-825-3220, 1-800-825-3221. Fax: 1-412-825-3224. E-mail: expo@pittcon.org.
Home page: www.pittcon.org.

Notes

[1] Davis, S., Botkin, J., The Monster Under the Bed: How Business is Mastering the Opportunity of Knowledge for Profit, Simon & Schuster: New York (1994).

Appendix B- Flow Diagram

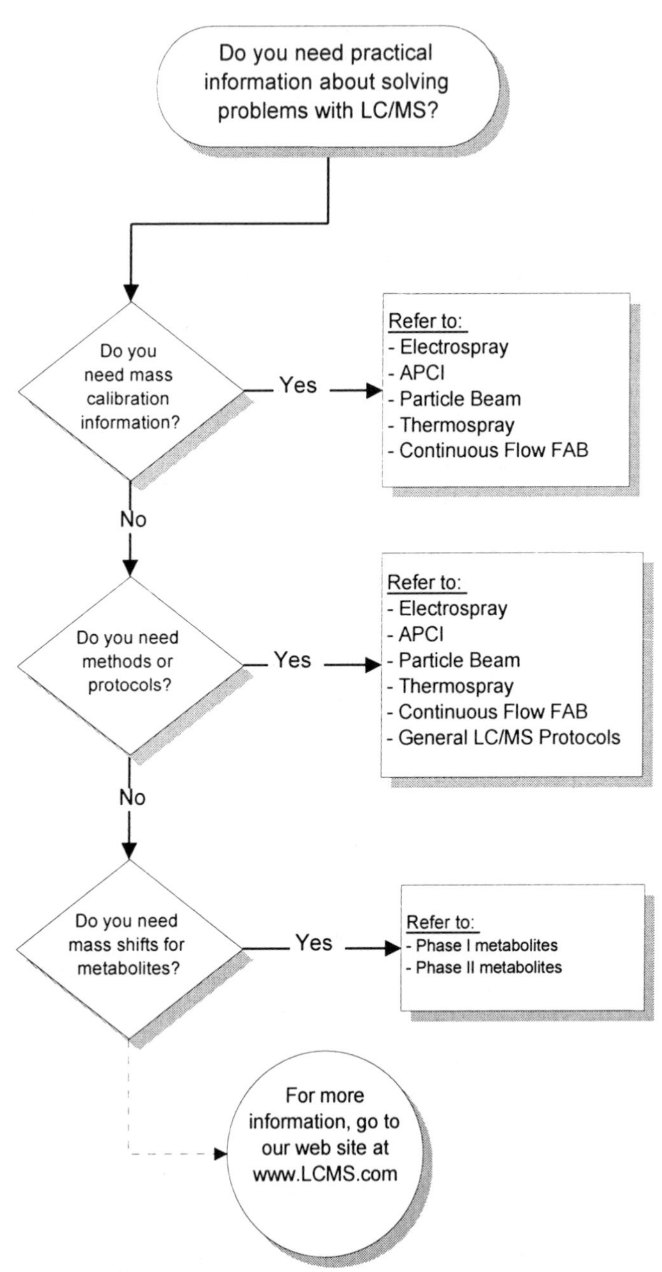

Appendix B

Practical Information for LC/MS

(or, Useful Information)

There are two "A"s required for successful LC/MS:
***Attitude**, and*
***Aptitude**.*
You need both.
Professor Jack Henion
Cornell University

You Can't Do Anything Without a Calibration!

The calibration[1] of the mass scale is one of the most important and routine procedures in the operation of all mass analyzers. In general, this is pat of system optimization. Remember, your mass spectral data is only as reliable as the calibration of the mass scale. All identification, interpretation, and quantification can be effected by errors in calibration, drift, and mass misassignment.

Various schemes exist to calibrate the mass scale in LC/MS. Choosing which scheme depends on the mass analyzer and the type of interface and samples you are running. All manufacturers provide recommended procedures to calibrate the mass scale of their instruments. Some modern instruments even have the ability to calibrate automatically. Here we include selected calibration standards.

Optimization of your LC/MS system involves tuning and calibrating with a reference compound that is suited for your intended experiments (e.g. high mass, high resolution, high scan speed.) Keep in mind that changing experimental parameters including tuning can affect peak shape and calibration. Remember that tune and calibration are linked. Don't throw in any arbitrary tune file without subsequently checking calibration.

Calibration: Electrospray[2]

Cesium Iodide Clusters

Cesium iodide-trifluoroacetic acid cluster ions generated with a pneumatic electrospray source is useful for positive and negative ionization in the 500-2100 dalton mass range.[3]

Table 1- CsI- Positive Ionization

Major +ES Mass Spectral Peaks for Cesium Iodide Clusters Quadrupole Instrument, $(Cs_{n+1}I_n)+$, n = 0 - 8					
n	Expected Nominal Mass	Relative Abundance[a]	n	Expected Nominal Mass	Relative Abundance[a]
0	133	53	5	1432	1
1	393	100	6	1692	2
2	653	4	7	1952	*
3	912	12	8	2211	*
4	1172	3			

*: indicates abundance <1%.

Table 2- CsI- Negative Ionization

Major -ES Mass Spectral Peaks for Cesium Iodide Clusters Quadrupole Instrument, $(Cs_nI_{n+1})^-$					
n	Expected Nominal Mass	Relative Abundanc[a]	n	Expected Nominal Mass	Relative Abundance[a]
0	127	100	5	1426	2
1	387	100	6	1686	*
2	647	30	7	1946	*
3	906	14	8	2205	*
4	1166	3			

*: indicates abundance <1%.

Sample Preparation: A 0.1 mole/L solution of CsI in aqueous acetonitrile. Other salts and acids (e.g., formic acid, rubidium iodide, cesium nitrate) were also reported to be useful.

Cytochrome C

The multi-charged spectrum of cytochrome C ($(M+H_n)^+$, n=7-19) is useful for tuning and mass calibration of positive ionization electrospray in the 650-1600 amu range.[4]

Table 3- Cytochrome C

Multi-charged +ES Mass Spectral Peaks for Cytochrome C* Quadrupole Instrument, Calculated MW 12,360.1					
Expected Nominal Mass	Expected Mass	Number of Charges (n)	Expected Nominal Mass	Expected Mass	Number of Charges (n)
652	651.54	19	1031	1031.02	12
688	687.68	18	1125	1124.65	11
728	728.07	17	1237	1237.02	10
774	773.51	16	1374	1374.36	9
825	825.02	15	1546	1546.02	8
884	883.87	14	1767	1766.74	7
952	951.79	13			

*: from horse heart, Sigma part#: C-7752 (Sigma, home page: www.sigma.sial.com).
Note: assumes H = 1.0070.

Myoglobin

The multi-charged spectrum of apomyoglobin (myoglobin minus the heme group) ($(M+H_n)^+$, n=9-24) is also useful for tuning and mass calibration of positive ionization electrospray in the 650-1600 amu range.[5]

Table 4- Myoglobin

Multi-charged +ES Mass Spectral Peaks for Apomyoglobin* Quadrupole Instrument, Calculated MW 16,951.49[6]					
Expected Nominal Mass	Expected Mass	Number of Charges (n)	Expected Nominal Mass	Expected Mass	Number of Charges (n)
707	707.32	24	1060	1060.48	16
738	738.03	23	1131	1131.11	15
772	771.53	22	1212	1211.83	14
808	808.02	21	1305	1304.97	13
849	848.58	20	1414	1413.63	12
893	893.19	19	1542	1542.05	11
943	942.76	18	1696	1696.16	10
998	998.15	17	1885	1884.51	9

*: from horse heart, Sigma part#: M1882. (Sigma, home page: www.sigma.sial.com).
Note: assume H = 1.0070, molecular weight of myoglobin = 16,951.49.

Quaternary Ammonium Salts

Quaternary salts (anion, M^+) are useful for tuning and mass calibration of positive ionization electrospray in the 70-500 amu mass range.

Table 5- Quaternary Ammonium Salts

Mass Spectral Peaks for Quaternary Ammonium Salts (+ES)			
Salt	CAS#	Mass anion (m/z)	Aldrich Chemical[a] (part #)
Tetramethylammonium Acetate	10581-12-1	74	24,507-0
Tetramethylammonium Chloride	75-57-0	74	T1,952-6
Tetraethylammonium Acetate	1158-59-7	130	20,558-3
Tetraethylammonium Chloride	56-34-8	130	11,304-2
Tetrabutylammonium Acetate	10534-59-5	242	33,599-1
Tetrabutylammonium Chloride	37451-68-6	242	34,585-7
Tetrapentylammonium Chloride	4965-17-7	298	25,896-2
Tetrahexylammonium Chloride	5922-92-2	354	26,383-4
Tetrahepthylammonium Bromine	4368-51-8	410	23,784-1
Tetraoctylammonium Bromine	14866-33-2	467	29,413-6

[a]Aldrich Chemical Co., home page: www.aldrich.sial.com.

Ultramark-1621

The electrospray spectrum of Ultramark-1621, a mixture of fluorinated phosphazenes, is useful for tuning and mass calibration of positive and negative ionization electrospray in the 900-2200 amu range.[7]

Table 6- Ultramark- Positive Ionization

Major +ES Mass Spectral Peaks for Ultramark-1621* Magnetic Sector Instrument, $(M+H)^+$					
Expected Nominal Mass	Expected Mass	Relative Abundance	Expected Nominal Mass	Expected Mass	Relative Abundance
922	922.010	2	1622	1621.965	70
1022	1022.004	7	1722	1721.959	70
1122	1211.997	40	1822	1821.952	40
1222	1221.991	50	1922	1321.946	20
1322	1321.964	70	2022	2021.940	7
1422	1421.978	80	2122	2121.933	2
1522	1521.972	100			

*: from PCR Inc., PO Box 1466, Gainesville, FL, 32602, USA, Telephone: 800-331-6313 or 904-376-8246, Fax: 904-371-6246. Part#: 11944-6.

Table 7- Ultramark- Negative Ionization

Major -ES Mass Spectral Peaks for Ultramark-1621* Magnetic Sector Instrument, $(M-CH_2(CF_2CF_2))^-$					
Expected Nominal Mass	Expected Mass	Relative Abundance	Expected Nominal Mass	Expected Mass	Relative Abundance
1106	1105.966	3	1706	1705.928	100
1206	1205.959	11	1806	1805.921	83
1306	1305.953	27	1906	1905.915	73
1406	1405.947	58	2006	2005.906	21
1506	1505.940	73	2106	2105.902	8
1606	1605.934	59			

*: from PCR Inc., Part#: 11944-6.

Sample Preparation: The sample was prepared by mixing a solution of 50 uL of Ultramark-1621 in 50 mL acetonitrile with a 50 mL solution of water, methanol, acetic acid (47/47/6, v/v). It was also noted that Ultramark-1621 is soluble in a variety of solvents. Other mass peaks were observed and were attributed to impurities and/or sodium adducts of Ultramark-1621.

Calibration: APCI

Perfluoro-amines

A mixture of perfluoro-amies (similar to compounds used for tuning and calibration of EI and CI sources, Ultramark 1621) have been proposed to be useful for APCI systems. At this time no spectra or specific components have been devolved.[8]

Water Clusters

Water clusters were one of the first compounds to be used to calibrate the mass analyzers for API sources. The spectrum of water clusters ((H2O)nH+ with n = 3-112) is useful for tuning and mass calibration of positive APCI in the 90-2000 amu range.[9] Eight cluster ions peaks were used for mass calibration in the 90-800 amu range.

Table 8- Water Clusters- Positive Ionization

Major +APCI Mass Spectral Peaks for Water Clusters Quadrupole Instrument, $((H_2O)_nH)^+$ with n=3-112					
Number of Water Molecules (n)	Expected Nominal Mass	Expected Mass	Number of Water Molecules (n)	Expected Nominal Mass	Expected Mass
5	91	91.061	25	451	451.272
10	181	181.113	30	541	541.325
15	271	271.166	35	631	631.377
20	361	361.219	40	721	721.430

Calibration: Particle Beam

Perfluoroacyl-Carbohydrates

A mixture of heptafluorobutyrate/pentafluoropropionate derivative of cellobiose has been reported to be non-contaminating and useful for tuning and mass calibration in the chemical ionization (+CI and -CI) mode only.[10] It is useful for mass calibration in the 1000-2000 amu mass range for magnetic sector and quadrupole instruments.

Table 9- Perfluoro....- Positive Ionization

Major +CI Mass Spectral Peaks for Perfluoro-Carbohydrates (Heptafluorobutyrate/Pentafluoropropionate Derivative of Cellobiose) Mix (3:1) heptafluorobutyrate/pentafluoropropionate derivative of cellobioseMagnetic Sector Instrument, Methane +CI, M^+					
Expected Mass	Relative Abundance[a]	Expected Mass	Relative Abundance[a]	Expected Mass	Relative Abundance[a]
307.0204	26	896.9673	23	1446.9601	*
320.9997	21	946.9641	11	1482.9581	*
418.9976	49	1118.9813	3	1496.9574	1
468.9944	100	1168.9781	4	1546.9542	2
518.9916	88	1184.9530	*	1596.9510	2
603.9811	12	1218.9749	4	1646.9478	1
604.9890	9	1232.9741	*	1690.9376	*
653.9781	18	1254.9560	*	1696.9446	*
654.9860	13	1268.9717	1	1740.9344	*
703.9749	10	1282.9709	1	1790.9312	*
704.9828	8	1332.9677	2	1840.9280	*
796.9737	10	1382.9645	2	1890.9248	*
846.9705	21	1432.9613	2		

*: indicates abundance <1%.

a: indicates temperature & source pressure dependence.

Other mass peaks were observed & were attributed to impurities.

Table 10- Perfluoro....- Negative Ionization

Major -CI Mass Spectral Peaks for Perfluoro-Carbohydrates (Heptafluorobutyrate/Pentafluoropropionate Derivative of Cellobiose) Mix (3:1) heptafluorobutyrate/pentafluoropropionate derivative of cellobiose Magnetic Sector Instrument, Methane -CI, M⁻					
Expected Mass	Relative Abundance[a]	Expected Mass	Relative Abundance[a]	Expected Mass	Relative Abundance[a]
307.9730	3	1279.9620	4	1643.9443	22
326.9714	31	1311.9537	6	1646.9478	12
338.9714	15	1329.9606	10	1659.9392	11
357.9698	15	1361.9505	7	1662.9427	12
376.968	100	1379.9592	11	1676.9383	4
388.9682	22	1393.9603	7	1689.9298	60
407.9666	26	1412.9587	7	1696.9446	5
426.9650	100	1426.9543	10	1709.9360	19
743.9434	*	1443.9571	16	1739.9266	74
793.9702	*	1446.9606	12	1759.9328	20
812.9686	1	1462.9555	10	1789.9234	60
843.9670	1	1476.9511	21	1809.9296	18
862.9654	2	1493.9539	28	1822.9210	4
890.9603	2	1496.9574	24	1839.9202	28
893.9638	1	1512.9523	17	1853.9194	5
912.9622	1	1526.9480	31	1859.9264	9
943.9606	*	1543.9507	42	1872.9178	3
990.9539	2	1546.9542	35	1889.9170	7

*: indicates abundance <1%.
a: indicates temperature & source pressure dependence.

Other mass peaks were observed & were attributed to impurities.

Perfluorotributylamine- (PFTBA)

Perfluorotributylamine (also referred to as PTA or FC-43) is a fluorinated amine. It is "the" standard tuning compound (EI, +CI, and -CI) and mass marker typically used with most quadrupole and ion trap instruments. It is useful for mass calibration in the 30-650 amu mass range.[11]

Table 11- PFTBA

Physical/Chemical Properties of PFTBA	Value
CAS#	311-98-7
Formula Weight	671.08
Nominal Mass	671
Formula	$(CF_3(CF_2)_3)_3N$
Boiling Point	170 - 180 °C
Density	1.88
Available from: PCR Inc.	Part# 18201-4
Available from: Scientific Instrument Services	Part# FC-43-x

PCR Inc., PO Box 1466, Gainesville, FL, 32602, USA;
telephone: 800-331-6313 or 904-376-8246; fax: 904-371-6246.
Scientific Instrument Services, Inc., 1027 Old York Road, Ringoes, NJ, 08551, USA; telephone: 908-788-5550; fax: 908-806-6631;
home page: www.sisweb.com.

Note: The reported relative abundance will differ within the same class of mass analyzers (e.g., quadrupoles) and between different mass analyzers (e.g., quadrupoles and magnetic sectors).

Table 12- PFTBA- Electron Ionization

Major EI Mass Spectral Peaks for PFTBA (FC-43) Quadrupole Instrument, $M^{+\cdot}$					
Expected Mass	Relative Abundance	Expected Mass	Relative Abundance	Expected Mass	Relative Abundance
50	10	131	48	314	2
69	100	132	2	326	*
70	1	145	1	352	1
76	*	150	3	376	*
81	*	164	1	414	3
93	1	169	4	415	*
95	*	176	2	426	1
100	18	181	2	464	3
101	*	214	1	502	6
112	*	219	51	503	*
113	*	220	2	614	1
114	8	226	1	615	*
119	13	264	13		
		265	*		

*: indicates abundance <1%.

Table 13- PFTBA- Chemical Ionization

Major +CI Mass Spectral Peaks for PFTBA (FC-43) (Methane, Source Chamber Pressure ~1 x 10-4 torr) Quadrupole Instrument, M^+					
Expected Mass	Relative Abundance[a]	Expected Mass	Relative Abundance[a]	Expected Mass	Relative Abundance[a]
69	5	376	*	576	10
70	*	377	1	577	*
131	2	414	100	614	25
132	*	415	8	615	2
219	20	464	2	652	70
220	1	502	25	653	8
314	1	503	2		
315	*				

*: indicates abundance <1%.

a: indicates temperature source pressure dependence.

Table 14- PFTBA- Electron Ionization

Major -CI Mass Spectral Peaks for PFTBA (FC-43) (Methane, Source Chamber Pressure ~2 x 10-4 torr) Quadrupole Instrument, M⁻					
Expected Mass	Relative Abundance[a]	Expected Mass	Relative Abundance[a]	Expected Mass	Relative Abundance
219	1	414	10	514	2
264	5	415	*	557	15
302	30	433	5	558	1
303	1	452	60	583	2
333	10	453	7	595	70
352	6	464	*	596	7
364	3	476	10	633	100
383	5	477	*	634	18

*: indicates abundance <1%.
a: indicates temperature source pressure dependence.

Perfluorotripentylamine- (PFTPA)

Perfluorotripentylamine (also referred to as FC-70) is another fluorinated amine used as a standard tuning compound (EI +CI and -CI) and mass marker. It is useful for mass calibration in the 30-780 amu mass range.[12]

Table 15- PFTPA

Physical/Chemical Properties of PFTPA	Value
CAS#	338-84-1
Formula Weight	821.16
Nominal Mass	821
Formula	$(CF_3(CF_2)_4)_3N$
Boiling Point	210-220 °C
Density	1.94
Available from: PCR Inc.	Part# 12206-9
Available from: Scientific Instrument Services	Part# FC-70- x

PCR Inc., PO Box 1466, Gainesville, FL, 32602, USA;
telephone: 800-331-6313 or 904-376-8246; fax: 904-371-6246.
Scientific Instrument Services, Inc., 1027 Old York Road, Ringoes, NJ, 08551, USA; telephone: 908-788-5550; fax: 908-806-6631;
home page: www.sisweb.com.

Table 16- PFTPA- Electron Ionization

Major EI Mass Spectral Peaks for PFTPA (FC-70) Quadrupole Instrument, $M^{+\cdot}$					
Expected Mass	Relative Abundance	Expected Mass	Relative Abundance	Expected Mass	Relative Abundance
50	*	131	10	269	20
69	100	132	*	270	1
70	1	151	1	314	3
93	1	169	3	364	*
100	10	181	14	514	*
101	*	182	*	526	*
114	3	219	1	564	*
119	12	231	*	602	*
120	1				

*: indicates abundance <1%.

Table 17- PFTPA- Chemical Ionization

Major -CI Mass Spectral Peaks for PFTPA (FC-70) (Methane, Source Chamber Pressure ~2 x 10-4 torr) Quadrupole Instrument, M^{-}					
Expected Mass	Relative Abundance[a]	Expected Mass	Relative Abundance[a]	Expected Mass	Relative Abundance[a]
219	2	483	2	602	3
281	1	502	3	614	7
314	2	514	13	626	2
333	5	515	1	633	1
352	5	526	1	664	4
383	1	533	5	676	1
412	2	534	*	683	1
414	3	552	100	733	2
433	4	553	10	745	5
450	1	564	4	783	15
452	4	576	1	784	2
476	2	583	2	833	1

*: indicates abundance <1%.

a: indicates temperature source pressure dependence.

Calibration: Thermospray

Polypropylene Glycol-425

Polypropylene glycol-425 (also referred to as PPG-425) is a standard thermospray tuning compound (+TS) and mass marker typically used with most quadrupole instruments. It is useful for mass calibration in the 150-750 amu mass range.[13]

Table 18- PPG

Physical/Chemical Properties of PPG-425	Value
CAS#	25322-69-4
Average Mass (M_n)	~425
Formula	$H(OCH(CH_3)CH_2)_nOH$
Viscosity	80 centistokes
Density (25 °C)	1.004
Available from: Aldrich Chemical	Part# 20,230-4

Aldrich Chemical Co., 1001 West Saint Paul Avenue, Milwaukee, WI, 53233, USA; telephone: 1-800-962-9591; fax: 1-800-962-9591; e-mail: aldrich@sial.com; home page: www.aldrich.sial.com.

Table 19- PPG- Positive Ionization

Positive Thermospray Ionization, Spectrum of PPG-425 (Possible Protonated & Adduct Ions)				
n	M	$(M+H)^+$	$(M+NH_4)^+$	$(M+Na)^+$
1	76	77	94	99
2	134	135	152	<u>157</u>
3	192	193	510	515
4	250	<u>251</u>	268	273
5	308	<u>309</u>	<u>326</u>	331
6	366	367	<u>384</u>	389
7	424	425	<u>442</u>	447
8	482	483	<u>500</u>	505
9	540	541	<u>558</u>	563
10	598	599	<u>616</u>	621
11	656	657	<u>674</u>	679
12	714	715	<u>732</u>	737

Note: Underlined number denotes masses observed in mass spectrum.

Table 20- PPG- Major Ions

Major +TS Mass Spectral Peaks for PPG-425 (Source Chamber Pressure ~2 x 10-4 torr) Quadrupole Instrument					
Expected Mass	Relative Abundance[a]	Expected Mass	Relative Abundance[a]	Expected Mass	Relative Abundance[a]
157	25	326	30	558	40
158	1	327	4	559	10
215	15	384	60	616	25
216	1	385	12	617	2
251	20	442	100	674	70
252	2	443	25	675	8
309	10	500	80	732	1
310	*	501	20	733	*

*: indicates abundance <1%.

a: indicates temperature source pressure dependence.

<u>LC/MS Conditions</u>: Flow 2 mL/min.; 1/1: methanol/100 mM ammonium formate. Thermospray tip: 257 °C; Thermospray Block: 200 °C. Source pressure: ~2 x10-4 torr.

<u>Sample Preparation</u>: A 1% solution (v/v) of PPG-425 dissolved in the mobile phase.

Other mass peaks were observed and were attributed to impurities and/or sodium or potassium adducts.

Ultramark 1621

Ultramark 1621 is a fluorinated phosphazine traditionally used for high mass calibration (EI) for magnetic sector instruments and also for electrospray LC/MS instruments (see above). Here it is used with the LC/MS interface, thermospray (+TS), on a 2000 amu quadrupole instrument. It is useful for mass calibration in the 921-2100 amu mass range.[14]

Table 21- Ultramark 1621

Physical/Chemical Properties of Ultramark-1621	Value
CAS#	na
Formula Weight	a mixture
Formula	$N_3P_3O_3R_6$ R = $-CH_2(CF_2-CF_2)_xH$ x = 1,2,3
Available from: PCR Inc.	Part# 11944-6

na: not available.
PCR Inc., PO Box 1466, Gainesville, FL, 32602, USA,
Telephone: 800-331-6313 or 904-376-8246, Fax: 904-371-6246.

Table 22- Ultramark 1621- Positive Ionization

Major +TS Mass Spectral Peaks for Ultramark-1621 (Source Pressure ~ 2 x 10^{-4} torr) Quadrupole Instrument, $(M+H)^+$					
Expected Mass	Relative Abundance	Expected Mass	Relative Abundance	Expected Mass	Relative Abundance
922	*	1322	100	1722	35
1022	45	1422	92	1822	5
1122	70	1522	95	1922	*
1222	82	1622	60		

*: indicates abundance <1%.
a: indicates temperature & source pressure dependence.

LC/MS Conditions: Flow 2 mL/min.; 1/1: acetonitrile/100 mM ammonium acetate. Thermospray tip: 260 °C; Thermospray Block: 200 °C. Source pressure: ~2 x10-4 torr.

Sample Preparation: Equal volumes of Ultramark 1621 and mobile phase. The mixture was shaken until an emulsion formed. A 50 uL aliquot of the emulsion was injected.

Calibration: Continuous Flow FAB

Cesium Iodide (CsI)

Cesium iodide is a standard FAB tuning compound (+CFF) and mass marker typically used with both magnetic sector and quadrupole mass analyzers. It is useful for mass calibration in the 130-5000 amu mass range.

Table 23- Ultramark 1621- Positive Ionization

Physical/Chemical Properties of Cesium Iodide	Value
CAS#	7789-17-5
Formula Weight	259.81
Formula	CsI
Melting Point	626 °C
Density	4.510
Available from: Scientific Instrument Services	Part# FAB-C2

Scientific Instrument Services, Inc., 1027 Old York Road, Ringoes, NJ, 08551, USA; telephone: 908-788-5550; fax: 908-806-6631; home page: www.sisweb.com.

Table 24- Ultramark 1621- Positive Ionization

Major +FAB Mass Spectral Peaks for Cesium Iodide Clusters $(Cs_{n+1}I_n)+$, n=0-8					
n	Nominal Mass	Exact Mass	n	Nominal Mass	Exact Mass
0	133	132.905	11	2991	2990.814
1	393	392.715	12	3251	3250.624
2	653	652.525	13	3510	3510.434
3	912	912.335	14	3770	3770.244
4	1172	1172.145	15	4030	4030.054
5	1432	1431.955	16	4290	4289.864
6	1692	1691.765	17	4550	4549.674
7	1952	1951.548	18	4809	4809.484
8	2211	2211.385	19	5069	5069.294
9	2471	2471.195	20	5329	5329.104
10	2731	2731.004			

*: indicates abundance <1%.

Methods & Protocols: Electrospray

LC/MS & LC/MS/MS

Mike Lee and his colleagues at Bristol-Myers Squibb Pharmaceuticals Research Institute describes approaches to make LC/MS techniques the cornerstone of accelerated drug development, and the identification of impurities, degradants and biomolecules.

(Lee, M.S., Kerns, E.H., Hail, M.E., Liu, J., Volk, K.J., "Recent applications of LC-MS techniques for the structure identification of drug metabolites and related compounds," LC-GC 15, pages 542-558, 1997).

LC/MS/MS for Protein Sequencing:

John Yates and his colleagues at the Department of Molecular Biotechnology, (Univ. of Washington) describe the use of electrospray (ES) LC/MS/MS to match spectral data to their corresponding translated nucleotide sequences from a protein database.

((a) Link, A.J., Eng, J., Yates, J.R., "Analyzing complex biological systems using micro-LC-ESI-MS-MS," *American Laboratory*, pages 27-30, July 1996. (b) Yates, J.R., Eng, J.K., McCormack, A.L., "Mining genomes: Correlating tandem mass spectra of modified and unmodified peptides to sequences in nucleotide databases," *Anal. Chem.* 67, pages 3202-3210, 1995. (c) Eng, J.K., McCormack, A.L.,Yates, J.R., "An approach to correlate tandem mass spectral data of peptides with amino acid sequences in a protein database," *J. Am. Soc. Mass Spectrom.* 5, pages 976-989, 1994.)

LC/MS Used to Study Protein Folding

David Booth at the Immunological Medical Unit (Royal Postgraduate Medical School) and Carol Robinson at the Oxford Centre for Molecular Sciences (University of Oxford) and colleagues describe the utilization of electrospray to study the aggregation of human lysozyme variants underlying amyloid fibrillogenesis.

(Robinson, C,V., Grob, M., Eyles, S.J., Ewbank, J.J., Mayhew, M., Hartl, F.U., Dobson, C.M., Radford, S.E., "Conformation of GroEL-bound a-lactalbumin probed by mass spectrometry," *Nature*, 372, pages 646-651 (1994). Booth, D.R., Sunde, M., Bellotti, V., Robinson, C.V., Hutchinson, W.L., Frase, P.E., Hawkins, P.N., Dobson, C.M., Radfor, S.E., Blake, C.C., Pepys, M.B., "Instability, unfolding and aggregation of human lysozyme variants underlying anyloid fibrillogenesis," *Nature* 385, pages 787-793.

LC/MS/MS in Clinical Diagnosis:

David Millington, Donald Chase and their associates at Duke University Medical Center describe semi-automated electrospray (ES) LC/MS/MS methods, based on isotope dilution, for the analysis of amino acids and acylcarnitines in human whole blood, plasma and urine. Complete metabolic profiles of target compounds were generated in less than one minute per sample. These semi-automated methods will allow the analysis of up to 500 samples per day. They describe one method for the diagnosis homocystinuria and other hypermethioninemias from dried blood spots on newborn screening cards. They predict that utilizing LC/MS/MS will successfully detect hypermethioninemias with very low rates for false positives and false negatives.

((a) Millington, D.S., Chace, D.H., Stevens, R.D., "Rapid, automated analysis of diagnostically important metabolites in physiological matrices by LC/ESI-MS/MS," *Proceedings of the 12th LC/MS Montreux Symposium*, Hilton Head Island, SC, November 1-2, 1995. (b) Chace, D.H., Hillman, S.L., Millington, D.S., Kahler, S.G., Adam, B.W., Levy, H.L., "Rapid diagnosis of homocystinuria and other hypermethioninemias from newborns' blood spots by tandem mass spectrometry," *Clin. Chem.* 42, pages 349-355, 1996.)

LC/MS Assay of Recombinant Product Drugs

Ragulan Ramanathan (from Washington University), Walter Zielinski (from the FDA) and coworkers describe a protocol to test the integrity of recombinant proteins and the consistency from batch to batch of their preparations. They used the techniques of ESI to assay the rate of H/D exchange of recombinant insulins. The assay could be conducted at one time point for a sample size of less than 2 ug.

(Ramanathan, R., Gross, M.L., Ziellinski, W.L., Layloff, T.P., "Monitoring recombinant protein drugs: A study of insulin by H/D exchange and electrospray ionization mass spectrometry," *Anal. Chem.* 69, pages 5142-5145, 1997.)

Methods & Protocols: APCI

Screening Small-Molecule Libraries with LC/MS:

Jack Henion and associates at New York State College of Veterinary Medicine describe a method combining immunoaffinity and atmospheric pressure ionization (APCI) LC/MS for screening a combinatorial library of 20-30 closely related benzodiazepines.

(Wieboldt, R., Zweigenbaum, J. Henion, J., "Immunoaffinity ultrafiltration with Ion Spray HPLC/MS for screening small-molecule libraries," *Anal. Chem.* 69, pages 1683-1691, 1997.)

High Speed LC/MS/MS in PK Studies:

Timothy Olah and colleagues at Merck Research Laboratories describe the use of LC with tandem MS for determining mixtures of drug candidates as part of a high-throughput bioanalysis involving pharmacokinetics (PK) and metabolic stability studies. This approach relies on (1) simple isolation methods, (2) isocratic chromatography, and (3) LC/MS/MS conditions of atmospheric pressure chemical ionization (APCI) and Reaction Monitoring that can be adapted and applied to numerous agents. In a single analysis, they were able to determine the plasma concentration of 12 drug candidates. Using this method the authors were able to screen more than 400 compounds in a 6 month period.

(Olah, T.V., McLoughlin, D.A., Gilbert, J.D., "The simultaneous determination of mixtures of drug candidates by liquid chromatography atmospheric pressure chemical ionization mass spectrometry as an *in vivo* drug screening procedure," *Rapid Commun. Mass Spectrom.* 11, pages 17-23, 1997.)

LC/MS/MS for Drug Analysis:

Gary Bowers and colleagues at GlaxoWellcome, Inc. describe a totally automated approach for combining solid phase extraction (SPE) with flow injection analysis and atmospheric pressure chemical ionization (APCI) LC/MS/MS to analyze several thousand samples in support of Phase I clinical studies.

(Bowers, G.D., Clegg, C.P., Hughes, S.C., Harker, A.J., Lambert, S., "Automated SPE and tandem MS without HPLC columns for quantifying drugs at the picogram level," *LC-GC* 15, pages 48-53, 1997.)

Methods & Protocols: Particle Beam

Alcohols Increase Response:

J. White and associates from Health and Safety Laboratory describe the use of various alcohols (methanol, ethanol, propanol, butanol, pentanol) to improve the response of particle beam (PB) LC/MS. They demonstrated that the post-column addition of propanol improved the mass spectral response for neburon 7-fold.

(White, J., Brown, R.H., Clench, M.R., " Particle beam liquid chromatography/mass spectrometry analysis of hazardous agricultural and industrial chemicals," *Rapid Commun. Mass Spectrom.* 11, pages 618-623, 1997.)

Micro Flow Rate LC/PB/MS for Pesticides:

Achille Cappiello and his associates at the Istituto di Scienze Chimiche (Universita di Urbino) describe a particle beam (PB) LC/MS method for the determination of 32 base/neutral and 13 acidic pesticides in water. They utilized a particle beam interface they developed which allows the introduction of a much lower mobile-phase flow rates, ~ 1 microliters/min., into the mass spectrometer ion source.

(Cappiello, A., Famiglini, G., Bruner, F., "Determination of acidic and basic/neutral pesticides in water with a new microliter flow rate LC/MS particle beam interface," *Anal. Chem.* 66, pages 1416-1423, 1994.)

Non-Volatile Compounds in Drinking Water:

Tom Behymer and associates from the EPA/USA describes the use of particle beam (PB) LC/MS for analyzing semivolatile contaminants, benzidines and nitrogen-containing pesticides, which may be present in drinking water or drinking water sources.

(Behymer, T.D., Bellar, T.A., Ho, J.S., Budde, W.L., "Method 553: Determination of Benzidines and Nitrogen-Containing Pesticides in Water by Liquid-Liquid Extraction or Liquid-Solid Extraction and Reverse Phase High Performance Liquid Chromatography/Particle Beam/Mass Spectrometry," pages 173-212, IN: Methods for the Determination of Organic Compounds in Drinking Water, Supplement II; US-EPA Report, EPA/600/R-92/129; Order No. PB92-207703, 270 pages, 92(23), Abstract No. 266,774, 1992.)

M.P. Additive to Enhance Signal Intensity:

Tom Bellar and his associates at the US EPA describe the enhancement of signal intensity in two instances of particle beam (PB) LC/MS during a reverse-phase gradient elution HPLC separation: (1) the chromatographic coelution of analytes and (2) when ammonium acetate was added to the mobile phase. They attributed this enhancement to both improved chromatographic efficiency and a phenomenon they described as the "carrier effect".

(Bellar, T.A., Behymer, T.D., Budde, W.L., "Investigation of enhanced ion abundances from a carrier process in high-performance liquid chromatography particle beam mass spectrometry," *J. Am. Soc. Mass Spectrom.* 1, pages 92-98, 1990.)

Polar Compounds of Environmental Interest:

Tom Behymer and his associates at the US EPA describe the use of two particle beam (PB) LC/MS systems as a possible general purpose, broad-spectrum analytical method for the determination of nonvolatile organic compounds in environmental samples. Instrument tuning and signal optimization, detection limits, linear calibration range and signal intensity were among the performance factors evaluated.

(Behymer, T.D., Bellar, T.A., Budde, W.L., "Liquid chromatography/particle beam/mass spectrometry of polar compounds of environmental interest," *Anal. Chem.* 62, pages 1686-1690, 1990.)

LC/PB/MS:

Jack Northington and co-workers at West Coast Analytical Service, Inc. describe the particle beam (PB) LC/MS retention times and detection limits of approximately 50 compounds. They analyzed amines, organometallics, and organic acids. Although the detection limits for the analytes were variable, many had detection limits of 5-50ng in a full scan (electron ionization) acquisition. These detection limits were equivalent to 0.25 - 2 micrograms/mL (in a 20 uL injection volume), which is similar to the sensitivities commonly required in the U.S. EPA Method 8270 using GC/MS.

(Northington, D.J., Hovanec, B.M., Shelton, M., "Particle beam LC-MS in environmental analysis," *Am. Env. Lab.* 2, pages 34-41, 1990.)

Methods & Protocols: Thermospray

Compounds of Environmental Interest:

Tom Bellar and Bill Budde from the EPA-USA describe a thermospray (TS) LC/MS method for 52 pesticides and other compounds of environmental interest.

(Bellar, T.A., Budde, W.L. "Determination of nonvolatile organic compounds in aqueous environmental samples using liquid chromatography/mass spectrometry," *Anal. Chem.* 60, pages 2076-2083, 1988.)

Pesticides in Ground Water:

Dietrich Volmer and Karsten Levsen describe a method that employs off-line and on-line solid-phase extraction and thermospray (TS) LC/MS analysis with time-scheduled selected ion monitoring for environmental monitoring for a series of 51 nitrogen- and phosphorus-containing pesticides. On column detection limits range from 40 - 600pg.

(Volmer, D., Levsen, K., "Mass spectrometric analysis of nitrogen-containing pesticides by liquid chromatography-mass spectrometry," *J. Am. Soc. Mass Spectrom.* 5, pages 655-675, 1994.)

Methods & Protocols: General LC/MS

Phosphate Buffers, No More a Problem

Alexander J. Debets of N.V. Organon (Oss, The Netherlands) modified a Dionex (Sunnyvale, CA) electrosuppression device to continuously remove bad ions (e.g., phophate) from the sample stream after the separation. Phosphate was released with hydroxide. This improved the detection of basic drugs with MS, and UV also. When the voltage was turned off, within 10 minutes the LC/MS interface was plugged up.

(Stevenson, R., editor's report from "PBA '97, The Era of Triple-Stage Rockerts," *Biotechnology Laboratory* 15, pages 4- 10, 1997).

Pictograms for MS & LC/MS:

Wolf Lehmann at German Cancer Research Center has proposed a visual documentation scheme for depicting various mass analyzers, CID schemes, ionization, LC/MS, MS and MS/MS modes.

(Lehmann, W.D., "Pictograms for experimental parameters in mass spectrometry," *J. Am. Soc. Mass Spectrom.* 8, pages 756-759, 1997).

Analytical Method Development & Validation:

Michael Swartz (Waters Corp.) and Ira Krull (Northeastern University) have written a book that covers the subject of method development and validation. They are also hosts of a quarterly column in LC-GC addressing the needs of chemist who validate analytical methods in a regulated environment.

(Swartz, M.E., Krull, I.S., "Analytical Method Development and Validation," Marcel Dekker: New York, 1997. Krull, I, Swartz, M., "Validation Viewpoint. Introduction: National and international guidelines," *LC-GC* 15 pages 534-540, 1997.)

Column Selection & Switching for LC/MS:

William Letter reviews the use of automatic column selection and column switching techniques for LC/MS.

(Letter, W.S., "Automated column selection and switching systems for HPLC," *LC-GC* 15, pages 508-512, 1997.)

HPLC Solvents:

Paul Sadek provides detailed coverage of HPLC solvents currently in use.

(Sadek, P.C., *The HPLC Solvent Guide*, John Wiley: New York 1997.)

Method Validation:

Ludwig Huber (Hewlett-Packard) has written a book to guide the analytical chemist through the entire validation process; validation of software, and the validation and verification of the entire system (computers and instrumentation).

(Huber, L., *Validation of Computerized Analytical Systems*, Interpharm Press: Buffalo Grove, 1995.)

Information: Mass Shifts

Table 25- Mass Shifts for Phase I Metabolites

Reaction	Phase I Modifications Example	Mass Change
Oxidation		
Aromatic Hydroxylation	Ar-H ⇒ Ar-OH	+15.9949
Aliphatic Hydroxylation	R-CH_2 ⇒ CHOH	+15.9949
N-Dealkylation	R-NH-CH_3 ⇒ R-NH_2	-14.0156
O-Dealkylation	R-O-CH_3 ⇒ R-OH	-14.0156
S-Dealkylation	R-S-CH_3 ⇒ R-SH	-14.0156
Epoxidation	R-HC=CH-R ⇒ R-CHOCH-R	+15.9949
Desulfuration	P=S ⇒ P=O	+15.9771
Sulfoxidation	R-S-R ⇒ R-SO-R	+15.9949
N-Hydroxylation	R-NH-CO-R ⇒ R-NOH-CO-R	+15.9949
Amine Oxidase	R-NO_2 ⇒ R-NH_2	-29.9741
Dehydrogenases		
Alcohol & Aldehyde Dehydrogenase	R-CH_2-OH ⇒ R-COH	-2.0156
	R-COH ⇒ R-COOH	+15.9949
	R-CH_2-OH ⇒ R-COOH	+13.9792
Hydratase		
Epoxide Hydratase	R-CH(O)CH-R ⇒ R-CH(OH)-CH(OH)-R	+18.0105

Table 26- Mass Shifts for Phase II Metabolites

Phase II Modifications		
Reaction	Example	Mass Change
Glucuronyl Transferase		
Aliphatic or Aromatic Alcohols	R-OH ⇒ R-O-Glu Ar-OH ⇒ Ar-O-Glu	+176.0320
Carboxylic Acids	R-COOH ⇒ R-COO-Glu	+176.0320
Sulfhydryl Compounds	R-SH ⇒ R-S-Glu	+176.0320
Amines	R-NH_2 ⇒ R-NH-Glu	+176.0320
Glutathione S-Transferase		
Alkene	R-CH=CH_2 ⇒ R-CH_2-CH_2-S-G	+309.0994
	R-CH=CH_2 ⇒ R-CH_2-CH_2-S-Mercapturic Acid	+162.0224
Aromatic Epoxide	Ar ⇒ Ar(O)	+15.9949
	Ar(O) ⇒ Ar-(OH)-S-G	+309.0994
	Ar(O) ⇒ Ar-(OH) S-Mercapturic Acid	+162.0224
Aliphatic Nitro	R-CH_2-NO_2 ⇒ R-CH_2-S-G	+262.0749
	R-CH_2-NO_2 ⇒ R-CH_2-S-Mercapturic Acid	+115.0217
Sulfotransferase		
Alcohols	R-OH ⇒ R-OSO_3H	+79.9568
Aromatic Amines	Ar-NH_2 ⇒ Ar-NH-SO_3H	+79.9568

Table 27- Mass Shifts for Phase II Metabolites

Reaction	Phase II Modifications Example	Mass Change
Amino Acid Conjugases		
Glycine	R-COOH ⇒ R-CO-Gly	+57.0214
Alanine	R-COOH ⇒ R-CO-Ala	+71.0371
Ornithine	R-COOH ⇒ R-CO-Orn	+114.0793
Methyl Transferase		
Amines	R-OH ⇒ R-OCH_3	+14.0156
Alcohols	R-OH ⇒ R-OCH_3	+14.0156
Carbamate	R_2N-CS-SH ⇒ R_2N-CS-CH_3	+14.0156
N-Acetyl Transferase		
Amines	R-NH_2 ⇒ R-NH-CO-CH_3	+42.0105
Hydrazine	R-NH-NH_2 ⇒ R-NH-NH-CO-CH_3	+42.0105
Sulfonamide	R-NH_2 ⇒ R-NH-CO-CH_3	+42.0105

Information: Symbols for LC/MS[15]

Table 28- Pictograms- Scan Modes

MS Scan Modes	
+ FAB	Positive CF-FAB, using a BE sector instrument; regular scan mode.
+ CI 281	Particle Beam or Moving Belt with a positive CI source, using a BE sector instrument; electric sector (E) SIM & magnetic sector (B) set at m/z 281 at full acceleration voltage.
+ ESI	Positive ESI, using a Q or QQQ instrument; regular scan mode.
+ APCI 762	Positive APCI, using a Q or a QQQ instrument; SIM of m/z 762.
EI	Particle Beam or Moving Belt with an EI source, using a Q or QQQ instrument; regular scan mode.
+ TS	Positive TS, using a Q or a QQQ instrument; SIM of more than one m/z value.
+ ESI 40	Positive ESI, using a Q or a QQQ instrument; skimmer-CID (sCID) at 40 V offset & regular scan mode.

Adapted with permission from reference 15.

Table 29- Pictograms- Scan Modes

MS Scan Modes	
-/+ ESI -/+	Dual polarity ESI programmed sCID, using a Q or a QQQ instrument; fragment ion detection in SIM mode in low mass region, positive ESI for molecular ion detection in high mass region using regular scan function. Polarity switch occurs when sCID offset is switched off.
-/+ ESI -/+/+	Dual polarity ESI programmed sCID, using a Q or a QQQ instrument; negative & positive polarity for fragment ion detection in scan mode, positive polarity for molecular ion detection in high mass region using regular scan mode. Polarity switch occurs when sCID offset is switched off.

Adapted with permission from reference 15.

Table 30- Pictograms- MS/MS Modes

MS/MS Scan Modes	
+ FAB 705 He 800	Positive CF-FAB, using a hybrid (BEQQ) instrument; product scan of m/z 705 (MS^2). He as collision gas at 800 V offset.
+ TS 760 Ar 25	Positive TS, using a QQQ instrument; product scan of m/z 760 (MS^2). Ar as collision gas at 25 V offset.

Adapted with permission from reference 15.

Table 31- Pictograms- MS/MS Modes

MS/MS Scan Modes	
+ ESI ▬ ✳ 184 ▬ He 25	Positive ESI, using a QQQ instrument; precursor scan of m/z 184. He as collision gas at 25 V offset.
+ APCI ▬ ✳ ▬ (98) Xe 20	Positive APCI, using a QQQ instrument; neutral loss scan for loss of mass 98. Xe as collision gas at 20V offset.
+ APCI 898 ✳ 800 Xe 20	Positive APCI, using a QQQ instrument; selected reaction monitoring for the fragmentation m/z 898 to 800 (SRM). Xe as collision gas at 20V offset.
+ ESI ✳ 50 ▬ ✳ He 30 211	Positive ESI, using a QQQ instrument; sCID (50 V offset) plus precursor ion scan for m/z 211. He as collision gas at 30V offset.
- ESI 465 ✳ He 25	Negative ESI, using a QQ-TOF (orthogonal extraction) instrument; product ion scan for m/z 465 (MS^2). He as collision gas at 25V offset.

Adapted with permission from reference 15.

Table 32- Pictograms- MS/MS Modes

MS/MS Scan Modes	
	Positive ESI, using an IT instrument; product scan of m/z 1000 (MS^2).
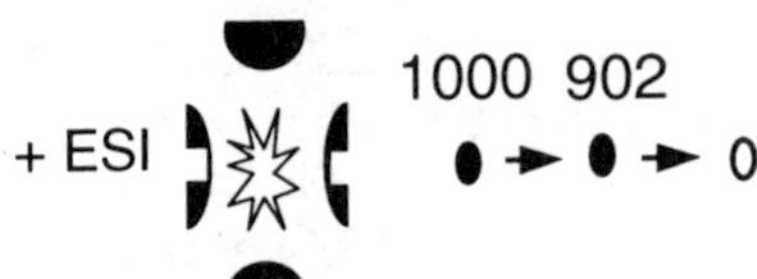	Positive ESI, using an IT instrument; product scan of m/z 902, formed by CID of m/z 1000 (MS^3).

Notes

[1] There is a lot of confusion when the word "calibration" is mentioned. Calibration in terms of chromatography typically refers to the quantitation curve, while in terms of mass spectrometry calibration also refers to the quantitation curve, but it also refers to the calibration of the mass scale, mass calibration.

[2] Other methods have been proposed and used, such as: Polytetramethylene ether glycol (Larsen, B.S., McEwen, C.N., "An electrospray ion source for magnetic sector mass spectrometers," *J. Am. Soc. Mass Spectrom.* 2, pages 205-211 (1991)); Poly(ethylene) and poly(propylene) oxides and their sulfates (Cody, R.B., Tamura, J., Musselman, B.D., "Electrospray ionization/magnetic sector mass spectrometry: Calibration, resolution, and accurate mass measurements," *Anal. Chem.* 64, pages 1561-1571 (1992)).

[3] (a) Anacieto, J.F., Boyd, R.K., Pleasance, S., Sim, P.G., Thibault, P., "Calibration techniques in IONspray mass spectrometry," *Proceedings of the 39th ASMS Conference on Mass Spectrometry and Allied Topics*, Nashville, TN, May 19-24, pages 1366-1367, 1991. (b) Anacieto, J.F., Pleasance, S., Boyd, R.K., *Org. Mass Spectrom.* 27, Pages 660-666, 1992. Calibration of ion spray mass spectra using cluster ions.

[4] Data acquired in the authors' lab.

[5] Data acquired in the authors' lab.

[6] Zaia, J., Annan, R.S., Biemann, K., Rapid Commun. Mass Spectrom. 6, pages 32-36, 1992. *The correct molecular weight of myoglobin, A common* calibrant for mass spectrometry.

[7] Moini, M., Rapid Commun. Mass Spectrom. 8, pages 711-714, 1994. *Ultramark 1621 as a calibration/reference compound for mass spectrometry. II. Positive- and negative-ion electrospray ionization.*

[8] Imatani, K., Werlich, M., Flanagan, M., Reinecke, E., Ong, A., "Novel calibration system for API LC/MS," Proceedings of the 45th ASMS Conference on Mass Spectrometry and Allied Topics, Palm Springs, California, June 1-5, 1997.

[9] Xu, Y., Crutchfield, S., "A pratical calibration method for liquid chromatograph-mass spectrometry systems," Proceedings of the 45th ASMS Conference on Mass Spectrometry and Allied Topics, Palm Springs, California, June 1-5, 1997.

[10] Hsu, F.F., Tyler, A.N., Sherman, W.R., Biol. Mass Spectrom. 20, pages 339-344, 1991. *Readily synthesized calibration compounds for quadrupole and magnetic instruments for use over the mass range to 2000 daltons.*

[11] Data acquired in the authors' lab.

[12] Data acquired in the authors' lab.

[13] Data acquired in the authors' lab.

[14] Data acquired in the authors' lab.

[15] Lehmann, W.D., "Pictograms for experimental parameters in mass spectrometry," J. Am. Soc. Mass Spectrom. 8, pages 756-759, 1997.

Appendix C- Flow Diagram

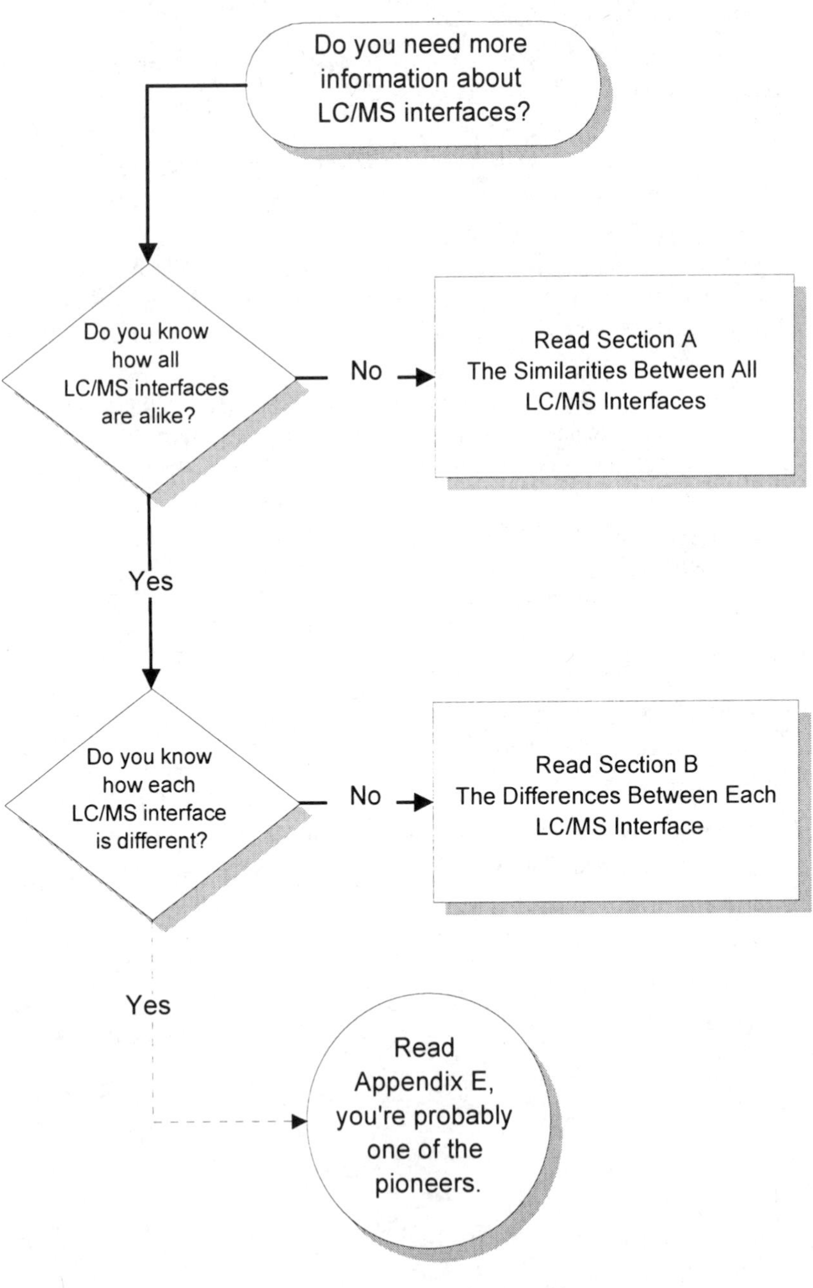

Appendix C

Interfaces- How do they work?

(or, Interface Fundamentals)

Similarities Between All Interfaces

To gain a appreciation for the operation of LC/MS interfaces, it is useful to understand the similarities between them. For example, all interfaces function to convert a dissolved solute (your analyte) into gas-phase ions. Table I shows the physical state-of-matter at each end of the LC/MS interface with a description of the conversion process that occurs in-between. Note the common processes.

All LC/MS interfaces have to accommodate the following processes; 1) evaporate liquids into gases, 2) ionize neutrals into charged species, and 3) evacuate large amounts chromatographic eluent to maintain the required vacuum for mass analysis. This process is further illustrated by Figure 1 showing the chemical pathway from one state (the LC) to the other (the MS). This model of LC/MS has been described as a "thermodynamic perspective."[1]

Table I- All LC/MS Interfaces Address This Change-of-State

Physical Parameter	LC / MS
1) State-of-Matter	liquid (evaporation) gas
2) Charge State	neutral[2] (ionization) ion
3) Pressure	760 Torr (evacuation) 10^{-5}-10^{-8} Torr

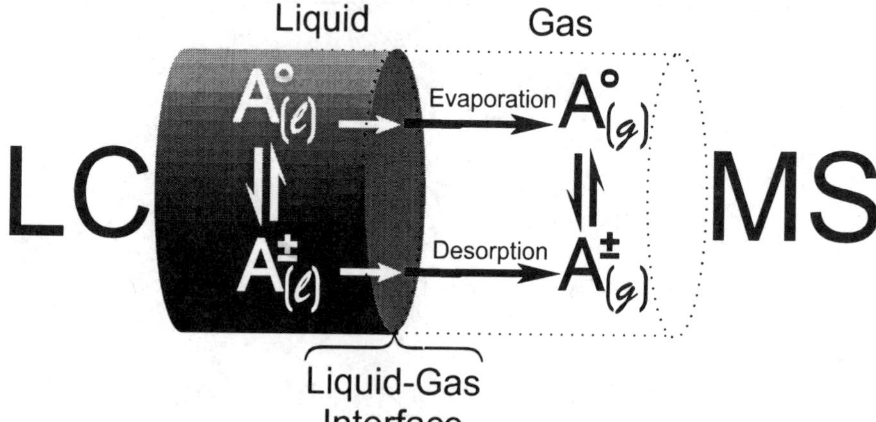

Figure 1- Change-of-state diagram that applies to all LC/MS interfaces. Whether the analyte is neutral in the liquid phase [$A^{o}_{(l)}$] or charged [$A^{\pm}_{(l)}$], the analyte must make its way to a surface to evaporate (or desorb to [$A^{\pm}_{(g)}$]) into the gas phase. Once in the gas phase, a neutral analyte [$A^{o}_{(g)}$] must be ionized to accommodate mass analysis.[3]

The model above is intended to illustrate the similarity between all LC/MS interfaces. This figure shows, in an idealized fashion, the pathways taken by all interfaces begins and ends at essentially the same place. Only the pathway in-between will create the seemingly subtle but important differences between one interface and another.

Four of the six interfaces we discuss utilize an aerosol generation process to augment evaporation or ion desorption. Aerosols are very important in LC/MS because they provide the surface area required for your analytes and the surrounding solvent to evaporate (or desorb) into the vapor phase. Evaporation is an interfacial phenomena, that is, only molecules at a surface will evaporate. The more surface; the more evaporation. From this point of view, it is understandable that developers of interfaces struggle relentlessly to improve their aerosol generation capabilities. Also note that electrospray is the most efficient aerosol generation process of all.[4]

Another important consideration is the energy requirements for this thermodynamic change-of-state. The energy requirements to facilitate the change-of-state are independent of path. The energy expenditures are similar for all interfaces. How is it, then, that each interface behaves so differently?

Differences Between Each Interface

Evaporation, ionization, and evacuation are common to all interfaces; however, the specific method or pathway of evaporation, ionization, and evacuation can be quite different with each interface. In this section we will mainly focus on these three processes to explain many of the advantages and limitations of one interface over another. The value of learning the similarities and differences between each technique at a more fundamental level should assist you in making selections between different techniques or developing sound strategies for interface optimization.

One important theme that we will continually repeat throughout this section is energy. Since all interfaces have essentially the same start- and end-states, by the first law of thermodynamics they will require about the same amount of energy to affect a change-of-state. A major distinction between these interfaces is found in the amount and type of energy that is used to drive the state change. For example, every chemist knows that high temperature will promote increased internal energy of a molecule through random processes and potentially cause degradation. Heating a molecule is clearly the less favorable pathway. The LC/MS technologies that minimized the amount of heat imparted to the analyte are the ones that have the least degradation.

Table II- Comparison Between the Different LC/MS Interfaces

Interface	Evaporation	Ionization	Evacuation
Electrospray	desorption at atmospheric pressure	ions in solution	staged reduction, sampling ions from atmospheric pressure
APCI	high temperature (at atmospheric pressure)	APCI	staged reduction, sampling ions from atmospheric pressure
Particle Beam	high temperature (at low pressure)	EI, CI,......	staged reduction, sampling neutral particles
Continuous Flow FAB	desorption	plasma from primary ion beam	direct introduction at low flow
Thermospray	high temperature	reagent ions in solution- (and gas-phase)	staged reduction, sampling ions from ca. 10 Torr
MALDI	desorption	plasma from high energy pulsed laser	direct introduction of solid sample

Fundamentals: Electrospray

Electrospray (ES) is the most unique interface in LC/MS because the ion generation process imparts very little internal energy into the analyte. This is because the energy driving the thermodynamic change-of-state is electrostatic potential energy from a high voltage applied between the electrospray needle and collector (Figure 2). This process is best understood by considering the circuit configuration below. The needle is one electrode, the collector is the other, respective oxidation and reduction processes are required in order for current to flow. The eluent from the liquid chromatograph is an integral part of the circuit. The current of the circuit is limited by the properties of the liquid flowing through the tube. Ions in solution migrate from the bulk to the surface of the liquid under the influence of the high field strength near the tip of the needle. Counterions are depleted from the surface of the liquid leaving a net charge. The characteristic "cone-jet" geometry is the result of the liquid accelerating toward the collection electrode. When the electrostatic forces exceed the inward forces from the surface tension a cone-jet is created. The highly charged surface of the liquid jet immediately disrupts into droplets that accelerate toward the collector electrode.

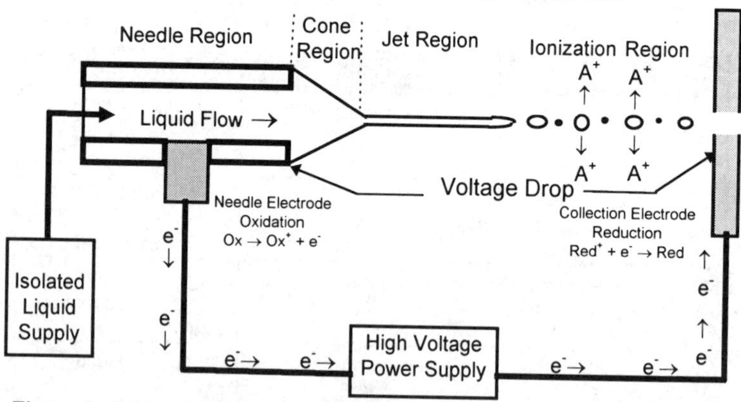

Figure 2- Schematic diagram of electrospray represented as an electrical circuit with the motion of the charged droplets completing the circuit. The high voltage applied to the needle electrode (in this case) drives the deformation of the liquid into a cone-jet geometry.[5] All variations of electrospray; including ion spray, turbospray, ultraspray, and nanospray conform to this basic circuit.

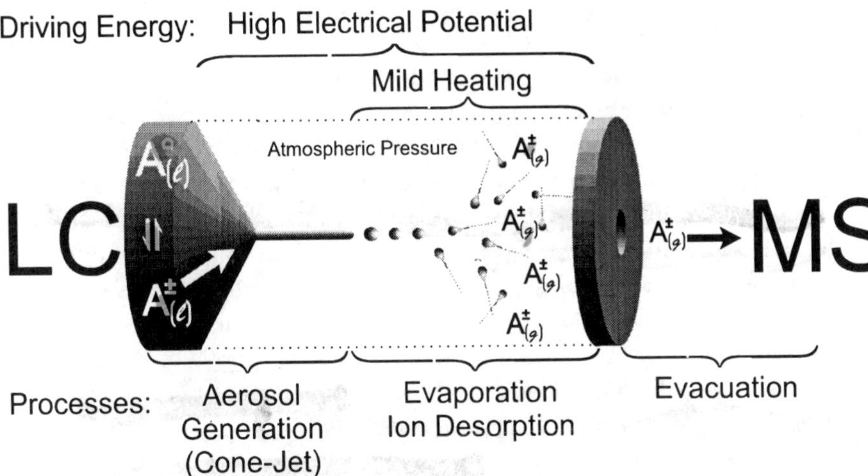

Figure 3- Schematic diagram of the mechanism of electrospray showing cone-jet formation, droplet formation, droplet evaporation and disruption, ion formation, and finally sampling into the vacuum system from atmospheric pressure.

The charged droplets will evaporate in the region between the needle and the collector (Figure 3). The excess charge on the droplet concentrates on the surface as the droplet evaporates. This increases the field around the droplet to the point where the surface deforms into tear-shaped emitters of smaller particles.[6] Depending on the initial droplet size, the particles leaving can either be streams of smaller particles that continue the process, or discrete solvated surface ions. With continued evaporation this tear-shaped geometry is stable until the charge is depleted, the analyte crystallizes into solid particles, or the droplets completely evaporate. The significance of this process is that it occurs at relatively low temperatures.

Stripped of their counterions, the surface ions do not have any difficulty leaving the surface. In fact, they are pushed off the surface by repulsion from other like-charged ions on the surface. The ions do not leave the surface as "naked" ions, rather, they carry with them their solvation sphere into the gas phase. At atmospheric pressure, collisions with the surrounding gases quickly desolvate the solvent-clustered ion resulting in a low internal energy quasi-molecular ion (or multicharged ion).

Fundamentals: APCI

APCI is the ideal complement to electrospray because of the common interfacing hardware for sampling ions from atmospheric pressure into the vacuum (Figure 5). This is where the similarities end. In stark contrast to the room temperature spray of ES, APCI utilizes a pneumatic nebulizer to create a fine spray that is directed into a heated region at 450 - 550°C. The high temperature is used to desolvate the droplets and subsequently evaporate the solutes into the vapor phase. Although high temperature is required to supply the heat of vaporization to the flowing solvent, the analyte is not directly exposed to these high temperatures. However, thermally labile species will decompose at the elevated temperatures of APCI.

As discussed in *Chapter 12*, neutral species are more easily vaporized than charged species. The analyst should always consider the charge state (acid-base equilibrium) of their analytes in solution when operating APCI. Once in the vapor phase, neutral solute molecules are ionized with reagents shown below (Figure 4).

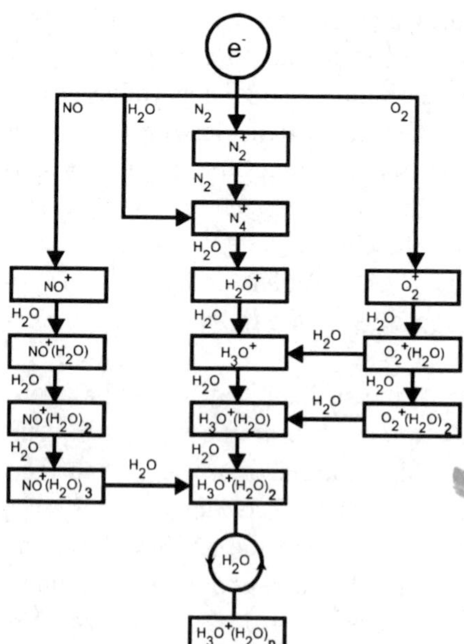

Figure 4. There is a significant benefit to ionization at atmospheric pressure because all species in the gas phase undergo significant collisions with surrounding gases. When ionized by discharge or beta-emitter source of electrons, the surrounding gases (N_2, H_2O, O_2, air) interact and cascade through the reactions shown to the left.[7] The major ions produced in air are N_2^+, O_2^+, H_2O^+, and NO^+. The primary reagent ion that interacts with gas phase neutral analytes is shown at the bottom, $H_3O^+(H_2O)_n$.

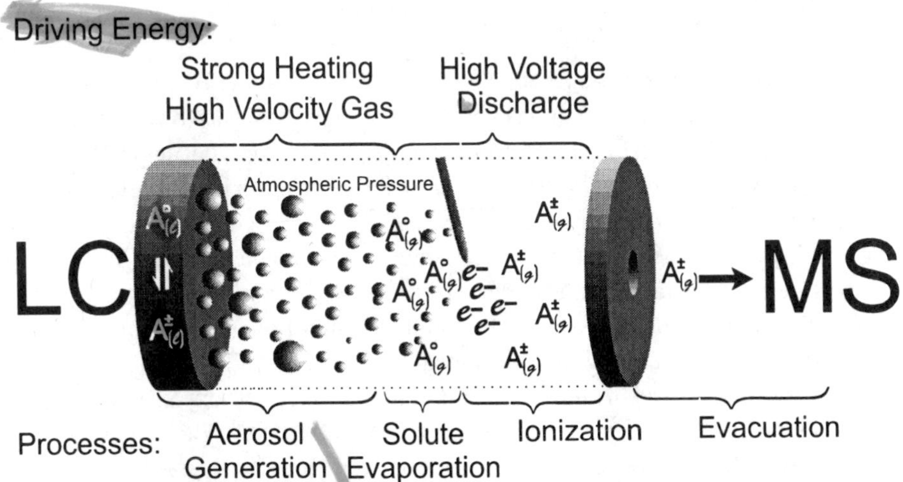

Figure 5- Schematic diagram of the mechanism of APCI showing aerosol generation, desolvation, solute evaporation, discharge and ionization, and finally sampling into the vacuum system from atmospheric pressure.

APCI can occur in a number of ways; including, proton transfer and charge exchange in the positive ion mode, and proton abstraction or electron capture in the negative ion mode.[8] Since water cluster ions are a major source of reagent ions, the proton affinity of these clusters relative to analyte ions will have a profound effect on sensitivity.[9] Below in Figure 6 is a graph of the sensitivity regions for APCI based on proton affinity.

Figure 6-Estimated Response of APCI relative to proton affinity. Note that the optimum response region is for analytes with proton affinity greater water. Also be aware that your mobile phase solvents will compete with your analyte for protons. Keep in mind proton affinity of <u>all</u> sample components, they all compete for the same protons.

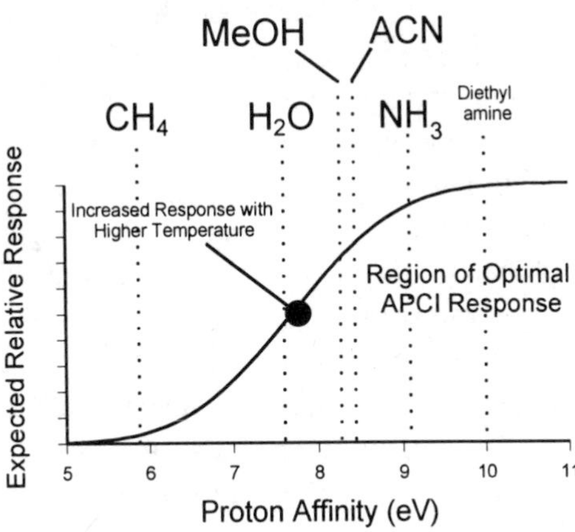

Fundamentals: Particle Beam

Contrary to the popular designation, particle beam (PB) is also an atmospheric pressure interface. At the time PB was developed, most of the interfaces under development were spraying inside the vacuum system.[10] The developers of PB realized that rapid evaporation of high flows of liquid could more efficiently be accomplished at atmospheric pressure. Figure 8 shows the decrease in the rate of evaporation as a function of pressure.[11] Particle beam operates in the same fashion to APCI in that higher flows are accommodated by nebulizing and desolvating the liquid flow. This is accomplished by heating either the nebulizer or the expansion region downstream from the nebulizer. Unlike APCI, PB typically uses helium because of its high thermal conductivity to assist in the desolvation process. Also different from APCI is the degree of heating required for PB. The objective of PB is to dry the liquid droplets into solute particles. Excessive heating will drive volatile analytes into the gas phase in the expansion region causing poor transport to the mass analyzer.

Once solvent-depleted solute particles have been formed, they are swept via viscous flow through a single exit aperture from the desolvation region into the vacuum system. The pressure drop across the aperture causes a high velocity aerosol beam, whereby the lighter solvent gases and helium expand radially, while the heavier solute particles travel on the beam axis through several skimmers until they reach the ion source region of a conventional mass spectrometer. Therefore, PB is considered a "transport" interface in that it simply accommodates the transport of enriched solute into almost any convention ion source.

Before ionization occurs, the particles must impact with the surface of the ion source and "flash vaporize."[12] Once in the vapor phase, solute molecules can be ionized with EI or CI processes. One important advantage of decoupling solute introduction and ionization processes is the ability to select specific chemical reagent gases (e.g. ammonia, methane, isobutane).

An electron micrograph of a particle beam collected in the ion source is shown in order to better visualize the process (Figure 9).

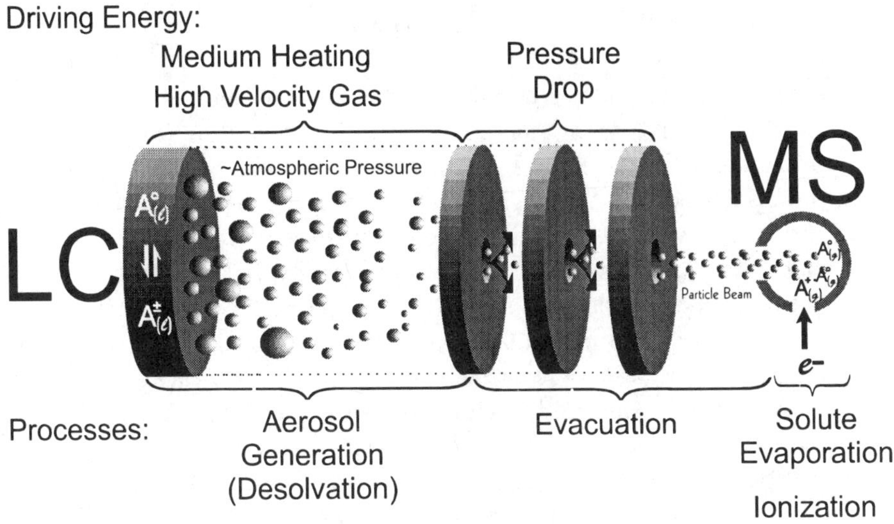

Figure 7- Schematic diagram of the processes associated with particle beam LC/MS showing aerosol generation, desolvation, sampling of the particle beam, solute evaporation in the ion source, and electron ionization to produce gas-phase ions.

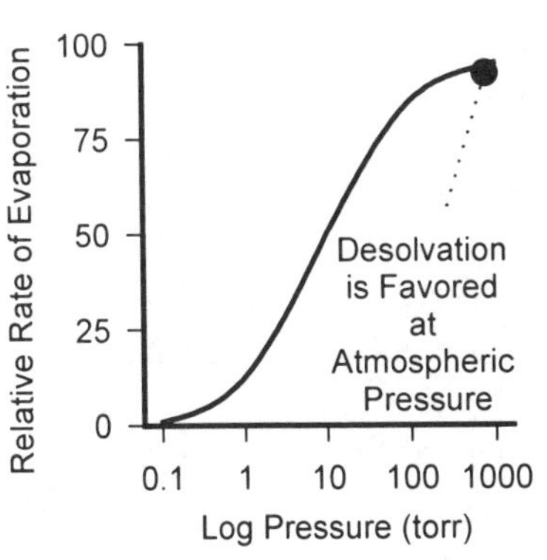

Figure 8- Rate of evaporation as a function of pressure. This shows that atmospheric pressure is the optimal pressure regime for high rates of evaporation.

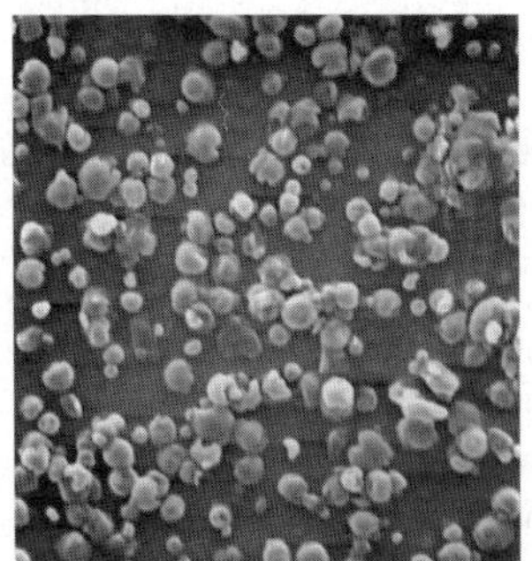

Figure 9- Electron micrograph of dry solute particles (of beta blocker drug atenolol) collected on the surface of an electron ionization source. The collected particles resulted from particle beam introduction into a cold ion source. In actual operating conditions the source would be several hundred degrees and the particles would "flash vaporize."

Fundamentals: Continuous Flow-FAB

The simplest interface is no interface. Continuous flow FAB (CFFAB) simply introduces sample directly into the high vacuum of the mass spectrometer through a capillary tube at low flow rates (low microliter/min). At these flows, the entire gas load from the evaporated flow can be accommodated by the source vacuum pumps of the mass analyzer. Typically, a matrix (glycerol, thioglycerol) is added to the mobile phase at 5-10%. No transport losses occur in CFFAB making it one of the most sensitive techniques available in LC/MS. In addition, CFFAB typically operates with a continuous flow of liquid and a continuous primary beam of ions, resulting in a continuous stream of secondary ions. Consequently, continuous introduction mass analyzers such as sectors and quadrupoles have been most commonly implemented.

The ionization process is accomplished by creating "primary" ions (or atoms)[13] at very high potential relative to the end of the capillary tubing. These ions can be created in the gas phase through discharge processes ($Xe^{o}{}_{(g)}$ going to $Xe^{+}{}_{(g)}$) or they can be created by boiling Cs pellets at high current producing Cs^{+}. The primary ions are accelerated to extremely high kinetic energy (5kV to 20kV) and focused at the tip of the capillary (Figure 10). The higher the voltage applied, the more stringent the vacuum requirements to avoid electrical discharge or arcing.

Ionization occurs when the energy from the primary ion partitions into the liquid. Figure 11 illustrates the disruption of the surface when a primary ion hits. This is conceptually described as a "billiard-ball" process where the incident energy is transferred through the liquid resulting in desorption of surface layers of the target.[14]

The sensitivity of CFFAB is highly dependent upon the matrix, solvents, and other sample components. Desorption, being an interfacial phenomena, requires that the analyte be at or near the surface when the primary ion hits. The analyst must consider aspects of surface activity in order to optimize response.[15] In CFFAB charged species tend to migrate to the surface. In this way we can equate electrospray and CFFAB in that analyte sensitivity is a competitive process for positions on the surface. Anything that can be done to the analyte to promote charge may enhance sensitivity (e.g. pH, alkali metals, surfactants, derivatize).

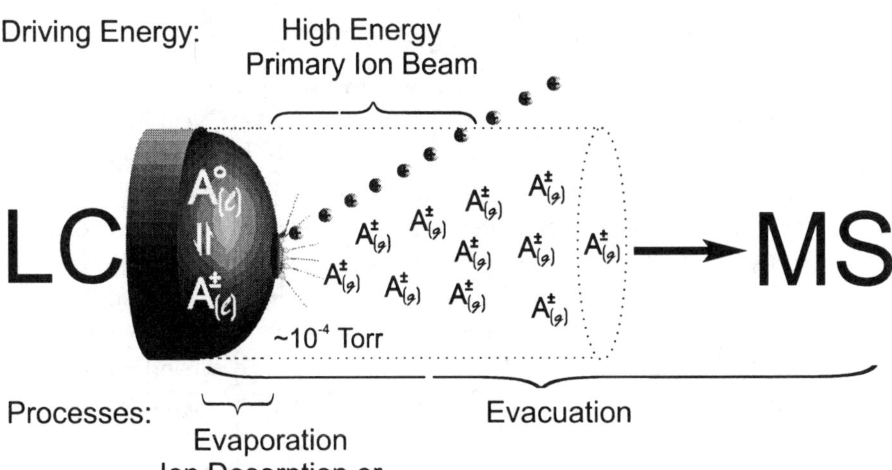

Figure 10- Schematic diagram of the processes associated with continuous flow FAB LC/MS showing the high energy primary ion beam intercepting with a liquid bead (containing matrix) at the exit of a direct liquid insertion probe. Secondary ions including analyte ions desorb from the liquid matrix into the gas phase.

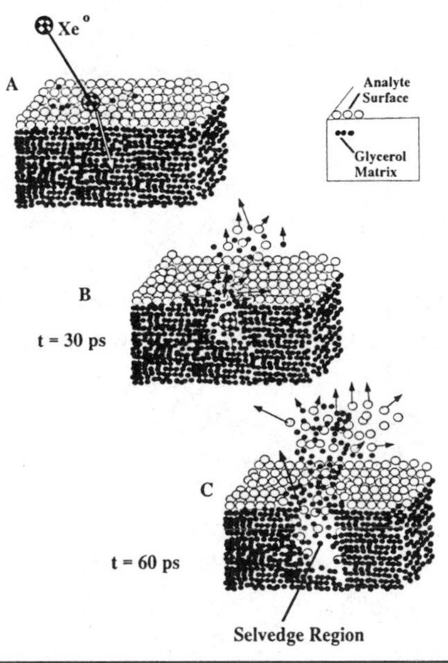

Figure 11- Desorption of analyte molecules during FAB ionization. (A) A xenon atom accelerated to 8 kV hits the matrix surface. (B) A thermal spike is generated in a region local to the Xe atom penetration. (C) Ion emission occurs from the selvedge region within a few seconds. (Reproduced with permission from reference 14)

Fundamentals: Thermospray

If electrospray is the most efficient aerosol generator in terms of surface area per volume, thermospray is the most efficient in terms of vaporization. Thermospray (TS) generates an aerosol by applying heat directly to the liquid while flowing through the tubing. This is much more efficient than other techniques at getting heat directly into the liquid, and consequent getting the liquid into the vapor phase. The mobile phase literally boils inside the nebulizer tubing (see Figure 13). The elegance of this technique was found in the clever method devised by Marvin Vestal and coworkers for controlling the exact amount of heat applied to the liquid.[16] By placing thermocouples on the nebulizer tubing to monitor the temperature of the tube, precise control of the exact amount of heat applied to the liquid was accomplished. This feedback is important to the performance of TS since the ion production is highly dependent upon the percent evaporation of the solvent inside the tubing.

Thermospray ionization is a unique in that it results directly from the aerosol generation process (Figure 12). When flowing liquid containing a conductive buffer (e.g. NH_4OAc) is boiled inside the capillary tubing, highly charged droplets are formed due to "static electrification."[17] As the droplets evaporate inside the tube, the increased charge density of the droplets promotes field induced "ion evaporation."[18] The ions produced by this desorption process may come from any performed ion in solution, including the buffer or analyte ions. This process is essentially the same as electrospray. The ions produced by ion evaporation serve as reagent ions for gas-phase neutrals to undergo chemical ionization. Neutral analytes present in solution can evaporate through thermal processes. From this vantage point, TS almost seems like a combination of ES and APCI, participating in both ion evaporation and chemical ionization in the same source.

Alternative ionization processes are utilized with TS to extend the applicability to normal phase separations and non-buffered solutions. This is accomplished by simply adding either a supply of electrons from a filament (for non-polar solvents) or a discharge (for nonbuffered samples).

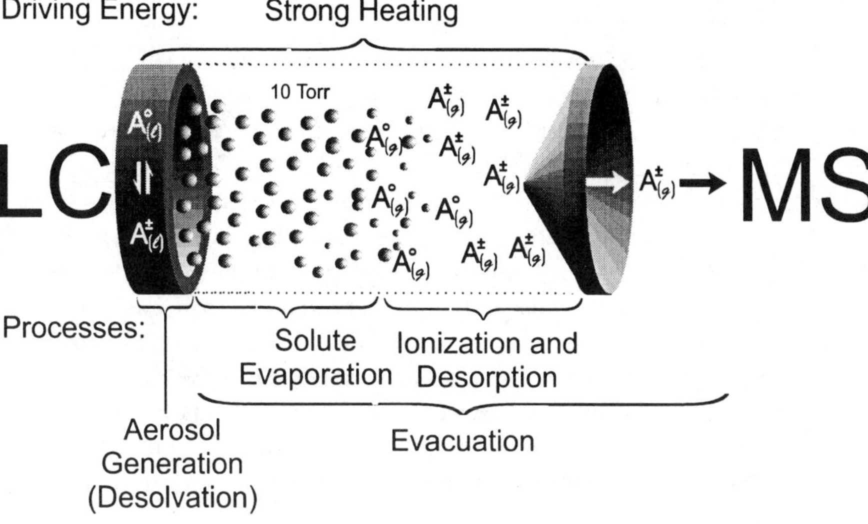

Figure 12- Schematic diagram of the processes associated with thermospray LC/MS showing the thermospray vaporizer, ion generation, and ion sampling into the mass spectrometer.

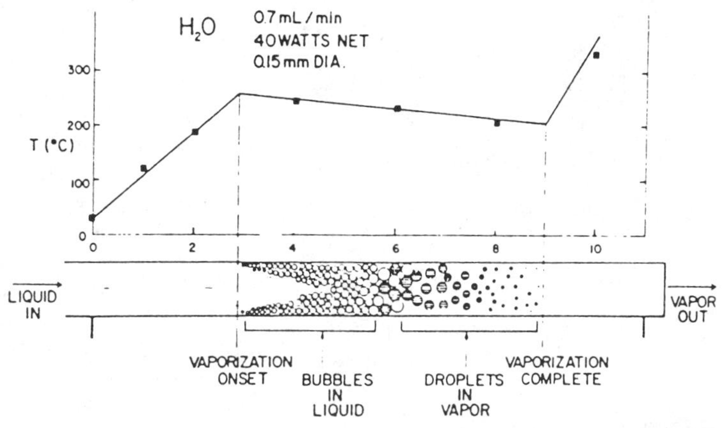

Figure 13- Temperature profile versus distance along a heated thermospray tube. Comparison between calculated (solid line) and measured (points) temperature profile at 0.7 mL/min together with a schematic representation of the model used in the calculation. (Reproduced from reference 16 with permission.)

Fundamentals: MALDI

Matrix Assisted Laser Desorption Ionization (MALDI) is included in this section because it is an important tool for problem solving with LC and MS.[19,20] Most of the applications of MALDI are not with on-line separations; however, with modern-day automation there is a very gray line between on- and off- line!

MALDI is yet another desorption technology similar in concept to FAB. Both techniques involve applying the sample to the tip of a probe and imparting significant amounts of energy into the surface of the sample in order to desorb and ionize analytes. Both techniques also employ a matrix to facilitate the desorption process. The difference between MALDI and FAB is in the type of energy applied to the surface, the matrix used, and the state of matter where the sample resides (see Figure 14).

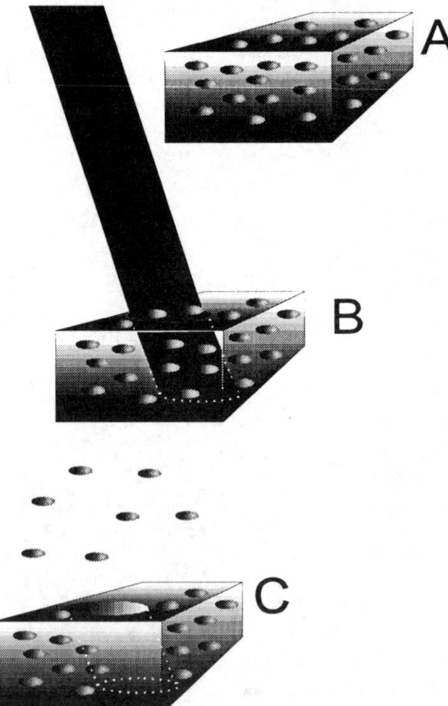

Figure 14- A representation of the desorption process in MALDI.
(A) The analyte is occluded within an VU-absorbing crystal lattice.
(B) The pulse from a high power laser penetrates the crystal rapidly vaporizing the UV-absorbing matrix.
(C) The result is a gas-phase analyte ion that has been desorbed from the target region. An empty crater remains as evidence of the pulse. This process can be repeated hundreds of times per second generation hundreds of craters.

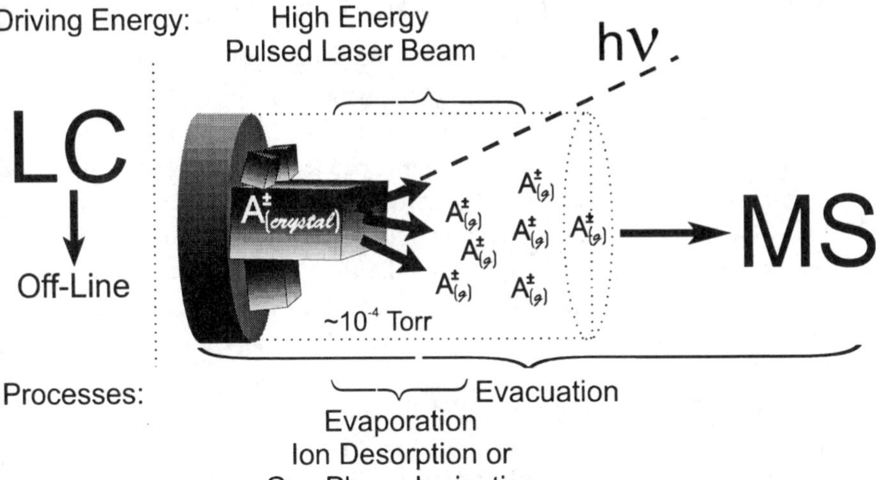

Figure 15- Schematic diagram of the processes associated with Matrix Assisted Laser Desorption Ionization showing the desorption of analyte trapped inside of matrix crystals. A high energy pulsed laser rapidly vaporizes the UV-absorbing matrix, desorbing matrix and analytes.

A high energy pulsed laser is used to excite the crystals on the surface of the target as shown in Figure 15. The laser is typically a nitrogen laser at 337 nm. The pulsed nature of this technique makes it ideal for coupling to pulsed source mass analyzers such as TOF and the various traps.

Desorption occurs when the matrix occluding the analyte has a strong absorbance band at the wavelength of the laser. Preferably your analyte does not. Typical matrices are conjugated structures that are solid at room temperature, but sublimes very easily when heated with the laser. Examples of matrices are sinapinic acid, nicotinic acid, and amino benzoic acid. They are usually very soluble in water.

A typical analysis involves apply a sample to a target already containing a matrix solution. The solution is dried until it crystallizes. The sample is then inserted into the path of the laser beam through a vacuum interlock. Multiple samples may be placed in individual wells on a single target making possible multiple analysis from one target.

Further Reading On Fundamentals

These two-page nutshells are certainly not intended to present of comprehensive treatment of the fundamental aspects of the various LC/MS interface technologies; rather, they have been include to give the reader a comparative and broader perspective on the operation of each interface. To justly and fully treat these topics we would have to devote another 500 pages of text and several years to this book. We have neither the time or energy to take on this task at the present time. We do however offer as encouragement some useful references that have taken the challenge and succeeded.

Most of the current research on LC/MS interface phenomena has been concentrated in the area of electrospray for obvious reasons; therefore, these references all apply to electrospray. Many of the cited references within this book also can be found in these books or compilations.

There are three books devoted to the topic of electrospray processes that we would recommend to those interested in this topic. These include:

- Michelson, D., *Electrostatic Atomization*, Adam Hilger (1990). This book on electrospray processes is 150 pages and covers the basic theory of cone-jet formation and Rayleigh instability.
- Bailey, A.G. *Electrostatic Spraying of Liquids*, John Wiley, New York, (1988). This 197 page book also covers the basic theory behind electrospray.
- Prewett, P.D., and Mair, G.L.R., *Focused Ion Beams from Liquid Metal Ion Sources*, John Wiley, New York (1991). This 320 page book is devoted to electrospray applications within the semi-conductor industry but has the most comprehensive treatment in a text. In addition, this book has our favorite cover upon which the authors names are written with an ion beam across the width of a human hair. Analytical chemist may learn something from their colleagues in the semi-conductor industry about controlling ion production and ion focusing.

In addition to books, there are several journals that have published dedicated issues to the topic of electrospray fundamentals.

- The first, *Journal of Aerosol Science*, 25/6, September 1994 is most likely the best compilation of theory about the spray process to date. For the serious reader, this is the best source of information about electrospray processes. This issue begins with a comprehensive review of electrospray literature by J.M. Grace and J.C.M. Marijnissen.
- For the mass spectrometry community, a recent issue of the *International Journal of Mass Spectrometry and Ion Processes*, 162, (1997). This excellent issue edited by Scott McLuckey and Gary Van Berkel contains 14 papers dealing with both solution phase and gas phase processes associated with electrospray. This issue should be a reference on the bookshelf of anyone serious about electrospray processes.
- There is a special edition of *Analytica Chimica Acta* planned for issue before the end of 1998 that deals with topics relating to the mechanism of electrospray. We look for this to be a valuable reference.

Notes

[1] Willoughby, R.C. and Sheehan, E.W., "The Physics of LC/MS: A Thermodynamic Perspective," Proceedings of the 42nd ASMS Conference on Mass Spectrometry and Allied Topics, Chicago, Illinois, May 29-June 3 (1994).

[2] The analyte can be either a neutral or an ion in solution. Neutral is included to stress the importance of ionization.

[3] This diagram is intended to show the dual pathway that a dissolved analyte can take into the gas phase, as a neutral or an ionic specie. Neutral species are more volatile than charged species because of the lower interaction of a neutral molecule with their surrounding solvents or counterions. Ionic interaction is the strongest. Below is a table summarizing intra- and intermolecular interactions that dominate the behavior of most molecules in LC/MS.

Table III- Interactions That Dominate LC/MS Behavior.

Type of Interaction	Energy (eV)	Fall off with:	Comments
Covalent Bond	3-5	n.a.	C-C, C=C
Ion-ion	0.4-4	1/r	less falloff
Hydrogen Bonding	0.1-0.3	n.a.	O, N
Ion-dipole	0.03-0.1	$1/r^2$	Affect solubility
Dipole-dipole	0.005-0.03	$1/r^6$	Affect solubility

[4] One way to measure aerosol efficiency is in terms of surface area generated per mass (or volume). Pneumatic aerosol generators cannot compete with electrospray in terms of this measure of efficiency. Unfortunately, the high efficiency of electrospray is limited to low flows rates (0-10 microliter/min); the regime where electrostatic forces produce a stable cone-jet.

[5] One of the first paper to point out the electrospray circuit relationship was:
Kebarle, P. and Tang. L., "From Ions in Solution to Ions in the Gas Phase," Anal. Chem. 65, pages 972-986 (1993).

[6] Gomez, A., Tang, K., "Charge and fission of droplets in electrostatic sprays," Phys. Fluids 6, pages 404-414, (1994)

[7] Huertas, M.L., Fontan, J., "Evolution Times of the Tropospheric Positive Ions," Atmospheric Environ. 9, 1018 (1875). As presented in the API book, PE Sciex, Toronto, Canada (The API Experts).
[8] Carroll, D.I., Dzidic, I., Horning, E.C., Stillwell, R.N. "Atmospheric Pressure Ionization Mass Spectometry," Appl. Spectrosc. Rev. 17, pages 337-406 (1981).
[9] Sunner, J., Nicol, G., Kebarle, P., Anal. Chem., 60, 1300 (1988). Sunner, Ikonomou, M.G., G., Kebarle, P., Anal. Chem., 60, 1308 (1988).
[10] Baldwin, M.A., McLafferty, F.W., "LC-MS Interface, I. The Direct Introduction of Liquid Solutions into a Chemical Ionization Mass Spectrometer," Org. Mass Spectrom. 7, pages 1111-1112 (1973).
[11] Willoughby, R.C., "Studies with an Aerosol Generation Interface for LC-MS," Ph.D. thesis, Georgia Institute of Technology (1983).
[12] Cotter, R.J., "Mass Spectrometry of Nonvolatile Compounds," Anal. Chem. 52, pages 1589A-1606A (1980).
[13] Many of the early developer in FAB utilized high energy neutral atom beams by charge stripping ions on their way to the target. FAB has since taken a broaden common usage for all beams of atoms and ions. The correct term for most CFFAB is liquid SIMS (secondary ion mass spectrometry). Barber, M., Bordoll, R.S., Elliott, G.J., Sedgewick, R.D., Tyler, A.N., "Fast Atom bombardment Mass Spectrometry," Anal. Chem. 54, pages 654A-657A (1982).
[14] Murphy, R.C., *Mass Spectrometry of Lipids*, Plenum Press: New York, pages 47-50 (1990).
[15] Ligon, W.V. "Evaluating the composition of liquid surfaces using mass spectrometry," IN: *Biological Mass Spectrometry* (Burlingame, A.L., McCloskey, J.A., eds), pages 61-76, Elsevier Science: Amsterdam (1990).
[16] Vestal, M.L., Fergusson, G.J., "Thermospray LC/MS Interface with Direct Electrical Heating of the Capillary," Anal. Chem. 57, pages 2373-2378 (1985).
[17] A complete theoretical treatment of droplet charging and evaporation is contain from pages 37 to 85 of the following text: Yergey, A.L., Edmonds, C.G., Ivor, A.S.L., and Vestal, M.L., *Liquid Chromatography/Mass Spectrometry: Techniques and Applications*, Plenum Press: New York (1989).
[18] Iribarne, J.V., Thomson, B.A., "On the Evaporation of Small Ions from Charged Droplets," Chem. Phys. 64, pages 2287-2294 (1976).
[19] Karas, M., Hillenkamp, F., Anal. Chem. 60, pages 2299-2301 (1988).
[20] Tanaka, K., Waki, H., Ido, Y., Akita, S., Yoshida, Y., Yoshida, T., Rapid Commun. Mass Spectrom. 8, pages 151-153 (1988).

Appendix D- Flow Diagram

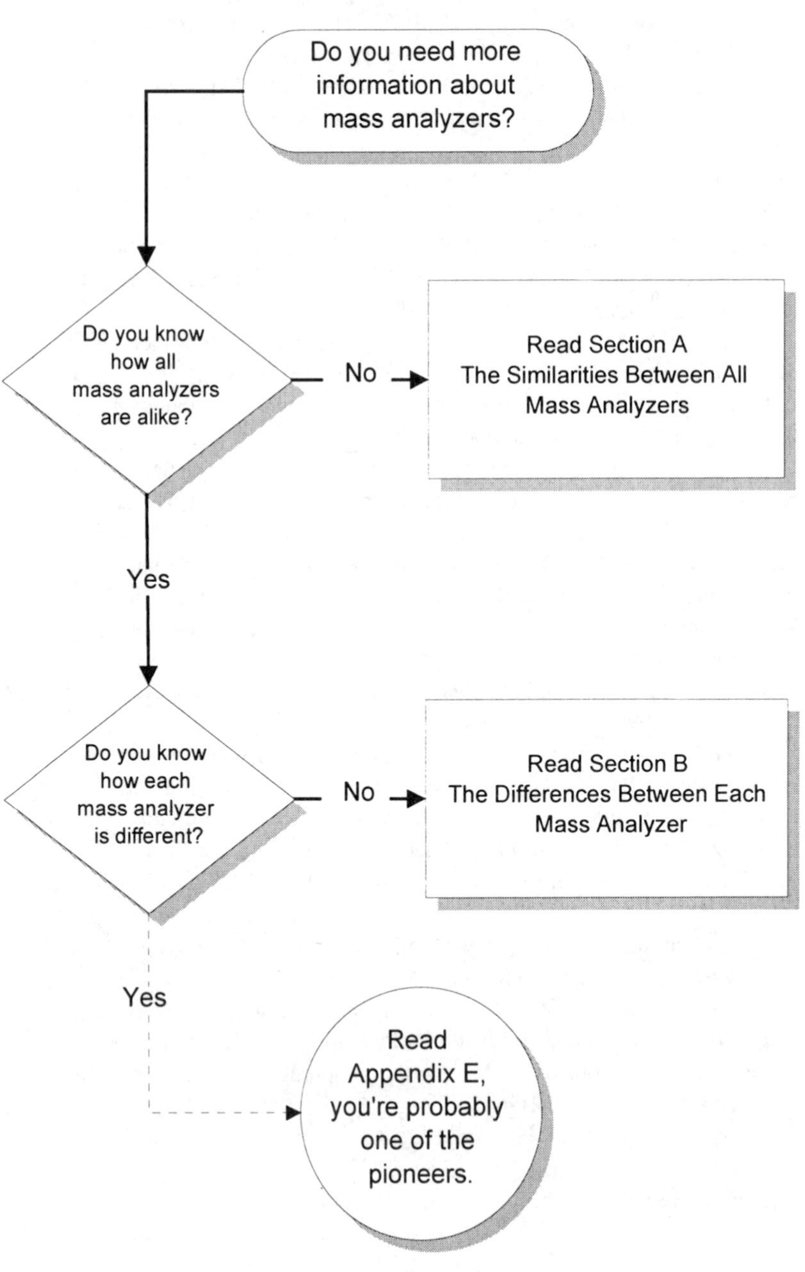

Appendix D

Mass Analyzers-How Do They Work?

(or, Mass Analysis Fundamentals)

Similarities Between All Mass Analyzers

As with interfaces, to gain a appreciation for mass analysis it is useful to understand the similarities between the various mass analyzers. In the introduction to mass analysis in *Chapter Two* we stated that all mass analyzers determine the mass of an ion, determine mass-to-charge ratio, measure gas-phase ions, and operate at low pressure. But how do they really work?

To gain insight into how analyzers operate we should consider the three primary processes associated with mass analysis; namely, ion production (source), ion separation (analyzer), and ion collection (detector). All analyzers must deal with these three processes in one fashion or another. All analyzers sample ions from an ion source. All analyzers resolve and separate the ions for subsequent detection. Mass analysis is all about controlling the motion of ions.

Table I is a general summary of the similarities between the analyzers. Remember, they all do basically the same thing, some just do it better than others.

Table I- All Mass Analyzers Have These Common Components.

Component	Key Processes
1) Source	Ion Production
2) Analyzer	Ion Separation
3) Detector	Ion Collection

Table II- Distance Between Collisions (MFP) vs. Pressure.[1]

Mean Free Path	Pressure Atm	Pressure torr	Pressure PASCAL	Density molecules/cm^3
10^{-6} meters	1	760	10^5	2.7×10^{19}
0.05 mm	10^{-3}	1	10^2	3.5×10^{16}
0.5 mm	10^{-4}	10^{-1}	10^1	3.5×10^{15}
0.5 cm	10^{-5}	10^{-2}	1	3.5×10^{14}
5 cm	10^{-6}	10^{-3}	10^{-1}	3.5×10^{13}
50 cm	10^{-7}	10^{-4}	10^{-2}	3.5×10^{12}
50 meters	10^{-9}	10^{-6}	10^{-4}	3.5×10^{10}

The motion of molecules is affected by pressure gradients, temperature gradients, electric fields, and magnetic fields. If we want to limit the influence of pressure and temperature on the motion of molecules in order to more precisely control their motion, we need a vacuum. Figure 1 shows a trajectory of a molecule at high and low pressure. Collisions (except in ion traps) have a random effect on the direction of an ion, thus limiting our ability to control their motion. The low pressure (vacuum) in a mass analyzer is required so that the ions will move from the ion source to the detector of the mass analyzer without being perturbed. If the distance between the source and detector is 10 cm, then your pressure has to support that travel distance without collision. Table II shows the relationship between travel distance (in terms of mean free path) and pressure. Designers of instruments must balance the gas load of the sample inlet with the pumping speed of the vacuum system to accommodate this travel distance. Pressure requirements for various analyzers are given below.

Table III- Pressure Requirements for Various Analyzers

Analyzer	Pressure (torr)	Mean Free Path
FTMS	$\leq 10^{-8}$	5 kilometers
Sector	$\leq 10^{-6}$	50 meters
TOF	$\leq 10^{-6}$	50 meters
Quadrupole	$\leq 10^{-4}$	50 cm
Ion Trap	$\leq 10^{-4}$	50 cm

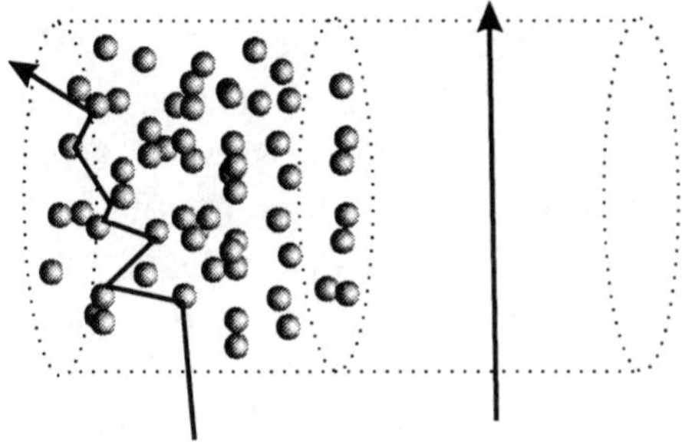

Figure 1- Effects of high and low pressure on ion motion.

At reduced pressures the motion of an ion is controlled by applying either an electric field or a magnetic field. The motion of ions in electric and magnetic fields is shown in Figure 2. When an ion is subjected to an electric potential, the ion will tend to move in a linear pathway that is in the direction normal to the equipotential plane (force and velocity vectors are the same direction). For plates, negative move to positive, and positive moves to negative. When an ion is subjected to a magnetic field, the ion will move in a circular pathway with the magnetic field being normal to the plane of motion. Note the velocity and force vectors are 90° apart causing the circular motion.

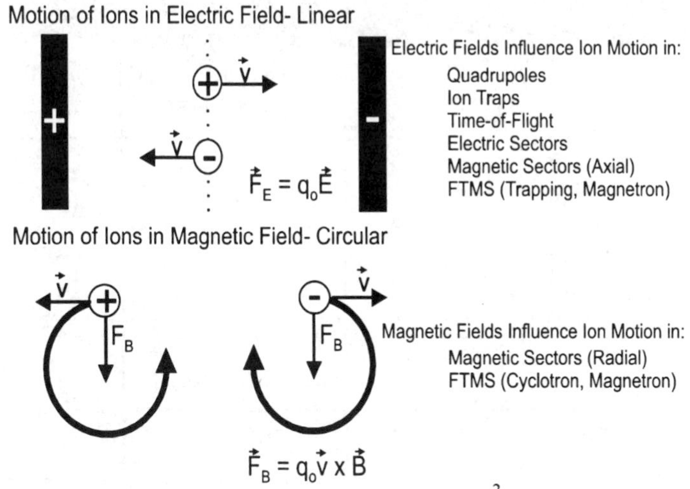

Figure 2- Motion of ions in electric and magnetic fields.[2]

Differences Between Mass Analyzers

Ion production, mass separation, and ion detection are common to all mass analyzers; however, how ions are introduced into an analyzer, how ions of different masses are separated, and how they are detected can be quite different with each specific mass analyzer. Some analyzers introduce ions into the mass analyzer as a continuous beam of ions; others as a pulse. Some analyzers are spectrometers that separate ions in space; while others are traps that separate ions in time. Some analyzers collect ion current on a surface, while others never collect the ions at all. The differences between each approach (Table IV) usually results in unique advantages and limitations.

The one theme we will be returning to with each analyzer is the mode of isolating unique mass-to-charge ratios (m/z). This ultimately results in our measurement of an intensity at an unique m/z (a mass-intensity pair). This section will acquaint you with how the different mass analyzers determine this mass-intensity pair. In addition, we will discuss aspects of MS/MS since it is virtually a prerequisite to problem solving in LC/MS today.

Table IV- Comparing the Different Mass Analyzers

Mass Analyzer	Produce Ions	Separate Ions	Detect Ions
Quadrupole	Continuous	Electronic Band Pass Filter	Current (direct)
Ion Trap	Pulsed	Voltage (rf)	Current (direct)
Time-of-Flight	Pulsed	Flight Time	Current (direct)
Sectors	Continuous	Magnetic Field (momentum)	Current (direct)
Fourier Transform	Pulsed	Frequency	Image Current (indirect)

Fundamentals: Quadrupoles

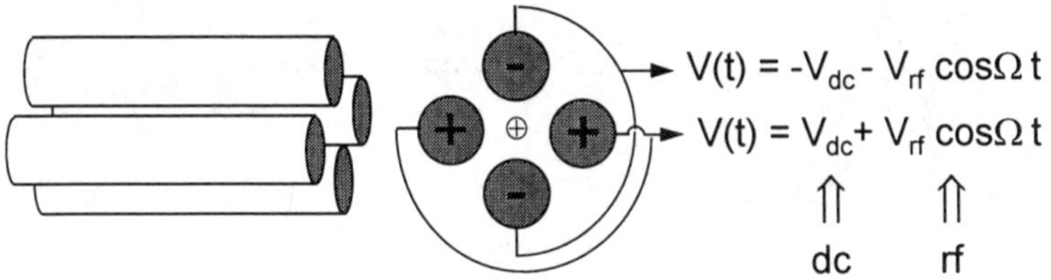

$V(t) = -V_{dc} - V_{rf} \cos\Omega t$

$V(t) = V_{dc} + V_{rf} \cos\Omega t$

dc rf

Measurement:

$$m/z = [k'/\Omega^2 r^2]\, V \qquad (1)$$

Variations:

1. Single quads
2. Triple quads (MS/MS)
3. Rf prequads ("ion pipes")
4. Hexapolar, octapolar lens

Figure 3- Schematic of a quadrupole mass analyzer with equation for m/z.

Quadrupole mass analyzers (Q) consist of a set of four electrodes, with either a circular or hyperbolic cross section. Opposing sets of rods have both a dc (direct current) and an ac (alternating current or a radio frequency) voltage component; one set positive, one set negative (see Figure 3).

For mass analysis, a continuous beam of ions enters one end of this assembly from the ion source, and exits the opposite end to be detected by a high voltage detector. Ions are filtered on the basis of their mass-to-charge ratio (See equation 1). Ions below and above a certain m/z value will be filtered out of the beam depending on the ratio of the dc and ac voltages.

The basis of this filtering can be explained by the way ions are influenced by opposing sets of rods. Ions between the two rods with a positive potential that are *above* a critical m/z value are transmitted through the center of the quadrupole assembly and on to the detector. This forms a high pass mass filter (see Figure 4).[3] Any ions between the two rods with the negative potential that are *below* a critical m/z value, will be transmitted through the center of the quadrupole assembly. This forms a low pass mass filter.

Combining both sets of rods into a quadrupole configuration overlaps the two mass filter regions and creates an area of mutual stability where ions of a certain mass-to charge ratio pass through the rods (forming a band pass filter, see top portion of Figure 5). Ions outside the bandpass region run into the rods.

The mass-to-charge ratios (m/z) of the ion that is passed through the mass filter is proportional to the voltage (V) applied to the rods (Equation 1). At higher voltages, a higher mass is allowed to pass. The typical transit time (from ion source to detector) is ~50-100 microseconds.

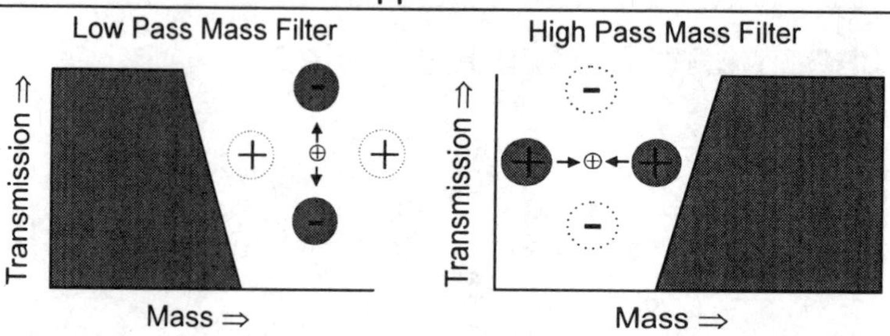

Figure 4- Negative rods create a low pass filter. Positive rods create a high pass filter.

Resolution for quadrupole mass filters (the width of the bandpass region) is dependent on the voltages applied to the opposing sets of rods. By decreasing the magnitude of the dc voltage in relationship to the rf, two things happen: (1) at the negative rods, the low pass mass filter is shifted to the right, to a higher mass; and (2) at the positive rods, the high mass filter is shifted to the left, to a lower mass. Putting the two situations together results in a wider bandpass (peak) and lower resolution (see Figure 5). For higher resolution the situation is reversed, that is, the bandpass region is made more narrow. Mass resolution is generally ~1,000 at mass 1,000.[4]

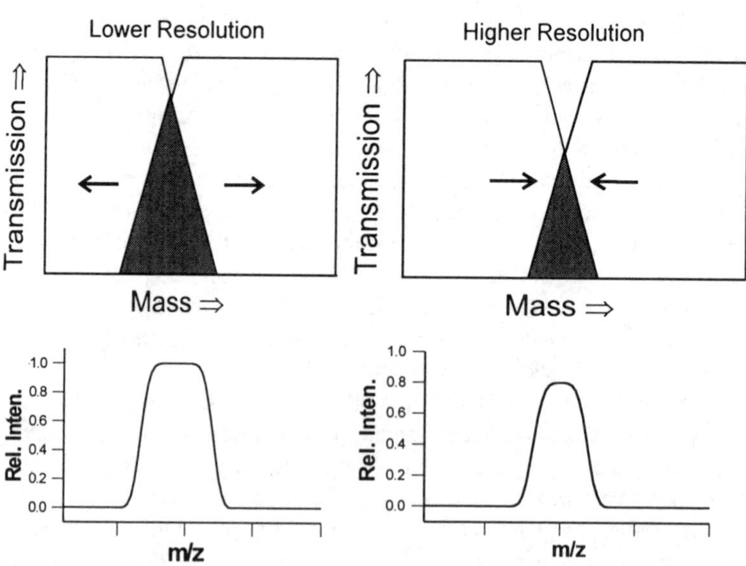

Figure 5- Low and high mass resolution. Note the higher transmission with lower resolution (left side of figure).

Scanning a quadrupole mass analyzer involves increasing (ramping) the amplitude of the dc and rf voltages at a constant ratio. Ramping changes the position of the bandpass region, thereby allowing different masses to be transmitted (see Figure 6 and 7).

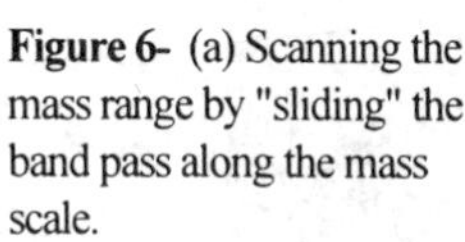

Figure 6- (a) Scanning the mass range by "sliding" the band pass along the mass scale.

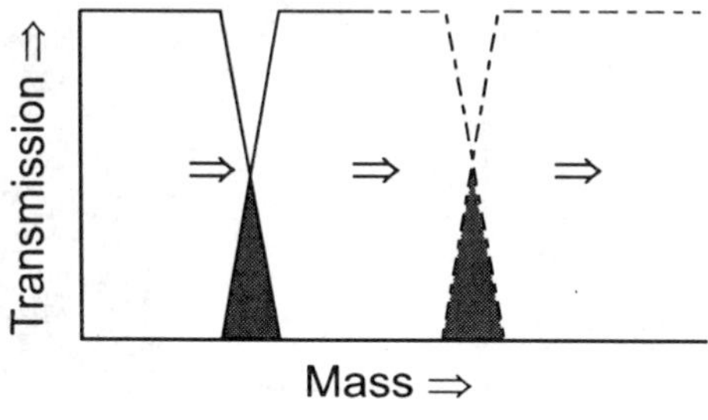

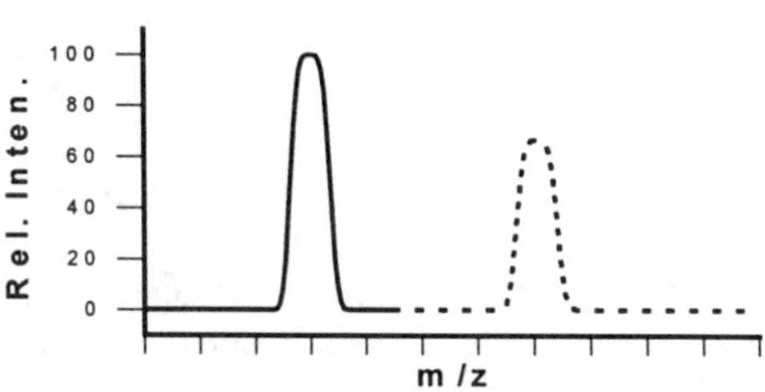

Figure 7- (b) The mass spectral output.

Performing selective ion monitoring (SIM) is accomplished by selecting a voltage (dc and rf ratio) that will allow one specific mass to pass through the mass analyzer (see Figure 8). By jumping from one voltage to another, a series of individual masses will pass through the quadrupole mass analyzer. The advantage being that you spend more time at that selected mass giving orders of magnitude sensitivity enhancement.

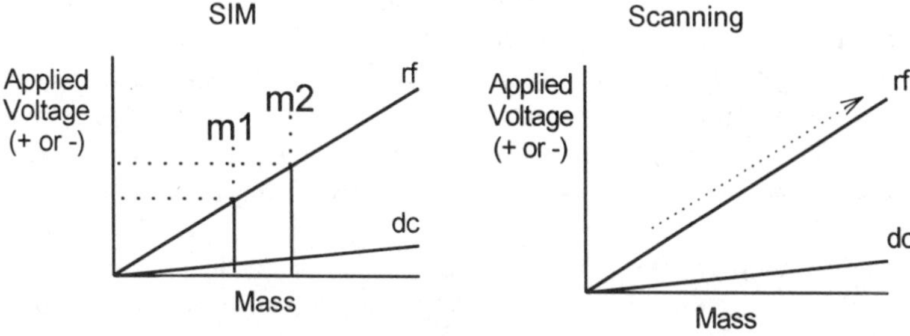

Figure 8- SIM: stepping from mass to mass at set rf/dc ratios; scanning the mass range by ramping the voltages, keeping the ratio of rf to dc constant throughout the scan.

A quadrupole assembly can be operated without the dc component of the voltage. This mode is referred to as the rf-only mode. With an rf only quadrupole, all ions above a critical m/z value are transmitted- a high pass mass filter (see Figure 9).

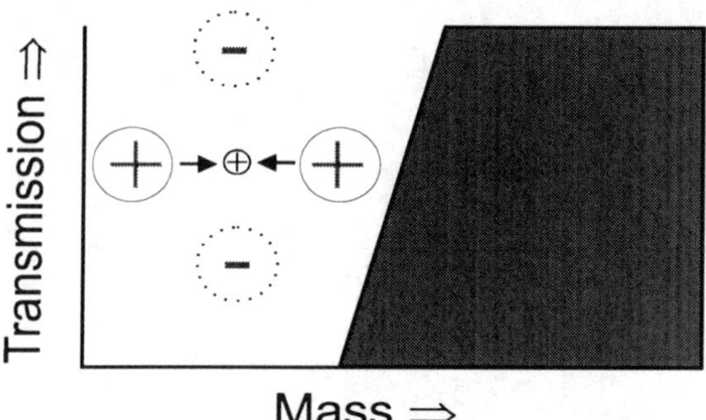

Figure 9- A rf-only quadrupole is a high pass mass filter.

Rf-only quadrupoles (hexapoles and octopoles) are used to transport ions from external sources, such as atmospheric pressure ionization sources (APCI and ES) to the mass analyzers. In this application they are commonly referred to as "ion pipes". They are also used as a means of transporting ions through the collision regions between mass analyzers (e.g., triple quadrupole) because of their high transport efficiency.

For MS/MS analysis three quadrupole mass filters can be configured together to form a tandem mass spectrometer. Instruments of this type are referred to as triple quadrupoles (or triple quads). The first and third quadrupoles are used for scanning while the middle quadrupole (an rf-only quadrupole) serves as a collision cell. With triple quadrupole mass spectrometry, the ions are passed from Q1 into a higher pressure Q2 (collision cell) where they undergo collisions with a background gas (e.g., argon, nitrogen). These low energy collisions transfer the kinetic energy from the incoming ion into internal energy. The increased internal energy results in unimolecular decomposition of the analyte ions. The fragment ions from decomposition are mass analyzed with the third quadrupole (Q3). The triple quadrupole mass spectrum typically shows prominent product ions with a small precursor ion. Quadrupoles are also used with other tandem mass analyzers as both scanning and rf-only devices. Examples are the "hybrid" instruments such as time-of-flight (Q-TOF) and sectors (EB-QQ).

An important consideration in triple quad is the low mass cutoff.[5] The low mass product ions from a high mass precursor ion may be lost below approx. 1/3rd of the mass of the precursor mass. Hexapoles and octopoles in the collision cell can reduce this value to about 1/6th. Be wary of low mass losses in triple quads as well as quadrupole traps.

Fundamentals: Ion Traps

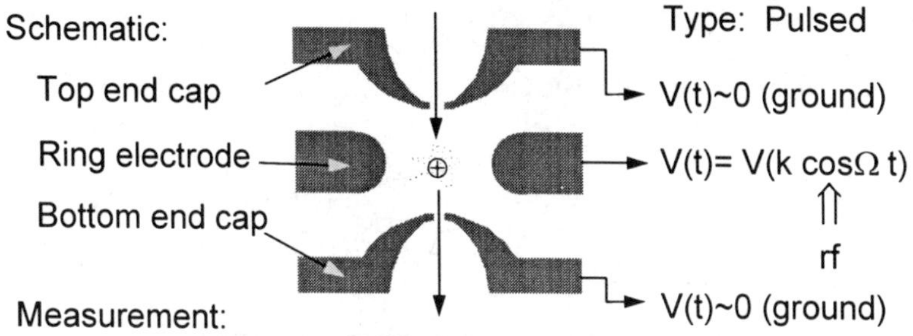

Schematic: Type: Pulsed

Top end cap — V(t)~0 (ground)

Ring electrode — V(t)= V(k cosΩ t)

⇑

rf

Bottom end cap — V(t)~0 (ground)

Measurement:

$$m/z = [k'/\Omega^2 r^2]\, V \qquad (2)$$

Variations:

1. Internal source (GC/MS)
2. External source (GC & LC/MS)
3. MS/MS/MS....
4. Zoom Scan

Figure 10- Schematic of an ion trap mass analyzer.

Conceptually, if you bent a linear rod into donut-shaped electrode and placed two hemisphere-shaped electrodes above and below the donut, you could create a three dimensional analogue of the linear quadrupole mass filter. This configuration (Figure 10) is called a "quadrupole" ion trap (IT).

With ion traps, the ions are held in a static position in contrast to linear quads that filter ions on a trajectory from point A to point B. Inside the trap, ions are subjected to oscillating electrical fields applied by *only* a radio frequency (rf) voltage applied to the ring electrode. The end caps are held near ground potential. Unlike quadrupoles, there is no dc component of the voltage. The radio frequency field traps the ions near the center of the ring electrode. This situation is analogous to the high pass mass filter (see Figure 11) for a quadrupole as discussed in the previous section.

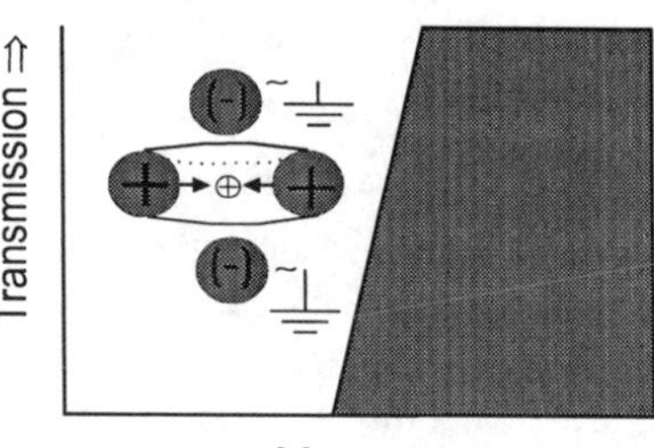

Figure 11- The ion trap behaves like a high pass filter. As the rf voltage on the ring electrode is ramped, low masses become unstable and exit the trap to the detector.

For mass analysis, a ions are formed outside of the trap (e.g., external source for APCI or ES) and guided up to the trap. A portion (or pulse) of the ions is sampled into the trap. Once inside the trap, the ions are confined by an oscillating rf field and "damped" into the center by a buffer gas, helium. In the process of trapping, the ions take on an oscillating frequency that is related to their mass-to-charge ratio (m/z).

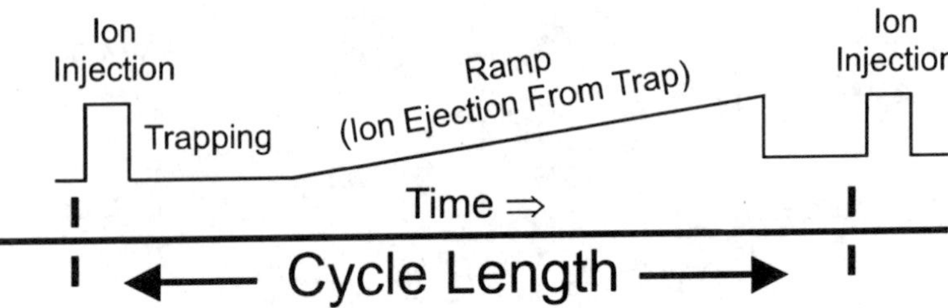

Figure 12- Cycle of ion injection, trapping, mass analysis (ion ejection).

To scan the mass range, the amplitude of the rf voltage on the ring electrode is ramped (equation 2). At the same time, a small rf voltage is applied to the end caps. As the ring electrode voltage changes, the frequency of ion stability changes as well. When the end cap frequency matches the resonance frequency of an ion, the ion will become excited into an oscillating motion until ejected along the axis of the endcaps. This resonance frequency is a function of mass. This cycle of events takes ~ 50-100 milliseconds. A typical scan cycle is shown in Figure 12. The band pass transmission model is shown in Figure 13. Upon ejection, the ions are typically detected by electron multipliers.

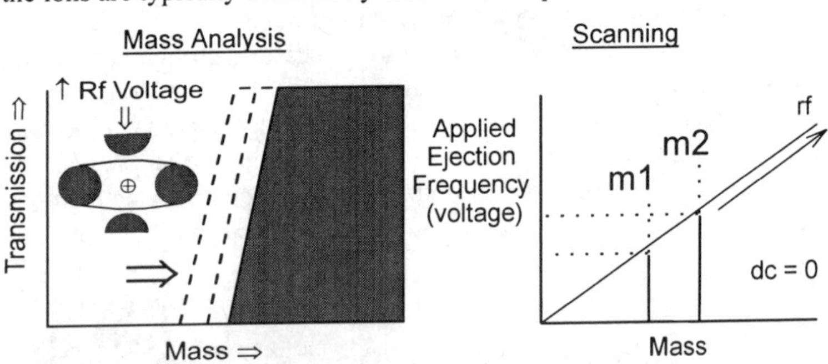

Figure 13- Mass analysis by ramping the rf voltage on the end caps.

In addition to scanning a type of selected ion monitoring (SIM) can be implemented with an ion trap. To do SIM, a specific ion from the external source is accumulated in the trap by ejecting all other ions during the accumulation period. After a period of time the ion is cooled and then ejected from the trap. This technique is typically used to enhance sensitivity by improving signal-to-noise for a specific ion or family of ions.

Resolution in an ion trap is maintained by the presence of a buffer gas-helium. As the ions collide with helium they are energetically "cooled". This causes ions of a specific mass to move towards a similar spatial point in the trap. As ions are ejected from the trap for mass analysis, ions of a specific mass all start from the same point and arrive at the detector at the same time. Mass resolution of the trap is similar to a linear quadrupole, ~1,000 at mass 1,000.[6]

Loss of resolution is noticed when the amplitude of the rf voltage for ejection is changed too fast (ramped too quickly), ions will fail to respond to the induced instability. This results in the loss of resolution. By slowing the rate at which the rf voltage is ramped, the resolution over a small mass range can be increased dramatically (e.g. $\sim10^6$). This method of increasing resolution by slowly scanning over a short mass range is called zoom-scan mode.

Traps can also be used for MS/MS.../MS analysis (referred to as MS^n). Unlike the triple quadrupole mass analyzer which performs MS/MS by placing the mass analyzers in tandem. Traps perform MS/MS experiments sequentially in time. Ions are selected for MS/MS analysis by selectively ejecting (with a radio frequency on the end caps) all other ions in the trap, leaving a specific precursor mass ion behind. A resonating frequency that corresponds to the isolated mass is then applied to the end caps (sometimes referred to as a tickle voltage). This voltage is typically a few percent of the amplitude of the ejection voltage for the particular mass. The precursor ion starts to oscillate and collide with the helium bath gas. After <u>numerous</u> collisions, the increased internal energy of the precursor ion induces fragmentation. Since the fragment ions (product ions) are not at the same resonating frequency as the parent, they "fall" back to the center of the trap through collisional damping with the helium gas. After a period of time the rf (ring) voltage is scanned causing the ejection of ions (precursor and product) from the trap.

For MS^n analysis the sequence of events starts out the same. First an ion is trapped and then fragmented. Instead of ejecting all of the ions, a selected product ion is trap while all the other product ions from the first precursor ion are ejected. This cycle can theoretically proceed indefinitely (e.g., MS^6).

As mentioned with quadrupoles, there is a low mass cutoff for traps as well. This would mean that product ions that are created in the trap during resonance excitation in MS/MS may not remain in the trap for detection.[7]

Fundamentals: Time-of-Flight

Schematic: Type: Pulsed

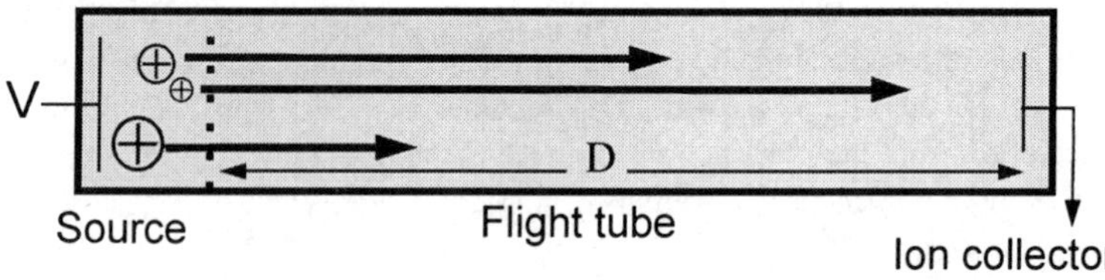

Measurement: $$m/z = [k\ V/D^2]\ t^2 \qquad (3)$$

Variations:

1. Reflectrons
2. Post Source Decay
3. Q-TOF
4. Trap storage

Figure 14- Schematic of a time-of-flight mass analyzer.

A time-of-flight (TOF) mass analyzer consists of simply a source, a flight tube, and a detector (see Figure 14). Typically for mass analysis in LC/MS, a continuous beam of ions, generated from an external source, such APCI or ES, are introduced into the source region either orthogonally (90-degrees to the flight tube) or on-axis with the flight tube. Ions entering orthogonally pass through the source. As they pass in front of the "repeller" they are pulsed out of the source into the flight tube to be detected by a high speed detector (e.g. microchanneltron plate or discrete dynode). Duty cycles approaching 100% are possible with optimal source design. For axial introduction, ions are introduced into the source region and collected. Ions are accumulated for a short period of time and then pulsed out of the source region into the flight tube. In both situations the mass-to-charge ratio (m/z) is related to the flight time of an ion through the flight tube (Equation 3). Lighter ions traverse the flight tube faster than heavier ions. The typical fight time for an ion is on the order of 5-100 microseconds.[8]

Since TOF mass analyzers determine mass based on their arrival time, they are in a sense, scanning all the time. SIM is not a practical alternative for TOF, nor would it have any advantage.

In a simple linear TOF mass analyzer, mass resolution is generally ~1,000.[9,10] This low mass resolution is attributed to the uncertainties in the time of formation (temporal), location (spatial) and the initial kinetic energy distribution of the ions in the source at the time they are pulsed out of the source region into the flight tube. Variations in initial source conditions will cause ions of the same mass to have different arrival times. This is the major cause of poor resolution (see Figure 15).

Two techniques are utilized to compensate for temporal, spatial, and kinetic distribution. For example, by delaying the extraction of the ions from the source, wide spatial and temporal distributions can be avoided. This technique is referred to as "delayed extraction." This is generally applied to MALDI sources where ions leave the target at wide energy distributions. Delaying allows the ions to undergo collisions before extraction.

Temporal Distribution

Two ions of the same mass formed at different times in the source arrive at the detector at different times.

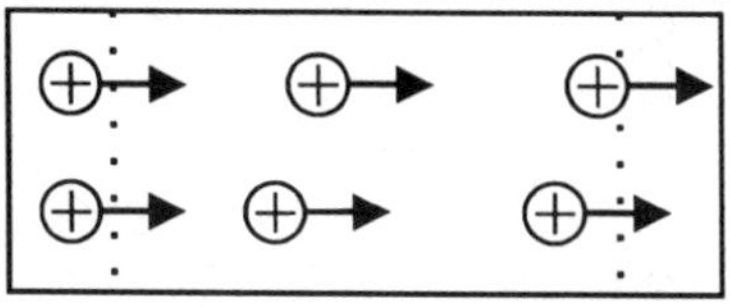

Spatial Distribution

Two ions of the same mass formed at different locations arrive at the detector at different times.

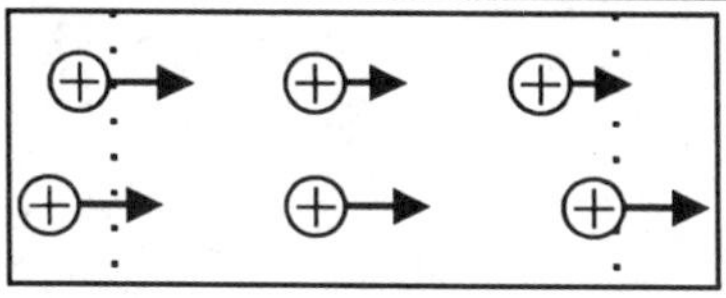

Kinetic Energy Distribution

Two ions of the same mass formed with different initial kinetic energies arrive at the detector at different times.

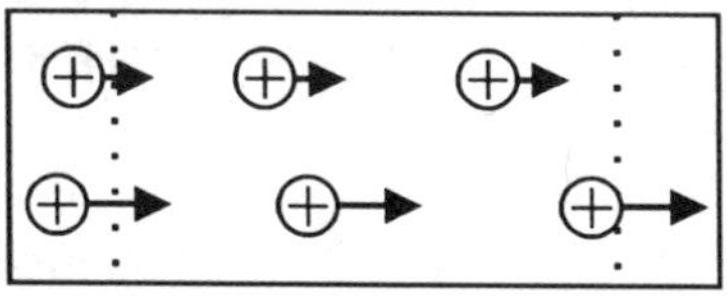

Two ions of the same mass with the same kinetic energies but initial velocities in the opposite directions arrive at the detector at different times.

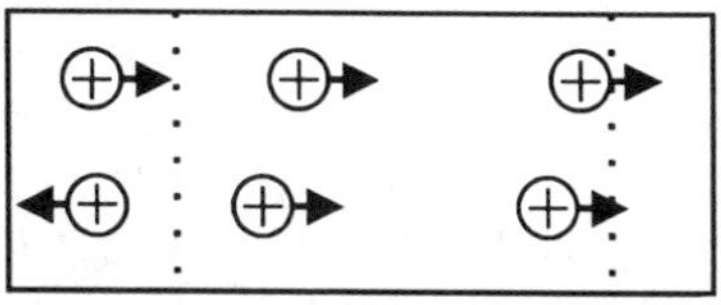

Figure 15- Influence of the time when an ion is formed, the place it is formed and the energy it is formed at on resolution (adapted from reference 11 with permission).

Alternatively, reflectrons (ion mirrors) are used to compensate for variations in energy distribution. Ions are reflected based on their forward kinetic energy. The more energetic the ion the deeper it penetrates the retarding field of the reflectron before being reflected. This allows an energetic ion, traveling a longer flight path, to arrive at the detector at the same time as the less energetic ions of the same mass.

TOF mass analyzers, alone, cannot practically be used for tandem mass spectrometry (MS/MS). For this reason, hybrid TOF instruments are often configured as the second mass analyzer for MS/MS. Configurations with quadrupoles (e.g., QQ-TOF) and hybrid sector instruments (e.g., EB-TOF) have improved scanning sensitivities due to the multiplexing ability of TOF.

Fundamentals: Sectors

Schematic:

Type: Continuous

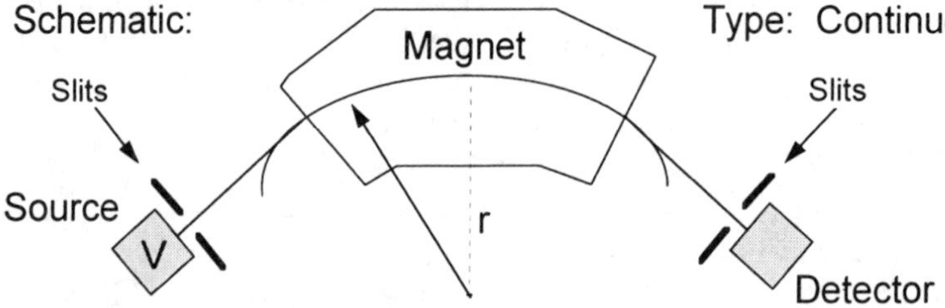

Measurement:

$$m/z = [k\ r^2/V]\ B^2 \qquad (4)$$

Variations:

1. EB
2. BE
3. EBQQ
4. Hybrids: EB-TOF, EB-IT

Figure 16- Schematic of a magnetic sector mass analyzer.

A magnetic sector mass analyzer consists of a strong magnet (see Figure 16). For mass analysis, a continuous beam of ions is generated in an ion source. For an ion to reach the high voltage detector and be recorded, it must be accelerated out of the source by a voltage (accelerating voltage V) towards the source slit. The ion must traverse a radius of curvature r through a magnetic field of strength B. The motion of an ion is determined by the balance of its angular momentum and the centrifugal force caused by the magnetic field.

The mass-to-charge ratio (m/z) of an ion is therefore based on its momentum.[12] By varying either the magnetic field (B) or the accelerating voltage (V), ions of different mass-to-charge ratios (m/z) are separated by the magnetic field (Equation 4). The ions are resolved from each other by dispersing them in space (translationally). The typical analysis time (from the source to the detector) is ~1 microsecond.

Resolution (the translational energy spread) is determined by changing the widths of the source and collector slits, and transmitting a narrow band of ion energies on to the detector. To provide more resolution, an energy analyzer (electrostatic sector, E) can be added in front of or after the magnetic sector. This combination of a magnetic and an electrostatic sector (e.g., EB) is designed to have velocity focusing properties that reduce translational energy spread. Mass resolution is generally on the order ~10,000 at mass 1,000.[13]

Scanning a magnetic sector is typically accomplished by scanning the mass range from high mass to low by varying the magnetic field strength exponentially. This has the advantage of producing mass spectral peaks of constant width.[14] To perform SIM (selected ion monitoring) a particular magnetic field strength is chosen to pass a particular ion or ions of interest.

For MS/MS analysis, ions are injected into a collision chamber to undergo collisions with a background gas (e.g., helium, argon). The type of fragmentation that results from these collisions depends on the velocity of the ion. Fragments from high velocity ions result in high energy collisions while low velocity ions result in low energy collisions.[15] The collision chambers are either field free areas between sectors (high energy collisions) or quadrupole collision cells (rf-only quadrupole, low energy collisions) after the sector but before the detector.

In high energy collisions ions collide with the background gas. Due to the low gas pressure (10^{-3} torr) and the high forward velocity of the ion there are very few if any collisions with the background gas. A collision, if it occurs, primarily involves a transfer of energy to the electrons of the molecule. Thus, a high energy collision of an ion can result in odd-electron precursors ions as well as odd-electron product ions.[16] Due to the fact that there are not many collisions, the MS/MS mass spectrum typically shows a prominent precursor ion and less product ion abundance when compared to low energy collisions.

In low energy collisions, ions are first decelerated before they enter the quadrupole collision cell. Inside the rf only quadrupole collision cell an ion experiences many collisions. These low energy collisions transfer energy into the bonds of the molecule (vibrational excitation). The fragments that result from these low energy collisions typically give enough information to characterize and identify the molecule. Sector instruments equipped with a quadrupole collision cell and quadrupole mass analyzer (EBQQ) can be used to examine fragments (product ions) from both high and low energy collisions.

Fundamentals: FTMS

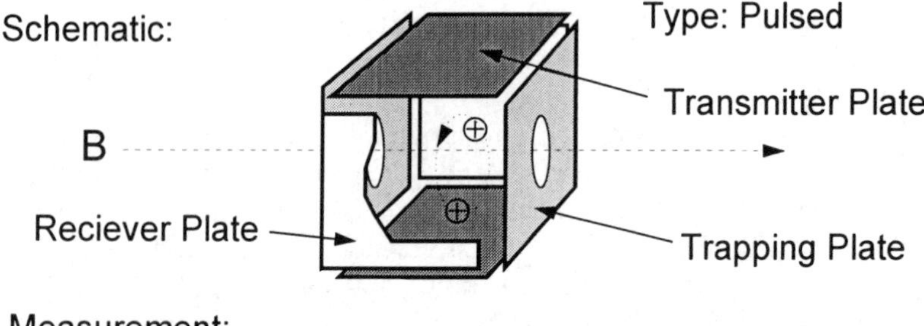

Measurement:

$$m/z = [B\ e] / \Omega \qquad (5)$$

Variations:

1. External source (LC/MS)
2. Internal source (EI, MALDI, LD)
3. Dual cell
4. Traps

Figure 17- Schematic of a fourier transform mass analyzer.

A Fourier Transform Mass Analyzer (FTMS) or ICR-MS (ion cyclotron resonance-MS) is also an "ion trap." It consists of a cubic cell (a trap) inside a strong magnetic field (B).[17] The cell has three distinct sets of plates: trapping, transmitter and receiver plates. For mass analysis, a continuous beam of ions is formed outside the cell in an external source such as APCI or ES. The ions are guided up to the trap and a portion of the beam is <u>pulsed</u> into the cell. Inside the cell the ions are constrained by a strong magnetic field (B) such that they move in circular orbits (cyclotron motion) in a plane perpendicular to the magnetic field (see Figure 17). The ions are also constrained in the cell by electric potentials applied to a set of trapping plates perpendicular to the magnetic field.

For mass analysis, radio frequency electric potentials are applied to the transmitter plates causing the trapped ions to be excited into larger circular orbits. As the excited ions pass near the receiver plates, the frequency of their passage is detected as an induced current called the "image" current. The frequency of an ion's motion is inversely related to its mass (equation 5).[18] But note, FTMS is unlike any other mass analyzers in that ion detection is non-destructive. You "listen" to the frequency of the ions. Signal-to-noise is improved by averaging many cycles before transforming and storing the data.

Once the ions have been detected, a radio-frequency pulse (a quench pulse) is applied to the cell to eject the ions before the next bundle of ions are collected into the cell. This cycle of trapping, excitation and detection can be repeated as fast as ~10 milliseconds (see Figure 18).

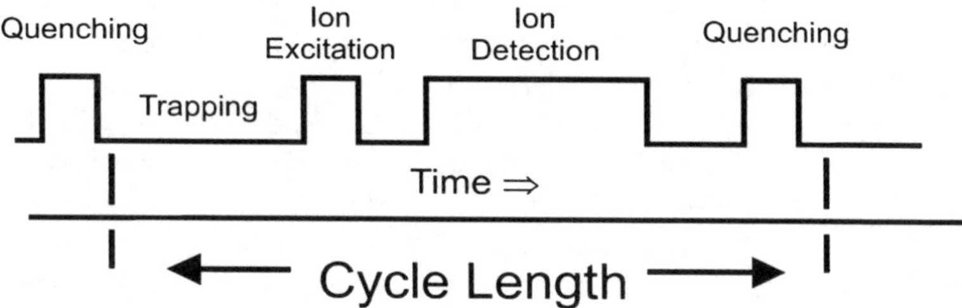

Figure 18- Cycle of mass FTMS mass analysis.

Time-Domain

Mass Domain

M_1 ⇒

M_2 ⇒

M_3 ⇒

$M_{1,\,2\,\&\,3}$ ⇒

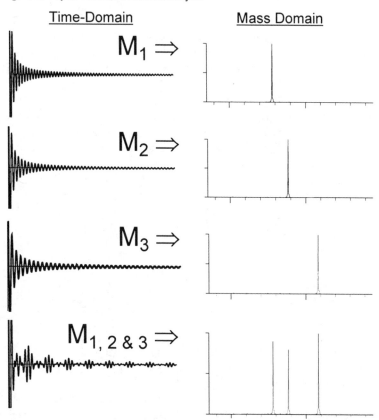

Figure 19- Time and mass domain for FTMS. As the ions oscillate in the cell, each mass will assume a unique frequency of rotation. That frequency is directly related to mass-to-charge ratio. Above is illustrated the frequency measurement on the left which is transformed into mass on the right.

Because there are ions of various masses present in the cell, the recorded frequency at the receiver plates is the sum of all the different frequencies. This time-domain signal of the image current can consist of hundreds of sine waves superimposed upon each other, each representing a unique mass. Figure 19 illustrates the time-domain signal for three individual masses and their corresponding mass spectra. If all three of these masses were present, the mass spectrum recorded by the instrument would be the sum of all three time-domain signals. Fourier transformation is used to convert this time-domain signal into a series of mass-intensity pairs to reconstruct a mass spectrum from frequency data.

Mass resolution is determined by the gas pressure in the cell and the strength of the magnetic field. At pressures $>10^{-6}$ torr there are too many collisions between the ions and the background gas. This causes the ions to be scattered which blurs the differences in the frequency of adjacent masses. For this reason the cells of FTMS instruments are operated at very low pressures; 10^{-8} - 10^{-9} torr.

Higher resolution can be obtained with stronger magnetic fields. A higher magnetic field is utilized to induce a higher frequency of the ions present in the cell. At higher magnetic field strength there is a proportional increase in the frequency of the ions. This has two effects; namely, ions resonating at higher frequencies (higher momentum) are scattered less when they do collide with background gas, and more importantly, increasing the magnetic field increases the frequency *difference* between two ions (see Figure 20 for an illustration). It is this increase frequency difference (or increase in the number of steps) between masses that accounts for the higher resolution. Mass resolution can be ~1,000,000 at mass 1,000.[19]

	Frequency	
Mass	0.5 Tesla	5 Tesla
CO	823.2107	8,232.107
N_2	823.1541	8,231.541
Difference	0.0566	0.566

Figure 20- Mass resolution of carbon dioxide (nominal mass = 28) and nitrogen (nominal mass = 28) at two different magnetic field strengths.

MS/MS analysis with FTMS is similar to that in the quadrupole ion trap. For example, ions are first isolated by ejecting all other ions in the cell through resonance excitation. Then, a pulse of gas is introduced into the cell and a "tickle voltage" is applied to the transmitter plates. By varying the amplitude of the resonating frequency, the ion (precursor ion) starts to oscillate and collide with any background gas in the cell resulting in the production of fragments (product ions). After a period of time the precursor and any product ions are excited into higher orbits and detected. Alternatively, photons from a laser (e.g., CO_2 laser) can be used to increase the internal energy of the ions causing fragmentation without the need for gas. For MS^n analysis the sequence of events of isolation, excitation and detection are repeated.

Notes

[1] Table values are from the ELQ-400 Operators Manual from Extrel Corporation. These are only single significant figure estimates to give the reader an appreciation for pressure and distance between the instrument parts. Mean free path is defined as the average distance between successive collisions.

[2] Where: B = magnetic field; E = electric field: F_B = magnetic force; F_E = electric force; q_0 = charge; v = velocity (adapted from Marshall, A.G., Grosshans, P.B., "Fourier transform ion cyclotron resonance mass spectrometry: The teenage years," Anal. Chem. 63, pages 215A-229A (1991)).

[3] Miller, P.E., Denton, M.B., "The quadrupole mass filter: Basic operating concepts," J. Chem. Educ. 7, pages 617-622 (1986), adapted from.

[4] Lambert, J.B., Shurvell, H.F., Lightner, D.A., Cooks, R.G., Introduction to Organic Spectroscopy, MacMillian Publishing: New York (1987). A mass resolution of 1,000 at mass 1,000 would allow you to discern one mass from another that differ by 1 u (dalton). For a further explanation of mass resolution see Chapter Three.

[5] Pedder, R., private communication. The ability of the collision cell to transmit a high mass efficiently precludes the efficient transmission of low mass fragment ion. If one optimizes transmission of a parent, there is a "cutoff" at low mass. You should keep this in mind when you experience analytes that disappear upon dissociation.

[6] Lambert, J.B., et al., ibid.

[7] Pedder, R., op cit.

[8] Cotter, R.J.(ed.), Time-of-flight mass spectrometry, ACS: Washington,D.C.(1994).

[9] Cotter, R.J., ibid.

[10] Lambert, J.B., et al., op cit.

[11] Cotter, R.J., "Time-of-flight mass spectrometry for the structural analysis of biological molecules," Anal. Chem. 64, pages 1027A-1039A (1992). adapted from.

[12] Chapman, J.R., Practical Organic Mass Spectrometry, John Wiley: Chichester (1985).

[13] Lambert, J.B., et al., op cit.

[14] Chapman, J.R., ibid.

[15] Gaskell, S.J., Reilly, M.H., "The complementary analytical value of high- and low-energy collisional activation in hybrid tandem mass spectrometry," Rapid Comm. Mass Spectrom. 2, pages 139-141 (1988).

[16] Chapman, J.R., op cit.

[17] Marshall, A.G., Grosshans, P.B., "Fourier transform ion cyclotron resonance mass spectrometry: The teenage years," Anal. Chem. 63, pages 215A-229A (1991).

[18] Johnston, M., "Fourier transform-mass spectrom., Part I", Spectroscopy 2, pages 14-17 (1988).

[19] Lambert, J.B., et al., op cit.

The Tree of LC/MS

Wilm & Mann (1996)
Bruins, Covey, & Henion (1987)
Whitehouse, Dreyer, Yamashita & Fenn (1985)
Horning et al. (1974)
Caprioli, Fan, & Cottrell (1986)
Blakley, Carmody, & Vestal (1980)
Willoughby & Browner (1984)
ES
APCI
Iribarne & Thomson (1976)
Ito, Takeuchi, Ishii, & Goto (1985)
Baldwin & McLafferty (1973)
Scott et al. (1974)
API
DLI
TRANSPORT
Tal'rose et al. (1969)
Dole et al. (1968)

Pioneers in LC/MS

(by Ross Willoughby)

This book is intended to assist scientists, students, and managers with real problem needs to become acquainted with the LC/MS technology, it's strengths and limitations. Today, there is very little risk, very little uncertainty about the future of LC/MS as a tool for problem solving. However, it wasn't always that way. Several decades ago, many thought the pursuit of this field was a waste of time. Liquid chromatography and mass spectrometry were "fundamentally incompatible techniques." "Why not simply collect fractions from the LC and run them with a probe?" Researchers couldn't get funded to do work in this field (including my graduate advisor). Instrument companies wouldn't invest R&D budgets to develop products in this field. No stable LC pumps were available to accommodate LC/MS. No one would even consider performing separations at microliters per minute. There were no commercial instruments that were either reliable or practical. This was clearly not a conducive environment for new technology development.

Fortunately there was a group of scientists who were too stubborn, too relentless, and maybe too naive to let these barriers stand in their way. The tree at the left is a representation of the growing and changing field of LC/MS and includes some (but certainly not all) of the pioneers who persevered to overcome these barriers.[1-13] A tree in my view is a better symbol of change (and time) than a simple timeline. It grows branches, sheds it's leaves, and grows a new set each season. Each new ring is added to the old, building on the past. This metaphor is ideal for technology. Some limbs die and fall off, others grow and flourish. It is my greatest hope that the newcomers to this field will not only consume the fruit of this tree but will nourish and help it sprout new branches.

One of my recent honors was having the opportunity to chair a half-day oral session on "Pioneers in LC/MS" at the annual ASMS meeting in Palm Springs, California in June of 1997. The distinguished speakers are shown in Figure 1. The response to this session was overwhelming with a standing room only, packed house. The speakers included Bruce Thomson (discussing ion evaporation and APCI), Jack Henion (Direct Liquid Introduction to IonSpray), Bill McFadden (out of retirement) (Moving Belt to Electrospray), Marvin Vestal (Thermospray), myself (a stand-in for Rick Browner [Figure 2] on Particle Beam), and John Fenn (Electrospray). (Video tapes of this historical session are available by contacting Mike Grayson at grayson@wuchem.wustl.edu.)

Each speaker was asked to relate their experiences in "the early days" in order that people coming into the field may gain an appreciation for how and why the developments occurred. Each speaker was asked to discuss the barriers that they encountered and to discuss the popular opinions of the day.

The following are brief excerpts from this session. It is my hope that the readers of this text will gain some insights and inspiration from the wisdom and experience of these icons of the field.

- Bruce Thomson- on Ion Evaporation, *"The ion evaporation model was a conceptual model that is probably overanalyzed today."*

- Bruce Thomson- on APCI, *"We sprayed through a heated tube to ensure that the droplets hit the walls because heat was difficult to get into the droplets."*

- Jack Henion- on the development of API, *"It required the development of a number of parallel technologies such as micro-LC, tandem mass spectrometry, API (Horning, et al.), and ion evaporation (Iribarne and Thomson work presented at Asilomar 1981)."*

- Jack Henion- on DLI, *"It was simply dumb determination that kept me going. I had to take a machine shop course to get a key for the shop in order to make glass tips. It took one year to get it to work."*

Figure 1- Speakers for the "Pioneers in LC/MS" oral session at the 45th ASMS Conference in Mass Spectrometry and Allied Topics, Palm Springs, California (June 3rd, 1997).

(l-r) Ross Willoughby, Bill McFadden, Marvin Vestal, Bruce Thomson, John Fenn, and Jack Henion.

Figure 2- A pioneer in particle beam and fundamental aerosol processes in LC/MS, Professor Rick Browner.

- Bill McFadden- on early influences on LC/MS, *"Everyone wanted EI. Everyone wanted one milliliter per minute flow rates. Should we be driven by customers? Or should we let the scientists do the work?"*
- Marvin Vestal- on CO_2 laser droplet cross beam experiments, *"We had the world's most expensive ice-maker. The laser would evaporate the droplets causing them to freeze. When we moved the laser over to the tip of the tube we got great evaporation."*
- Marvin Vestal- on Thermospray discovery, *"Cal Blakley walked over and asked- When did the filament burn out?"*
- John Fenn- on Malcolm Dole's experiments, *"Malcolm knew they needed a bath gas to supply enthalpy of evaporation."*
- John Fenn- on concurrent flow of gas, *"It was doomed to failure because it produced highly solvated ions."*

It was both entertaining and educational to listen to these great gentlemen. Each made a special point of explaining their work as a collective effort of many people; students, instrument companies, and research collaborations. Each pointed out that the advancements in LC/MS were not the result of a single breakthrough, but a progression of experiments, one improvement leading to the next.[14]

Portraits of many of the pioneers is included on the Dedication Page of this text and Figures 1 and 4. Of special note, is the inclusion of the great Victor Tal'rose of the Russian Academy of Science on the Dedication Page. He was the first to couple LC to MS; he certainly won't be the last.

Figure 3- Leaders in the development of LC/MS at the 13th Montreux Symposium held at Montreux, Switzerland November 13-15, 1996.
Back row- (l-r) Jan van der Greef, Paul Goodley, Wilfried Niessen
Front row- (l-r) Jack Henion, Dai Games, Patrick Arpino

Figure 4- One of the great spirits in LC/MS, Professor Fred McLafferty. He has contributed in one way or another to virtually every technique. Of particular note is his early work with direct liquid introduction (DLI).

Notes

[1] Baldwin, M.A., McLafferty, F.W., Org. Mass Spectrom. 7, 1111 and 1353 (1973).

[2] Blakley,C.R., Carmody, J.J., and Vestal, M.L., Adv. Mass Spectrom. 8B, 1616 (1980).

[3] Bruins, A.P.,Covey, T.R., Henion, J.D., Anal. Chem. 59, 2642 (1987).

[4] Caprioli, R.M., Fan, T., and Cottrell, J.S., Anal. Chem. 58, 2949 (1986).

[5] Dole,M.,Mack,L.L.,Hines,R.l.,Mobley,R.C.Ferguson,L.D., Alice, M.B., J Chem. Phys. 49, 2240 (1968).

[6] Horning, E.C., Carroll, D.I., Dzidic, I., Haegele, K.D., Horning, M.G., Stillwell, R.N., J. Chromatogr. Sci. 12, 725 (1974).

[7] Iribarne, J.V., Thomson, B.A., J. Chem. Phys. 64, 2287 (1974).

[8] Ito,Y.,Takeuchi, D.,Ishii, D., Goto, M., J.Chromatogr. 346, 161 (1985).

[9] Scott, R.P.W., Scott, C.G., Munroe, M., Hess, J., J. Chromatogr. 99, 395 (1974).

[10] Tal'roze, V.L., Skurat, V.E., Karpov, G.V., Russ. J. Phys. Chem. 43, 214 (1969).

[11] Wilm, M.S., and Mann, M. Anal. Chem. 68: pages 1-8 (1996).

[12] Willoughby, R.C., Browner, R.F., "Monodisperse Aerosol Generation Interface for Coupling Liquid Chromatography with Mass Spectroscopy," Anal. Chem. 56, pages 2626-2631 (1984).

[13] Yamashita, M. and Fenn, J.B., J. Phys. Chem. 88, pages 4451(1984) and Whitehouse, C.M., Dreyer, R.N., Yamashita, M., and Fenn, J.B., Anal. Chem., 57, 675 (1985).

[14] In our view the best and most organized accounting of the detailed experiments during the development of LC/MS is found in Chapters 4 through 12 of *Liquid Chromatography-Mass Spectrometry,* by W.M.A. Niessen and J. van der Greef, Marcel Dekker, New York (1992). We recommend this book for more information on the "pioneers."

Index

Order Form

Fax orders: 1-412-963-6882

Telephone orders: Call 1-412-963-6881
(Have your AMEX, Discover, VISA or MasterCard ready)

On-line orders: E-mail to: success@LCMS.com
Internet at: www.LCMS.com

Postal orders: Global View Publishing
P.O. Box 111384
Pittsburgh, PA 15238

Please Send:

- ❒ The book: *A Global View of LC/MS*
- ❒ Information: About on-site training

I understand that I may return any books for a full refund-for any reason, no question asked.

Company Name: ______________________________

Name: ______________________________

Address: ______________________________

City: ____________ State: _____ Zip: ______ - ________

Telephone: 1- (____)________________

Book Price:
$49.95

Book Shipping:
$5.00 for shipping within the US
$7.50 for shipping outside of the US

Payment:

❒ Cheque
❒ Credit Card: ❒VISA, ❒MasterCard, ❒Optima, ❒Amex, ❒Discover

Card number:______________________________

Name on card: ____________________ Exp. date:____/_____